Audrey D. Levine

Audrey D. Levine

Comparative Reservoir Limnology
and Water Quality Management

Developments in Hydrobiology 77

Series editor
H. J. Dumont

Comparative Reservoir Limnology and Water Quality Management

Edited by

M. Straškraba, J. G. Tundisi & A. Duncan

KLUWER ACADEMIC PUBLISHERS

DORDRECHT / BOSTON / LONDON

Library of Congress Cataloging-in-Publication Data

```
Comparative reservoir limnology and water quality management / edited
   by M. Straškraba, J.G. Tundisi, and A. Duncan.
       p.    cm. -- (Developments in hydrobiology ; 77)
     Originated at the International Conference on Reservoir Limnology
   and Water Quality Management held Aug. 1987 at České Budějovice.
     Includes bibliographical references and index.
     ISBN 0-7923-1919-2 (acid free paper)
     1. Reservoir ecology.  2. Limnology.  3. Water quality.
   I. Straškraba, Milan.  II. Tundisi, J. G.  III. Duncan, A. (Annie)
   IV. Series.
   QH541.5.R4C66   1992
   628.1'32--dc20                                              92-26377
```

ISBN 0-7923-1919-2

Published by Kluwer Academic Publishers,
P.O. Box 17, 3300 AA Dordrecht, The Netherlands.

Kluwer Academic Publishers incorporates
the publishing programmes of
D. Reidel, Martinus Nijhoff, Dr W Junk and MTP Press.

Sold and distributed in the U.S.A. and Canada
by Kluwer Academic Publishers,
101 Philip Drive, Norwell, MA 02061, U.S.A.

In all other countries, sold and distributed
by Kluwer Academic Publishers Group,
P.O. Box 322, 3200 AH Dordrecht, The Netherlands.

Printed on acid-free paper

All Rights Reserved
© 1993 Kluwer Academic Publishers

No part of the material protected by this copyright notice may be reproduced or utilized in any form or by any means, electronic or mechanical, including photocopying, recording, or by any information storage and retrieval system, without written permission from the copyright owners.

Printed the Netherlands

Contents

Introduction vii

Comparative Reservoir Limnology
Chapter I. A test of hypotheses relating to the comparative limnology and assessment of eutrophication in semi-arid man-made lakes
    by J. A. Thornton & W. Rast 1
Chapter II. Limnology and management of reservoirs in Brazil
    by J. G. Tundisi, T. Matsumura-Tundisi & M. C. Calijuri 25
Chapter III. Problems in reservoir trophic-state classification and implications for reservoir management
    by O. T. Lind, T. T. Terrell & B. L. Kimmel 57
Chapter IV. Limnology of a subalpine pump-storage reservoir
    by B. Kiefer, F. Schanz & D. Imboden 69

Mathematical Models and New Techniques
Chapter V. A hierarchy of mathematical models: towards understanding the physical processes in reservoirs
    by B. Henderson-Sellers 93
Chapter VI. Modelling of physical, chemical and biological processes in Polish lakes and reservoirs
    by J. Uchmański, W. Szeligiewicz & M. Loga 99
Chapter VII. Sedimentation and mineralization of seston in a eutrophic reservoir, with a tentative sedimentation model
    by J. A. Gálvez & F. X. Niell 119
Chapter VIII. Impacts of growth factors on competitive ability of blue-green algae analyzed with whole-lake simulation
    by O. Varis 127
Chapter IX. Design of limnological observations for detecting processes in lakes and reservoirs
    by J. Kettunen 139
Chapter X. Remote sensing estimation of total chlorophyll pigment distribution in Barra Bonita Reservoir, Brazil
    by E. M. L. M. Novo, C. Z. F. Braga & J. G. Tundisi 147

Reservoir Water Quality Management
Chapter XI. Succession of fish communities in reservoirs of Central and Eastern Europe
    by J. Kubečka 153
Chapter XII. Framework for investigation and evaluation of reservoir water quality in Czechoslovakia
    by M. Straškraba, P. Blažka, Z. Brandl, P. Hejzlar, J. Komárková, J. Kubečka, I. Nesměrák, L. Procházková, V. Straškrabová & V. Vyhnálek 169

Conclusions
Chapter XIII. State-of-the-art of reservoir limnology and water quality management
    by M. Straškraba, J. G. Tundisi & A. Duncan 213
Index 289

Introduction

At ever increasing rates, reservoirs are providing the main source for water supply, in developed and developing countries of the world. This is because the ever increasing demand for water can no longer be satisfied from the usual sources of ground water, lakes and rivers. Building dam reservoirs for water storage for various uses is complicated enough to warrant the use of innovative scientific and engineering approaches such as mathematical modelling, systems engineering and large systems methodology. However, but efficient management of scarce water resources for an ever expanding human population needs insight into and understanding of the biology of the situation – how living organisms interact within reservoir ecosystems to process freely cycling chemicals into their basic elements. Biological management is better as it produces fewer unpleasant chemical surprises. Quite a detailed knowledge of the underlying physical, chemical and biological processes and their interactions is necessary for such a purpose.

Reservoirs are, at present, one of the main anthropogenic impacts on the hydrological cycle and their construction produced also several direct and indirect effects in the aquatic and terrestrial systems. The total reservoir surface is now estimated as 590.000 km^2 (or 0.3% of all continents) and their maximum volume by regulation of rivers attains 5.900 km^3 (Lvovich *et al.* 1990). Figure 1 shows how these values were growing in the past and how they are distributed over the globe.

Furthermore, the study of such dynamic systems as reservoirs is contributing and will contribute to a better understanding of basic problems in ecology – such as the succession of communities in fast changing systems, the colonization patterns and the pulse effects. An important scientific background for water quality management of reservoirs, lakes and rivers is also obtained. The scientific understanding of reservoirs as near-natural systems manipulated by man, their interactions with the watershed and upstream/downstream regions is adding new dimensions to the systems approach in ecosystem functioning and management.

The idea of this book originated at the International Conference on Reservoir Limnology and Water Quality Management at České Budějovice in August 1987, where, unusually, limnologists and water quality engineers met together to discuss theoretical and applied aspects of the mutual inter-dependence of their disciplines. The Proceedings of the Conference (Straškrabová *et al.* 1989) could not cover all the scientific knowledge disclosed at the Conference. Moreover, some contributions needed further elaboration and some reservoir investigations were not presented at the Conference so that this book was prepared with the intention of furthering our understanding of the inter-relationships between theoretical limnology and management of water quality. Basic limnological properties or reservoirs are determined by geographical location, size, shape, water depth, throughflow, outflow depth and so on but decided upon by man for society's needs. Man too is responsible for deteriorating water quality of reservoir source water but it will be by management of limnological processes that improvements can be made.

The book is divided into four parts. The first part is devoted to a comparison of reservoirs from different geographical regions, with different trophic states and having different functions. The coverage is not exhaustive and covers only some regions and some kinds of usages. This part shows that regional differences are rather profound and further theoretical analysis of these differences would enable us to make more effective use of the knowledge accumulated.

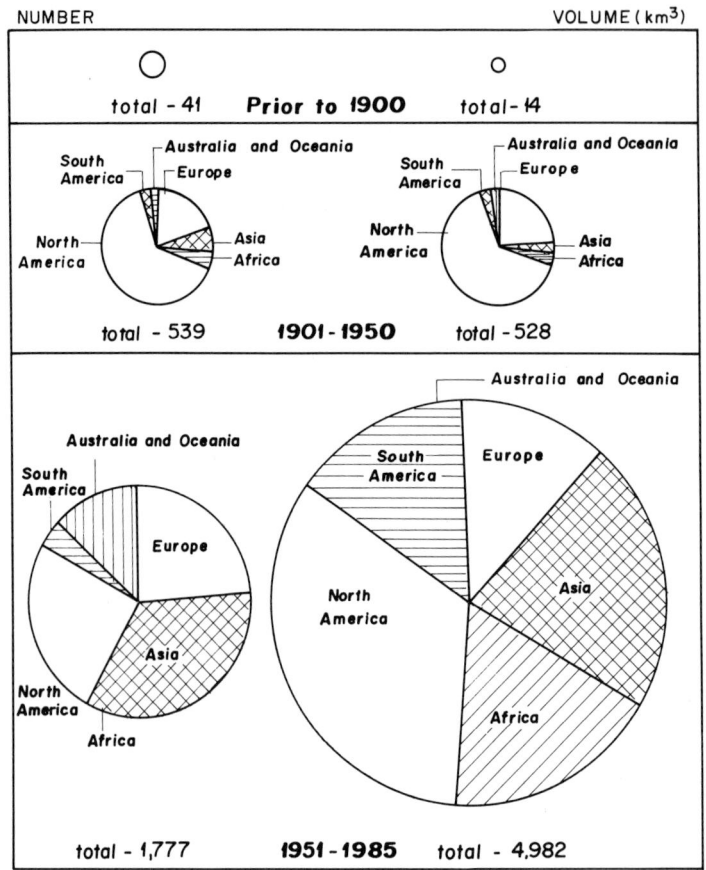

Fig. 1. The number and volume of large reservoirs during the period prior to 1900 up to 1985.

Classification of reservoirs by trophic state is still complicated by other differences due to geography, hydrology and usage which need further study. Ecosystems are multi-factorial systems which, in reservoirs, include management by man for his own purposes. One Swiss pump-storage reservoir provides a good example of a highly variable system which has been described using the technique of ecosystem modelling which encompasses its physics, chemistry and biology.

This modelling methodology is introduced extensively in the second part of the book, together with other new techniques of advanced statistical analysis and remote sensing. Models of individual processes are presented which include the physics of reservoirs, sedimentation of seston and competition of cyanobacteria with algae as well as whole ecosystem models. Models are used in other ways such as for planning the time and space design for sampling reservoirs in order to optimise costs and manpower.

Although problems of reservoir management come up in the first two parts of the book, part three focuses specifically on how to manage improvements in water quality. Fish communities are generally recognised as good indicators of the environmental health of a water body but can themselves directly influence water quality by their activities. One chapter classifies the fish communities in reservoirs of Central and East Europe and interprets differences as stages in the ecological succession of these man-made lakes. Another chapter dealing with methods in the study of water quality which, although largely based upon Czechoslovak experience, are generally applicable to all European and most temperate reservoirs and may be beneficial to limnologists and water quality engineers from other geographical regions.

The final part of the book is an attempt at summarising the present state of knowledge on reservoir limnology and management of water quality. It tries to bring out shortcomings and inadequacies. It is clear that our present level of knowledge is inadequate for proper management of water quality in such dynamic systems as reservoir ecosystems which are subject to pressures from society, to management for a particular function and to natural forces. It becomes evident that the application of brute-force and expensive technology at one site is no longer feasible because it shifts environmental problems elsewhere, downstream or to different regions, and creates other problems less amenable to solution. We need deeper insight into how reservoirs function as a part of nature which involves thinking more globally in order to approach the achievement of a sustainable development of mankind on earth.

This book is aimed at water quality engineers and theoretical limnologists to an equal degree but also to all our un-named colleagues involved with reservoirs for the benefit of mankind.

The authors would like to acknowledge the support of the activities connected with editing this book by the following institutions: IUNU, UNEP and ICSU for financial support of trips of J.G. Tundisi to Europe and Czechoslovakia respectively in May and September 1990 and December 1991 and British Council for financial support of a trip by M. Straškraba to England. In addition, we are thankful to Mrs. Dagmar Šusterová (České Budějovice) for carefully preparing the final Word-Perfect version of all manuscripts for this book.

M. Straškraba, J. G. Tundisi & A. Duncan
November 1991

References

Straškrabová, V., Z. Brandl, B. Henderson-Sellers, O. T. Lind, V. Sládeček & J. F. Talling, 1989. Proceedings of the International Conference on Reservoir Limnology and Water Quality. Arch. Hydrobiol. Beih. Ergebn. Limnol. 33, 975pp.

Chapter I

A test of hypotheses relating to the comparative limnology and assessment of eutrophication in semi-arid man-made lakes

J. A. Thornton[1] & W. Rast[2]
Present addresses: [1] *S.E. Wisconsin Regional Planning Commission, P.O. Box 1607, Waukesha, Wisconsin 53187-1607, U.S.A.;* [2] *United Nations Environment Programme, Water & Lithosphere Unit, P.O. Box 30552, Nairobi, Kenya*

Key words: reservoirs, limnology, eutrophication, arid regions, nutrients, plankton, water quality management

Abstract

In this paper we present selected morphological and limnological data on 113 man-made lakes in the semi-arid climatic zones around the world. These data are used to independently test certain hypotheses framed by Thornton & Rast (1989); to wit, semi-arid zone man-made lakes comprise a defined lake class having specific criteria which differ from those of temperate zone waterbodies. Specifically, we test the effect of the large catchment area : lake surface area ratio on turbidity, nutrient status, and phytoplankton standing crop and community composition, and assess the relevance of eutrophication threshold values previously proposed. Our results confirm our hypotheses concerning the morphological configuration of semi-arid zone waterbodies and the effects on the limnology and enrichment response of these lakes. We present an outline of a decision tree defining baseline conditions in semi-arid zone reservoirs relative to temperate zone waterbodies, which has significant implications for lake and reservoir management and the blanket application of water quality standards across climatic zones.

1. Introduction

Thornton & Rast (1987) stated the southern African semi-arid lake paradigm as the existence of significant differences in the limnology and eutrophication responses of man-made lakes from this region compared to those of the north temperate zone, based on the occurrence of high turbidities, extreme hydrological variability and relatively stable climatological factors affecting the growing season. We subsequently attempted to quantify these factors using data from 89 reservoirs in southern Africa and the southwestern United States (Thornton & Rast, 1989). To date, our studies suggest there was sufficient support for our hypotheses of significant differences, at least in terms of degree of response, to warrant a more rigorous examination of the hypotheses. In this paper, we draw on a worldwide data base, compiled from the scientific literature and from the Texas Natural Resources Information System, to critically examine this semi-arid man-made lake paradigm.

2. Materials and methods

2.1. The semi-arid lake paradigm

The semi-arid man-made lake paradigm was conceived in southern Africa, for reasons of geography, relative scientific isolation and economics, some twenty years ago. The components of the descriptive model, set out in tabular form by Thornton & Rast (1989) and summarised in Table I, evolved slowly and the complete paradigm has only recently been stated in full by Davies & Walmsley (1985). Nevertheless, the perceived differences between northern and southern hemispheres have provided, and continue to provide, much of the motivation for limnological research in the region.

Table I. Distinguishing features of lakes in the semi-arid zones of North America and southern Africa (median values shown), compared to temperate-zone lakes of the northern hemisphere (after Thornton & Rast, 1989).

Feature	Units	Semi-arid	Temperate
Turbidity	SDT (m)	1.1	1.0–10.0
Seasonal inflow	TW (yr)	0.9	1–100
Drought Cycle	CV rainfall (%)	65–114	30
Mixing regime	–	mono- / poly-mictic, non stratified	dimictic
Morphology	z (m)	6.7	1.5 → 20.0
	AD:A	183	<20
Drawdown	% capacity	up to 90%	minimal
Hydroclimate	Temperature (°C)	10 – 30	0 – 20
Oxygen depletion	% of lakes studied	73%	few (?)
Nutrient tolerance	N:P	3.5	12.0
	Limiting nutrient	P	N
Algal dominance	–	Diatoms or Chrysophytes	Cyanophytes
Algal biomass	CHA ($\mu g l^{-1}$)	8	3
Productivity	(g C $m^{-3} d^{-1}$)	<30	<0.3

Geographically, much of southern Africa is subtropical, giving rise to highly seasonal rainfall and relatively stable (aseasonal) temperatures. These factors in turn lead to a high coefficient of variation of annual river flow that, combined with the lack of perennial rivers and natural lakes, has led to the need for, and dependence upon, man-made lakes to meet the water requirements of industry, agriculture and domestic living (Alexander, 1985; DWA, 1986). Geography, too, has kept the area relatively isolated from water resources professionals in Europe and North America, due partly to the distances from these centres to southern Africa. This isolation has stimulated the examination from first principles of eutrophication assessment in the local context – this assumption of major difference from the classical, north temperate paradigm that characterises lake studies in southern Africa has been omitted from similar studies in North America (Thornton & Rast, 1987). And, finally, economics have played a large part in ensuring that public monies could be channelled into water research; monies that are not readily available elsewhere in Africa where the GDP is a fraction of that of the more developed countries in the south of the continent (Thornton, 1987a). Combined with this, international programmes such as the IBP have also contributed financially and otherwise to the stimulation of a resident scientific community in the developed countries (Thornton, 1984).

While recent studies have suggested many of the statements of difference from classically-defined phenomena that comprise the southern African paradigm are more matters of degree than of substance (and hence of little scientific interest from the point of view of generating new theories of lake functioning in these regions; Davies & Walmsley, 1985), the differences that have been identified have implications for eutrophication assessment and management. For example, our identification of eutrophication threshold values for total phosphorus and chlorophyll-a of 50 $\mu g\, l^{-1}$ and 14 $\mu g\, l^{-1}$, respectively, can have substantial implications in terms of lake restoration activities costing millions of dollars, compared to the lower concentrations considered desirable in temperate zone paradigms (Thornton & Rast, 1989). With this in mind, Thornton & Rast (1987; 1989) subsequently extended the southern African paradigm to the semi-arid American southwest with some success, and have hypothesised a more widespread applicability of the model throughout the semi-arid zones. For these reasons, therefore, we have compiled an independent data base from which to further test our hypotheses.

2.2. The data base

Unlike our previous investigation where we were only able to select two similar semi-arid areas (having similar areas, populations, climates and elevations; Thornton & Rast, 1989), the data base of 113 man-made lakes in the present study is

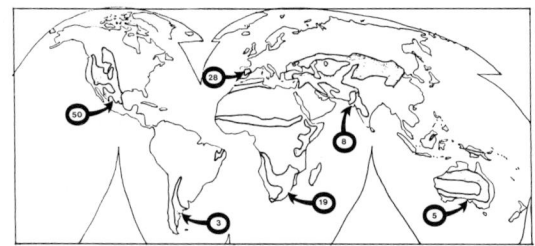

Fig. 1. Semi-arid areas of the world showing the relative distribution of numbers of lakes included in this study.

drawn from much more diverse areas (Fig. 1). Two primary sources of information were used; namely direct extraction of raw data on 50 Texan man-made lakes (in addition to the 17 other impoundments included in the original study) from the Texas Natural Resources Information System, and indirect extraction of data from the literature. The latter process was initiated via a bibliographic search using the WATERLIT base, and via personal contacts at the International Symposium on Reservoir Limnology and Water Quality held in České Budějovice, Czechoslovakia, in June 1987. For these reasons, the current data base is not as complete in all respects nor as consistent, as the data base used in our preliminary study. Thus, our remarks concerning methodological, analytical and procedural variability in the data set are even of greater applicability to this paper.

Generally, we attempted to compile selected morphological data (volume, surface area, maximum and mean depths, drainage basin area and water residence times) and limnological data (total and soluble reactive phosphorus (TP, SRP), total inorganic nitrogen, chlorophyll-a, dominant phytoplankton class, algal growth-limiting nutrient, Secchi disc transparency and phosphorus loading rate) for each of the 113 reservoirs included in this study (Table II). Commonly, total inorganic nitrogen (TIN) was estimated as the sum of the concentrations of nitrate, nitrite and ammonium; the algal growth limiting nutrient was estimated for the Texas lakes by a TIN:TP ratio greater than 7.2 equating to P-limitation; and phosphorus loading data omitted due to a lack of gauged flows.

Table II. Morphological and limnological features of the lakes used in the present study (Study = 1) and that of Thornton & Rast (1989) (Study = 2), V = volume (million m³), A = surface area (km²), AD = catchment area (km²), Z = maximum depth (m), $\bar{z}$ = mean depth (m), TW = water residence time (y), TP = total phosphorus, SRP = soluble reactive phosphorus, TIN = total nitrogen, CHA = chlorophyll-a, LN = limiting nutrient, SDT = Secchi disc transparency (m), LP = phosphorus loading rate (g/sq m/y), TS = trophic state, *PHYTO* = dominant phytoplankton class. Units are μg/l unless specified, AD:A and TN:TP ratios calculated.

LAKE	STUDY	AREA	V	A	AD	Z	$\bar{z}$
ANAHUAC	1	TEXAS	–	21	–	–	–
AQUILLA	1	TEXAS	–	13	–	18.3	4.9
ARLINGTON	1	TEXAS	–	9	–	15.5	61
AUSTIN	1	TEXAS	–	7	–	–	–
BARDWELL	1	TEXAS	–	143	–	13.1	3.7
BENBROOK	1	TEXAS	–	15	–	22.6	7
BRADY CREEK	1	TEXAS	–	–	–	–	–
BRIDGEPORT	1	TEXAS	–	52	–	26.2	7.6
BROWNWOOD	1	TEXAS	–	29	–	–	–
BUFFALO SPRINGS	1	TEXAS	–	1	–	–	–
CARANCAHUA	1	TEXAS	–	–	–	–	–
AMON CARTER	1	TEXAS	–	6	–	–	–
CEDAR CREEK	1	TEXAS	136	–	–	–	–
CHOKE CANYON	1	TEXAS	104	–	–	–	–
PAT CLEBURNE	1	TEXAS	–	6	–	–	–
CISCO	1	TEXAS	–	2	–	–	–
COLEMAN	1	TEXAS	–	8	–	–	–
FALCON	1	TEXAS	–	347	–	–	–
O.C. FISHER	1	TEXAS	–	22	–	–	–
FT. PHANTOM HILL	1	TEXAS	–	17	–	–	–
GEORGETOWN	1	TEXAS	–	5	–	25.9	–
GRAHAM	1	TEXAS	–	12	–	–	–
GRANBURY	1	TEXAS	–	35	–	22	5.5
GRANGER	1	TEXAS	–	18	–	–	–
GRAPEVINE	1	TEXAS	–	30	–	19.8	10.9
HORDS CREEK	1	TEXAS	–	2	–	–	–
HOUSTON COUNTY	1	TEXAS	–	6	–	–	–
HUBBARD CREEK	1	TEXAS	–	61	–	–	–
INKS	1	TEXAS	–	3	–	18.3	7
LEON	1	TEXAS	–	6	–	–	–

Table II. Continued.

LAKE	STUDY	AREA	V	A	AD	Z	z̄
LEWISVILLE	1	TEXAS	–	93	–	24.4	6.1
LIMESTONE	1	TEXAS	–	55	–	12.8	4.9
MARBLE FALLS	1	TEXAS	–	3	–	9.1	–
NASWORTHY	1	TEXAS	–	6	–	–	–
NAVARRO MILLS	1	TEXAS	–	20	–	14.9	3
PALO PINTO	1	TEXAS	–	11	–	–	–
PROCTOR	1	TEXAS	–	18	–	–	–
RAY HUBBARD	1	TEXAS	–	91	–	–	–
RED BLUFF	1	TEXAS	–	47	–	–	–
SABINE	1	TEXAS	–	–	–	–	–
SOMERVILLE	1	TEXAS	–	46	–	–	–
E.V. SPENCE	1	TEXAS	–	60	–	–	–
SWEETWATER	1	TEXAS	–	3	–	–	–
J.B. THOMAS	1	TEXAS	–	31	–	–	–
TOWN	1	TEXAS	–	2	–	5.5	3.4
WACO	1	TEXAS	–	29	–	–	–
WEATHERFORD	1	TEXAS	–	5	–	11.9	5.2
WHITE RIVER	1	TEXAS	–	8	–	–	–
WHITE ROCK	1	TEXAS	–	4	–	–	–
WORTH	1	TEXAS	–	14	–	7.6	1.8
UTAH	2	SW USA	832	396	6410	4.3	2.1
OTTER CREEK	2	SW USA	65	10	914	11.3	6.3
SHADOW MOUNTAIN	2	SW USA	21	5	411	11	3.8
GREEN MOUNTAIN	2	SW USA	191	8	1543	75.3	22.2
CHERRY CREEK	2	SW USA	16	3	994	7.6	5.2
DILLON	2	SW USA	313	12	855	60.9	24.6
BLUE MESA	2	SW USA	1138	36	8836	104.2	31.1
NAVAJO	2	SW USA	2108	63	8380	39.6	33.3
BLUE WATER	2	SW USA	47	7	514	17.4	6.7
CONCHAS	2	SW USA	456	38	19189	43	11.8
UTE	2	SW USA	133	16	28837	25	8
ELEPHANT BUTE	2	SW USA	2707	148	76135	47.9	18.3
ALAMOGORDO	2	SW USA	150	18	11350	19.8	8.1
MACMILLAN	2	SW USA	48	23	43931	7.9	2.1
MEREDITH	2	SW USA	1066	66	41497	26.2	16
TEXOMA	2	SW USA	2053	360	87104	28.7	5.7
CADDO	2	SW USA	231	132	6974	3	1.8
COLORADO CITY	2	SW USA	39	6	827	8.2	6
STAMFORD	2	SW USA	66	19	913	10.6	3.5
KEMP	2	SW USA	330	62	5340	13.1	5.3
EAGLE MOUNTAIN	2	SW USA	233	36	5055	15.8	6.4
GARZA LITTLE ELM	2	SW USA	564	93	4205	20.4	6
LAVON	2	SW USA	563	86	1949	11.9	6.5
TAWAKONI	2	SW USA	1144	148	1809	20.4	7.7
TEXARKANA	2	SW USA	373	126	8791	4.6	3
LAKE OF PINES	2	SW USA	308	75	2126	10.1	4.1
PALESTINE	2	SW USA	508	103	2098	15.2	4.9
TRINIDAD	2	SW USA	9	2	–	2.4	3.1
WHITNEY	2	SW USA	773	95	67684	30.2	8.1
POSSUM KINGDOM	2	SW USA	897	80	58324	44.2	11.2
BELTON	2	SW USA	545	50	9181	25.6	10.8
STILLHOUSE HOLLOW	2	SW USA	291	26	3387	31.1	11.3
SAM RAYBURN	2	SW USA	3574	463	8475	16.8	7.7
SAN ANGELO	2	SW USA	142	21	3834	8.5	6.5
TWIN BUTES	2	SW USA	229	36	6557	13.1	6.3
BUCHANAN	2	SW USA	1222	93	80844	40.2	13.1

Table II. Continued.

LAKE	STUDY	AREA	V	A	AD	Z	$\bar{z}$
JOHNSON	2	SW USA	172	25	93965	24.4	6.7
TRAVIS	2	SW USA	1447	76	98756	58.5	18.9
BASTROP	2	SW USA	20	3	21	13.4	5.6
CANYON	2	SW USA	471	33	3685	39	14
MEDINA	2	SW USA	313	22	1619	36.3	13.9
AMISTAD	2	SW USA	4330	262	318566	61	16.5
CORPUS CHRISTI	2	SW USA	372	88	43060	16.8	4.2
LIVINGSTONE	2	SW USA	2158	334	42615	21.3	6.5
HOUSTON	2	SW USA	180	49	7275	6.1	3.7
FORT SUPPLY	2	SW USA	90	7	3861	23.5	11.9
OOLOGAH	2	SW USA	679	119	11118	24.4	5.7
GRAND LAKE O CHEROKEES	2	SW USA	2051	188	26483	50	10.9
KEYSTONE	2	SW USA	819	106	162567	22	7.7
ELLSWORTH	2	SW USA	115	22	618	18.3	5.1
EUFAULA	2	SW USA	4189	414	97544	26.5	10.1
THUNDERBIRD	2	SW USA	147	24	637	27.7	6
FORT COBB	2	SW USA	105	16	738	19.2	6.4
ALTOS	2	SW USA	162	25	5439	23.5	6.4
ARBUCKLE	2	SW USA	89	9	316	28.6	9.4
WISTER	2	SW USA	37	16	2554	13.4	2.3
KARIBA	2	STH.AFR.	160368	5100	514892	120	31.4
ROBERTSON	2	STH.AFR.	490	81	3800	22.6	6
MCILWAINE	2	STH.AFR.	250	26	2227	27.4	9.4
PRINCE EDWARD/HENRY HALLAM	2	STH.AFR.	13	3	793	18	4
JOHN MACK	2	STH.AFR.	21	5	2300	17.5	4.3
CACTUSPOORT	2	STH.AFR.	3	1	1279	16.4	3.9
LOWER UMGUSA	2	STH.AFR.	1	1	474	16	5.1
UPPER UMGUSA	2	STH.AFR.	3	1	461	13	3.9
LINDLEYSPOORT	2	STH.AFR.	14	2	704	22.2	8
BOSPOORT	2	STH.AFR.	19	4	1080	14.3	5.2
RUST DER WINTER	2	STH.AFR.	28	5	1147	20	5.8
LOSKOP	2	STH.AFR.	180	17	12285	36	9.6
NEW DORINGSPOORT	2	STH.AFR.	109	13	3627	36	8.1
TONTELDOOS	2	STH.AFR.	1	0.1	55	10.5	3.9
BRONKHORSTSPRUIT	2	STH.AFR.	58	8	1263	19.5	6.7
RIETVLEI	2	STH.AFR.	13	2	479	17.2	6.3
ROODEPLAAT	2	STH.AFR.	42	4	684	43	10.6
BUFFELSPOORT	2	STH.AFR.	11	1	119	23	7.7
HARTBEESPOORT	2	STH.AFR.	193	20	4112	32	9.6
OLIFANTSNEK	2	STH.AFR.	14	3	492	13.6	6.3
BLOEMHOF	2	STH.AFR.	1273	228	107911	18	4.5
P.K. LE ROUX	2	STH.AFR.	2930	128	–	73	23
ALBERT FALLS	2	STH.AFR.	293	24	1644	24.6	12.4
NAGLE	2	STH.AFR.	23	2	2535	38.1	15.2
HAZELMERE	2	STH.AFR.	22	2	381	30.6	10.8
VERNON HOOPER	2	STH.AFR.	7	1	786	16.2	8.8
HENLEY	2	STH.AFR.	6	1	238	18.8	8.2
MIDMAR	2	STH.AFR.	177	16	917	22.3	11.4
NAHOON	2	STH.AFR.	6	1	473	18.4	5.9
LAING	2	STH.AFR.	22	2	913	37.5	10.2
BRIDLEDRIFT	2	STH.AFR.	76	6	1176	40.9	11.8
H.F. VERWOERD	2	STH.AFR.	5952	364	70749	68.7	16.3
EBRO	1	SPAIN	540	62	466	30	8.5
ABUILAR DE CAMPOO	1	SPAIN	247	16	546	42	21.9
PORMA	1	SPAIN	318	11	250	72	27.5

Table II. Continued.

LAKE	STUDY	AREA	V	A	AD	Z	$\bar{z}$
LINARES DEL ARROYO	1	SPAIN	58	5	756	35	7.7
CUERDO DEL POZO	1	SPAIN	178	17	380	36	11.8
TRANQUERA	1	SPAIN	83	5	–	46	15.8
BUENDIA	1	SPAIN	1520	80	3342	70	19.8
ENTREPENAS	1	SPAIN	891	34	3829	68	23.8
EL VADO	1	SPAIN	57	3	426	57	19
EL ATAZAR	1	SPAIN	426	–	10	125	39
SANTILLANA	1	SPAIN	91	10	236	33	6.1
SAN JUAN	1	SPAIN	162	6	1790	67	25.2
GURGUILLO	1	SPAIN	217	9	1050	77	25.8
CAZALEGAS	1	SPAIN	26	5	3993	5	0.6
TORCON	1	SPAIN	4	1	205	30	3.3
GOAJARAZ	1	SPAIN	25	1	–	42	15.6
PENARROYA	1	SPAIN	48	4	950	30	9.8
ALARCON	1	SPAIN	1112	65	2918	47	16.3
GENERALISSIMO	1	SPAIN	228	12	4264	87	22.2
LORIGUILLA	1	SPAIN	71	3	4716	35	20.4
EL VELLON	1	SPAIN	45	4	–	47	10.4
GUADALMENA	1	SPAIN	346	12	–	80	27.8
GUADALEN	1	SPAIN	173	17	1281	49	9.8
TRANCO DE BEAS	1	SPAIN	500	18	358	93	31.2
RUMBLAR	1	SPAIN	137	6	583	53	17.3
JANDULA	1	SPAIN	322	13	2158	86	23.5
CENAJO	1	SPAIN	472	16	2637	84	23.1
TALAVE	1	SPAIN	42	1	757	36	22.9
HOME	1	AUSTRALIA	3070	202	15275	41.5	15.2
MULWALA	1	AUSTRALIA	116	44	27300	13.7	2.6
MOONDARA	1	AUSTRALIA	23	–	–	–	–
BURLEY GRIFFIN	1	AUSTRALIA	33	7	–	–	4.7
GRINNINDERA	1	AUSTRALIA	–	–	–	–	3.5
PROSPECT	1	AUSTRALIA	50	5	–	24	9.7
MOUNT BOLD	1	AUSTRALIA	30	2	–	41	13.2
EURRAGORANG	1	AUSTRALIA	2057	75	9000	105	27.4
ITEZHI-TEZHI	1	AFRICA	–	–	–	–	–
DA GAMA	1	AFRICA	14	1	62	23.5	10.6
KLIPKOPJE	1	AFRICA	12	2	78	11.5	5.1
WITKLIP	1	AFRICA	13	2	64	15	7
RUSFONTEIN	1	AFRICA	75	11	940	22.1	6.5
ARMENIA	1	AFRICA	14	4	858	12.2	3.6
ERFENIS	1	AFRICA	224	33	4750	30.5	6.7
CHONGWE	1	AFRICA	–	–	–	–	–
SWARTWATER	1	AFRICA	4	1	19	18	6.2
NOORA	1	AFRICA	161	15	1772	16	10.7
OMTATA	1	AFRICA	255	2	–	125.5	–
XONXA	1	AFRICA	143	13	1487	17	11
LUSUSI	1	AFRICA	157	11	1300	35.9	14.7
BONKOLO	1	AFRICA	7	–	88	–	–
VAAL	1	AFRICA	2122	291	38505	36	9
WELBEDACHT	1	AFRICA	39	15	15245	17.7	2.6
CHELMSFORD	1	AFRICA	198	34	830	18.6	5.8
KAFUE	1	AFRICA	740	809	15300	33	1
CAHORA BASSA	1	AFRICA	52000	2660	1200000	157	20
SAN ROQUE	1	ARGENTINA	–	0.4	25	–	0.02
LOS MOLINOS	1	ARGENTINA	–	0.3	21	–	0.01
EMBALSE DEL RIO TERCERO	1	ARGENTINA	–	0.7	46	–	0.02
DAK PATTHAR	1	INDIA	–	–	–	18	0.2

Table II. Continued.

LAKE	STUDY	AREA	V	A	AD	Z	z̄
RIHAND	1	INDIA	10600	466	1333	76.2	23.9
MANSAGAR	1	INDIA	–	–	–	–	–
SATPURA	1	INDIA	–	10	–	5	–
GANDHISAGAR	1	INDIA	–	660	23025	62	–
SANPNA	1	INDIA	–	5	32	8	–
TAWA	1	INDIA	–	200	5983	46	–
RAILWAY (NAINPUR)	1	INDIA	–	5	–	5	–

LAKE	STUDY	AREA	LN	SDT	LP	TS	TN:TP	AD:A
ANAHUAC	1	TEXAS	N	0.1	–	–	1.0	–
AQUILLA	1	TEXAS	P	1.1	–	–	8.5	–
ARLINGTON	1	TEXAS	N	0.7	–	–	5.0	–
AUSTIN	1	TEXAS	P	1.5	–	–	12.1	–
BARDWELL	1	TEXAS	N	0.6	–	–	4.7	–
BENBROOK	1	TEXAS	N	0.9	–	–	2.5	–
BRADY CREEK	1	TEXAS	N	0.7	–	–	2.3	–
BRIDGEPORT	1	TEXAS	N	1	–	–	3.1	–
BROWNWOOD	1	TEXAS	N	1.1	–	–	3.0	–
BUFFALO SPRINGS	1	TEXAS	P	0.7	–	–	14.4	–
CARANCAHUA	1	TEXAS	–	0.6	–	–	0.4	–
ANON CARTER	1	TEXAS	N	0.9	–	–	3.7	–
CEDAR CREEK	1	TEXAS	N	1.1	–	–	2.5	–
CHOKE CANYON	1	TEXAS	–	0.8	–	–	1.2	–
PAT CLEBURNE	1	TEXAS	N	0.4	–	–	3.6	–
CISCO	1	TEXAS	N	1.5	–	–	9.0	–
COLEMAN	1	TEXAS	N	1.2	–	–	5.4	–
FALCON	1	TEXAS	–	1.3	–	–	4.3	–
O.C. FISHER	1	TEXAS	–	–	–	–	–	–
FT. PHANTOM HILL	1	TEXAS	N	0.5	–	–	1.7	–
GEORGETOWN	1	TEXAS	P	2.4	–	–	6.3	–
GRAHAM	1	TEXAS	N	0.7	–	–	2.2	–
GRANBURY	1	TEXAS	N	1.4	–	–	5.8	–
GRANGER	1	TEXAS	N	0.6	–	–	7.3	–
GRAPEVINE	1	TEXAS	N	0.9	–	–	6.0	–
HORDS CREEK	1	TEXAS	N	0.8	–	–	3.0	–
HOUSTON COUNTY	1	TEXAS	P	1.8	–	–	12.5	–
HUBBARD CREEK	1	TEXAS	N	1.2	–	–	7.8	–
INKS	1	TEXAS	N	1.3	–	–	3.5	–
LEON	1	TEXAS	N	1.2	–	–	3.6	–
LEWISVILLE	1	TEXAS	N	0.8	–	–	5.7	–
LIMESTONE	1	TEXAS	N	1.2	–	–	5.9	–
MARBLE FALLS	1	TEXAS	N	1.3	–	–	8.0	–
NASWORTHY	1	TEXAS	N	0.5	–	–	1.6	–
NAVARRO MILLS	1	TEXAS	P	0.5	–	–	12.5	–
PALO PINTO	1	TEXAS	N	0.6	–	–	2.6	–
PROCTOR	1	TEXAS	N	0.6	–	–	2.0	–
RAY HUBBARD	1	TEXAS	N	1	–	–	4.5	–
RED BLUFF	1	TEXAS	N	0.9	–	–	1.6	–
SABINE	1	TEXAS	–	0.8	–	–	4.4	–
SOMERVILLE	1	TEXAS	N	0.7	–	–	2.0	–
E.V. SPENCE	1	TEXAS	N	1.5	–	–	4.0	–
SWEETWATER	1	TEXAS	N	0.8	–	–	2.2	–

Table II. Continued.

LAKE	STUDY	AREA	LN	SDT	LP	TS	TN:TP	AD:A
J.B. THOMAS	1	TEXAS	N	0.4	–	–	2.1	–
TOWN	1	TEXAS	P	1.6	–	–	12.8	–
WACO	1	TEXAS	P	0.6	–	–	15.2	–
WEATHERFORD	1	TEXAS	N	0.8	–	–	1.7	–
WHITE RIVER	1	TEXAS	N	1.4	–	–	4.6	–
WHITE ROCK	1	TEXAS	N	0.5	–	–	5.0	–
WORTH	1	TEXAS	N	0.7	–	–	2.5	–
UTAH	2	SW USA	P	1.2	0.5	H	2.4	16.4
OTTER CREEK	2	SW USA	N	1.1	0.6	E	0.6	97.4
SHADOW MOUNTAIN	2	SW USA	N	1	1.4	M	2.0	94.2
GREEN MOUNTAIN	2	SW USA	P	1	0.8	O	4.0	192.9
CHERRY CREEK	2	SW USA	N	1.2	1.5	E	0.7	331.3
DILLON	2	SW USA	P	0.5	0.6	O	4.4	71.3
BLUE MESA	2	SW USA	N	1	2.7	M	2.1	245.4
NAVAJO	2	SW USA	P	1.2	2.5	M	1.4	133.0
BLUE WATER	2	SW USA	P	1.2	0.1	M	3.9	73.4
CONCHAS	2	SW USA	P	1.1	0.3	M	2.0	505.0
UTE	2	SW USA	P	1.1	0.7	M	1.9	1802.3
ELEPHANT BUTE	2	SW USA	N	1.2	17.4	E	1.3	514.4
ALAMOGORDO	2	SW USA	P	1.2	1.5	E	2.0	630.6
MACMILLAN	2	SW USA	P	1.2	1	E	0.5	1912.2
MEREDITH	2	SW USA	N	1.1	6.4	M	3.3	628.7
TEXOMA	2	SW USA	N	1.1	4.1	E	3.8	242.0
CADDO	2	SW USA	N	1.1	0.7	E	1.3	52.8
COLORADO CITY	2	SW USA	N	1.2	0.2	E	2.1	137.8
STAMFORD	2	SW USA	N	1.2	1.4	E	0.8	48.1
KEMP	2	SW USA	P	1.1	0.2	E	4.8	86.1
EAGLE MOUNTAIN	2	SW USA	N	1.2	1.5	M	2.9	140.7
GARZA LITTLE ELM	2	SW USA	P	1.2	1.2	E	8.4	45.2
LAVON	2	SW USA	P	1.2	1	E	2.9	22.7
TAWAKONI	2	SW USA	N	1.1	0.9	E	2.2	12.2
TEXARKANA	2	SW USA	N	1.2	2.1	E	1.1	69.8
LAKE OF PINES	2	SW USA	N	1.1	0.9	E	2.9	28.3
PALESTINE	2	SW USA	P	1.1	0.7	E	5.8	20.4
TRINIDAD	2	SW USA	N	1.2	–	E	0.3	–
WHITNEY	2	SW USA	N	1	1.2	E	4.3	712.5
POSSUM KINGDOM	2	SW USA	N	1	2.7	M	3.0	729.1
BELTON	2	SW USA	P	0.9	4	M	11.6	183.6
STILLHOUSE HOLLOW	2	SW USA	P	1	0.4	M	8.9	130.3
SAM RAYBURN	2	SW USA	P	1.1	0.3	E	5.2	18.3
SAN ANGELO	2	SW USA	N	1.2	0.1	E	1.4	182.6
TWIN BUTTES	2	SW USA	P	1.1	0.04	E	8.6	182.1
BUCHANAN	2	SW USA	P	1	1.1	E	6.9	869.3
JOHNSON	2	SW USA	P	1.1	1.4	E	10.0	3758.6
TRAVIS	2	SW USA	P	0.9	0.6	M	13.9	1299.4
BASTROP	2	SW USA	N	1	0.3	M	4.1	7.0
CANYON	2	SW USA	P	1	0.1	M	60.0	73.6
MEDINA	2	SW USA	P	0.9	1	M	38.5	1215.9
AMISTAD	2	SW USA	N	1.2	2.4	E	1.2	489.3
CORPUS CHRISTI	2	SW USA	N	1.1	6.4	E	2.9	127.6
LIVINGSTONE	2	SW USA	N	1.2	6.1	E	2.7	148.5
HOUSTON	2	SW USA	N	1.2	6.1	E	2.7	148.5
FORT SUPPLY	2	SW USA	N	1.2	0.4	E	1.9	551.6
OOLOGAH	2	SW USA	P	1.2	3.5	E	9.8	93.4
GRAND LAKE O CHEROKEES	2	SW USA	P	1.1	6.9	E	8.5	140.9
KEYSTONE	2	SW USA	N	1.2	16.5	E	5.1	1533.7

Table II. Continued.

LAKE	STUDY	AREA	LN	SDT	LP	TS	TN:TP	AD:A
ELLSWORTH	2	SW USA	P	1.1	0.2	E	1.9	28.1
EUFAULA	2	SW USA	N	1.2	4.2	E	5.0	235.6
THUNDERBIRD	2	SW USA	P	1.1	0.3	E	5.6	26.5
FORT COBB	2	SW USA	P	1.1	0.6	E	2.9	46.1
ALTOS	2	SW USA	N	1.1	0.4	E	1.5	217.6
ARBUCKLE	2	SW USA	P	1.1	0.4	M	3.3	35.1
WISTER	2	SW USA	N	1.2	2.9	E	2.9	159.6
KARIBA	2	STH.AFR.	P	–	0.1	O	–	101.0
ROBERTSON	2	STH.AFR.	P	1.2	0.1	M	–	46.9
MCILWAINE	2	STH.AFR.	N	1.2	3	E	0.7	85.7
PRINCE EDWARD/HENRY HALLAM	2	STH.AFR.	P	1.5	–	E	–	264.3
JOHN MACK	2	STH.AFR.	–	0.4	3	M	14.3	460.0
CACTUSPOORT	2	STH.AFR.	–	–	9.3	M	–	1279.0
LOWER UMGUSA	2	STH.AFR.	–	–	11	E	–	474.0
UPPER UMGUSA	2	STH.AFR.	N	–	11	E	–	461.0
LINDLEYSPOORT	2	STH.AFR.	P	0.4	1.6	M	–	352.0
BOSPOORT	2	STH.AFR.	P	0.8	5.3	H	16.3	270.0
RUST DER WINTER	2	STH.AFR.	P	1.4	0.4	M	2.2	229.4
LOSKOP	2	STH.AFR.	P	0.8	0.9	E	4.5	722.6
NEW DORINGSPOORT	2	STH.AFR.	–	0.2	0.3	M	5.2	279.0
TONTELDOOS	2	STH.AFR.	N	1.4	1.3	O	–	550.0
BRONKHORSTSPRUIT	2	STH.AFR.	N	0.3	0.3	M	5.3	157.9
RIETVLEI	2	STH.AFR.	N	2.9	15.8	H	0.2	239.5
ROODEPLAAT	2	STH.AFR.	N	3.5	11.1	H	5.9	171.0
BUFFELSPOORT	2	STH.AFR.	P	1.4	0.2	M	–	119.00
HARTBEESPOORT	2	STH.AFR.	P	0.8	17.5	H	1.3	205.6
OLIFANTSNEK	2	STH.AFR.	P	0.3	0.2	M	14.3	164.0
BLOEMHOF	2	STH.AFR.	N	0.7	7.9	E	–	473.3
P.K. LE ROUX	2	STH.AFR.	–	0.3	–	M		
ALBERT FALLS	2	STH.AFR.	P	2.5	0.02	O	2.3	68.5
NAGLE	2	STH.AFR.	P	2.5	0.2	O	17.5	1267.5
HAZELMERE	2	STH.AFR.	P	0.8	0.1	M	8.5	190.5
VERNON HOOPER	2	STH.AFR.	P	1.2	4.4	E	17.0	786.0
HENLEY	2	STH.AFR.	P	1	0.2	M	32.1	238.0
MIDMAR	2	STH.AFR.	P	1.3	0.05	M	9.3	57.3
NAHOON	2	STH.AFR.	P	0.4	1	E	1.2	473.0
LAING	2	STH.AFR.	P	0.2	13.8	E	1.1	456.5
BRIDLEDRIFT	2	STH.AFR.	–	0.1	2.4	E	1.3	196.0
H.F. VERWOERD	2	STH.AFR.	N	0.1	0.6	M	–	194.4
EBRO	1	SPAIN	–	3.8	–	O	–	7.5
ABUILAR DE CAMPOO	1	SPAIN	–	4.4	–	M	–	34.1
PORMA	1	SPAIN	–	5.2	–	O	–	22.7
LINARES DEL ARROYO	1	SPAIN	–	1.1	–	M	–	151.2
CUERDO DEL POZO	1	SPAIN	–	1.7	–	E	–	22.4
TRANQUERA	1	SPAIN	–	4.5	–	O	–	–
BUENDIA	1	SPAIN	–	4	–	O	–	41.8
ENTREPENAS	1	SPAIN	–	3.5	–	O	–	112.6
EL VADO	1	SPAIN	–	2.8	–	M	–	142.0
EL ATAZAR	1	SPAIN	–	6.2	–	M	–	–
SANTILLANA	1	SPAIN	–	4.4	–	E	–	23.6
SAN JUAN	1	SPAIN	–	3	–	E	–	298.3
GURGUILLO	1	SPAIN	–	2.6	–	E	–	116.7
CAZALEGAS	1	SPAIN	–	0.6	–	E	–	798.6
TORCON	1	SPAIN	–	1.9	–	E	–	205.0
GOAJARAZ	1	SPAIN	–	2.1	–	M	–	–

Table II. Continued.

LAKE	STUDY	AREA	LN	SDT	LP	TS	TN:TP	AD:A
PENARROYA	1	SPAIN	–	7.6	–	M	–	237.5
ALARCON	1	SPAIN	–	3.2	–	M	–	44.9
GENERALISSIMO	1	SPAIN	–	4.4	–	O	–	353.3
LORIGUILLA	1	SPAIN	–	4.3	–	O	–	1572.0
EL VELLON	1	SPAIN	–	3.2	–	E	–	–
GUADALMENA	1	SPAIN	–	8.9	–	O	–	–
GUADALEN	1	SPAIN	–	1	–	M	–	75.4
TRANCO DE BEAS	1	SPAIN	–	2.9	–	O	–	19.9
RUMBLAR	1	SPAIN	–	4.2	–	M	–	97.2
JANDULA	1	SPAIN	–	2.4	–	M	–	166.0
CENAJO	1	SPAIN	–	2.4	–	M	–	164.8
TALAVE	1	SPAIN	–	2.1	–	O	–	757.0
HOME	1	AUSTRALIA	–	1	1.2	E	2.7	75.6
MULWALA	1	AUSTRALIA	N/P	0.5	9.4	E	1.1	620.5
MOONDARA	1	AUSTRALIA	–	0.9	–	E	–	–
BURLEY GRIFFIN	1	AUSTRALIA	–	–	3.5	E	7.8	–
GRINNINDERA	1	AUSTRALIA	–	–	1.1	M	17.6	–
PROSPECT	1	AUSTRALIA	–	–	1.1	M	61.4	0
MOUNT BOLD	1	AUSTRALIA	–	–	6.1	M	14.4	0
EURRAGORANG	1	AUSTRALIA	–	–	2.9	–	–	120
ITEZHI-TEZHI	1	AFRICA	–	–	–	O		
DA GAMA	1	AFRICA	N	5	–	O	–	62
KLIPKOPJE	1	AFRICA	N	1.9	–	O	–	39.0
WITKLIP	1	AFRICA	N	4.8	–	O	–	32.0
RUSFONTEIN	1	AFRICA	P	–	–	E	–	85.5
ARMENIA	1	AFRICA	P	–	–	O	–	214.5
ERFENIS	1	AFRICA	P	–	–	M	–	143.9
CHONGWE	1	AFRICA	–	–	–	E	–	–
SWARTWATER	1	AFRICA	–	2.1	–	M	76.7	19.0
NOORA	1	AFRICA	–	–	–	M	–	118.1
OMTATA	1	AFRICA	–	–	–	M	–	–
XONXA	1	AFRICA	–	–	–	M	–	114.4
LUSUSI	1	AFRICA	–	–	–	M	–	–
BONKOLO	1	AFRICA	–	–	–	M	–	–
VAAL	1	AFRICA	N	0.1	–	E	–	132.3
WELBEDACHT	1	AFRICA	N	–	–	M	–	1016.3
CHELMSFORD	1	AFRICA	N	–	–	M	–	24.4
KAFUE	1	AFRICA	–	–	–	E	–	189.1
CAHORA BASSA	1	AFRICA	–	–	–	M	–	451.1
SAN ROQUE	1	ARGENTINA	–	1	–	–	–	–
LOS MOLINOS	1	ARGENTINA	–	–	–	–	–	–
EMBALSE DEL RIO TERCERO	1	ARGENTINA	–	–	–	–	–	–
DAK PATTHAR	1	INDIA	–	0.6	–	E	–	–
RIHAND	1	INDIA	–	0.1	–	O	–	2.9
MANSAGAR	1	INDIA	–	–	–	E	–	–
SATPURA	1	INDIA	–	0.8	–	O	–	–
GANDHISAGAR	1	INDIA	–	1.7	–	M	–	34.9
SANPNA	1	INDIA	–	0.6	–	E	–	6.4
TAWA	1	INDIA	–	1.3	–	M	–	29.9
RAILWAY (NAINPUR)	1	INDIA	–	–	–	E	–	–

Table II. Continued.

LAKE	STUDY	AREA	TW	TP	SRP	TIN	CHA	PHYTO
ANAHUAC	1	TEXAS	–	171	98	170	8	–
AQUILLA	1	TEXAS	–	30	10	255	–	–
ARLINGTON	1	TEXAS	–	21	5	105	8.5	–
AUSTIN	1	TEXAS	–	10	5	121	2	–
BARDWELL	1	TEXAS	–	50	25	235	8.5	–
BENBROOK	1	TEXAS	–	40	10	98	10	–
BRADY CREEK	1	TEXAS	–	30	10	70	8	–
BRIDGEPORT	1	TEXAS	–	26	10	80	2	–
BROWNWOOD	1	TEXAS	–	13	5	65	2	–
BUFFALO SPRINGS	1	TEXAS	–	70	28	1010	30	–
CARANCAHUA	1	TEXAS	–	28	13	10	–	–
AMON CARTER	1	TEXAS	–	39	10	146	10	–
CEDAR CREEK	1	TEXAS	–	40	10	100	13	–
CHOKE CANYON	1	TEXAS	–	80	17	98	–	–
PAT CLEBURNE	1	TEXAS	–	59	12	215	4	–
CISCO	1	TEXAS	–	10	10	90	2	–
COLEMAN	1	TEXAS	–	14	5	75	2	–
FALCON	1	TEXAS	–	22	5	95	4	–
O.C. FISHER	1	TEXAS	–	50	–	–	–	–
FT. PHANTOM HILL	1	TEXAS	–	38	23	65	6	–
GEORGETOWN	1	TEXAS	–	20	5	125	3	–
GRAHAM	1	TEXAS	–	50	10	110	4	–
GRANBURY	1	TEXAS	–	13	10	75	7	–
GRANGER	1	TEXAS	–	40	5	290	4	–
GRAPEVINE	1	TEXAS	–	26	10	155	3	–
HORDS CREEK	1	TEXAS	–	20	10	60	2	–
HOUSTON COUNTY	1	TEXAS	–	20	10	250	4.5	–
HUBBARD CREEK	1	TEXAS	–	20	–	155	2	–
INKS	1	TEXAS	–	20	5	70	6	–
LEON	1	TEXAS	–	18	10	65	5	–
LEWISVILLE	1	TEXAS	–	40	15	229	3.5	–
LIMESTONE	1	TEXAS	–	30	8	178	3	–
MARBLE FALLS	1	TEXAS	–	20	5	160	6	–
NASWORTHY	1	TEXAS	–	40	5	65	14	–
NAVARRO MILLS	1	TEXAS	–	36	10	450	6	–
PALO PINTO	1	TEXAS	–	55	10	145	3	–
PROCTOR	1	TEXAS	–	33	10	65	25	–
RAY HUBBARD	1	TEXAS	–	50	10	224	8	–
RED BLUFF	1	TEXAS	–	40	10	65	11	–
SABINE	1	TEXAS	–	54	16	235	4	–
SOMERVILLE	1	TEXAS	–	40	10	78	20	–
E.V. SPENCE	1	TEXAS	–	20	5	80	7	–
SWEETWATER	1	TEXAS	–	30	8	65	8	–
J.B. THOMAS	1	TEXAS	–	52	12	110	6	–
TOWN	1	TEXAS	–	23	–	295	2	–
WACO	1	TEXAS	–	23	10	350	6	–
WEATHERFORD	1	TEXAS	–	42	10	73	4	–
WHITE RIVER	1	TEXAS	–	14	10	65	2	–
WHITE ROCK	1	TEXAS	–	78	12	390	15	–
WORTH	1	TEXAS	–	33	10	82	6	–
UTAH	2	USA	2.5	132	12	320	72	BG
OTTER CREEK	2	USA	1.4	67	30	40	11	D
SHADOW MOUNTAIN	2	USA	0.2	20	3	40	5	D
GREEN MOUNTAIN	2	USA	0.4	10	2	40	5	D
CHERRY CREEK	2	USA	3.6	54	7	40	23	D
DILLON	2	USA	1.3	9	2	40	3	D

Table II. Continued.

LAKE	STUDY	AREA	TW	TP	SRP	TIN	CHA	PHYTO
BLUE MESA	2	USA	0.9	19	5	40	6	BG
NAVAJO	2	USA	1.4	36	13	50	2	CR
BLUE WATER	2	USA	5.8	36	12	140	3	BG
CONCHAS	2	USA	3	20	4	40	3	CR
UTE	2	USA	1.4	21	4	40	3	CR
ELEPHANT BUTE	2	USA	2.2	83	52	110	6	D
ALAMOGORDO	2	USA	1.7	25	3	50	5	CR
MACMILLAN	2	USA	0.2	97	9	45	14	D
MEREDITH	2	USA	8.6	21	9	70	3	G
TEXOMA	2	USA	0.4	42	18	160	12	CR
CADDO	2	USA	0.1	55	13	70	14	BG
COLORADO CITY	2	USA	3.9	42	12	90	12	BG
STAMFORD	2	USA	2.5	73	12	60	18	D
KEMP	2	USA	2.8	23	7	110	10	BG
EAGLE MOUNTAIN	2	USA	0.9	24	8	70	5	D
GARZA LITTLE ELM	2	USA	0.9	45	18	380	14	BG
LAVON	2	USA	1.4	63	18	180	5	D
TAWAKONI	2	USA	3.2	46	13	100	10	D
TEXARKANA	2	USA	0.1	106	30	120	19	D
LAKE OF PINES	2	USA	0.6	31	11	90	12	CR
PALESTINE	2	USA	1.3	31	10	180	10	D
TRINIDAD	2	USA	–	389	240	110	24	G
WHITNEY	2	USA	0.8	28	8	120	6	BG
POSSUM KINGDOM	2	USA	1	23	9	70	9	BG
BELTON	2	USA	1.6	16	7	185	8	G
STILLHOUSE HOLLOW	2	USA	1	18	10	160	3	CR
SAM RAYBURN	2	USA	1.4	29	9	150	6	BG
SAN ANGELO	2	USA	11	98	11	140	24	BG
TWIN BUTTES	2	USA	14.9	29	9	250	8	BG
BUCHANAN	2	USA	1.7	36	12	250	8	BG
JOHNSON	2	USA	0.1	42	13	420	8	G
TRAVIS	2	USA	1	18	7	250	5	D
BASTROP	2	USA	5.2	22	7	90	12	G
CANYON	2	USA	1	10	6	450	2	CR
MEDINA	2	USA	2.7	10	4	600	12	CR,BR
AMISTAD	2	USA	3.8	13	9	500	2	D
CORPUS CHRISTI	2	USA	0.5	113	50	130	19	D
LIVINGSTONE	2	USA	0.3	196	128	559	16	D,BG
HOUSTON	2	SW USA	0.01	97	36	260	16	D
FORT SUPPLY	2	SW USA	1.6	70	14	135	9	G
OOLOGAH	2	SW USA	0.3	59	31	580	5	CR
GRAND LAKE O' CHEROKEES	2	SW USA	0.3	87	38	740	6	BG
KEYSTONE	2	SW USA	0.1	136	96	690	21	D
ELLSWORTH	2	SW USA	1.4	37	9	70	8	CR
EUFAULA	2	SW USA	0.8	81	29	405	4	D
THUNDERBIRD	2	SW USA	14.6	27	9	150	8	D
FORT COBB	2	SW USA	7.6	38	12	110	14	D
ALTUS	2	SW USA	2.1	41	10	60	14	BG
ARBUCKLE	2	SW USA	3.2	20	8	70	7	D
WISTER	2	SW USA	0.04	80	16	230	4	D
KARIBA	2	STH.AFR.	4	–	20	20	–	D
ROBERTSON	2	STH.AFR.	1.1	–	10	30	9	BG
MCILWAINE	2	STH.AFR.	0.8	82	40	60	12	BG
PRINCE EDWARD/HENRY								
HALLAM	2	STH.AFR.	0.2	–	80	490	–	G
JOHN MACK	2	STH.AFR.	0.2	58	10	830	9	–

Table II. Continued.

LAKE	STUDY	AREA	TW	TP	SRP	TIN	CHA	PHYTO
CACTUSPOORT	2	STH.AFR.	0.03	–	5	–	–	–
LOWER UMGUSA	2	STH.AFR.	0.2	–	170	600	–	BG
UPPER UMGUSA	2	STH.AFR.	0.6	–	170	600	–	BG
LINDLEYSPOORT	2	STH.AFR.	0.1	–	30	530	4	BG
BOSPOORT	2	STH.AFR.	0.2	72	40	1170	12	BG
RUST DER WINTER	2	STH.AFR.	0.2	37	10	80	4	–
LOSKOP	2	STH.AFR.	0.3	44	20	200	11	BG
NEW DORINGSPOORT	2	STH.AFR.	0.8	88	30	460	3	BG
TONTELDOOS	2	STH.AFR.	0.03	–	10	190	1	–
BRONKHORSTSPRUIT	2	STH.AFR.	0.3	66	20	350	5	BG
RIETVLEI	2	STH.AFR.	1	800	390	160	10	G
ROODEPLAAT	2	STH.AFR.	1.5	75	50	440	13	BG
BUFFELSPOORT	2	STH.AFR.	0.4	–	10	110	7	G
HARTBEESPOORT	2	STH.AFR.	0.8	750	500	1000	52	BG
OLIFANTSNEK	2	STH.AFR.	0.2	49	20	700	3	BG
BLODEMHOF	2	STH.AFR.	0.6	–	100	820	23	–
P.K. LE ROUX	2	STH.AFR.	–	–	20	470	3	BG
ALBERT FALLS	2	STH.AFR.	1.5	13	1	30	5	D
NAGLE	2	STH.AFR.	0.1	12	1	210	3	G
HAZELMERE	2	STH.AFR.	0.3	48	10	410	6	BG
VERNON HOOPOER	2	STH.AFR.	0.1	81	20	1380	16	D
HENLEY	2	STH.AFR.	0.1	24	2	770	4	Y
MIDMAR	2	STH.AFR.	1	15	10	140	2	D
NAHOON	2	STH.AFR.	1.6	560	70	690	1	BG
LAING	2	STH.AFR.	0.4	1050	100	1180	1	BG
BRIDLEDRIFT	2	STH.AFR.	0.5	1140	50	1490	1	BG
H.F. VERWOERD	2	STH.AFR.	0.5	–	20	3340	–	BG
EBRO	1	SPAIN	–	–	3	297	1	Y
AGUILAR DE CAMPO	1	SPAIN	0.8	–	6	626	3	Y,D
PORMA	1	SPAIN	1.5	–	3	393	3	Y
LINARES DEL ARROYO	1	SPAIN	0.5	–	5	1312	8	D,BG
CUERDO DEL POZO	1	SPAIN	–	–	3	32	7	Y
TRANQUERA	1	SPAIN	0.2	–	6	4393	2	Y
BUENDIA	1	SPAIN	3.5	–	27	2022	1	Y
ENTREPENAS	1	SPAIN	1.1	–	5	1933	1	D
EL VADO	1	SPAIN	0.3	–	5	480	1	–
EL ATAZAR	1	SPAIN	1.3	–	19	680	2	D
SANTILLANA	1	SPAIN	0.8	–	5	695	1	D,G
SAN JUAN	1	SPAIN	0.2	–	10	745	13	D,BG
GURGUILLO	1	SPAIN	0.5	–	12	886	14	D,BG
CAZALEGAS	1	SPAIN	–	–	135	1300	2	D
TORCON	1	SPAIN	–	–	7	136	2	D,G
GUAJARAZ	1	SPAIN	1.1	–	6	4133	3	Y,D
PENARROYA	1	SPAIN	–	–	4	4345	1	Y,D
ALARCON	1	SPAIN	2.6	–	1	921	2	Y,G
GENERALISSIMO	1	SPAIN	1.4	–	5	3333	1	Y
LORIGUILLA	1	SPAIN	–	–	3	2600	3	Y
EL VELLON	1	SPAIN	1.2	–	86	1333	1	D,G
GUADALMENA	1	SPAIN	1.2	–	125	2446	1	Y,D
GUADALEN	1	SPAIN	1	–	97	500	1	Y,G
TRANCO DE BEAS	1	SPAIN	1.9	–	139	367	1	Y
RUMBLAR	1	SPAIN	–	–	9	393	1	Y,G
JANDULA	1	SPAIN	–	–	109	9266	3	Y,G
CENAJO	1	SPAIN	1.2	–	2	2866	2	Y,D
TALAVE	1	SPAIN	–	–	9	5918	3	Y
HUME	1	AUSTRALIA	0.5	22	5	60	5	D

Table II. Continued.

LAKE	STUDY	AREA	TW	TP	SRP	TIN	CHA	PHYTO
MULWALA	1	AUSTRALIA	0.01	33	10	37	10	D
MOONDARA	1	AUSTRALIA	–	–	–	–	–	–
BURLEY GRIFFIN	1	AUSTRALIA	–	45	–	350	6	BG
GRINNINDERA	1	AUSTRALIA	–	21	–	370	5	–
PROSPECT	1	AUSTRALIA	0.1	22	–	1350	3	BG
MOUNT BOLD	1	AUSTRALIA	0.4	99	–	1430	3	BG
BURRAGORANG	1	AUSTRALIA	–	–	–	–	3	D,G
ITEZHI-TEZHI	1	AFRICA	–	–	250	30	2	BG
DA GAMA	1	AFRICA	–	–	6	400	3	–
KLIPKOPJE	1	AFRICA	–	–	5	300	10	–
WITKLIP	1	AFRICA	–	–	6	400	4	–
RUSFONTEIN	1	AFRICA	–	–	50	1300	97	BG
ARMENIA	1	AFRICA	–	–	–	–	1	BG
ERFENIS	1	AFRICA	–	–	–	–	1	BG
CHONGWE	1	AFRICA	–	–	160	280	11	BG
SWARTWATER	1	AFRICA	0.7	3	1	230	6	G
NOORA	1	AFRICA	–	–	60	610	–	–
UMTATA	1	AFRICA	–	–	10	698	–	–
XONXA	1	AFRICA	–	–	20	360	–	–
LUBUSI	1	AFRICA	–	–	60	1000	–	–
BONKOLO	1	AFRICA	–	–	40	590	–	–
VAAL	1	AFRICA	–	–	40	130	2	BG
WELBEDACHT	1	AFRICA	–	–	–	520	–	–
CHELMSFORD	1	AFRICA	–	–	–	130	–	–
KAFUE	1	AFRICA	–	–	950	1000	–	–
CAHORA BASSA	1	AFRICA	–	–	210	130	–	–
SAN ROQUE	1	ARGENTINA	–	–	70	5.5	–	–
LOS MOLINOS	1	ARGENTINA	–	–	–	–	–	–
EMBALSE DEL RIO TERCERO	1	ARGENTINA	–	–	–	–	–	–
DAK PATHAR	1	INDIA	–	–	35	75	40	BG
RIHAND	1	INDIA	1.2	–	27	330	–	–
MANSAGAR	1	INDIA	–	–	4240	2000	7	BG
SATPURA	1	INDIA	–	–	10	0.1	1	BG
GANDHISAGAR	1	INDIA	–	–	10	0.1	10	G,BG
SANPNA	1	INDIA	–	–	0.1	0.1	10	BG
TAWA	1	INDIA	–	–	10	0.1	1	–
RAILWAY (NAINPUR)	1	INDIA	–	–	0.1	0.1	10	BG

All data cited were from studies of at least one year in duration, with sampling frequencies varying from weekly to bimonthly. Most lakes were sampled fortnightly, however, and this data base can be considered to be reasonably robust from that point of view. Certainly, the Texas lake data are more authoritative than the NES data employed during our preliminary investigations (Thornton & Rast, 1989). Thus, on the whole, the data can be assessed as at least equivalent to our previous data base, insofar as being representative of the regions concerned. This is despite the greater potential for analytical variability in this data set as mentioned above.

Statistical analysis was carried out using the techniques of Tukey (Erickson & Nosanchuk, 1979) using the spreadsheet programme LOTUS 123 (Lotus Development Corporation, 1986).

3. Results

Table III presents a statistical summary of the data, highlighting some of the distinctive features of the limnology of these semi-arid lakes. The lakes have a large catchment : surface area ratio (AD:A), which underlies the importance of catchment effects on their limnology and their potential for

Table III. Comparison of semi-arid lake data gathered during the present study to similar data gathered by THORNTON & RAST (1989) upon which they based their hypotheses concerning these lakes, and to data gathered under the OECD eutrophication programme (OECD 1982). Median values and interquartile ranges are shown for present study and that of THORNTON & RAST (1989); median values and ranges are shown for OECD (1982) data.

Parameter	Symbol	Units	This study	Thornton & Rast	OECD
Volume	V	10^6 m^3	42< 157< 346	37< 180< 679	–
Area	A	km^2	5< 13< 34	5< 23< 88	03< 6.6<5800
Catchment area	AD	km^2	426<1281<4264	855<3387<11350	–
Max. depth	Z_{max}	m	17< 30< 49	15< 22< 36	–
Mean depth	$\bar{z}$	m	5< 10< 20	5< 7< 11	1.7< 14.3< 313
Retention time	TW	y	0.5< 1.1< 1.3	0.3< 0.9< 1.7	0.01< 1.2< 700
Area ratio	AD:A	–	85< 166<1572	114< 143< 239	–
Total P	TP	µg l^{-1}	20< 20< 42	23< 42< 81	3< 47< 750
Ortho-P	SRP	µg l^{-1}	5< 10< 20	9< 12< 30	1< 16< 891
Total N	TIN	µg l^{-1}	80< 229< 610	70< 160< 470	263<1244<6095
Chlorophyll	CHA	µg l^{-1}	2< 4< 8	4< 8< 12	0.3< 8.4< 89
Secchi disk	SDT	m	0.7< 1.2< 2.4	1.0< 1.1< 1.2	0.8.< 3.3< 28.3
P load	LP	g m^{-2} y^{-1}	–	0.3< 1.3	0.02< 1.2< 80
N:P ratio	TN:TP	–	2.3< 4.3< 7.3	1.1< 3.0< 13.3	–

nonpoint source nutrient enrichment. This is reflected in the high percentage (70%) of enriched impoundments, defined on the basis of observer-assessed trophic status. It is also reflected in the relatively short water retention times (TW; median value = 1.1 y) recorded for these lakes. Many of these lakes are also moderately shallow, having a median value of 300 m for the maximum depth and a median value of just under 10 m for the mean depth.

The physico-chemical and biological effects of the large catchment:surface area ratios are reflected in the enriched nature of the impoundments, as noted earlier. Median total phosphorus concentrations exceed the 25 µg l^{-1} commonly used to define enrichment in the temperate zone (Vollenweider, 1968). Much of the phosphorus is in particulate form (determined as the difference between total and soluble reactive phosphorus). High concentrations of suspended particulates are common in semi-arid zone impoundments as demonstrated by a median Secchi disc transparency value of 1.2 m. Such turbidities result from the high erosivity in the large catchments, which leads to soil erosion (Lal, 1985). Nitrogen:phosphorus (TIN:TP) ratios are low (median value = 4.3), although, where such limitation to algal growth was reported, most (78%) of the study lakes were nitrogen limited. Chrysophytes dominated the algal flora in most of the study lakes, although cyanophytes dominated in many others. Chlorophyll-a concentrations were generally low, however (median value = 4 µg l^{-1}).

The lakes involved in this study were subject to significant intra-annual variations in volume, surface area, water residence time and depth, which affected their chemical and biological parameters (Margalef et al., 1976; Walker & Hillman, 1982; Cullen & Smalls, 1981; Obrdlík, 1987; Walmsley & Toerien, 1977; Grobbelaar, 1979; Stegmann et al., 1981; Du Preez, 1985; Salter, 1985; Marshall, 1984; Bonetto & Di Persia, 1984; Bhargava, 1984; Singh, 1985; Gopal et al., 1984; Unni, 1985). These variations are essentially seasonal and can lead to significant variations in phytoplankton species composition and biomass despite favourable year-round growing conditions in many of the reservoirs, as well as evaporative concentration effects in the chemical environment of the waterbodies.

4. Discussion

4.1. Semi-arid lake classification

Comparison of the data given in Table II with that of Thornton & Rast (1989) suggests there are a number of common features characteristic of lakes in the semi-arid zone, which distinguish them from lakes in other climatic regions (Table III).

Although there are obviously internal difference between the lakes included in the two semi-arid lake data bases, these are in part due to the nature of the data sources. The data included in our previous study were predominantly from eutrophication studies, and hence could have been biased to some degree toward nutrient enriched systems. This bias was not present in the present data, as a result, these data permit a more objective assessment of the relative numbers of lakes subject to nutrient enrichment and their characteristics. The similarities between the two data bases are obvious and outweigh the slight differences, thereby validating our previous conclusion regarding the representativeness of the data used to formulate our models.

Comparing man-made lakes of the semi-arid zone with the classical lake model of the north temperate zone (Table III) suggests the features identified above do in fact constitute a clearly-defined class of lakes. This confirms our conclusion that the differences between the semi-arid zone and the temperate zone are based on other factors, such as natural variations (e.g., geology and runoff; Grobler & Silberbauer, 1985), although we cannot completely eliminate the effect of induced variability (e.g., differences due to sampling and analysis). The OECD (1982) identified the fact that semi-arid zone lakes may respond differently to external stimuli (e.g., phosphorus loading) due to peculiar inflow and throughflow regimes, turbidity (in the case of Mt. Bold Reservoir) and morphometry. In some cases, these reservations were extended to other waterbodies as well. Thus, before proceeding further, we must question if it is realistic to compare our semi-arid lake paradigm (based on man-made lakes) to the classical temperate lake paradigm (based on natural lakes).

This question can be approached on several levels and in a number of ways. First, the comparison is justified in terms of the fact that, in countries such as the United States with a range of waterbodies across several climatic zones, there is no legislative distinction made between types of lakes in the application of water laws. As pointed out in our description of the southern African paradigm, this often results in the blanket application of standards without regard for limnology. Second, man-made lakes in the semi-arid zone perform the same socioeconomic functions as natural lakes in the more water rich areas of the world and can be considered even more valuable resources due to the general scarcity of water. Thus, there is a greater need to manage these systems wisely; wise management demands understanding. Since our limnological understanding of waterbodies is based largely on the temperate lake paradigm, it becomes exceedingly relevant to examine this paradigm in a 'local' context. Finally, many studies, including the OECD programme, have demonstrated the applicability of research methodologies and management techniques between climatic regions, and hence between types of lakes (e.g., OECD, 1982, Thornton & Walmsley, 1982; Walmsley & Thornton, 1984; Ryding & Rast, 1989; Symoens et al., 1981; Davies & Walmsley, 1985).

On another level, in addition to the waterbodies of the "Shallow Lakes and Reservoirs Project", the OECD (1982) study included reservoir data in other regional projects without distinction. Where data points were excluded, there were clearly identified reasons for such exclusions other than a distinction between man-made and natural lakes; to wit, lakes that were subject to artificial manipulation such as aeration, lakes that were limited by light or nitrogen (non-phosphorus limited waterbodies), and lakes that were subjected to heavy point source pollutant inputs, internal loading or severe constriction along their longitudinal axes (OECD, 1982, p. 116). The empirical statistical relationships developed for the various component projects within the OECD programme were also shown to fit well into a combined model (OECD, 1982). This suggests that temperate reservoirs react in much the same manner as temperate lakes and, while this latter assumption should be examined in more detail, it does suggest the validity of our comparison on both conceptual and scientific levels.

4.2. Characteristics of semi-arid zone lakes

The primary distinguishing characteristic of semi-arid zone lakes is the large catchment : surface area ratio required to maintain a minimum sustainable water yield from these reservoirs (Alexander, 1985). This single feature has major implications for the limnology of the resultant waterbodies that lead to manifestations throughout the aquatic environment. Semi-arid zone catchments are commonly comprised of well-leached, organic-poor soils which are extremely susceptible to erosion (Lal, 1985). This is aggravated by the seasonal nature of the rainfall events, which commonly occur in the form of concentrated storm fronts having a high erosivity potential (Lal, 1985). Over the relatively large catchments, the result is high suspended particulate loads carried by rivers and the occurrence of turbid waterbodies (Davies & Walmsley, 1985). Deposition of this material in waterbodies can significantly reduce the capacities of impoundments in the region and alter their morphology and catchment characteristics.

Such alterations in the catchments affect the nutrient budgets of the waterbodies in various ways; namely, in terms of the total nutrient input from diffuse sources, in terms of the quantities of particulate nutrients in the systems, and in terms of their flushing rates. The former implies there is likely to be an higher input of nutrients to semi-arid impoundments than to temperate impoundments having smaller catchment areas, despite the possibility that nutrient export from the organic-poor soils of the semi-arid zone catchments might be lower than from the richer soils of the temperate zone. Although there are relatively few data with which to test this contention, in a review of nonpoint source literature from the African sub-region, Thornton (in print) suggests that nutrient export from these catchments is not significantly lower than in the temperate zone. Hence, simply by virtue of the size of the catchment area, there is a greater probability that semi-arid zone lakes will be 'enriched' when compared to similar sized lakes elsewhere. Generally our data support this concept, with over half of the study lakes being meso- or eutrophic (Table II).

The accentuated effect of nonpoint source inputs from the catchment area in semi-arid zone lakes is further magnified by the need to conserve water in these regions. This leads to the reintroduction of waste waters into natural water courses (DWA, 1986, Thornton, 1987b). This necessity has aggravated the occurrence of cultural eutrophication in these regions (Toerien *et al.*, 1975, for example). Most countries in the semi-arid zone are only now beginning to come to terms with this problem of nutrient enrichment (Ouano *et al.*, 1978, Alabaster, 1980). However, Thornton (1987a) has pointed out many socioeconomic issues that inhibit the promulgation and enforcement of water quality regulations in these largely underdeveloped areas.

A further artfact of the high turbidities and nutrient loads in the semi-arid zone lakes is the occurrence of relatively high concentrations of particulate phase nutrients in the study lakes. While our data only refer to phosphorus, median values of total phosphorus exceed median values of orthophosphorus by a factor of 3 (Table II). This is roughly twice the mean for temperate zone lakes (OECD, 1982) and is consistent with our hypothesis. While a large percentage of this particulate phosphorus is bioavailable (Thornton, 1979; Grobler & Davies, 1979), the presence of inorganic suspensoids/turbidity does restrict the depth of light penetration, and therefore the percentage of the water column conducive to algal growth (Robarts, 1979; 1984). In fact, Stegmann (1982) has shown that primary production in turbid reservoirs is greatly compressed into the upper few centimetres of the surface waters, which restricts the availability of particulate nutrients to algae in these lakes except under the most turbulent conditions. Melack (1985) cites evidence of enhanced bacterial activity in turbid lakes at the expense of phytoplankton, and notes that cases of mutual flocculation of phytoplankton and bacteria with clays have been reported by some investigators. These observations may explain the relatively low median chlorophyll-a concentration of 4 μg l^{-1} recorded in the study lakes (Table II).

The low N:P ratios also contribute to the biotic responses of semi-arid zone reservoirs. Median values of just over 4 : 1 have been reported (Table II). Smith (1982) suggests such a low ratio favours the dominance of cyanophytes over other classes of algae in these systems. Similarly, Robarts *et al.*, (1986) note that modifications to the light regime, due to inorganic turbidity effects should also favour cyanophyte dominance. Our data suggest that bluegreen algae do, in fact, dominate in the southern African and Indian impoundments. However, algal communities are dominated elsewhere by chrysophytes and diatoms (Table II). Whitford & Schumacher (1973) suggest these classes of algae favour cooler water and/or lower light intensities more than the cyanophytes which frequently occur in warmer waters. Certainly, differences in the climatic zone classification (semi-arid mid-latitude versus semi-arid tropical) might account for these differences in algal class dominance. Cyanophytes dominate in the tropical regions (southern African and India) and the other classes dominate in the mid-latitude regions (United States, Spain, Australia and Argentina) (Table II).

The low N:P ratio observed in most semi-arid zone lakes favours nitrogen limitation of phytoplankton growth. This is generally confirmed in Table II for lakes other than the Texas waterbodies (where nutrient limitation was calculated from N:P ratios). This is in contrast to our previous findings which suggested that despite low N:P ratios (median value = 3) most semi-arid lakes in the southwest United States and southern Africa were phosphorus-limited. Limiting nutrients were determined by bioassay for the lakes for which data were available (cf. Toerien *et al.*, 1975, NES, 1977), although the NES does note their results must be used with caution because of delays between sample collection and bioassay analysis. Generally, however, the few data on nutrient limitation included in the present study (excluding Texas lake data, $n = 10$) make the resolution of this issue problematical. Thus, we cannot comment meaningfully on the use of nutrient limitation as a semi-arid lake characteristic, although low N:P ratios do appear to be general within the region and do appear to influence phytoplankton growth (e.g., Table II and above discussion).

Finally, the morphological implications of fluctuating water levels and siltation on flushing rates can also modify the response of the lakes in the semi-arid zone to nutrient loading. Median flushing rates of just over one year will not greatly reduce algal growth through washout on an annual basis. However the highly seasonal nature of the inflows can do so at times during an annual cycle (Thornton, 1986).

4.3. Comparative limnology

In order to examine the differences in limnology between our semiarid lake data base and the (largely) temperate lake OECD data base, we plotted our data relative to the statistical

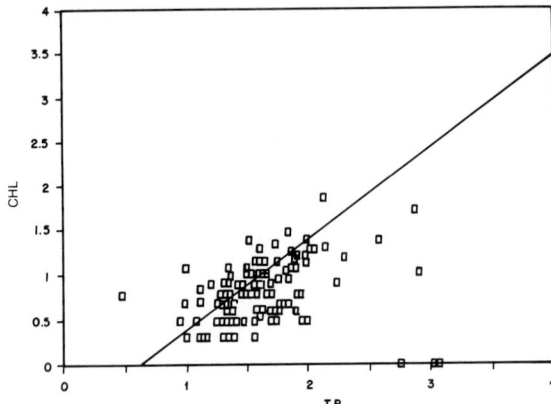

Fig. 2. Scatter plot of chlorophyll-a (CHL) and total phosphorus (TP) concentrations recorded from 133 semi-arid zone reservoirs. Log-log plot in mg m^{-3}; OECD (1982) line of best fit shown.

relationships derived by the OECD (1982). These plots (Figs. 2 to 8) identify the major points of difference between waterbodies in these climatic zones and highlight some similarities between their limnology in a graphical way. Generally, the phosphorus-chlorophyll relationships found in the OECD study apply to the semi-arid zone lakes of our data base (Figs. 2 and 3). This is consistent with various regional studies, such as those of Walmsley & Thornton (1984) and Archibald & Lee (1981), who validated these relationships in southern

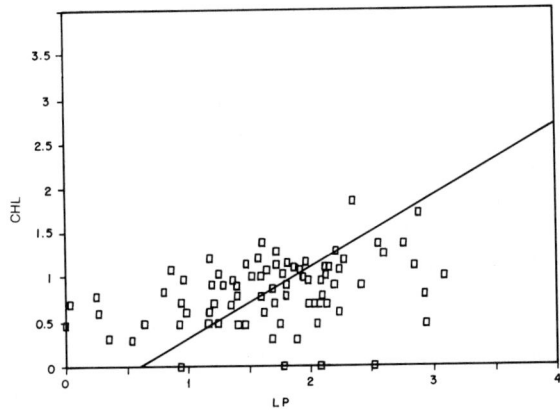

Fig. 3. Scatter plot of chlorophyll-a concentrations (CHL) and phosphorus loading rates (LP) recorded from 183 semi-arid zone reservoirs. Log-log values in mg m^{-3} and g m^2 y^{-1} for CHL and LP respectively; OECD (1982) line of best fit shown.

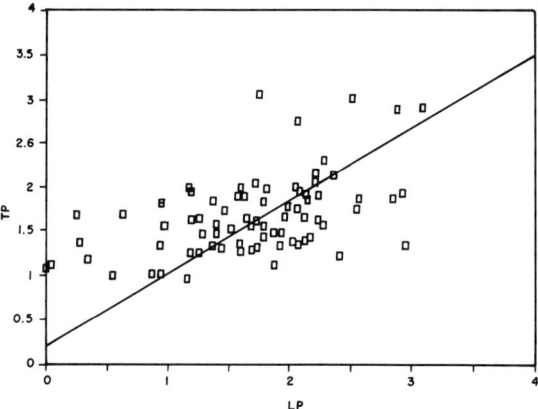

Fig. 4. Scatter plot of total phosphorus (TP) in mg m^{-3} and phosphorus loading (LP) in g m^{-2} y^{-1} recorded from 133 semi-arid zone reservoirs. Log-log values; OECD (1982) line of best fit shown.

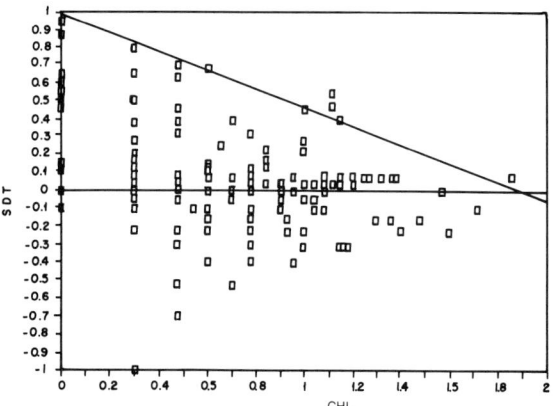

Fig. 5. Scatter plot of Secchi disc transparency (SDT) in m and chlorophyll (mg m^{-3}) recorded from 175 semi-arid zone reservoirs. OECD (1982) line of best fit shown on the log-log plot.

Africa and Texas, respectively. Similarly, the phosphorus concentration – phosphorus loading relationship also applied in these waterbodies (Fig. 4). Thornton & Walmsley (1982) also previously validated this relationship (e.g., the Vollenweider model) in southern Africa, but noted that the model tended to overpredict by a relatively constant margin (cf., Twinch *et al.*, 1986). These results were confirmed by a complementary study by Jones & Lee (1984) using the USA-OECD models (Rast and Lee, 1978). Thus, it appears the limnological relationships governing the behaviour of phosphorus in a waterbody, and its transfer to primary producers, are common to all climatic zones studied. Indeed, Chapman & Thornton (1986) found there were numerous similarities between the chemical behaviour and biotic responses of lakes, estuaries and oceans, ascribing these similarities to the essential sameness of the underlying biochemistry of the organisms involved, and to the physical properties of the medium in which they occur. These similarities, however, do not address the differences in eutrophication threshold values and the composition of the total phosphorus fraction that may exist between regions.

Figures 5 and 6 begin to identify these differences between climatic zones. The OECD relationships between Secchi disc transparency and both phosphorus and chlorophyll concentrations overestimate the depth of light penetration in semi-arid zone lakes by a substantial amount. These figures suggest there is a relatively high degree of inorganic turbidity in these waterbodies compared to temperate zone waters (Fig. 5); a fact confirmed in the reviews in Davies & Walmsley (1985). Likewise, the relationship between phosphorus and Secchi disc transparency in the temperate zone is predicated upon lower inorganic turbidities (i.e., the statistical relationship is indirect in this case, being based on the relationship between phosphorus and chlorophyll concentrations; Figs. 2 and 6). These relationships are not inconsistent with our semi-arid lake paradigm in that the observed relationship between phosphorus and chlorophyll concentrations is maintained in this

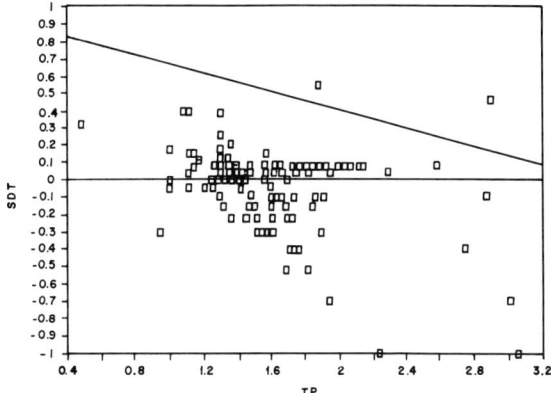

Fig. 6. Log-log scatter plot of Secchi disc transparency (SDT in m) and total phosphorus (TP in mg m^{-3}) recorded from 133 semi-arid zone reservoirs. OECD (1982) line of best fit shown.

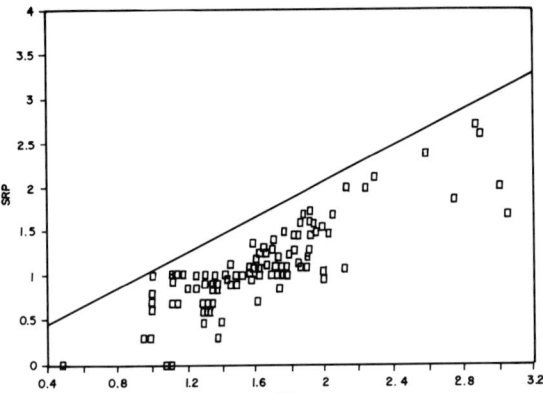

Fig. 7. Log-log scatter plot of soluble reactive (SRP) and total phosphorus (TP) concentrations in mg m^{-3} recorded from 133 semi-arid zone reservoirs. Isotonic line shown.

model (e.g., for any given unit of phosphorus, there is likely to be a unit of chlorophyll produced by the primary producers of both zones; however, the light penetration in semi-arid zone lakes will be less due to the presence of additional suspensoids from the larger catchment areas, in addition to algal turbidity). The presence of a higher concentration of inorganic particulates is confirmed in part in Fig. 7 which shows a much lower orthophosphate concentration for a given level of total phosphorus than is common in the temperate lake paradigm (OECD, 1982, p. 40). Fewer of our data points approach the isotonic line than do those from the temperate zone, suggesting a higher particulate phosphorus concentration in these waters.

A plot of Secchi disc transparency against observer-assessed trophic state (Fig. 11) is less clear than the chlorophyll and phosphorus concentration frequency plots (Figs. 9 and 10), but has similar importance for lake management. Like the OECD (1982, p. 84) relationship, Secchi disc transparency clearly decreases with an increase in trophic status from oligotrophic to mesotrophic. At high trophic levels, however, the relationship breaks down, with the eutrophic and hypertrophic waterbodies becoming less turbid by often substantial degrees; to wit, the data on the five hypertrophic systems included in our combined data base (Table II) have a median depth of light penetration of approximately 10 m. At first glance, we found this situation to be anomalous, until we examined two features of this figure. The first was that our semi-arid lakes were more turbid than the temperate lakes used to define the OECD distribution. This fact has been noted above as one of the distinguishing characteristics of semi-arid zone lakes. The second, and related, factor is the relationship between water clarity and increasing trophic status in turbid reservoirs. Grobler *et al.* (1983; 1987) clearly showed that increasing salination associated with cultural enrichment in the Vaal River system resulted in increased flocculation of particulates and improved water clarity or, as our data show, clearer water at high levels of enrichment. This relationship has been extended to a number of other South African waterbodies, including several of the reservoirs included in this survey (Grobler *et al.*, 1983); from the foregoing, it would appear to be a general feature of turbid lakes. Obviously, other factors such as basin geology and geomorphology play a role in the incidence and likelihood of turbid conditions in these reservoirs (Lal, 1985), and will modify the general application of the relationship (as shown by the breadth of the curves in Fig. 11 at the higher trophic levels). However, our data strongly suggest that turbidity as reflected in Secchi disc transparency has limited usefulness as a trophic state index in semi-arid waterbodies.

The relationship between total nitrogen and total phosphorus concentrations is also different from that of the temperate zone (Fig. 8). The resultant distribution of the data suggests an almost inverse relationship (compare our Fig. 8

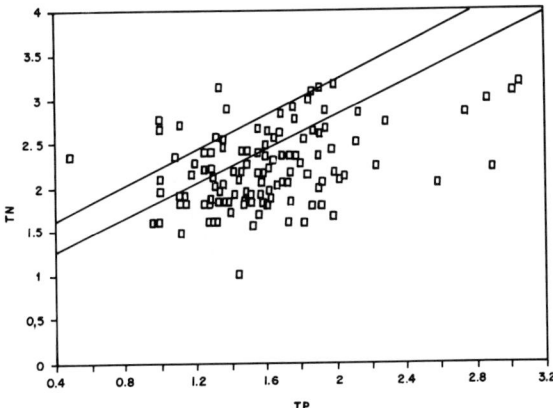

Fig. 8. Log-log scatter plot of total phosphorus (TP) and total nitrogen (TIN) concentrations in mg m^{-3} recorded from 133 semi-arid zone reservoirs. N:P ratios of 7 : 1 and 15 : 1 comparable to the OECD (1982) presentation are shown on the plot.

with Fig. 4.1 of the OECD, 1982). The low N:P ratio in semi-arid zone lakes was highlighted above as a characteristic of these lakes.

Figures 2 to 8 and the data contained in Table III confirm there are significant differences between temperate zone and semi-arid zone waterbodies. These differences relate primarily to the distinguishing features of the semi-arid zone reservoirs studied. However, there are also similarities in the nutrient loading-response area of the limnology of these lakes which suggest the basic principles remain valid in the semi-arid zone. Thus, our comments regarding the importance of our findings for lake management are reenforced. The implications of these issues are discussed below.

5. Implications for eutrophication management

The foregoing discussion confirms our previous conclusion that there are real differences in degree between the potential responses of semi-arid zone reservoirs to external stimuli compared to those of temperate zone waterbodies. Thus, while the fundamental processes might be unchanged between climatic zones, these distinguishing features can have major implications for the management and assessment of eutrophication in the lakes of the semi-arid zone. This is particularly relevant in terms of common eutrophication indicators, such as water transparency, chlorophyll concentration, ambient phosphorus concentration, and the presence of indicator organisms (e.g., cyanophytes). Our data confirm that semi-arid reservoirs are subject to enrichment and exhibit eutrophication or enrichment responses. In fact, our data suggest enrichment is a more common occurrence in these reservoirs for both natural and cultural reasons than in the temperate zone. Thus, we reiterate our injunction that application of temperate lake eutrophication paradigms to semi-arid zone lakes *must be done with caution*. For example, plotting chlorophyll-a concentration and phosphorus concentration frequencies against observer-assessed trophic status (Figs. 9 and 10) confirms our previous hypotheses concerning the elevated threshold values of these indices in semi-arid climates; to wit, concentrations of 60 μg l^{-1} and 12 μg l^{-1}, respectively.

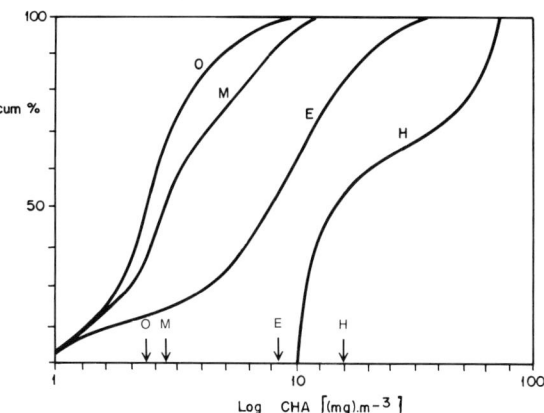

Fig. 9. Smoothed frequency plot of the mean chlorophyll-a concentrations recorded from the study lakes plotted against observer-assessed trophic status. Arrows indicate approximate mid-point of trophic boundaries in semi-arid zone lakes.

The larger data base afforded by this study suggests slight adjustments of our previously postulated eutrophication thresholds may be required. The total phosphorus threshold value indicated by our current data base suggests a slight upward adjustment of the 50 μg l^{-1} threshold to 60 μg l^{-1} might be warranted. Such an adjustment is marginal, and possibly an artfact of the apparent reluctance of observers to distinguish between eutrophy and hypertrophy. This failure results in a very broad phosphorus concentration peak (Fig. 10). Similarly, a downward adjustment of the chlorophyll threshold concentrations is also indicated (Fig. 9). Nevertheless, in both cases, the

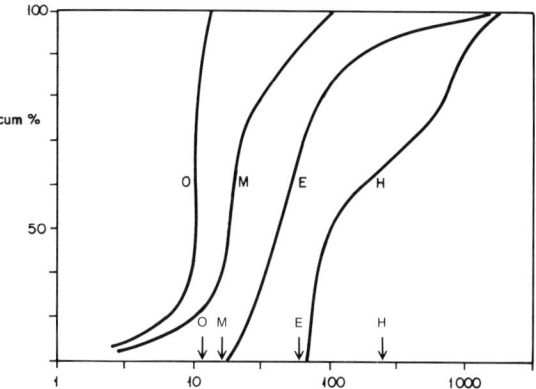

Fig. 10. Smoothed frequency plot of the mean total phosphorus concentrations recorded from the study lakes plotted in terms of the observer-assessed trophic state. Arrows indicate approximate mid-point of trophic boundaries in semi-arid zone lakes.

the concentration thresholds suggested from our data base exceed commonly accepted threshold values based on temperate zone data (OECD, 1982, p. 83–84). This phenomenon is consistent with the semi-arid zone lake paradigm set out above.

Thus, in terms of lake management, it is essential that these threshold values be incorporated into legislated water quality criteria on a regional basis, in order to provide a realistic eutrophication management response appropriate for the 'natural' conditions expected in the waterbodies being managed.

6. Conclusions

Our studies have shown that the difference in degree between semi-arid zone reservoirs and temperate lakes identified in our preliminary analysis (Thornton & Rast, 1989) are confirmed on the basis of an independent set of data compiled from the literature and other monitoring studies. The principle distinguishing feature of semi-arid zone lakes, their large AD:D ratio, underlies the decreased depth of light penetration, increased (particulate) phosphorus concentration, and higher algal biomass commonly found in these waterbodies. This feature also leads to a higher frequency of enrichment in these lakes, relative to temperate zone lakes such as those included in the OECD study. This fact, combined with regional microclimatic differences, affects algal species dominance at the class level. These differences, whilst not based on differing biogeochemical mechanisms, do affect broadly-based legislated water quality criteria which we feel cannot be applied realistically across different climatic zones. While our studies are as yet incomplete for lakes in the humid tropics and extreme latitudes, we can begin to summarise our present findings in terms of a 'decision tree' setting out 'base line' conditions in the semi-arid zones versus the temperate zone of classical limnology (Fig. 12), the ramifications of which are highly significant for eutrophication management in the semi-arid zone.

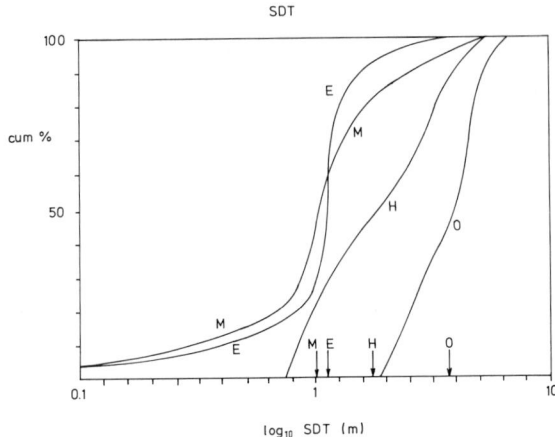

Fig. 11. Smoothed frequency plot of the Secchi disk transparency reported for the study lakes, plotted in terms of observer-assessed trophic state. Arrows indicated mid-point of trophic boundaries in semi-arid zone lakes.

Acknowledgements

We would like to thank the following persons for their assistance in this investigation: Dr. Juan Armengol for providing access to the Spanish reservoir data; Texas Natural Resources Information System for providing access to the data on the Texas reservoirs; Andrew Darroch for assistance with the computer analysis of our data base; Margaret Thornton for sketching out the decision tree; and our many colleagues who contributed to the concepts generated in this paper through discussion, provision of data, and comment on the manuscript. This paper is a contribution to the Zandvlei Working Group of the City of Cape Town's Inland Waters Management Team and is published with the permission of the City Planner, City of Cape Town. The opinions expressed in this paper are those of the authors and do not necessarily reflect the opinions of their organisations.

References

Alabaster, J. S., 1980. Draft review of the state of aquatic pollution of East African inland waters. CIFA Rep. 80/8. FAO, Rome.

Alexander, W. J. R., 1985. Hydrology of low latitude southern hemisphere land masses. Hydrobiologia 125: 75–83.

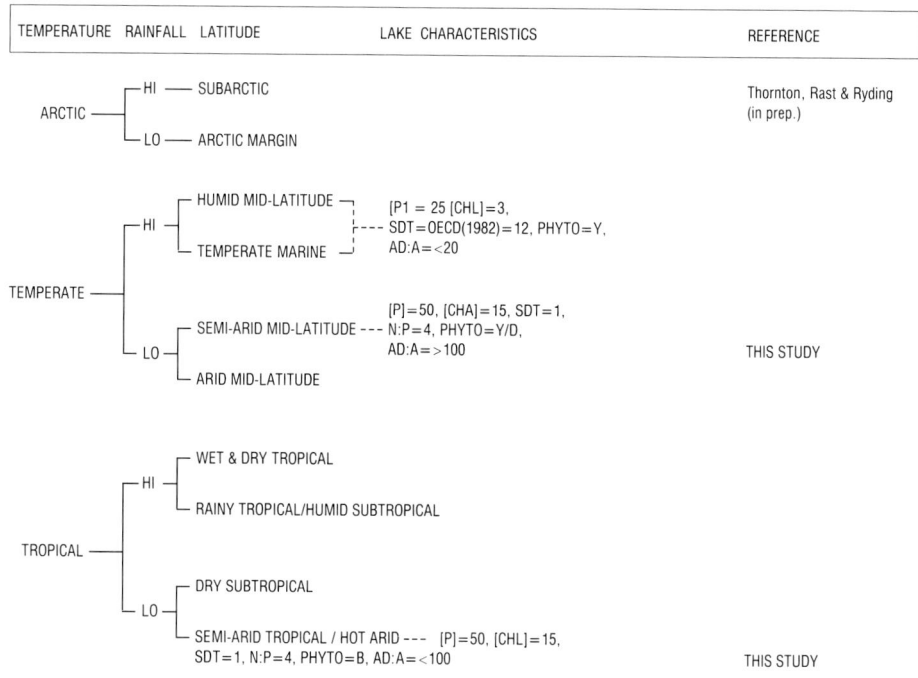

Fig. 12. Partial decision tree summarising median base line conditions expected and eutrophication threshold values of phosphorus concentrations, chlorophyll-a concentrations, Secchi disc transparency and dominant algal class for semi-arid zone and temperate lakes.

Archibald, E. M. & G. F. Lee, 1981. Application of the OECD eutrophication modelling approach to Lake Ray Hubbard, Texas. J. Amer. Water Works Assoc. 73: 590–599.

Bhargava, D. S., 1984. Equalizing effect of impounding at Dak Patthar along Yamuna. Institute of Engineers (India) Journal 65: 31–34.

Bonetto, A. A. & D. H. Di Persia, 1984. The San Roque Reservoir and other man-made lakes in the central region of Argentina. In: F. B. Taub (ed.), Lakes and Reservoirs. Ecosystems of the World 23: 541–556.

Chapman, P. & J. A. Thornton, 1986. Nutrients in aquatic ecosystems: An introduction to similarities between freshwater and marine ecosystems. J. Limnol. Soc. sth. Afr. 12: 2–5.

Cullen, P. & I. Smalls, 1981. Eutrophication in semi-arid areas: the Australian experience. Water Quality Bull. 6: 79–91.

Davies, B. R. & R. D. Walmsley, 1985. Perspectives in Southern Hemisphere Limnology. Developments in Hydrobiol. 28. Junk, The Hague.

Du Preez, A. L., 1985. The chemical composition of Transkei River water. Water S.A. 11: 41–47.

DWA (Department of Water Affairs), 1986. Management of the Water Resources of the Republic of South Africa. Government Printer, Pretoria.

Erickson, B. H. & T. A. Nosanchuk, 1979. Understanding Data. Open University Press, London.

Gopal, B., R. K. Trivedy & P. K. Goel, 1984. Influence of water hyacinth cover on the physico-chemical characteristics of water and phytoplankton composition in a reservoir near Jaipur (India). Int. Revue ges. Hydrobiol. 69: 859–865.

Grobbelaar, J. U., 1979. Early observations of some limnological characteristics of three Orange Free State impoundments. J. Limnol. Soc. sth. Afr. 5: 47–50.

Grobler, D. C. & E. Davies, 1979. The availability of sediment phosphate to algae. Water S. A. 5: 114–122.

Grobler, D. C. & M. J. Silberbauer, 1985. The combined effect of geology, phosphate sources and runoff on phosphate export from drainage basins. Water Res. 19: 975–981.

Grobler, D. C., D. F. Toerien & J. S. De Wet, 1983. Changes in turbidity as a result of mineralisation in the lower Vaal River. Water S.A. 9: 110–116.

Grobler, D. C., D. F. Toerien & J. N. Rossouw, 1987. A review of sediment/water quality interaction with particular reference to the Vaal River system. Water S. A. 13: 15–22.

Jones, R. A. & G. F. Lee, 1984. Application of OECD eutrophication modelling approach to South African dams (reservoirs). Water S. A. 10: 109–114.

Lal, R., 1985. Soil erosion and sediment transport research in tropical Africa. Hydrol. Sci. J. 30: 239–256.

Lotus Development Corporation, 1986. 1-2-3 Release 2.01. Lotus Corp., Cambridge.

Margalef, R., D. Planas Mont, J. Armengol Bachero, A. Vidal Celma, N. Prat Fornells, A. Guiset Serra, J. Toja Santillana & M. Estrada Miyares, 1976. Limnologia de los embalses Espanoles. Ministerio de Obras Publicas, Madrid.

Marshall, B. E., 1984. Predicting ecology and fish yields in

African reservoirs from pre-impoundment physico-chemical data. CIFA Tech. Pap. 12. FAO, Rome, 36pp.

Melack, J. M., 1985. Interactions of detrital particulates and plankton. Hydrobiologia 125: 209–220.

NES (National Eutrophication Survey), 1977. Working papers (various). US EPA, Corvallis.

Obrdlik, P., 1987. Chlorophyll-a concentration in the Lower Kafue River and Chongwe River Basins. J. Limnol. Soc. sth. Afr. 13: 58–61.

OECD (Organisation for Economic Cooperation and Development), 1982. Eutrophication of Waters: Monitoring, Assessment and Control. Organisation for Economic Cooperation and Development, Paris.

Ouano, E. A. R., B. N. Lohani & N. C. Thanh, 1978. Water Pollution Control in Developing Countries. Asian Institute of Technology, Bangkok.

Rast, W. & Lee, G.F., 1978. Summary Analysis of the North American (U.S. portion) OECD Eutrophication Project: Nutrient load-lake response relationships and trophic state indices. Ecol. Res. Series, No. EPA-600/3-78-008, U.S. Environmental Protection Agency. 454 p.

Robarts, R. D., 1979. Underwater light penetration, chlorophyll-a and primary production in a tropical African lake (Lake McIlwaine, Rhodesia). Arch. Hydrobiol. 86: 433–444.

Robarts, R. D., 1984. Factors controlling primary production in a hypertrophic lake (Hartbeespoort Dam, South Africa). J. Plankton Res. 6: 91–105.

Robarts, R. D., T. Zohary, F. Ojeda, C. C. E. Lovengreen & R. D. Walmsley, 1985. Observations on the underwater light field of two hypertrophic South African impoundments. J. Limnol. Soc. sth. Afr. 11: 78–81.

Ryding, S.-O. & W. Rast, 1989. The Control of Eutrophication of Lakes and Reservoirs. UNESCO/MAB Vol. 1. UNESCO, Paris and Parthenon Publishing Group, 314pp.

Salter, L. F., 1985. Preliminary study of water quality in the Kafue Flats, Zambia. S. Afr. J. Sci. 81: 529–531.

Singh, R. K., 1985. Limnological observations on Rihand Reservoir (Uttar Pradesh) with reference to the physical and chemical parameters of its water. Int. Revue ges. Hydrobiol. 70: 857–875.

Smith, V. H., 1982. The nitrogen and phosphorus dependence of algal biomass in lakes: An empirical and theoretical analysis. Limnol. Oceanogr. 27: 1101–1112.

Stegmann, P., 1982. Some limnological aspects of a shallow, turbid man-made lake. Ph.D. Diss., University of the Orange Free State, Bloemfontein.

Stegmann, P., A. J. H. Pieterse, D. F. Toerien, M. T. Seaman & B. C. W. Van Der Waal, 1981. A preliminary limnological survey of Swartwater Dam (Qwa Qwa). Water S.A. 7: 16–27.

Symoens, J. J., M. J. Burgis & J. J. Gaudet, 1981. The Ecology and Utilization of African Inland Waters. UNEP Rep. and Proc. Ser. 1. – UNEP, Nairobi, 141pp.

Thornton, J. A., 1979. Some aspects of the distribution of reactive phosphorus in Lake McIlwaine, Rhodesia: Phosphorus loading and abiotic responses. J. Limnol. Soc. sth. Afr. 5: 65–72.

Thornton, J. A., 1984. Strategic Approaches to Marketing Water Research in South Africa. MBA thesis, University of South Africa, Pretoria.

Thornton, J. A., 1986. Nutrients in African lake ecosystems: Do we know all? J. Limnol. Soc. sth. Afr. 12: 6–21.

Thornton, J. A., 1987a. Aspects of eutrophication management in tropical/sub-tropical regions. J. Limnol. Soc. sth. Afr. 13: 25–43.

Thornton, J. A., 1987b. A review of some unique aspects of the limnology of shallow southern African man-made lakes. GeoJournal 14: 339–352.

Thornton, J. A., in print. The assessment and control of nonpoint pollution of aquatic systems in tropical and subtropical Africa and other lesser developed regions. In: W. Rast, S.-O. Ryding, M. Holland, G. Jolankai & J. A. Thornton (eds.), Assessment and Control of Nonpoint Source Pollution of Aquatic Systems. A Practical Approach. UNESCO/MAB Ser.11 Parthenon Press, London.

Thornton, J. A. & W. Rast, 1987. Application of eutrophication modelling techniques to man-made lakes in semi-arid southern Africa. In: S. J. Nix & P. E. Black (eds.), Monitoring, Modeling and Mediating Water Quality. American Water Resources Association, Bethesda, MD: 547–558.

Thornton, J. A. & W. Rast, 1989. Preliminary observations on nutrient enrichment of semi-arid man-made lakes in the Northern and Southern Hemisphere. Lake and Reservoir Mgmt. 5 (2): 59–66.

Thornton, J. A. & R. D. Walmsley, 1982. Applicability of phosphorus budget models to southern African man-made lakes. Hydrobiologia 89: 237–245.

Toerien, D. F., K. L. Hyman & M. J. Bruwer, 1975. A preliminary trophic status classification of some South African impoundments. Water S.A. 1: 15–23.

Twinch, A. J., P. J. Ashton, J. A. Thornton & F. M. Chutter, 1986. A comparison of phosphorus concentrations in Hartbeespoort Dam predicted from phosphorus loads derived near the impoundment and in the upper catchment area. Water S.A. 12: 51–56.

Unni, K. S., 1985. Comparative limnology of several reservoirs in central India. Int. Revue ges. Hydrobiol. 70: 845–856.

Vollenweider, R. A., 1968. Scientific fundamentals of the eutrophication of lakes and flowing waters, with particular reference to nitrogen and phosphorus as factors in eutrophication. OECD Tech. Rep. DAS/SCI/68.27. Organisation for Economic Cooperation and Development, Paris, 182pp.

Walker, K. F. & T. J. Hillman, 1982. Phosphorus and nitrogen loads in waters associated with the River Murray near Albury-Wodonga, and their effects on phytoplankton populations. Aust. J. Mar. Freshwater Res. 33: 223–243.

Walmsley, R. D. & J. A. Thornton, 1984. Evaluation of OECD-type phosphorus eutrophication models for predicting the trophic status of southern African man-made lakes. S. Afr. J. Sci. 80: 257–259.

Walmsley, R. D. & D. F. Toerien, 1977. The summer conditions of three eastern Transvaal reservoirs and some considerations regarding the assessment of trophic status. J. Limnol. Soc. sth. Afr. 3: 37–42.

Whitford, L. A. & G. J. Schumacher, 1973. A Manual of Fresh-Water Algae. Sparks Press, Raleigh, 324pp.

Chapter II

Limnology and management of reservoirs in Brazil

J. G. Tundisi[1], T. Matsumura-Tundisi[2] & M. C. Calijuri[1]
[1] *Centre for Water Resources & Applied Ecology, School of Engineering, São Carlos, University of São Paulo, Brazil;* [2] *Limnological Laboratory, Department of Ecology & Evolutionary Biology, Federal University of Sao Carlos, Brazil*

Key words: South America, Brazil, tropics, reservoirs, mechanisms of functioning, management, ecotechnology, eutrophication

Abstract

In this paper the authors report and discuss limnological information for reservoirs in Brasil including physical, chemical and biological data in order to provide an insight into some basic mechanisms and interrelationships of forcing functions, biological and biogeochemical factors useful for management.

The introduction of ecotechnological measures to control, reduce eutrophication, suspended mater input and to recover water quality, is very urgent.

Basic theoretical problems of reservoirs structure and function are discussed as well as the introduction of new techniques such as satellite image to improve management.

1. Introduction

The size and complexity of the natural hydrographic network of South America crucially influences any control of the water resources in the continent and development towards multiple usage. Brazil, especially, is a reservoir-oriented country. The presence of large and artifical water bodies in the river basins of the continent not only alters radically their limnology and ecology but also results in major economic and social change. Thus the construction of reservoirs in particular regions is a topic of great significance for the whole continent.

In South America, both of the two largest river basins, the Amazon and La Plata (Figs. 1, 2), have been regulated during the last 50 years by the construction of reservoirs along their courses. The La Plata Basin, which supports several hydroelectric power plants, is sub-divided by the sub-basins of tropical and sub-tropical tributaries (Fig. 2). Along the major tributaries of the Paraná, and Uruguay have been built large and small reservoirs which have been the subject of several studies (Arcifa *et al.*, 1981; Bonetto, 1976; 1977; Branco & Rocha, 1977; Calijuri, 1988; Henry *et al.*, 1985; Pedrozo *et al.*, 1988; Tundisi, 1977; Tundisi, 1988a, b).

The hydroelectric potential of the major tributaries of the Amazon River (Fig. 3) has been estimated to be as much as 100 GW (Fig. 4). There are plans to exploit this potential (Eletrobrás, 1987) for Brazil in general since the water resources of the south-east are almost exhausted. Some studies on Amazonian reservoirs have been published together with pre-impoundment evaluations (Barrow, 1983; 1987; Garzon, 1983; Goodland, 1977; 1978a, b, c; Heide, 1976; 1982; Junk & Melo, 1987; Monosowski, 1984). Table I gives some information on the geographical co-ordinates, reservoir area, drainage area and the name of the host river of four existing and six projected Amazonian reservoirs which has been taken from Branski *et al.* (1988). These are very large water bodies up to 2360 km² in area and with drainage : reservoir area ratios in the existing reservoirs ranging from over 1000 (Coaracy Nunes) to 8 (Balbina).

Another large river in north-east Brazil is the S. Francisco River (Fig. 5) whose dam reservoirs form a significant water resource. These were built for multiple usages, to provide hydroelectric power

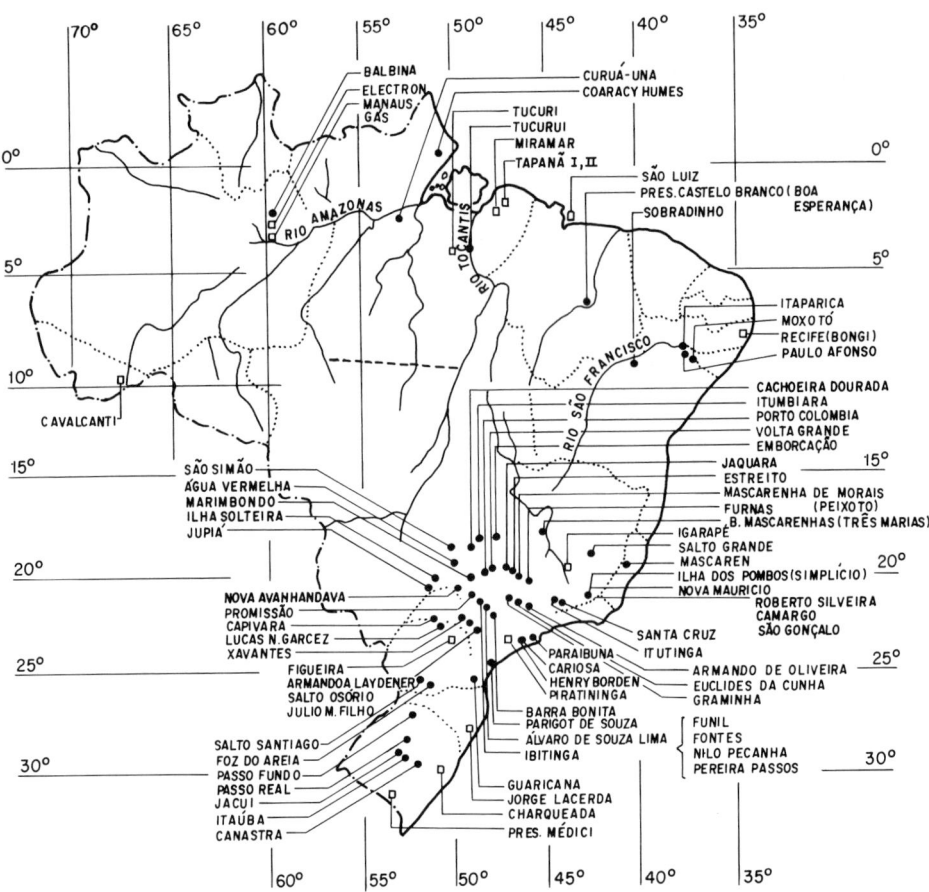

Fig. 1. Geographical distribution of reservoirs in Brasil showing the major groups in the La Plata Basin, the Amazon and the S. Francisco. From Eletrobras (1987).

and water for irrigation as well as to regulate navigation (CHESF, 1987).

Small reservoirs in the northeast of Brasil were built with the purpose of water storage and biomass production (Fig. 6).

All these large and small reservoirs pose ecological, limnological, economic and social problems, whether within their river basin or encompassing the very large range of latitude. Comparative limnological studies provide invaluable information for managing the reservoirs as the focal point within their river basin but this must include environmental data collected by regular monitoring or special research studies both before and after the construction of the reservoir.

This paper provides a summary of what limnological and ecological information is available for Brazilian reservoirs in relation to their physical, chemical and biological characteristics. Other aspects will also be considered: the nature of the main forcing functions, the influence of morphometry and how operational procedures can affect the river ecology upstream and downstream of the reservoir. Other objectives of this paper are to understand the functioning of the reservoir as an aid to management as well as to clarify what are its main environmental impacts and management problems.

Most of the results considered come from reservoirs built in the basins of the Amazon River and La Plata River. The series of large dams built on the larger tributaries within these basins pose problems for theoretical ecology and limnology in relation to processes, forcing functions and operational systems and patterns of reservoir

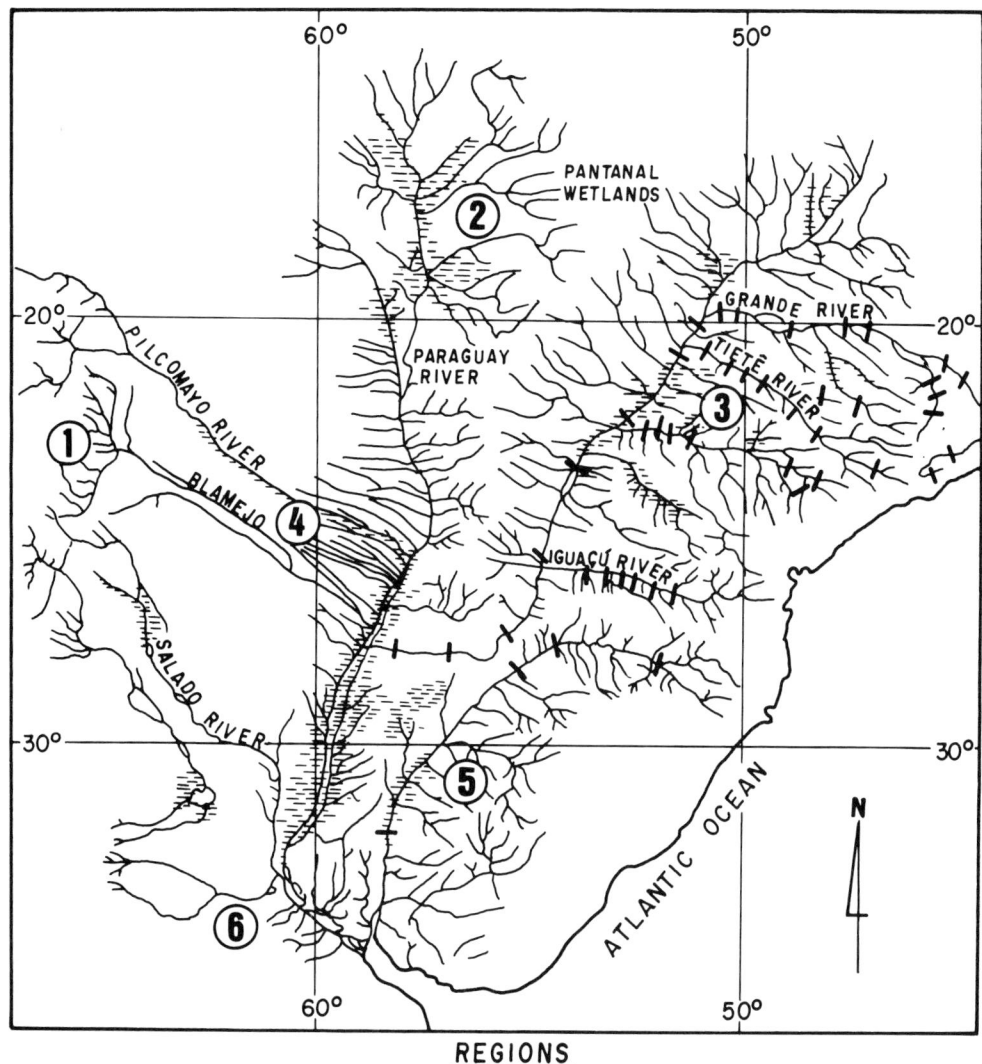

Fig. 2. Geographical distribution of reservoirs in the La Plata Basin.

ageing. The approach adopted here of comparative analysis may provide a useful framework for indicating or predicting the present status or future developments and may form a database for modelling reservoir processes (Margalef, 1975; 1976; Margalef *et al.*, 1976; Tundisi, 1990).

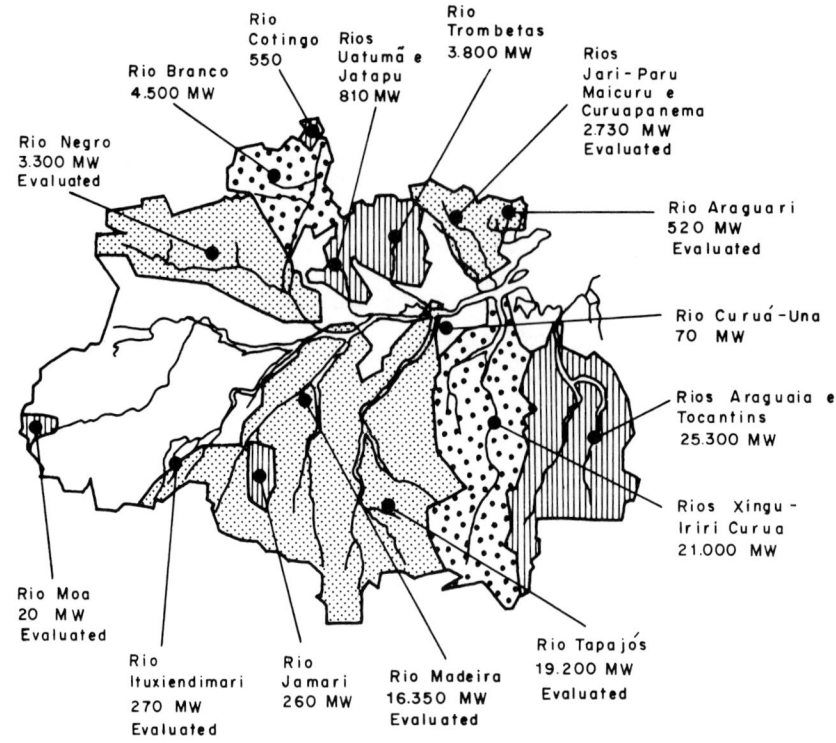

Fig. 3. The hydroelectric potential of the Amazon Region. (Junk & Mello, 1987, authorized).

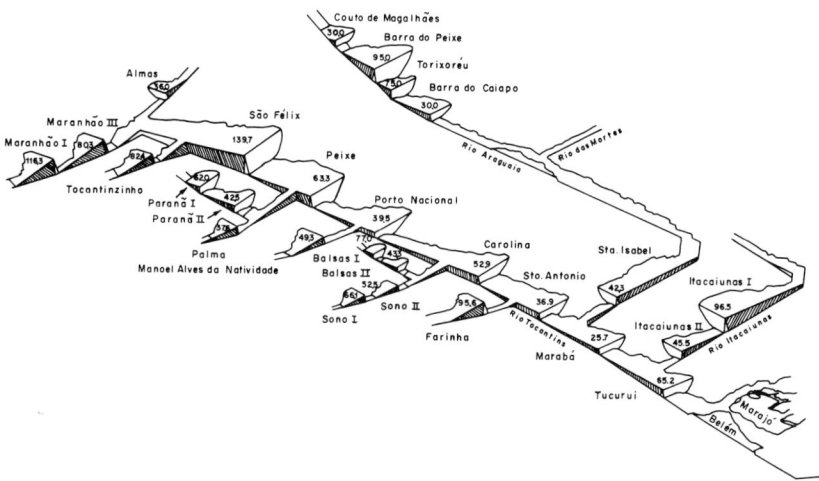

Fig. 4. A planned cascade of reservoirs along the Tocantins River, Amazonian (From Junk & Mello, 1987).

Table I. Existing and planned hydroelectric projects in legal Amazon (up to year 2010).

Project	River	Coordinates	Reservoir area (km²)	Drainage area (km²)		
				River basin	River basin reservoir area	Contribution to reservoir basin
Coaracy Nunes*	Araquari	00° 53' N 51° 15' W	23.1	30,800	1,333.33	24,200
Curuá-Una (CELPA)*	Curuá-Una					
Tucuruí	Tocantins	03° 55' S 49° 41' W	2,160	767,000	355.09	691,200
Balbina*	Uatumã	01° 55' S 59° 28' W	2,360	70,600	29.92	19,260
Samuel*	Jamari	08° 45' S 63° 25' W	579	29,700	51.30	15,280
Manso	Manso	14° 52' S 55° 48' W	427	10,890	25.50	9,364
Cachoeira	Trombetas	01° 03' S 57° 03' W	918	133,930	145.89	50,560
Jiparaná	Jiparaná	09° 44' S 61° 52' W	957	73,900	77.22	47,300
Kararaô	Xingú	03° 07' S 51° 46' W	1,225	509,000	415.51	477,000
Barra Do Peixe	Araguaia	16° 17' S 52°37' W	1,020	382,000	374.51	17,310
Couto Magalhães	Araguaia	17° 10' S 53°08' W	4,420	382,000	8,642.53	5,170

* Existing reservoirs from Branski *et al.* (1988)

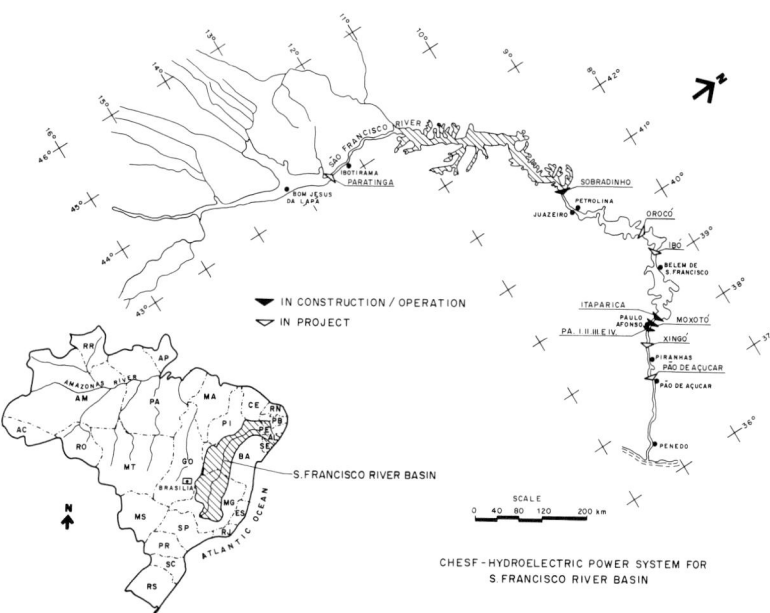

Fig. 5. A cascade of reservoirs along the S. Francisco River (from CHESF, 1987).

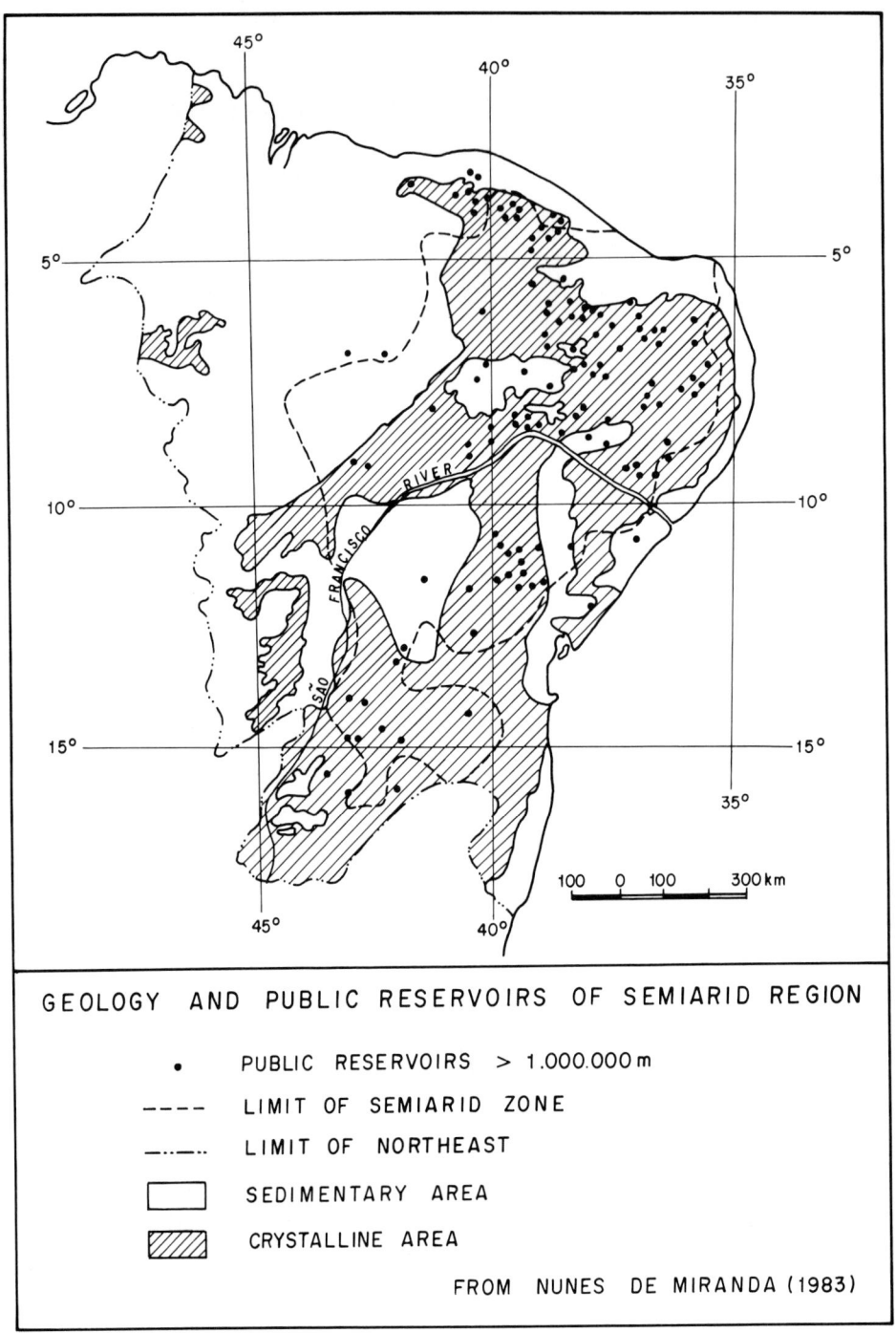

Fig. 6. Geographical distribution of reservoirs in the semi arid region of North East of Brasil in the S. Francisco River Basin (From Nunes De Miranda, 1983).

2. General characteristics of the reservoirs

2.1. Morphometry

Most of the reservoirs in Brazil have a dendritic shape. Fig. 7 illustrates the simple dendritic morphometry of Volta Grande Reservoir and Fig. 8 of the much more complex shape of the Furnas Reservoir. This type of complex morphometry greatly influences the ecology of the reservoir since the retention time of the water in its various compartments is much longer compared with that of the central channel. Some reservoirs are relatively shallow with much lower average depths than their maximal depths of about 20–25 m, due to the low gradients in the rivers so that very large areas are inundated. Shallowness, combined with a simple morphometry, can encourage a permanent mixing regime since the kinetic energy of the wind is sufficient to generate full vertical mixing and such a reservoir can be considered as a single unit for either management or modelling purposes. In reservoirs with more complex shapes, there may be several compartments with inlets which are useful from a management point of view. However, the highly complex morphometry of the Amazonian reservoirs is associated with poor vertical circulation of water and the development of anoxia in the hypolimnion.

2.2. Operational procedures

Operational procedures for the various functions in mulitple usage reservoirs influence greatly their limnology. For example, the operation of hydroelectric reservoirs produced several changes in the water releases and correspondingly on the limnological functioning throughout the year. This affects retention time, which is a crucial factor for the quality of reservoir water. In general, the upstream reservoirs have longer retention times (6–12 months) compared with downstream ones (30–40 days). This affects the development and succession of the phytoplankton and zooplankton community (De Fillipo, 1987; Tundisi & Matsumura-Tundisi, 1990). Reservoirs linked together in a cascade are normally treated in a single operational procedure which not only affects the ecology of all the reservoirs but also has consequences downstream of the cascade.

2.3. Seasonality

In some reservoirs, a combination of periodicity in local rainfall and man's control of the system produces a well marked seasonal cycle in water storage, retention time and nutrient concentration with consequences on primary production of

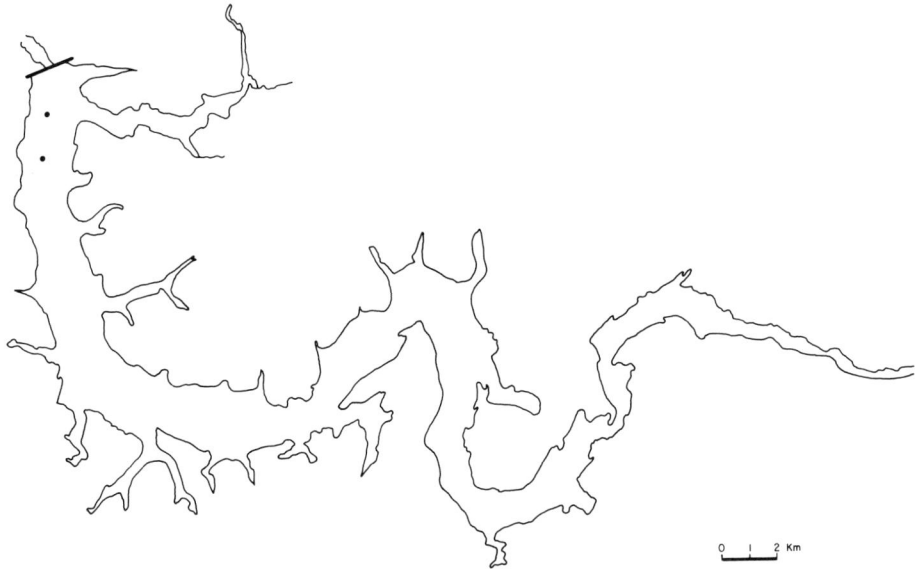

Fig. 7. Simple morphometry of the Volta Grande Reservoir on the Rio Grande, in the headwaters of Paraná River, upper La Plata Basin.

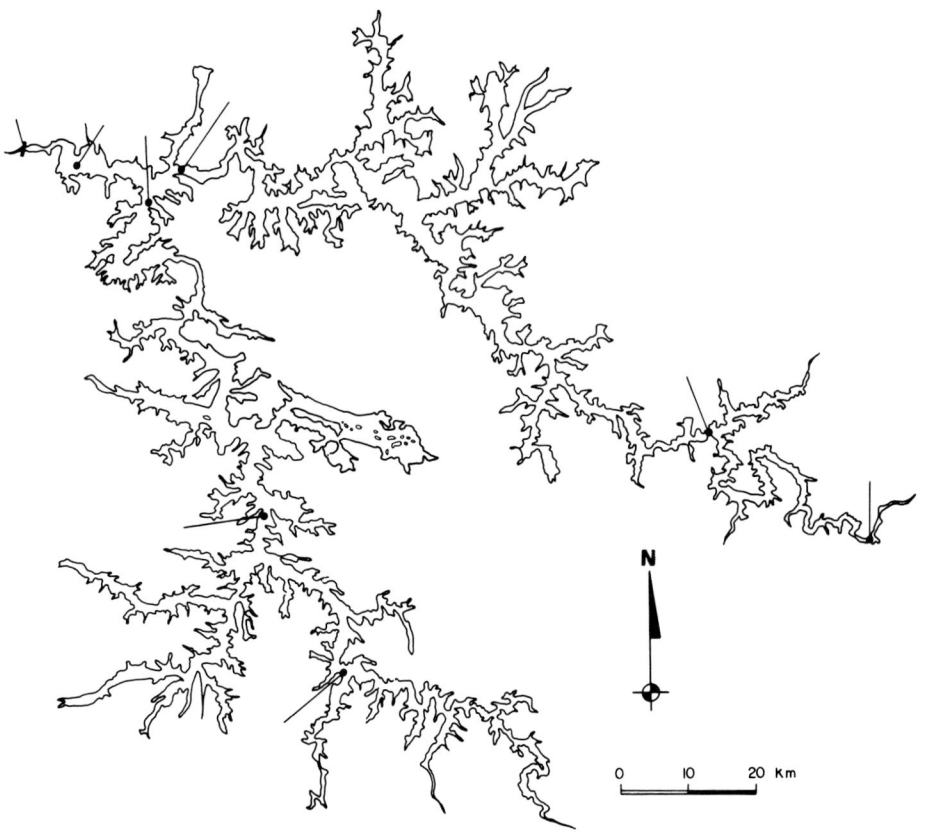

Fig. 8. Complex morphometry of the Furnas Reservoir on the same Rio Grande River upstream.

phytoplankton, on zooplankton and macrophytes. Such seasonal cycles may also have several effects above and below the reservoir: drying up of wetlands, reduction of volume, increase of retention time and reduced levels of dissolved oxygen downstream. Figs. 9 and 10 illustrate this for Barra Bonita and Tucuruí Reservoir respectively. The increasing of conductivity with long retention time is also shown in Fig. 9.

Rainfall occurring seasonally may also introduce a pulse effect into the river reservoir system. Suspended inorganic material will be transported out during the rainy period which results in a strong pulse interfering with the primary production of the phytoplankton and with the development of other organisms. There may also be some interference with biogeochemical cycles due to the introduction of particles onto which phosphate is adsorbed.

3. Physical and biogeochemical characteristics of the reservoirs

3.1. Light penetration

Several factors can affect light penetration in the reservoirs. For example, during periods of intense rainfall, there occurs a greatly increased contribution from the catchment of suspended organic and inorganic particles which will increase light attenuation and may change the light composition. Calijuri (1988) has shown that most of the increased attenuation is produced by the inorganic particles in turbulent and well-mixed reservoirs. This is illustrated for Barra Bonita Reservoir in Figs. 11 and 12. The annual variation in Secchi Disc transparency (Fig. 11) shows low depths (low light penetration) during periods of high rainfall, and short retention times. That is, rain in the catchment introduces a pulse of

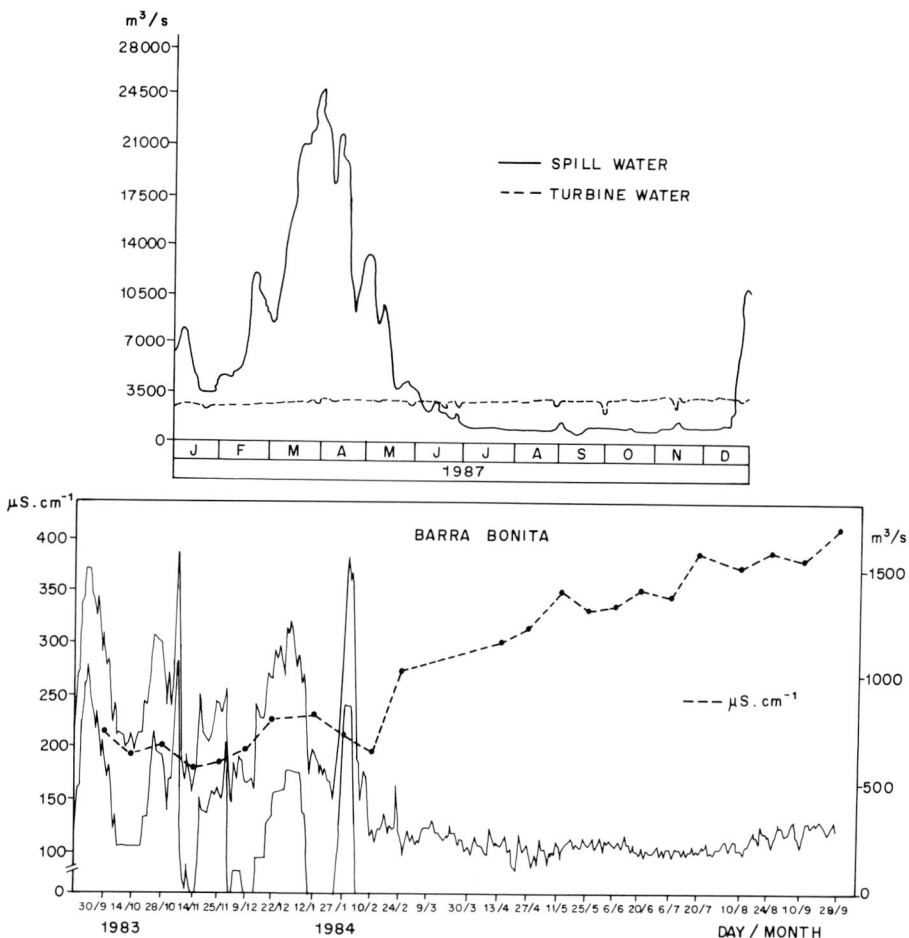

Fig. 9. A comparison of the annual variation in outflow rates in 1987 for Tucuruí Reservoir (Amazonia) and for 1983/84 for Barra Bonita Reservoir (B. Tieté River, upper La Plata Basin) Note how conductivity increases with longer water retention in Barra Bonita Reservoir.

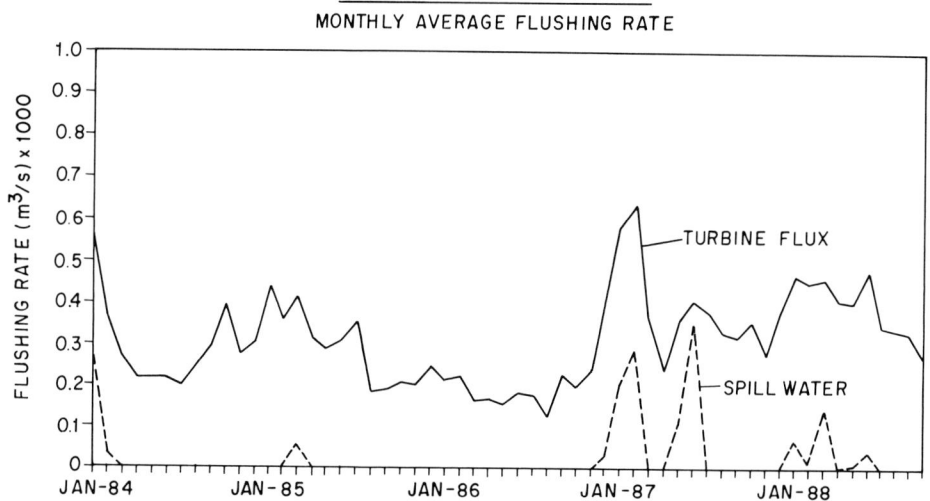

Fig. 10. The variation in outflow rates in Barra Bonita Reservoir (Tietê River) from 1984–1988.

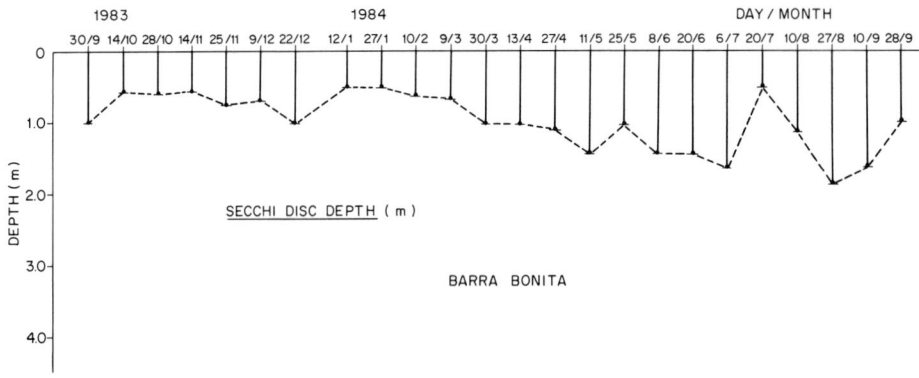

Fig. 11. The annual variation in Secchi disc depth during 1983/1984 in Barra Bonita Reservoir (S. Paulo). Note the increase during June and August. Decrease in Secchi depth during July was due to wind mixing.

suspended inorganic particles which greatly increases the vertical extinction coefficients. Table II shows, for Barra Bonita Reservoir, that it is the absorbance due to the suspended particles that contributes most. Table III compares the vertical extinction coefficients for five reservoirs measured during July 1989. All of them are similarly very high and mostly due to attenuation by suspended particles.

Water blooms of the cyanobacterial species *Microcystis aeruginosa, Anabaena spiroides* and *Anabaenopsis* sp. can also affect the light attentuation during periods of long retention times and low turbulence. In Amazonian reservoirs, water transparency is relatively high but the brown colour of the water, due to dissolved organic substances, both changes the light spectrum of the underwater radiation and permits the red end of visible light to penetrate deeper.

3.2. Thermal structure and circulation of the reservoirs

In a study of twenty-three reservoirs, Tundisi (1981) concluded that most were polymictic. Arcifa *et al.* (1981) classified three as warm monomictic and five as polymictic amongst eight studied reservoirs but later re-classified one of the warm monomictic ones as oligomictic. The development of warm monomictism is associated with the morphometric condition of a deep valley without wind action. Polymictism is largely due to the combination of wind action and appropriate morphometry and can vary in intensity. However, structural factors such as the position of the turbine water intakes and the elevation of the dam spillway gates can contribute to the thermal patterns. For example, Tundisi (1984) reports upon a thermal stratification that developed in the main channel of the reservoir because the spillway was sited half way up the dam wall: this was referred as

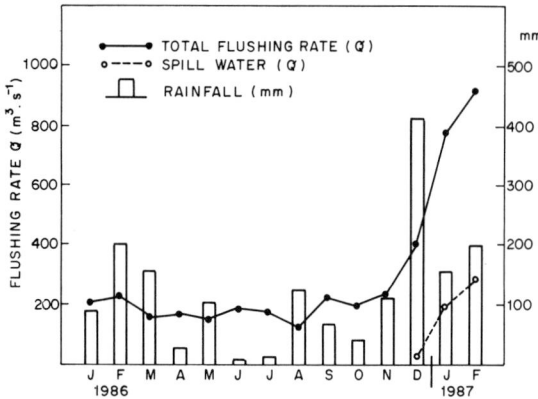

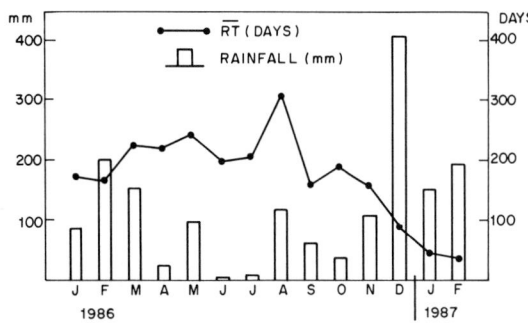

Fig. 12. The relationship between the monthly rainfall, total flushing rate and the retention period in Barra Bonita Reservoir (Tietê River) during 1986/1987.

Table II. Average values of vertical attenuation coefficients for Barra Bonita Reservoir and the phytoplanktonic and suspended matter components.

Barra Bonita Reservoir	Particulate material	Secchi depth	k	k_t	k_w	$k_c \cdot C$	k_x
Winter	4.4	1.00	1.70	1.44	0.04	0.06	1.34
Summer	10.7	0.50	4.54	4.54	0.06	0.20	4.28

k = vertical contrast attenuation coefficient
k_t = diffuse attenuation coefficient for total irradiance.
k_w = diffuse attenuation per meter for pure water.
k_c = specific attenuation for phytoplankton.
C = concentration of chlorophyll in the euphotic zone (mg m^{-3}).
k_x = attenuation by substances other than phytoplankton or water.
According to Kirk (1986) and Morris (1981)

a 'hydraulic stratification', lately confirmed also for some Spanish reservoirs.

Climate has an important influence upon the thermal patterns of reservoirs and diurnal changes may be more important than seasonal ones in some tropical reservoirs in relation to thermal stratification and vertical circulation. In Broa Reservoir, Simonato (1986) demonstrated that the diurnal influence was fundamental for stratification and destratification sequences and suggests that it may also be important for biogeochemical cycling. Although nocturnal cooling was insufficient in the man-made Brokopondo Lake to lower the epilimnetic temperatures, Heide (1982) thought that the nictemeral cycle did enhance vertical exchange (of water masses).

The Amazonian reservoirs stratify for longer periods than above and are mostly warm monomictic water bodies but with short periods of complete mixing. This is illustrated for Tucurui Reservoir in Figs. 13 and 14 for temperature and dissolved oxygen. Heide (1982) investigated the stratification of Brokopondo Reservoir during its filling and distinguished three main zones:

(i) the turbulent riverine zone with dissolved oxygen levels that were similar to those in the inflowing streams;
(ii) the recently flooded bank zone with trees present in which turbulence was reduced and a rapid stratification was followed by oxygen depletion;

Table III. Secchi depth, euphotic zone and absorption of Ph.A.R. by suspended matter and total chlorophyll in the five reservoirs (July – Winter – 1989). Units given in Table II.

Day	Reservoir	Secchi depth (m)	Zeu (m)	k	k_t	k_w	$k_c \cdot C$	k_x
25/07/89	Barra Bonita	1.90	5.00	4.74	0.92	0.02	0.11 / 11.96%	0.79 / 85.87%
26/07/89	Bairiri	1.40	4.50	6.43	1.05	0.02	0.18 / 17.14%	0.85 / 80.95%
26/07/89	Ibitinga	2.00	4.75	4.50	0.97	0.01	0.14 / 14.43%	0.82 / 84,54%
27/07/89	Promissao	2.00	4.75	4.50	0.98	0.02	0.19 / 19.39%	0.77 / 78.57%
27/07/89	Nova Avanhandava	3.00	8.00	3.00	0.59	0.02	0.06 / 10.17%	0.51 / 86.44%

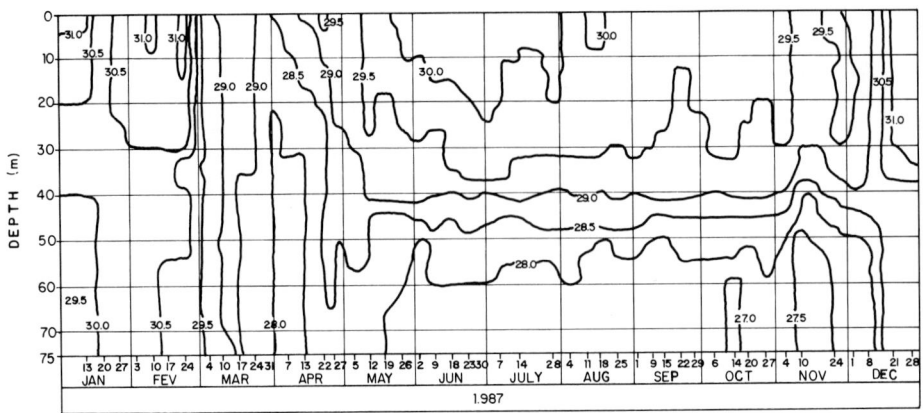

Fig. 13. The seasonal time course of the depth distribution of temperature at station n_1 in Tucuruí Reservoir (Amazonia) during 1987.

(iii) a lacustrine zone, with greater depths of water, in which a strong stratification is accompanied by oxygen depletion and anoxic deep water layers;

(iv) the wetlands zone associated with the reservoirs.

Such a horizontal zonation is typical for Amazonian reservoirs where projecting dead trees from the flooded forest significantly reduce both wind action and water turbulence and prevent streaming (Baxter, 1977; Ploskey, 1983; 1985). These findings are confirmed by more recent results for the new Amazonian Reservoirs of Balbina and Samuel (Matsumura-Tundisi *et al.*, 1991 and personal information from Mr. Roberto from Balbina Reservoir).

Although there are few studies on the physical stability of South American reservoirs, Henry & Tundisi (1988) conclude from their study on the two shallow Reservoirs of Broa and Rio Pardo that morphometry was the main factor influencing differences in stability illustrated in Figs. 15 and 16. The figures show stronger fluctuation during the year which is typical of shallow reservoirs. An increase in the stability ocurred in the summer for

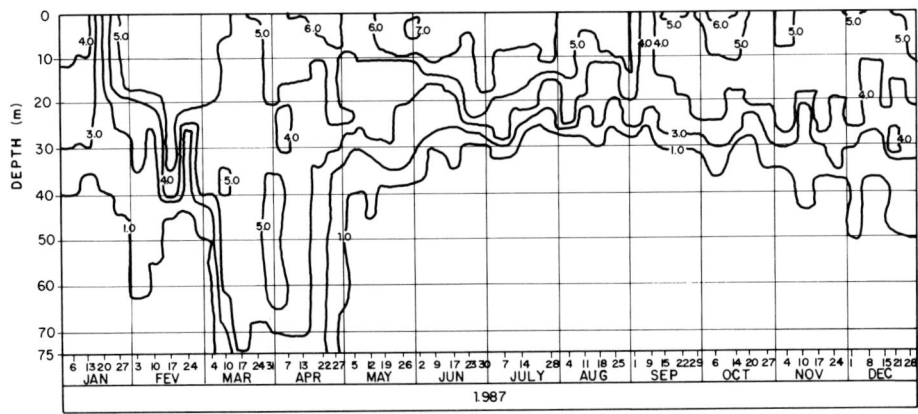

Fig. 14. The seasonal time course of the depth distribution of dissolved oxygen concentrations at station n_1 in Tucuruí Reservoir (Amazonia).

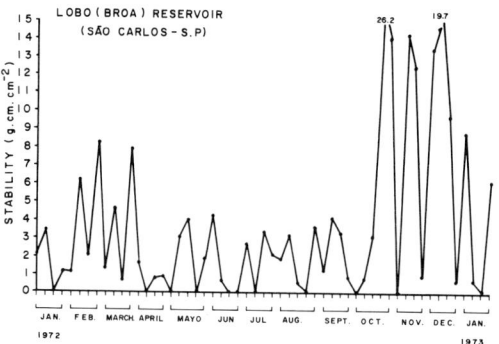

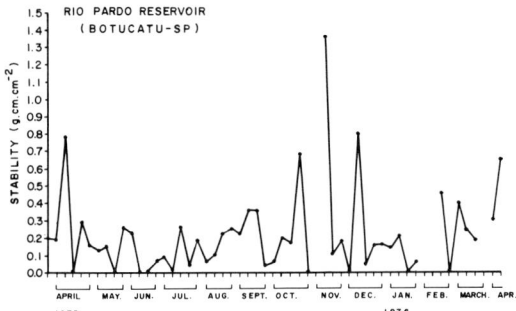

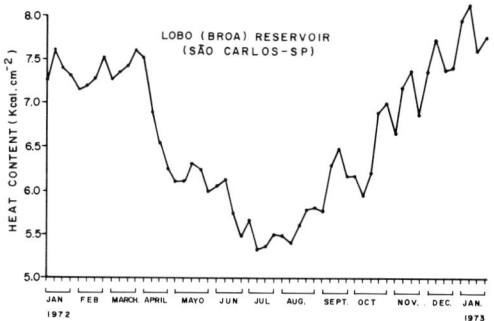

Fig. 15. The annual time course of heat content and stability in the shallow Rio Pardo Reservoir (Botucatu, S. Paulo) during 1975/1976.

Fig. 16. The annual time course of heat content stability content in the Lobo (Broa) Reservoir (S. Carlos), during 1975/1976.

both reservoirs, but due to shallowness and wind effects this has no long lasting ecological consequences, related with oxygen concentration and distribution of organisms. Wind and precipitation provide external energy that promotes circulation and mixing as, for example, when short periods of stratification develop after heavy rain which contributes with denser, colder water from the tributaries (unpublished results of Tundisi for Broa Reservoir during the southern summer). This contrasts with the situation during the southern winter, when the coupling of increased wind stress and resistance against wind decreased as the consequence of lower temperatures promotes homogeneity in many reservoirs.

The chemical and biological consequences of these thermal patterns are several. In polymictic reservoirs, homogeneous vertical profiles of dissolved oxygen and particles of inorganic and organic substances are usual. Reynolds (1987) comments upon the uniform vertical distribution of the phytoplankton in these reservoirs. Even such heavy filamentous forms as *Melosira italica* and *Melosira granulata* may be homogeneously distributed in the water column during periods of intense circulation. In Amazonian reservoirs, the development of an oxycline is common during the period of stratification which may have deleterious biological and chemical consequences downstream. Where the reservoir outlet is located halfway up the dam, as in Furnas Reservoir, a 'hydraulic stratification' may develop, giving vertical profiles typical of a normally stratified system. This is illustrated in Fig. 17 for the nitrite, nitrate and ammonia depth distributions in Furnas Reservoir.

3.3. Water chemistry and the major biogeochemical cycles

The chemical composition of reservoir water is dependent upon the hydrogeochemical nature of the watershed, the chemistry of the river water, the internal chemical processing within the reservoir

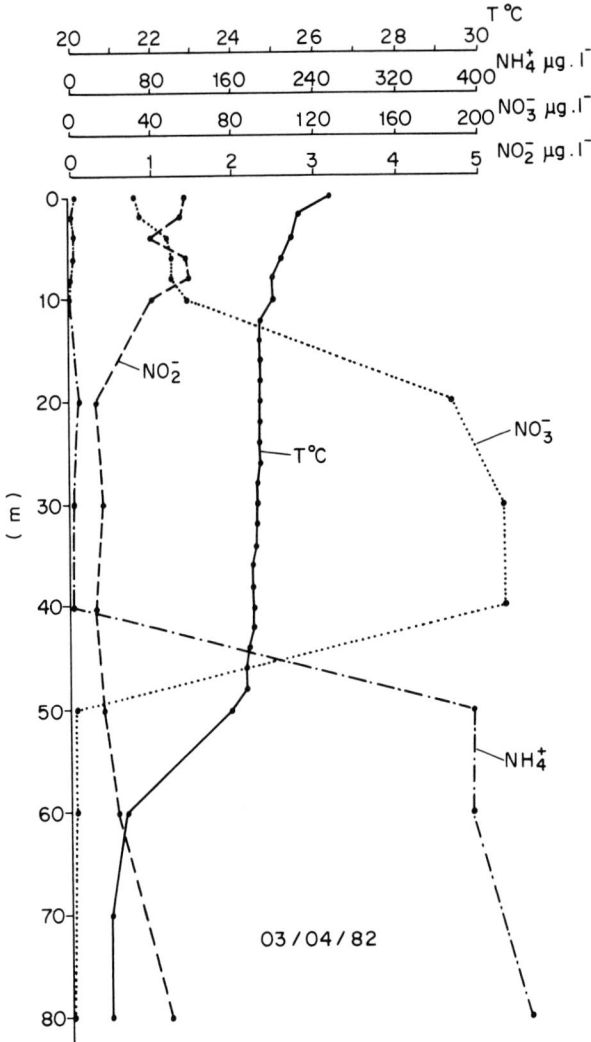

Fig. 17. An example of hydraulic stratification in Furnas Reservoir (Rio Grande, upper La Plata Basin) illustrated by depth profiles of temperature and construction of inorganic nitrogen species.

itself after the water level stabilizes, the soil usage (large and small scale agriculture with sugar cane, corn and other crops with heavy use of fertilizers and pesticides/herbicides) common for La Plata Basin Reservoirs.

In general, the levels of dissolved inorganic nutrients in the Reservoirs of the La Plata Basin are low, except for those that are very eutrophic. This is despite chemical inputs from external sources such as erosion in the upper Paraná River basin due to certain agricultural practices or the discharge of untreated sewage and industrial effluents (at 25 m^3 sec^{-1}) from S. Paulo City into the Tieté River Basin. Pedrozo *et al.* (1988) suggests that this discrepant result of low phosphate concentrations in the upper Paraná River coexisting with intense human activity is because the phosphate is trapped in the reservoirs. The same authors estimate the rate of erosion in the lateritic soils of southeastern Brazil is as high as 20 tons (ha·y)$^{-1}$ so that further eutrophication is inhibited by the precipitation of ferric phosphate onto the bottom. After mixing, this may be very fast (Whitaker, 1987).

A possible source of nitrogen depletion might be the denitrification processes occurring in the wetlands associated with the reservoirs or in the

Table IV. Ionic composition of five reservoirs in S. Paulo State (1979 Survey).

Reservoir	Cond [a]	Alc [b]	pH	SO_4^- [c]	Cl^- [c]	Ca^+ [c]	Mg^+ [c]	Na^+ [c]	K^+ [c]	C/A
Barra Bonita	103	0.65	7.91	0.24	0.18	0.34	0.21	0.44	0.08	1.00
Bairiri	100	0.66	7.72	0.21	0.18	0.34	0.21	0.44	0.08	1.02
Ibitinga	91	0.61	7.92	0.15	0.15	0.30	0.18	0.36	0.07	1.00
Promissao	81	0.63	7.93	0.12	0.12	0.31	0.18	0.33	0.08	1.03
Avanhandava	60	0.63	7.84	0.17	0.15	0.33	0.19	0.38	0.08	1.03

[a] – $\mu S\ cm^{-1}$,
[b] – $meq\ l^{-1}$,
[c] – $mg\ l^{-1}$,
C/A – cations/anions

littoral macrophyte stands, where there are high concentrations of ammonia (up to 1000 $\mu g\ l^{-1}$) and low levels of dissolved oxygen at night. In Barra Bonita Reservoir, Esteves (1983) found very high levels of phosphate in the sediments (513.6 ppm), suggesting that the reservoir sediments are also functioning as important nutrient traps. Some recent unpublished determinations showed very high levels of ammonia and dissolved inorganic phosphate in the sediment interstitial water (Whitaker, 1987).

Tables IV and V show the ionic composition of water from five reservoirs from the middle Tietê River (data from a 1979 survey) and the chemical composition of the Amazonian Reservoirs of Tucuruí and Samuel Reservoirs in the Solimóes and Negro Rivers.

Although there is little information available on the effects of decomposition of cyanobacterial blooms on dissolved oxygen levels or inorganic nutrient cycles, high levels of dissolved oxygen at the surface were detected by Tundisi (unpublished) during the day and in the presence of dense populations of *Microcystis aeruginosa*. The decomposition of organic matter from these blooms when senescent are probably an important source of dissolved organic carbon and the concentrations that can be attained are shown in Table VI for Barra Bonita Reservoir during a *Microcystis* bloom.

Biogeochemical cycles in the Amazonian reservoirs are strongly related to the decomposition of vegetation and anoxic conditions in the hypolimnion. The biogeochemical changes that follow the succesional stages during the filling of Amazonian reservoirs have been studied by Heide (1982) and Matsumura-Tundisi *et al.* (1991). Both

Table VI. Dissolved organic carbon DOC for Barra Bonita Reservoir, S. Paulo at period of highest residence time (August 1987) (data supplied by Dr. I. Ota, Water Research Institute, Nagoya University, Japan DOC was determined on Total Organic Carbon Analyses Toc – 500 Shimadzu Co. Japan)

Tietê River station:	$9.30 \pm 0.37\ mg\ C\ l^{-1}$
Piracicaba River station:	$5.29 \pm 0.07\ mg\ C\ l^{-1}$
Tietê and Piracicaba River (intermediary station):	$11.82 \pm 0.16\ mg\ C\ l^{-1}$
Station near the dam:	$18.24 \pm 0.34\ mg\ C\ l^{-1}$

Table V. Mean concentration for some chemical parameters in the Tucuruí Reservoir and Solimoes and Negro Rivers [a] (Amazonia). From Branski *et al.* (1988)

	Conductance ($\mu S\ cm^{-1}$)	$N(NO_3)$ ($\mu g\ l^{-1}$)	$P(PO_4)$ ($\mu g\ l^{-1}$)	pH	Ca + Mg + Na + K ($mg\ l^{-1}$)	Ca + Mg + Na + K (meq l^{-1})	
Tucuruí Reservoir [a]	56	41	39	6.8	10.4	0.44	0.11
Solimoes [b] River	57	48	11.6	6.9	10.6	0.45	0.12
Negro [c] River	9	36	5.8	5.1	1.03	0.0197	0.025

[a] Tucuruí data from Eletronorte, 1986
[b] Solimoes and Negro River data from Junk & Howard-Williams (1984). Time and depth average data in "M_1" station from May 1985 to March 1986
[c] Ditto, annual average data

the redox potential and horizontal gradients are related to changes in the biogeochemical conditions in the transitional riverine zone between the river and the reservoir and to the intensity of mixing processes in the main channel.

Exposure of bottom sediments to high air temperatures during large changes of water level also plays a role (as a forcing function) in biogeochemical cycles of tropical reservoirs. The annual pattern of drying and flooding provides a rich supply of inorganic and organic nitrogen and phosphorus which is responsible for the dense phytoplankton that develops in recently flooded shallow waters. This is exploited in the Reservoirs of S. Francisco River where a seasonal agriculture is commonly practiced on land exposed at low water levels and this, in turn, enhanced nutrient enrichment on re-flooding. Wetlands developing in the riverine transitional zones may be responsible for high losses of nitrogen, carbon and sulphur. Tundisi & Wada (1990) measured high amounts of N^{15} both at Barra Bonita (La Plata) and Samuel Reservoir (Amazon) and detected the evolution of CH_4 and H_2S, thus contributing to processes of denitrification and release of methane and sulphur to the atmosphere. This may be particularly important for Amazonian reservoirs due to the large volumes of hypolimnetic water, which are mostly anoxic (for a 35 m deep water column only the first 3 m have dissolved oxygen up to 6 mg l^{-1} and 32 m are anoxic (Matsumura-Tundisi et al., 1991).

4. Community processes in the reservoirs

4.1. Primary production

In most of the reservoirs, the primary production of phytoplankton predominates over that of the periphyton and macrophytes throughout the year and, in toto, indicates a moderate level of biological productivity. The level of planktonic primary production is lower in the flow-through reservoir systems compared with the values of up to 4000 mg C m^{-2} d^{-1} measured in eutrophic reservoirs by the ^{14}C technique (Tundisi, 1983). The level of primary production and chlorophyll-a biomass of the phytoplankton was measured in 23 reservoirs in the São Paulo State by Tundisi (1981) and these values are listed in Table VII, together with their coordinates and altitudes. Tropical lakes have higher levels of primary production. For example, Hill & Rai (1982) give a range of 100–8000 mg C m^{-2} d^{-1} for tropical lakes and Bonetto (1982) report a lower range, 50–800 mg C m^{-2} d^{-1} for a series of shallow lakes in the Argentina River Bermejo Basin, in which the seasonal variation in primary production is strongly correlated with water level fluctuations, concentrations of suspended solids and nutrient enrichment.

Although the vertical profiles in primary production varied amongst the reservoirs studied, the commonest pattern showed light inhibition at the surface together with a sub-surface photosynthetic maximum, as shown by Findenegg (1966). An important limiting factor to phytoplanktonic production in the reservoirs of southern Brazil is the low level of phosphorus as nitrogen is available as NO_3—N in very high concentrations (Henry et al., 1985; Salas, 1983; Tundisi et al., 1977; 1988; Henry, 1990).

Other factors which limit primary production are the high levels of inorganic suspended particles in the water during the rainy season and excessive growth of macrophytes such as *Eichhornia crassipes* and *Eichhornia azurea* (which was observed in Reservoirs of the Tietê River in São Paulo State). In contrast, the production of macrophytes and periphyton are far more important as food for herbivores and detritivores in the Amazonian reservoirs. In many compartments of these, there exist food chains supported largely by the decomposition of the flooded forest and its associated aquatic flora exploiting available surfaces. Such developments have been reported also for temperate reservoirs with drowned vegetation (Applegate & Mullan, 1967; Clafin, 1968; Cowell & Hudson, 1967).

The seasonal cycle of primary production in the reservoir, whether due to the phytoplankton, periphyton or macrophytes, is climatically influenced as rain and wind interferes with the thermal structure and nutrient cycles of the water body (Bonetto et al., 1987; Tundisi et al., 1977; Tundisi, 1983; 1986; 1989). It is also affected by variations in the retention period associated with the hydrological cycle, rainfall and fluctuations in water level. At periods of low precipitation, many littoral areas may dry up with consequential loss of macrophyte biomass.

Table VII. Average primary productivity (four samplings at different seasons during 1979), assimilation ratio, and chlorophyll-a for 23 reservoirs in S. Paulo State. From Tundisi (1983).

Reservoirs	Lat. (S)	Long. (W)	Alt. (m)	Prim. prod. (mg Cm^{-2}d^{-1})	CHA (mg m^{-3})	Ass. ratio [a]
Barra Bonita	22° 29'	48° 34'	430	398.27	15.9	2.56
Bairiri	22° 06'	48° 45'	442	521.85	20.3	2.64
Ibitinga	21° 45'	48° 50'	460	483.94	29.8	2.16
Promissao	21° 24'	49° 47'	410	584.08	68.7	0.83
Salto De Avanhandava	21° 13'	49° 46'	360	268.74	14.9	1.60
Capivara	22° 37'	50° 22'	520	188.67	11.7	3.40
Rio Pari	22° 51'	50° 32'	420	105.19	13.3	1.43
Salto Grande	22° 53'	49° 59'	405	102.80	5.7	2.07
Xavantes	23° 08'	49° 43'	400	193.79	20.8	0.95
Piraju	23° 11'	49° 16'	571	100.94	11.9	0.91
Jurumirim	23° 11'	49° 16'	571	103.05	9.7	1.02
Rio Novo	23° 06'	48° 55'	755	60.87	12.1	0.79
Limoeiro	21° 27'	47° 01'	650	225.89	22.3	2.26
Euclides Da Cunha	21° 36'	46° 54'	700	25.99	3.8	0.96
Graminha	21° 32'	46° 38'	800	582.98	34.4	0.94
Estreito	20° 32'	47° 24'	1000	126.71	25.1	0.61
Jaguara	20° 11'	47° 25'	536	154.08	22.3	0.70
Volta Grande	20° 05'	48° 02'	510	340.23	31.7	1.22
Porto Colombia	20° 10'	48° 48'	500	318.86	40.2	1.00
Marimbondo	20° 18'	49° 11'	390	262.10	37.5	0.80
Água Vermelha	19° 58'	51° 18'	452	232.47	32.5	0.80
Ilha Solteira	20° 24'	51° 21'	356	248.35	20.2	1.73
Jupiá	20° 58'	51° 43'	260	301.61	15.5	2.15

[a] Assimilation ratio [mg C(mg CHA h)$^{-1}$]

Figures 18 and 19 show the seasonal cycles of planktonic primary production and chlorophyll-a biomass during 1983–84 in two southern reservoirs, Lobo (Broa) and Barra Bonita, and in the Amazonian Tucuruí Reservoir. In Broa Reservoir, the primary production varies more throughout the year than the chlorophyll-a level due to rainfall and wind effects but was more stable compared with the situation in Barra Bonita Reservoir in which both production and chlorophyll-a varied by more than one order of magnitude. The highest values are associated with low flushing rates and long retention times (Fig. 9), high Secchi Disc depths (Tables II and III) and high levels of nutrients (Table IV). The high flushing rates during late 1983 and early 1984 in this reservoir (Fig. 9) reduces the chlorophyll-a levels, but not the primary production rates, to those attained by Broa Reservoir (Figs. 18 and 19). However, the seasonal distribution of cladoceran and copepod densities in Barra Bonita suggests that grazing pressures may also influence the phytoplankton biomasses (Fig. 20). In the tropical Tucuruí Reservoir, the large variations in primary production are associated with rainfall and the flushing out of nutrients from the small inlets or compartments. In this reservoir, the primary production levels are high for the planktonic biomass available because the assimilation ratio is high compared with those reported for the southern reservoirs in Table VII.

4.2. Composition and succession of reservoir communities

The community changes that occur from a river to a reservoir ecosystem after filling have been documented for many temperate reservoirs and for a few from the tropics and sub-tropics. The composition of the reservoir communities is influenced by several factors such as what is happening in the river upstream, the flow-through regime of the reservoir, the nature of the reservoir substrate, the severity of water level fluctuations

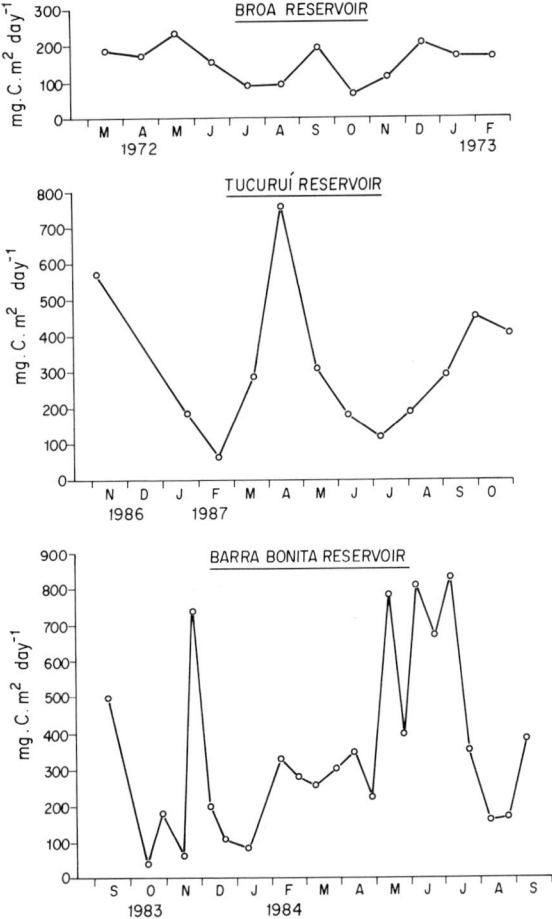

Fig. 18. A comparison of seasonal cycle of daily rates of phytoplanktonic primary production at Broa, Tucuruí and Barra Bonita Reservoirs.

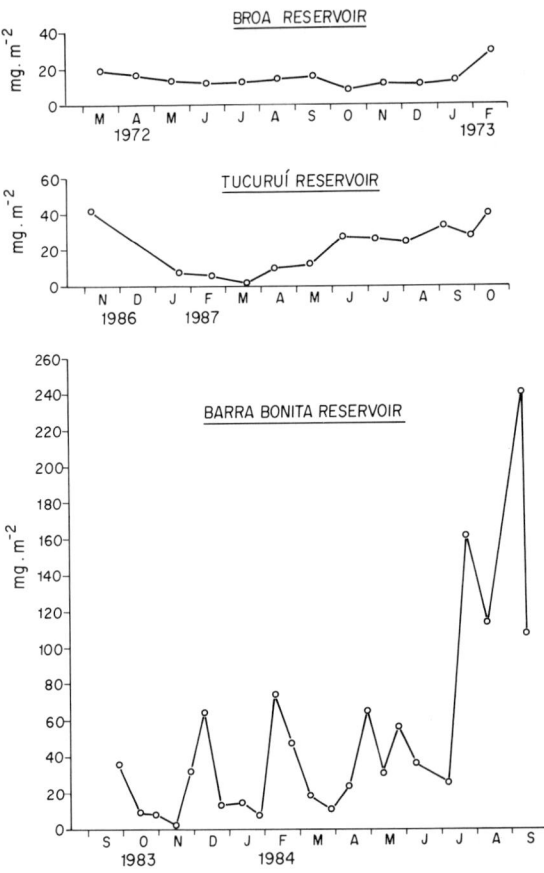

Fig. 19. A comparison of the seasonal cycle of planktonic chlorophyll-a concentrations at Broa, Tucuruí and Barra Bonita Reservoirs.

and the chemistry of the water. Organisms are transported into the reservoirs from lakes or reservoirs that exist upstream of the recipient water body, whose chemical status is wholly determined by the chemistry of the watershed. Short retention times and high flushing rates in the reservoir impoverish the species composition of the planktonic community and reduces their biomasses as well as lowers the level of primary production by mixing algae into the aphotic zone. Benthic plants and animal are adapted to different kinds of littoral and bottom sediments and dependent for food on inputs of organic matter from littoral macrophytes, which are very vulnerable to extreme water level fluctuations.

In the La Plata Basin, many of the reservoirs originated by the inundation of agricultural areas in which, previously, chemical fertilizers, pesticides and herbicides had been applied intensively. The inevitable deterioration in reservoir water quality after filling will impoverish the species diversity of the flora and fauna (by killing or lowering the reproductive capacity of the vulnerable species), greatly enhance (by fertilization) or reduce (by application of herbicides) the reservoir's biological productivity and interfere with the food web interactions of the phytoplankton, zooplankton and fish. In the Amazonian reservoirs, the inundated forest provides very extensive substrate for the development of macrophytes, periphyton and zoobenthos.

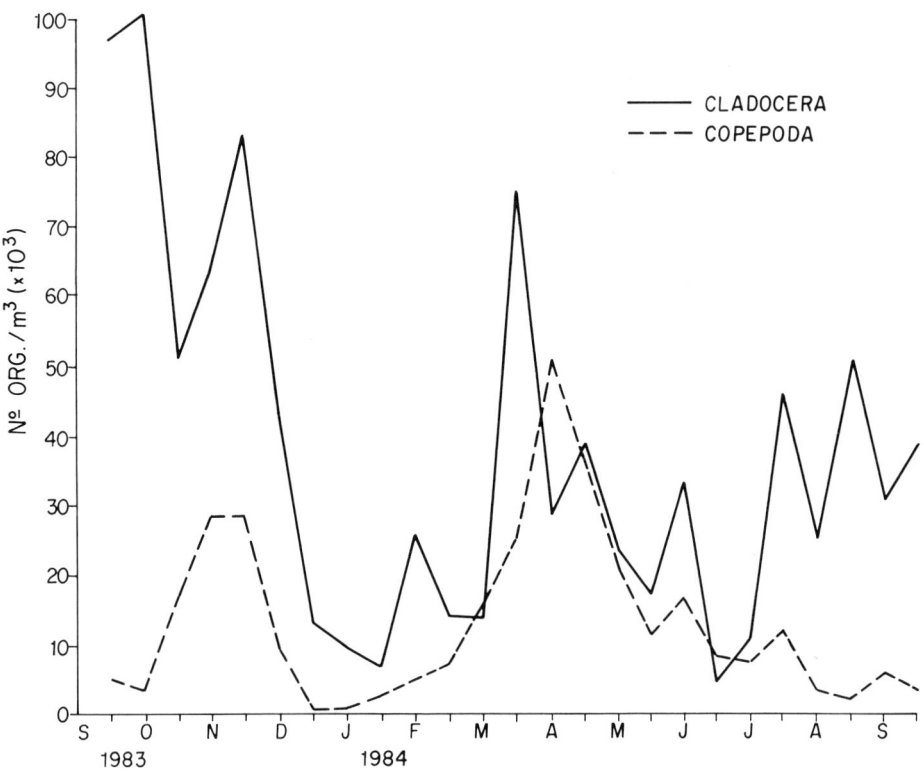

Fig. 20. The seasonal cycle of total copepod and total cladoceran densities during 1983/1984 in Barra Bonita Reservoir (Tieté River, S. Paulo)

The species composition of the phytoplankton is known for many reservoirs and that for Barra Bonita Reservoir is listed in Table VIII (Aranha, 1990). In general, species of green algae, cryptophytes and diatoms predominate in the oligotrophic and mesotrophic reservoirs where the diatoms, *Melosira italica* and *Melosira granulata*, are particularly important. A windy regime helps to keep these heavy algae in suspension. Several authors have studied the effect of wind on their vertical distribution (Lima *et al.*, 1979; Tundisi, 1982; 1983) and Nakamoto *et al.* (1976) describe the synchronous growth of *M. italica* with the rainy season in Broa Reservoir. In the eutrophic reservoirs, the Cyanobacteria predominate. In the upper Paraná Reservoirs, the common forms are *Microcystis aeruginosa*, *Anabaena* sp. and *Anabaenopsis* sp. Bonetto *et al.* (1987) report the presence of *Anabaena spiroides* and *Aphanocapsa delicatissima* in the Rio Bermejo Reservoir.

Table IX lists the common zooplanktonic species recorded for oligotrophic and eutrophic reservoirs in the São Paulo State. All but one of the 34 species of rotifers listed can be classified as cosmopolitan or cosmotropical forms in their distribution, in the sense of Green (1972a), whereas *Keratella americana* is one of 20 species listed by Green as being known only from the Americas. It is interesting, however, that only seven of the rotifer species in Table IX were also reported by Green (1972b) for five meander lakes of the Rio Suiá Missú in the Mato Grosso of Central Brazil; in these lakes, 24 species were cosmopolitan, 6 species were cosmotropical and 5 species belonged to the Americas group. The picture presented by the list of cladoceran species in Table IX appears to be similar in that cosmopolitan and cosmotropical species are present, apart from *Bosmina hagmanni* which is confined to the Americas (Dussart *et al.*, 1984; Green, 1972a).

Ecological studies on the zooplankton of Broa Reservoir in São Paulo State have shown that *Argyrodiaptomus furcatus* was the dominant and largest planktonic herbivore where it was patchily

Table VIII. Composition of phytoplankton of a eutrophic reservoir (Barra Bonita, S. Paulo State). From Aranha (1990).

CYANOPHYCEAE
  Chroococcales
Aphanocapsa roeseana
Chroococus sp.
Microcystis flos-aquae
Microcystis aeruginosa
Microsystis robusta
Microcystis psuedofilamentosa

  Nostocales
Oscillatoria spp.
Anabaena flos-aquae
Anabaena spiroides
Anabaena circinalis

CHLOROPHYCEAE
Ankistrodesmus spp.
Coelastrum spp.
Crucigenia spp.
Monoraphidium spp.
Paradoxia multiseta

ZYGNEMAPHYCEAE
Closterium spp.
Cosmarium spp.
Desmidium spp.

EUGLENOPHYCEAE
Phacus orbicularis

BACILLARIOPHYCEAE
Cyclotella sp.
Melosira granulata
Melosira granulata angustissima
Melosira granulata fo. *curvata*
Navicula sp.

CRYPTOPHYCEAE
Cryptomonas tetraptyrenoidosa
Cryptomonas paraptyrenoidifera
Chroomonas acuta
Rhodomonas minuta

DINOPHYCEAE
Peridinium spp.

Table IX. Common pelagic species of zooplankton found in the reservoirs of S. Paulo State and Amazon Region. Data from Samuel Reservoir kindly supplied by Miss Miriam H. Bueno Falotico, those for Balbina Reservoir by Dr. Ivan de Haro Moreno.

ZOOPLANKTON OF S. PAOLO STATE RESERVOIRS
ROTIFERA

Ascomorpha ovalis
Asplanchna sieboldi
Brachionus angularis
Brachionus calyciflorus
Brachionus caudatus
Brachionus patulus
Brachionus falcatus
Cephalodella gibba
Collotheca sp.
Cohochilus unicornis
Conochiloides coenobasis
Euchlanis dilatata
Filinia terminalis
Filinia longiseta f. saltator
Hexarthra mira
Keratella cochlearis
Keratella americana

Keratella lenzi
Keratella tropica
Lecane bulla
Lecane dorysa
Lecane lunaris
Mangredium eudactylotum
Polyarthra vulgaris
Ptygura libera
Synchaeta stylata
Synchaeta pectinata
Testudinella patina
Testudinella mucronata
Trichocerca capucina
Trichocerca similis
Trichocerca bicristata
Trichocerca elongata
Trichocerca cylindrica

Table IX. Continued.

CLADOCERA

Bosmina hagmani
Bosmina tubicens
Bosmina longirostris
Bosminopsis deitersi
Daphnia gessneri
Daphnia ambigua

Ceriodaphnia cornuta
Ceriodaphnia silvestri
Diaphanosoma birgei
Diaphanosoma spinulosum
Diaphanosoma brevireme
Moina minuta

COPEPODA

Argyrodiaptomus furcatus
Notodiaptomus iheringi
Notodiaptomus conifer
Notodiaptomus spinifer
Mesocyclops longisetus
Mesocyclops kieferi

Metacyclops mendocinus
Microcyclops anceps
Scolodiaptomus corderoi
Thermocyclops decipiens
Thermocyclops minutus

ZOOPLANKTON OF SAMUEL RESERVOIR IN THE FILLING PHASE

ROTIFERA

Synanterina spinulosa
Filinia longiseta var. *saltator*
Filinia longiseta var. *limnetica*
Polyarthra vulgaris
Conochilus coenobasis
Conochilus unicornis
Lecane bulla
Lecane lunaris
Brachionus quadridentatus
Brachionus patulus var. *macrocantus*
Brachionus zahnesi

Brachionus sp.
Platyas quadricornis
Hexarthra sp.
Lepadella patella
Ascomorpha ecaudis
Asplanchna sieboldi
Trichocerca similis
Mytilina bisulcata
Philodia sp.

Collotheca sp.
Manfredium eudactilotum

CLADOCERA

Moina reticulata
Moina minuta
Moina micrura
Ceriodaphnia cornuta cornuta
Ceriodaphnia cornuta intermedia
Ceriodaphnia cornuta rigaudi
Ceriodaphnia laticaudata
Daphnia laevis

Diaphanosoma breviremes

Bosminopsis deitersi
Bosmina longirostris
Euryalona orientalis
Simocephalus serulatus

Moinodaphnia macleayi
Scapholeberis fryeri
Grimaldina brazzai
Disparalona dadayi

COPEPODA

Argyrodiaptomus n.sp.
Notodiaptomus n.sp.
Mesocyclops brasilianus
Mesocyclops longisetus

Thermocyclops decipiens
Thermocyclops minutus
Microcyclops anceps

Table IX. Continued.

PROTOZOA

Votticella sp.	*Centrops acureata*
Arcella vulgaris	

INSECTA

Chaoborus sp.

CRUSTACEA

Conchostraca

ZOOPLANKTON OF BALBINA RESERVOIR IN THE FILLING PHASE

ROTIFERA

Brachionus patulus v. *macracanthus*	*Lecane ungulata*
Brachionus zahniseri	*Lecane quadridentata*
Keratella americana	*Mononumata* sp.
Platyias quadricornis	*Trichocerca similis*
Euchlanis sp.	*Polyarthra vulgaris*
Lecane curvicornis nitida	*Asplanchna*
Lecane luna	*Testudinella patina*
Lecane leontina	*Sinantherina spinosa*
Lecane lunaris	*Filinia longiseta var. limnetica*
Lecane ludwigi	*Filinia* sp.
Lecane signifera	*Collotheca* sp.
	Catenularia

CLADOCERA

Daphnia gessneri	*Bosminopsis deitersi*
Ceriodaphnia cornuta cornuta	*Diaphanosoma brevireme*
Ceriodaphnia cornuta rigaudi	*Diaphosoma* sp.
Moinodaphnia macleayi	*Chydorus* sp.
Moina reticulata	*Alona* sp.

COPEPODA

Notodiaptomus amazonicus	*Thermocyclops decipiens*
Rhacodiaptomus calamensis	*Thermocyclops minutus*
Mesocyclops brasilianus	

CRUSTACEA

Conchostraca: *Cyclestheria hislopi*

---

distributed with the highest concentrations, largest biomass and highest production located in station II where the macrophytes were densest (Matsumura-Tundisi & Tundisi, 1976; Rocha, 1978; Rocha *et al.*, 1982; Rocha & Matsumura-Tundisi, 1984). This probably explains Rocha's findings that the copepod's population density was inversely related to the biomass of nannoplankton, whose low abundance at station II (Watanabe, 1981) is interpreted by Tavares & Matsumura-Tundisi (1984) as being due to zooplanktonic grazing pressure. The dense areas of macrophytes offer *Argyrodiaptomus furcatus* a refuge from predation by the characid fish, *Astyanax fasciatus*, whose young and adult stages feed selectively upon the copepod (Barbosa & Matsumura-Tundisi, 1984). A high level of fish predation may also explain the zooplankton composition at the limnetic sampling station of Billings Reservoir which consisted of two species of cyclopoid copepods (39% of the zooplankton) and rotifers (36%); cladoceran species were only significant in the littoral sampling station (Sendacz, 1984). Studies of intrazooplanakton predation at Barra Bonita Reservoir (Matsumura-Tundisi *et al.*, 1991) showed that *Mesocyclops longisetus* feeds on *Brachiomus calyciflorus* and on *Ceriodaphnia cornuta*. *Mesocyclops kieferi* predates also on *Brachionus calyciflorus*. Based on experimental data and field sampling, the authors concluded that *Mesocyclops* has a significant impact on *Brachionus calyciflorus* population. This might explain the inverse relationship in the abundance of *Mesocyclops* and *Brachionus*. Low preference for *Ceriodaphnia cornuta* seems to relate to prey size. The coexistence of three species of *Mesocyclops* in the reservoir is probably due to a different seasonal incidence, or by size differences as in the case of *Mesocyclops longisetus* which is much bigger. The population of this species is low, and it is possibly limited by fish predation, according to the authors. Thus fish predation may have a significant qualitative and quantitative impact on the zooplankton community of these la Plata Basin Reservoirs.

For the Amazonian reservoirs the situation differs. In such extreme ecological conditions during the filling phase fish predation is very low or inexists, therefore masses of zooplankton specially cladocerans are observed (Matsumura-Tundisi *et al.*, 1991; and Tundisi, J. G. unpublished observations at Samuel Reservoir).

Table IX lists the zooplankton composition for Samuel Reservoir during the filling phase.

The commonest species of reservoir macrophytes are listed in Table X. In many of the reservoirs, the macrophytes exhibit a zoned distribution, with rooted forms (*Pontederia* sp.) found at the silted river entrance into the reservoir, followed further out by the presence of submerged species (*Mayaca fluviatilis*) and then by the floating *Nymphoides indica* nearest to the open water. The temporal as well as spatial succession of the reservoir macrophytes depends on various factors such as the reservoir's initial trophic status, the available substrate or degree of deposited sediment, the speed of local currents and the type of colonization strategy. For example, *Eichhornia* spp. dominate the eutrophic reservoirs in the La Plata Basin whereas *Mayaca fluviatilis* demands conditions of high water transparency. The spatial distribution of macrophytes is influenced by sediments brought in by the river tributaries. Often the first species to colonise are the free floating *Eichhornia crassipes* and *E. azurea* which are then succeeded by *Pistia stratiodes*, and later by *Typha dominguensis* in the regions with riverine sediments. Occasionally *Salvinia molesta* may appear, too, in the latter stages.

Table X. Composition of macrophyte species from the La Plata Basin Reservoirs.

Brachiaria nutrica	*Salvinia minima*
Brachiaria sp.	*Typha dominguensis*
Mayaca fluviatilis	*Ludwigia decurrens*
Eichhornia crassipes	*Myriophyllum brasiliensis*
Eichhornia azurea	*Ceratophyllum* sp.
Pistia stratiodes	*Pontederia* sp.
Polygonum spectrabile	*Nymphoides indica*
Salvinia curriculata	

5. Interactions between forcing functions and biogeochemical processes

The main forcing functions controlling the seasonal cycle, species succession and the time course of biogeochemical processes in South American reservoirs are rain, wind, outflow rates and retention times (Tundisi *et al.*, 1988; Tundisi, 1989; 1990).

Rainfall acts as a forcing function by introducing nutrients from the catchment, either in a soluble or particulate form whereas wind-mixing in reservoirs is fundamental to their limnological functioning, such as the biogeochemical processing in the iron and manganese cycles or keeping in suspension the photosynthesising populations of heavy diatoms like the *Melosira* spp. These two forcing functions operate at different times of year. In the La Plata Reservoirs, wind action is most significant during the winter whereas the effects of rain is more important in summer. The seasonal switching of these two forcing functions influences which algal species predominate and thus controls the succession pattern (Tundisi & Matsumura-Tundisi, 1990) or may even change the time scales of phytoplanktonic growth and succession (De Fillipo, 1987).

Outflow rates and retention times are operationally managed in relation to the local hydrological cycle. Short retention times will select for small planktonic algal species (nanoplankton and picoplankton) whose rapid reproduction can replace loss of biomass. Long retention times of 3–6 months favour larger colonial forms with flotation mechanisms such as some Cyanobacteria. In complicated reservoirs with many compartments, the seasonal cycle of algal succession has its own time course in each compartment and out of phase with the main channel. In shallow ones, like Broa Reservoir, rainfall influences the seasonal cycle in summer and wind in winter.

In Amazonian reservoirs, outflow rates and retention times are also key factors which acclerate the decomposition of the organic matter (by providing oxygen) in the vast inundated areas (1500–2000 km^2) as well as quickening the aging process of these reservoir ecosystems. In recent planning procedures, it was considered that a 40-day retention period was sufficient to ensure both the fast decomposition of the flooded forests and the rapid recovery of reservoir water quality within a time period of 7–10 years and despite the vast areas involved.

Because of the importance of these forcing functions to the limnology of reservoirs, management of the reservoirs for water quality must take account of when in the reservoir's seasonal cycle to change retention times. This is already a normal practice in the Amazonian Tucuruí Reservoir, namely to enhance downstream levels of dissolved oxygen by increasing outflow rates. Several authors (Legendre & Demers, 1984;

Margalef, 1975 amongst others) have described effective hydrodynamic methods for controlling phytoplanktonic succession. Blooms of *Microcystis aeruginosa* have been controlled by increasing spill water outflows.

6. The aging of reservoirs in Brazil

Several papers have been published on the kinds of changes that take place during the years that follow the filling of reservoirs mostly from the temperate region (Aggus, 1969; Armengol, 1977; Benson, 1973; Cowell & Hudson, 1967; Mills & Schiavone, 1982; Neel, 1967; Ploskey, 1983) and two from the tropics (Balon & Coche, 1974; Heide, 1982). Balon & Coche (1974) followed the changes in the African man-made Lake Kariba and Heide (1982) in Lake Brokopondo from the humid tropics. For Brazilian reservoirs, Matsumura-Tundisi has recent information on the filling phase of the Amazonian Samuel Reservoir (Matsumura-Tundisi *et al.*, 1991), and Bonetto (1976, 1977a, b), Tundisi (1983, 1986) and Tundisi (1986) have initiated long-term studies to follow the evolution of reservoirs in Argentina and Brazil (La Plata Basin).

Changes occur in the chemical composition of the water and in the biomass and diversity of the living organisms which are responses to events occurring in the catchment area and inputs into the reservoir. Margalef (1975) interprets the aging process as an accumulation of information. In the study on Brokopondo Reservoir, Heide (1982) showed that the initial inundation of the forest was immediately followed by oxygen depletion and high fish mortalities in the reservoir. High fish mortalities also occurred in the lower parts of Samuel Reservoir (Amazonia). During the transitional phase (in Brokopondo and in Samuel), Euglenophyceae and Gastrotricha developed in the plankton together with mucilaginous green algae and a good representation of planktonic crustaceans and rotifers. The chemical and biological events in the Amazonian reservoirs were largely determined by the decomposition of the drowned vegetation and not by the quality of the input water. Subsequently, the longer-term succession changed from Euglenophyceae to Chlorococcales and Desmidiales.

In the tropical Tucuruí Reservoir, filling was followed by a large increase in particulate organic matter of detrital origin from the forest and, later, by high biomasses of fish but produced by few species. The food web is based upon the allochthonous detrital input from the forest rather than on autotrophic production, even though both macrophytic and periphytonic growth is very intensive after the maximal filling has been achieved. A period of macrophyte dominance persists until nutrient depletion takes place. Heide (1982) even suggests that an oligotrophic condition returns once the organic matter has been completely decomposed. Such a pattern of events could be normal in all tropical reservoirs where rain forest is inundated and there is evidence from dark fixation experiments that indicates high levels of heterotrophic activity. Results from Tucuruí Reservoir suggest that it takes 10 years after filling to achieve complete decomposition of the inundated tropical rain forest. However, this depends upon the operational retention times. Relatively short retentions of 40 days produce fast flows and rapid decomposition of the organic matter in the main channel. In the side compartments, re-cycling of minerals will be slower than in the main channel, except in the event that the compartmental organic matter is exported to the region of main flow. Thus, the process of aging, and its related succession, may be dependent upon the kind of exchanges that may occur between the various side bays and the main channel.

Figure 21 attempts to summarise the organisation of the food web in the Tucuruí Reservoir at the initial phase in the aging process and illustrates the key role of organic detritus from the forest in supporting the detritivorous invertebrates (like *Macrobrachium amazonicum*) which generate the initial increased fish biomass (of species like *Cichlia ocelaris)* that is so characteristic of these reservoirs.

Patterns of aging are rather different for the reservoirs in the La Plata Basin because of problems related to eutrophication and toxicity arising from the disposal of untreated effluents from domestic and industrial sources and from runoffs from agricultural land on which chemical fertilizers had been applied intensively. How severe are these problems depends upon the hydrological regime, water level fluctuations and variation of

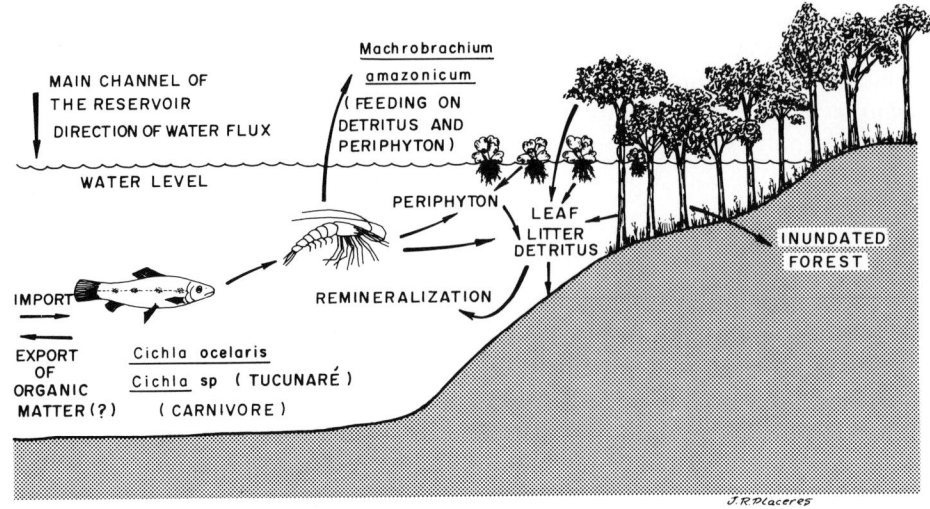

Fig. 21. A schematic representation of a detritus supported food web in one of the compartments of Tucuruí Reservoir (Amazonia).

retention time. Blooms of *Microcystis aeruginosa* appear as soon as the retention time is prolonged; ammonia concentrations may increase up to 1 mg l^{-1}; conductivity levels can change from 150 μS cm^{-1} to 400 μS cm^{-1} as shown for Barra Bonita Reservoir (Tundisi & Matsumura-Tundisi, 1990).

Eutrophication of the southern reservoirs seems to affect the ratio of calanoid : cyclopoid copepods in the zooplankton. Cyclopoid copepods, such as *Thermocyclops minutus* and *Thermocyclops decipiens* predominate over calanoids, largely *Argyrodiaptomus furcatus* in the eutrophic reservoirs (Matsumura-Tundisi & Tundisi, 1976). In general, it is notable that cyclopoid copepods are the most abundant group of the zooplankton of eutrophic reservoirs of southern Brazil. In conditions of hypertrophy, as in the small Taquaral Lake, no crustacean zooplankton existed.

The S. Francisco River Basin Reservoirs also suffer from problems of eutrophication and toxicity caused by inputs of chemical fertilizers, pesticides and herbicides. In this semi-arid region of Brazil, the irrigation system is well developed and the waste water is discharged into reservoirs, together with silt, high levels of nitrogen and phosphorus and the degradation products of various kinds of toxic substances.

In summary, the aging process is different in the three main hydrographic basins. In the Amazon reservoirs, the initial organic load from the inundated forests is decomposed at a speed depending upon retention time, mean depth and the area of forest inundated. In the La Plata Reservoirs, aging is associated with using the reservoirs to dispose of man's waste water in an untreated form, whether it originates domestically, from industry or from his agriculture. In the S. Francisco Reservoirs, there are the same influences but with the addition of siltation from the irrigation system.

A special situation exists with the reservoirs in the upper Paraná Basin which are arranged in a cascade so that the possibility exists of the upper ones contributing nutrients to the downstream ones thus accelerating their eutrophication. The main process which slows down the accumulation of nutrients downstream is full depth circulation of water so that phosphate (for example) is mainly trapped and accumulated in the interstitial water and particles of the sediments and exists at low levels only in the water column. Thus, despite the high population densities and high degree of industrialisation that exist in the upper course of the Paraná River, there is no sign of accleration eutrophication in the lower reservoirs. What happens to toxic chemicals is another problem, however, which urgently needs some research attention. The pulses of suspended particles and chemicals arising from rain in the catchment may transport adsorbed toxic chemicals in increasing

Table XI. Management programs for LOBO (Tietê River Basin), Tucuruí (Tocantins River Basin) and Itaparica (S. Francisco River Basin) Reservoirs.

	LOBO (BROA) Reservoir	TUCURUÍ Reservoir	ITAPARICA Reservoir
Water quality problems	Maintenance of present water quality for recreation, domestic use. Modelling for water quality in progress. Mass balance in progress	Improvement of water quality for turbine protection, fisheries and general use. Modelling of water quality in progress. Water Blooms in compartments.	Maintenance of water quality for irrigation, fisheries, domestic use. No modelling. Eutrophic reservoir. Water blooms.
Population	Maintenance of present level of activity (small scale agriculture, recreation). Environmental education system in progress.	Increase in the number of jobs, agricultural development. Relocation. Improvement of sanitary conditions.	Increase in the number of jobs. Fish farming, small scale agriculture. Relocation. Prevention of disease.
Protection and conservation measures (implemented or needed)	Maintenance of spatial heterogeneity (gallery forests, macrophytes). Prevention of eutrophication. Protection of rivers. Monitoring of rivers. Reforestation with native species.	Program of regional integration and insertion of the project. Protection of Islands. Marginal areas. Protection of downstream areas. Control of water quality downstream.	Regional integration. Protection of fish fauna. Reforestation. Protection of downstream areas.
Multiple uses of the reservoir. Implemented or planned.	Recreation, small scale fisheries, limited, irrigation, scientific activities, fish production. Hydroelectricity. Training in environmental sciences. Limnology. Used for environmental education. (Implemented).	Fisheries, hydroelectricity, transportation. Scientific studies, training. Recovery of flooded wood. (Implemented).	Hydroelectricity, fisheries. Fish farming, irrigation. Limited transportation. (Partially implemented)
Other general measures for correction and protection. Preventive and correctives actions.	Area of environmental protection. Management program in implementation. Legal actions. Partially implemented.	Establishment of protected areas around the reservoir. Control of water quality downstream and support of populations downstream. Regional insertion partially implemented.	Construction of farm villages. Improvement of sanitary conditions downstream. Partially implemented.

amounts and interfere in an unknown way with both the seasonal cycles and long-term developments of reservoir biology and limnology.

7. Problems of reservoir management in Brazil

The proliferation of reservoirs in different regions of Brazil has generated problems of management associated with demands for multiple usage and the urgent need to integrate their functioning into the regional life, with its characteristic ecological, economic and social conditions. Because regions do differ, so management objectives and plans will also differ widely. Any theoretical understanding of how reservoirs function must be based on fundamentals of limnology and must generate practical procedures applicable to reservoirs with diverse morphometry and hydrogeochemistry, sited in different climatic zones and surrounded by catchment area with different levels of human activity. Table XI is an attempt to summarise some management programmes for three reservoirs differing in size and age and situated in very different river basins. This table suggests some possible management activities for three different

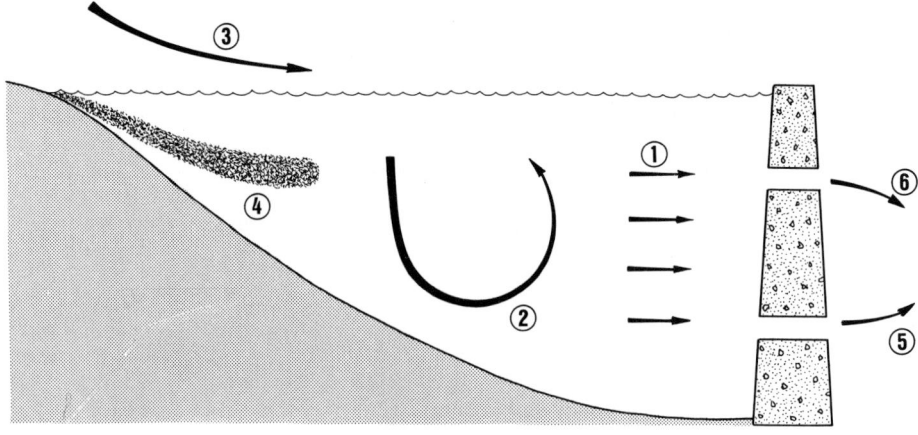

1 - FLUSHING RATE WHICH DETERMINES RETENTION TIME.
2 - WIND GENERATED CIRCULATION.
3 - INPUTS FROM PRECIPITATION AND RUNOFF (NON POINT SOURCE).
4 - RIVER DISCHARGE WITH DEPTH, VELOCITY AND TRANSPORT COMPONENTS.
5 - TURBINE FLOW.
6 - SPILL OUTFLOW.

Fig. 22. A schematic representation of the major forming functions in a generalized reservoir in Brazil.

reservoirs of various size, and also in different latitudes. The extent of this management will depend on spatial/temporal scale and the degree of regional location of the reservoirs. Their management depends mainly on the hydroelectric company that exploits them. Recently Eletrobras, the Brazilian main organization for energy exploitaition, issued regulations and recommendations for management of existing reservoirs. Also management alternatives for new reservoirs to be built were recommended. Research results were most helpful in the design of this recommendation.

8. Interactions between research and management

One of the aims of reservoir research is to produce ideas that are helpful for the management of reservoirs and their watersheds. It is hoped that the present paper's analysis of various reservoir case studies from a wide range of latitudes within Brazil will help to provide a conceptual framework for management, to improve operational strategies of existing reservoirs and to assist in the planning of new ones. Figure 22 illustrates the dominating processes that shape the structure and functioning of the reservoir ecosystem and which we have termed 'forcing functions'.

At our present stage in understanding, any thinking about the management of reservoirs must consider the following subjects:

(a) How to control water quality in reservoirs, especially in a cascade series, and how to slow down the process of eutrophication?
(b) What alternatives are available to multiple usage of reservoirs? Hino (1988) lists these as storage of drinking water, transport, recreation, fisheries, aquaculture, irrigation, tourism.
(c) How to regulate human activity in watersheds? What techniques are available to reduce loading levels, whether point or non-point sources?
(d) How to exploit the food resources without damaging the environment? For example, by increasing the natural fish populations, by stocking, by reservoir fish culturing, by aquaculture of shrimps and algae.
(e) How to improve the environmental quality of damaged reservoirs and watersheds?

9. Summary and conclusions

1. Studies on the ecology and limnology of large reservoirs from three major river basins, the Amazon, S. Francisco and La Plata, have advanced understanding on how they function.

This understanding should now be consolidated into a theoretical framework useful for present management and future planning.

2. Certain characteristics of these reservoir systems (size, morphometry, retention time, outflow rate) pose problems for their study (e.g. sampling) and control (e.g of water quality). Most of the reservoirs are polymictic by virtue of their shallowness and only a few are monomictic. The seasonality of the reservoirs are driven by the periodicity of rain (wet and dry periods); the major forcing functions of retention time and water level fluctuations originate from this seasonality. Most reservoirs are not simple in shape but dendritic and compartmentalised.

3. Spatial heterogeneity in terms of chemistry and biology is either longitudinal (along the line of flow), horizontal (associated with the above compartmentalisation) and vertical (large in the stratifying Amazonian reservoirs or coastal ones). The unidirectional water fluxes which change with operation procedures are a source of spatial heterogeneity and may enhance biological diversity.

4. Recent data suggests that diversity of species in reservoirs is greater than in lakes of the same volume and approximate morphometry.

5. The underwater light regime is greatly influenced by high levels of suspended inorganic particles in many reservoirs and by extensive growths of cyanobacterial blooms in highly eutrophic ones. Suspensions of inorganic particles are greatly increased during the rainy period and so produce a pulse into the reservoir.

6. Phytoplankton contributes most to the reservoirs' primary production but the production of macrophytes and periphyton is quantitatively important in certain reservoirs. Amazonian reservoir ecosystems are based upon the large detrital source from decomposing forest vegetation, macrophytes and periphyton, at least in the first years of inundation.

7. Pulse effects are important in these reservoirs and are expressed as large concentrations of suspended particles, surface runoff, wind-generated turbulence, rain and water level fluctuations. The frequency of pulses influences algal succession in the phytoplankton. Operation of the reservoirs may introduce another pulse effect.

8. The large size and morphometric complexity of Brazilian reservoirs demands a research methodology that suits the conditions. Some of the techniques adopted are extensive sampling combined with remote sensing; the use of fixed sampling sites and regular sampling period; the coupling of field studies with limnological experimentation.

9. The cascades of reservoir which are typical of several river systems in Brazil need special investigations in relation to their vulnerability to eutrophication, cumulation of toxic chemical deteriorating water quality since the cascade is operated as a single reactor unit.

10. Reservoirs of the Amazon River Basin pose particular problems such as the consequence of the intensive decomposition of the drowned forests, severe deterioration in water quality upstream and downstream of the reservoirs and large-scale production of gases (methane, hydrogen sulphide, carbon dioxide). Limnological studies have shown that water quality downstream can be improved by operating the spill water for oxygenation. Most of the Amazonian reservoirs are stratified with an anoxic hypolimnion.

11. Thinking about management of these man-made water bodies should also include considering ways of:
    (i) exploiting them for human food resources (such as fisheries, fish and shrimp culture) and
    (ii) tailoring multiple usages to conditions in the various geographical regions.

    Some effort should be put into developing simulation models of the eutrophic reservoir ecosystem to help with managerial thinking about future strategies or to provide some predictions on the time-course of eutrophication in the cascade reservoirs. Other urgent problems involve:
    (i) defining measures for controlling deleterious human activities in the catchment areas,
    (ii) getting on with re-planting the riparian forests with native species of trees and
    (iii) taking steps to reduce erosion.

12. The important subjects for future limnological

research on Brazilian reservoirs are comparative studies and a continuation of work on seasonal cycles, short-term fluctuations and nictemeral variations. Experimental limnology applied to a better knowledge of biological processes (interactions, such as predator-prey relationships, life cycles of species, grazing pressure over phytoplankton, toxicity response and biological indicators) is also urgently needed. Such scientific developments will help to understand the dynamic ecosystem structure of reservoirs.

Acknowledgements

The authors are grateful for financial support to: FAPESP (S. Paulo State Research Foundation): Grant numbers: 1290/78 initiative to J. G. Tundisi, 1360/83; 0653-2/85; 1957-3/88, Eletronorte (for partial support of field studies of Amazonian Reservoirs Tucuruí and Samuel, Eletrobras (for support of field visits in the S. Francisco River Basin Reservoirs), CESP (for logistic support of the field studies at Middle Tieté River Reservoirs), Organization of American States (for financial support of reservoir studies in Brasil), and to CNPg (National Research Council of Brasil) for partial support of reservoir studies. This is contribution number 1, thematic project, FAPESP (Process number 91-0612-5).

References

Aggus, L. R., 1969. Bottom Fauna Development in Beaver Reservoir, Northwest Arkansas, During the Period of Filling, 1964–1966. Ph.D. Thesis. Auburn, Auburn University, 105pp.

Applegate, R. L. & J. W. Mullan, 1967. Standing crops of dissolved organic matter, plankton, and seston in a new and an old Ozark Reservoir. In: Reservoir Fishery Resources Symposium. Presented by the Reservoir Committee of the Southern Division, American Fisheries Society, Washington, D.C.: 517–530.

Arcifa, M. S., C. G. Froehlich & S. M. Gianesella-Galvao, 1981. Circulation patterns and their influence on physico-chemical and biological conditions in eight reservoirs in Southern Brazil. Verh. Int. Ver. Limnol. 21: 1054–1059.

Armengol, J., 1977. Tipologia de los embalses espanoles. In: Seminário sobre Medio Ambiente y Represas. Tomo I: 3–13.

Balon, E. K. & A. G. Coche (eds.), 1974. Lake Kariba, a Man-Made Tropical Ecosystem in Central Africa. Junk, The Hague, 767pp.

Barbosa, P. M. M. & T. Matsumura-Tundisi, 1984. Consumption of zooplanktonic organisms by *Astyanax fasciatus* Cuvier, 1819 (Osteichthyes, Characidae) in Lobo (Broa) Reservoir, Sao Carlos, SP, Brazil. In: J. H. Dumont & J. G. Tundisi (eds.), Tropical Zooplankton. Junk, The Hague: 171–181.

Barrow, C. J., 1983. The environmental consequences of water-development in the tropics. In: Doi Jin Bee (ed.), Natural Resources in Tropical Countries. Singapore University Press, Singapore: 439–476.

Barrow, C. J., 1987. The environmental impacts of the Tucuruí Dam on the middle and lower Tocantins River Basin, Brazil, Regulated rzivers. Research & Management 1: 49–60.

Baxter, R. M., 1977. Environmental effects of dams and impoundments. Ann. Rev. Ecol. Syst. 8: 255–283.

Benson, N. G., 1973. Evaluating the effects of discharge rates, water levels, and peaking on fish populations in Missouri River main stem impoundments. In: W. C. Ackerman, F. G. White, E. B. Worthington & J. L. Ivens (eds.), Man-made Lakes: Their Problems and Environmental Effects. American Geophysical Union, Washington: 683–689.

Bonetto, A. A., 1975. Hydrologic Regime of the Paraná River and its influence on Ecosystems. In: A. D. Ahasler (ed.), Coupling of Land and Water Systems. Springer-Verlag Berlin: 175–197.

Bonetto, A. A., 1976. Water Quality of the Paraná River. Dir. Constr. Portuarias, INCYTH – PNUD – ONU, Argentina, 202pp. (In Spanish).

Bonetto, A. A., 1977a. Reservoirs and their ecological projections, problems and perspectives. In OAS, Seminar on the Environment and Reservoirs, I OAS, Montevideo, Uruguay: 14–34. (In Spanish).

Bonetto, A. A., 1977b. Los lagos de represa y suas projeciones ecologicas. In: Seminario Sobre Medio Ambiente y Represas, Tomo I: 13–34.

Bonetto, A. A., H. P. Castello & I. R. Wais, 1987. Stream regulation in Argentina, including the superior Paraná and Paraguay Rivers. Regulated rivers. Research & Management 1: 95–109.

Bonetto, C., 1982. Producion primaria del fitoplancton, concentracion de pigmentos, materia organica y nutrientes en la caracterizacion limnologica de los cuerpos de agua regionales del nordeste argentino. Ph.D. Thesis. Univ. Nacional de Buenos Aires, 185pp.

Branco, S. M. & A. A. Rocha, 1977. Pollution, Protection and Multiple Uses of Reservoirs. Edgar Blücher CETESB, Sao Paulo, Brazil, 185pp. (In Portugese).

Branski, J. M., J. P. De Avila, R. L. L. Lopez, A. J. T. Goncalves & J. G. Tundisi, 1988. Environmental Impact Assessment for the Proteira Hydroelectric Project. Paper presented in the 2nd Workshop River Reservoir Basin Ecosystem Approach to Environmental Sound Management of Inland Waters. Bangkok, Thailand, January 15–21, 1988.

Calijuri, M. C., 1988. Respostas Fisioecologicas da Comunidae Fitoplanctonica e Fatores Ecologicos em Ecossistemas com Diferentes Estágios de Eutrofizacao. Doctorate Thesis,

University of S. Paulo, School of Engineering, 293pp.

Claflin, T. O., 1968. Reservoir aufuwuchs on inundated trees. Trans. Am. Fish. Soc. 104: 524–525.

CHESF (Companhia Hidroelétrica do Sao Francisco), 1987. O empreendimento hidrelétrico de Itaparica e seus aspectos ambientais (Project Plan for Itaparica Reservoir). Hidroservice, 59pp.

Cowell, B. C. & P. L. Hudson, 1967. Some environmental factors influencing benthic invertebrates in two Missouri River Reservoirs. In: Reservoir Fishery Resources Symposium. Presented by the Reservoir Committee of the Southern Division, American Fisheries Society, Washington, D.C.: 541–555.

De Fillipo, R., 1987. Climatologia, Hidrologia e Sucessao do Fitoplancton na Represa de Barra Bonita, Médio Tieté. MsC. Thesis, UFSCar, 150pp.

Dussart, B. H., C. H. Fernando, T. Matsumura-Tundisi & R. J. Shiel, 1984. A review of systematics, distribution and ecology of tropical freshwater zooplankton. In: H. J. Dumont & J. G. Tundisi (eds.), Tropical Zooplankton. Junk, The Hague: 77–91.

Eletrobrás, 1987. Projeto Brasil, 2010.

Esteves, F. De A., 1983. Levels of phosphate, calcium, magnesium and organic matter in the sediments of some Brazilian reservoirs and implications for the metabolism of the ecosystems. Arch. Hydrobiol. 2: 129–138.

Findenegg, I., 1966. Phytoplankton und Primärproduktion einiger Ostschwizerischer Seen und des Bodensees. Schweiz. Z. Hydrol. 28: 148–172.

Garzon, C. E., 1983. Water Quality Management Strategies for the Alto Sinu Hydroproject. Preliminary proposal for Correlca, Colombia. Dartmouth College, Department of Engineering, Hanover, New Hampshire, 10pp.

Goodland, R. J. A., 1977. Environmental optimization in hydrodevelopment of tropical forest regions. In: R. D. Panday (ed.), Man-made Lakes and Human Health. Suriname, Paramaribo, University of Suriname: 10–20.

Goodland, R. J. A., 1987a. Environmental Assessment of the Tucuruí Hydroproject Rio Tocantins, Amazonia, Brazil. Eletronorte S.A., D.F., 168pp.

Goodland, R. J. A., 1978b. Environmental Assessment of the Hydroelectric Project Rio Tocantins, Amazonia. Survival International Review 3: 11–14.

Goodland, R. J. A., 1978c. Environmental Recconaissance of the Tucuruí Hydroproject Amazonia, Brazil. Eletronorte S.A., 141pp.

Green, J., 1970. Freshwater ecology in the Mato Grosso, Central Brazil. I. The conductivity of some natural waters. J. Nat. Hist. 4: 289–299.

Green, J., 1972a. Freshwater ecology in the Mato Grosso, Central Brazil II. Association of Cladocera in meander lakes of the Rio Suiá Missu. J. Nat. Hist. 6: 215–227.

Green, J., 1972b. Ecological studies on crater lakes in West Camerron zooplankton of Barombi Mbo, Mboandong, Lake Kotto and Lake Soden. J. Zool. 166: 283–301.

Green, J., 1972c. Latitudinal variation in associations of planktonic Rotifera. J. Zool. London 167: 31–39.

Heide, J. Van der, 1976. Hydrobiology of the man-made Brokopondo Lake. Brokopondo Research Report, Suriname 2, 95pp.

Heide, J. Van der, 1982. Lake Brokopondo-Filling Phase Limnology of a Man-Made Lake in the Humid Tropics. Amsterdam, 428pp.

Henry, R., K. Hino, J. G. Gentil & J. G. Tundisi, 1985. Primary production and effects of enrichment with nitrate and phosphate on phytoplankton in the Barra Bonita reservoir (State of S. Paulo, Brazil). Int. Revue ges. Hydrobiol. 70: 561–573.

Henry, R. & J. G. Tundisi, 1988. O conteudo em calor e a estabilidade em dois reservatorios com diferentes tempos de residencia. In: J. G. Tundisi (ed.), Limnologia e Manejo de Represas. Série Monografias em Limnologia. Vol. I (tomos 1 e 2), ACIESP/CRHEA/EESC/USP: 295/322.

Henry, R., 1990. Estructura espacial e temporal do ambiente físico-químico e análise de alguns processos ecologicos na Represa de Jurumirim (Rio Paranápanema) e sua baccia hidrogfáfica. DsC. Thesis, UNESP, Botucatu, 242pp.

Hill, G. & H. Rai, 1982. A preliminary characterization of the tropical lakes of the Central Amazon by comparison with polar and temperate systems. Arch. Hydrobiol. 96: 97–111.

Hino, K., 1988. Eletronorte: Environmental studies and experiences developed in the Amazon region. In: Expert Group Workshop on River/Lake Basin Approach to Environmentally Sound Management of Water Resources. ILEC/UNCRD/UNEP, 28pp.

Junk, W. & W. D. Howard, 1984. Ecology of aquatic macrophytes in Amazonia. In: H. Sioli (ed.), The Amazon – Limnology and Landscape Ecology of a Mighty Tropical River and its Basin. Junk, Dordrecht: 269–293.

Junk, W. J. & J. A. S. Nunes De Mello, 1987. Impactos ecologicos das represas hidrelétricas na bacia Amazonica brasilileira. Tübingen Geographische Studien 95: 357–387.

Kirk, J. T. O., 1983. Light and Photosynthesis in Aquatic Ecosystems. Cambridge Univ. Press, Cambridge, 401pp.

Legendre L. & S. Demers, 1984. Towards dynamic biologiccal oceanography and limnology. Can. J. Fish. Aquatic Sci. 41: 2–19.

Lima, W. D., J. G. Tundisi & M. A. Marins, 1979. A systematic approach to the sensitivity of *Melosira italica* (Ehr.) Kutz. Rev. Bras. Biol. 39: 559–563.

Margalef, R., 1975. Typology of reservoirs. Verh. Int. Verein. Limnol. 19: 1811–1816.

Margalef, R., 1976. Life forms of phytoplankton as survival alternatives in an unstable environment. Oceanologia Acta 1: 493–510.

Margalef, R., D. Planas, J. Armengol, S. G. Vidal, C. A. Alberto, S. G. Alberto, J. T. Santillana & M. E. Miyares, 1976. Limnologia de los Embalses Espanoles. Direccion General de Obras Hidraulicas, 421pp.

Matsumura-Tundisi, T., & J. G. Tundisi, 1976. Plankton studies in a lacustrine environment. I. Preliminary investigations on the seasonal cycle of zooplankton. Oecologia (Berlin) 25: 265–270.

Matsumura-Tundisi, T., J. G. Tundisi, A. A. Saggio, A. L. Oliveira Neto & E. L. G. Espindola, 1991. Limnology of Samuel Reservoir (Braxil, Rondonia) in the filling phase. Verh. Int. Verein. Limnol. 24: 1482–1487.

Mills, E. L. & A. Schiavone, 1982. Evaluation of fish communities through assessment of zooplankton populations and measures of lake productivity. N. Am. J. Fish. Manage. 24: 14–27.

Monosowski, E., 1984. Tucuruí Dam in the Amazon: development at environmental cost. IIUG preprint 84-8. International Institute for Environmental and Society WZB, Berlin, 18pp.

Morris, I., 1980. The Physiological Ecology of Phytoplankton. Blackwell Scientific Publisher, Oxford, 625pp.

Nakamoto, N., M. A. Marins & J. G. Tundisi, 1976. Synchronous growth of a freshwater diatom *Melosira italica* under natural environment. Oecologia (Berlin) 23: 179–184.

Neel, J. K., 1967. Reservoir eutrophication and dystrophication following impoundment. In: Reservoir Fishery Resources Symposium. Presented by Reservoir Committee of the Southern Division, American Fisheries Society, Washington, D.C.: 322–332.

Nunes De Miranda, 1983. Geology and Distribution of Reservoirs in the Northeast of Brazil. DNOCS, 27pp.

Pedrozo F., A. A. Bonetto & Y. Zalocar, 1988. A comparative study on phosphorus and nitrogen transport in the Paraná, Paraguay and Bermejo Rivers. In: J. G. Tundisi (ed.), Limnologia e Manejo de Represas. Série Monografias em Limnologia. Vol. I. Tomo 1. EESC-USP/CRHEA/ACIESP: 91–117.

Ploskey, G. R., 1983. A review of the effects of water level changes on reservoir fisheries and recommendations for improved management. Tech. Rep. U.S. Army Eng. Waterways Exp. Stn. (E-83-3), 83pp.

Ploskey, G. R., 1985. Impacts of preimpoundment cleaning on reservoir ecology and fisheries in the United States and Canada. FAO Fishery Technology Paper, FAO, Rome, 258pp.

Reynolds, C. S., 1987. The response of phytoplankton communities to changing lake environments. Schweiz. Z. Hydrol. 49: 220–236.

Rocha, O., 1978. Flucuacao sazonal e distribuicao da populacao de Diaptomus furcatus. Sars. (Copepoda Calanoida) na Represa do Lobo (Broa) Sao Carlos, Sp. Dissertacao de Mestrado. Instituto de Biociencias da USP, Sao Paulo, 147pp.

Rocha, O., T. Matsumura-Tundisi & J. G. Tundisi, 1982. Seasonal fluctuation of *Argyrodiaptomus furcatus* population in Lobo Reservoir, Sao Carlos, SP, Brazil. Tropical Ecology 23: 134–150.

Rocha, O. & T. Matsumura-Tundisi, 1984. Biomass and production of *Argyrodiaptomus furcatus*, a tropical calanoid copepod in Broa Reservoir Southern Brazil. In: H. J. Dumont & J. G. Tundisi (eds.), Tropical Zooplankton. Junk, The Hague: 307–311.

Salas, H. J., 1983. Resumen del Segundo Encuentro del Proyecto Regional: Desarollo de Metodologícas Simplificadas para la Evaluacion de Eutroficacion en los Cálidos, antes Lagos Tropicales. OPS, Centro Panam. de Ing. Sanitaria y Ciencias del Ambiente, 33pp.

Sendacz, S., 1984. A study of the zooplankton community of Billings Reservoir-Sao Paulo. In: H. J. Dumont & J. G. Tundisi (eds.), Tropical Zooplankton. Junk, The Hague: 121–127.

Simonato, A., 1986. Ciclos Diurnos de Fatores Ecologicos na Represa do Lobo (Broa). MsC. Thesis, Federal University of S. Carlos, 143pp.

Tavares, L. H. S. & T. Matsumura-Tundisi, 1984. Feeding in adult females of *Argyrodiaptomus furcatus* (Sars, 1901), Copepoda Calanoida, of Lobo Reservoir (Broa) Sao Carlos, Sao Paulo, Brazil. In: H. J. Dumont & J. G. Tundisi (eds.), Tropical Zooplankton. Junk, The Hague: 15–23.

Tundisi, J. G., 1977. Producao primária, "standing-stock", francionamento do fitoplancton e fatores ecologicos em ecossistema lacustre artifical (Represa do Broa, Sao Carlos). DsC. Thesis F.F.C.L.R.P., USP, 410pp.

Tundisi, J. G., 1981. Typology of reservoirs in Southern Brazil. Verh. Int. Verein. Limnol. 21: 1031–1039.

Tundisi, J. G., 1982. Shallow waters in South America: present knowledge and perspectives for future research and management. Ecosystem Dynamics in Wetlands and Shallow Water Bodies 1: 63–67.

Tundisi, J. G., 1983. A review of basic ecological processes interacting with production and standing-stock of phytoplankton in lakes and reservoirs in Brazil. Hydrobiologia 100: 223–243.

Tundisi, J. G., 1984. Estratificacao hidraulica em reservatórios e suas consequencias ecológicas. Cienc. Cult. 36: 1498–1504.

Tundisi, J. G., 1986. Limnologia de represas artificiais. Boletim de Hidráulica e Saneamento, No. 11, 46pp.

Tundisi, J. G. (ed.), 1988a. Limnologia e Manejo de Represas. Série Monografias em Limnologia. Vol. I. Universidade de Sao Paulo, Escola de Engenharia de Sao Carlos, Centro de Recursos Hídricos e Ecologica Aplicada, Sao Carlos Tomo 1, 506pp. Tomo 2, 432pp.

Tundisi, J. G., 1988b. The Lobo (Broa) ecosystem and reservoirs in Brazil. In: Expert Group Workshop on River/Basin Approach to Environmentally Sound Management of Water Resources. UNEP/ILEC/UNCRD, Nagoya and Otsu, 28pp.

Tundisi, J. G., 1989. Management of reservoirs in Brazil. In: S. E. Jørgensen & R. A. Vollenweider (eds.), Guidelines of Lake Management. Principles of Lake Management. Vol. 1: 155–169.

Tundisi, J. G., 1990. Key factors of reservoir functioning and geographic aspects of reservoir limnology. Chairman's overview. Arch. Hydrobiol. Beih. Ergeb. Limnol. 33: 645–646.

Tundisi, J. G. & T. Matsumura-Tundisi, 1986a. Eutrophication processes and trophic state for 23 reservoirs in S. Paulo State, Southern Brazil. Fourth Brazil/Japan Symposium on Sciences and Technology. Supplementary Volume, Publicat. Academy of Sciences, S. Paulo State, 26pp.

Tundisi, J. G. & T. Matsumura-Tundisi, 1986b. Limnological studies in lakes and reservoirs in Brazil. Report to Organization of American States, 146pp. (unpublished).

Tundisi, J. G. & T. Matsumura-Tundisi, 1990. Limnology and eutrophication of Barra Bonita Reservoir, S. Paulo State, Southern Brazil. Arch. Hydrobiol. Beih. Ergeb. Limnol. 33: 661–678.

Tundisi, J. G., T. Matsumura-Tundisi, O. Rocha, J. G. Gentil & N. Nakamoto, 1977. Primary production, standing-stock of phytoplankton and ecological factors in a shallow tropical reservoir (Represa do Broa, S. Carlos, Brazil). Sem. Medio

Ambiente y Represas 1: 138–172.

Tundisi, J. G., T. Matsumura-Tundisi, K. Hino & M. C. Calijuri, 1988. The Lobo (Broa) ecosystem and reservoirs in Brazil. Expert Group Workshop on River Lake Basin Approach to Environmentally Sound Management of Water Resources. UNEP/ILEC/UNCRD, Nagoya and Otsu, Japan, February 1988, 28pp.

Tundisi, J. G., T. Matsumura-Tundisi, M. C. Calijuri & E. M. L. M. Novo, 1991. Comparative limnology of five reservoirs in the Middle Tieté River, S. Paulo State, Brazil. Verh. Int. Verein. Limnol. 24: 1489–1496.

Tundisi, J. G. & E. Wada, 1990. Gas exchange in tropical aquatic ecosystems. (Abstract). Intecol, Japan, Yokohama: 1.

Watanabe, T., 1981. Fluctuacao sazonal e distribuicao espacial do nano e microfitoplancton na Represa do Lobo (Broa) Sao Carlos, SP. Dissertacao de Mestrado. PPG-ERN – UFSCar, 158pp.

Whitaker, V. A., 1987. Ciclo sazonal das espécies químicas do ferro no reservatorio do Lobo (Broa). MsC. Thesis, University of S. Paulo, School of Engineering at S. Carlos, 152pp.

Chapter III

Problems in reservoir trophic-state classification and implications for reservoir management

O. T. Lind,[1] T. T. Terrell,[2] & B. L. Kimmel[3]
[1] *Department of Biology, Baylor University, Waco, Texas 76798-7388, USA;* [2] *Cooperative Unit Center, Fish and Wildlife Service, US Department of Interior, Washington, D.C. 20240, USA;*
[3] *Environmental Sciences Division, Oak Ridge National Laboratory*, Oak Ridge, Tennessee 37831-6351, USA*

Key words: reservoirs, limnology, water quality management, eutrophication, classification

Abstract

Trophic-state classification methods developed for natural lakes are applied routinely to impoundments for management purposes. However, the trophic-state classification of a reservoir can often depend more on the variable selected for the basis of the classification or on the classification method used, than on the 'true' trophic status of the impoundment. Additionally, the application of common trophic-state classification schemes to reservoir ecosystems is often inappropriate because of features common to river impoundments that are usually not present in natural lakes. While reservoir limnologists may be aware that reservoir trophic-state classification is problematic, water-resources managers are not, and consequently, many reservoirs may be ineffectively or inappropriately managed.

1. Introduction

Trophic-state classification is a useful tool for managing lakes. The quantity of energy captured, stored, and potentially used in a lake ecosystem (i.e., as reflected by its trophic status) can be estimated relatively simply by applying empirical relationships based on variables such as primary productivity, algal biomass, water transparency, and/or nutrient concentrations. The trophic-state classification of lakes in a region or in a water-management district is useful for establishing lake management strategies and for setting priorities for lake restoration efforts.

Trophic-state classification is also applied routinely to impoundments for similar management purposes. However, our data and those of other researchers (e.g., Walker, 1984; Placke, 1983) indicate that the trophic-state classification of a reservoir can often depend more on the classification method employed or the variable selected as the basis of that classification than on the 'true' trophic status of the water body. Our experience with reservoirs in the United States leads us to conclude that the application of common trophic-state classification schemes to reservoir ecosystems is often inappropriate because of features common to reservoirs that are usually not present in natural lakes.

We think that the trophic-state classification of reservoirs is problematic for two primary reasons:
1. An uncertain relationship exists between nutrient supply and primary production in reservoirs. This uncertainty results from:
   (i) discrepancies between the estimated nutrient supply and the actual nutrient availability for primary production, and
   (ii) the influence of other factors (e.g., short water-residence time, limited light availability), unrelated to nutrient availability, on primary productivity.
2. Reservoirs commonly possess a high degree of spatial and temporal heterogeneity in the environmental factors that control primary

---

*Operated by Martin Marietta Energy Systems, Inc. under Contract No. DE-AC05-84OR21400 with the US Department of Energy.

productivity and, thus, influence trophic status. Commonly, an uplake-to-downlake gradient in trophic status exists within an individual reservoir. The spatially heterogeneous and temporally dynamic nature of impoundments makes it difficult to adequately characterize the trophic status of a reservoir.

The purpose of this paper is to illustrate the primary sources of potential error that are inherent in the trophic-state classification of reservoirs. Armed with this information, the individual responsible for trophic-state classification of reservoirs will be able to perform his task more accurately and with more insight, and the water resources manager will be able to prioritize and implement appropriate reservoir management strategies more effectively.

2. Methods

We employed several different classification schemes (Table I) to estimate the trophic status of a number of reservoirs located throughout the United States. The trophic-state classifications used are based on the following variables: nitrogen and phosphorus concentrations or loading rates (Sakamoto, 1966a; Vollenweider, 1968; 1976; USEPA, 1974; Carlson, 1977), phytoplankton production (Wetzel, 1975), chlorophyll concentrations (Sakamoto, 1966b; USEPA, 1974; Carlson, 1977), water transparency (USEPA, 1974; Carlson, 1977), and phytoplankton taxa ratios (Nygaard, 1949; Stockner, 1971).

We used our own data for Lake Roosevelt ($R = 40$), Washington, Belton Reservoir ($R = 623$), Texas, and Sam Rayburn Reservoir ($R = 521$), Texas. Additionally, we used data on reservoirs included in the US Environmental Protection Agency National Eutrophication Survey (USEPA NES) (USEPA, 1975). Because short water-residence time (i.e., rapid flushing rate) appears to be a primary factor influencing algal growth response to nutrient supply (Soballe & Kimmel, 1987) and, thereby, trophic status, we selected impoundments with short R (generally < 50 d) for our analysis. In addition to using published physical-chemical data on these reservoirs, we also calculated the ratio of araphid to centric (A/C) diatoms (Stockner, 1971) and the diatom index (Nygaard, 1949) from June 1983 data for Lake Roosevelt, June 1980 data for Lake Belton, and June 1972 data for the NES Reservoirs.

Table I. Variables used in various trophic-state classification methods.

Variable	Source	Oligotrophic	Mesotrophic	Eutrophic
Total phosphorus (μg l^{-1})	Sakamoto (1966b)	2–20	10–30	10–90
	Vollenweider (1968)	5–10	10–30	30–100
	USEPA (1974)	<10	10–20	>20
Inorganic nitrogen (mg m^{-3})	Vollenweider (1968)	200–400	300–650	500–1500
Chlorophyll-a (mg m^{-3})	Sakamoto (1966a)	0.3–2.5	1–15	5–140
	USEPA (1974)	<7	7–12	>12
Phytoplankton biovolume (cm^3 m^{-3})	Vollenweider (1968)	1	3–5	10
Araphid/centric diatoms	Stockner (1971)	<1	1–2	>2
Diatom index	Nygaard (1949)	0.0–0.3		0.0–1.75
Secchi depth (m)	USEPA (1974)	>3.7	2.0–3.7	<2.0
Trophic State Index (for TP, SD)	Carlson (1977)	<40	40–50	>50

3. Results and discussion

3.1. Variation in the trophic-state classification of reservoirs

The trophic-state classification of an impoundment often varies depending on the variable or variables used to derive the classification. For example, 33 of the 100 largest impoundments in Texas fall in a different trophic-state category according to Carlson's trophic-state index (TSI, Carlson, 1977) when the TSI is derived from average summer chlorophyll data versus total phosphorus data. When annual chlorophyll data rather than total phosphorus data are used to derive Carlson's TSI, 44 of the 100 Texas impoundments are classified differently (Texas Department of Water Resources, 1984). As estimated by the USEPA (1974) classification scheme, Lake Roosevelt is categorized as oligotrophic using chlorophyll-a data and as eutrophic using total phosphorus data (Table II).

There is also considerable variation in trophic-state classifications when a single variable is used to calculate different trophic-state indices for the same impoundment (Table II). For example, Lake Roosevelt is classified as oligo-mesotrophic by the Sakamoto (1966a) method and as eutrophic by the USEPA (1974) method, using total phosphorus as the basis of both classifications. Similarly, Sam Rayburn Reservoir can be classified as ultra-oligotrophic if primary production or inorganic nitrogen is the variable used, as mesotrophic if total phosphorus is used, and as eutrophic if areal-based phosphorus loading is the basis of the estimate. Belton Reservoir has an even greater

Table II. Variation in trophic-state classification of selected reservoirs based on variable chosen[1] or classification system[2] used. Data from USEPA (1975) unless otherwise indicated.

Variable	Value	Ultra	Oligo	Meso	Eutro	Poly
			Lake Roosevelt, WA (Nigro et al., 1983)			
TP	16		S_2	C V_2	E	
IN	239		V_1			
C_a	5		E	S_1,C		
PBv	2		V_1			
A/C	3.7				St	
DI	0.27		N			
SD	3.6		E,C			
			Sam Rayburn Reservoir, TX (Lind, 1979)			
TP	15		S_2	C,E,V_2		
IN	77	V_2				
C_a	3.5		E	S_1,C		
PP	40	W				
PL	0.6				V_3	
SD	1.9			E,C		
			Belton Reservoir, TX (Lind, 1984, Unpubl.)			
TP	122				E,C	V_2
IN	92	V_2				
C_a	8.4			E,S_2 C		
A/C	55.1				St	
SD	2.9		E	C		
			Clyde Pond, VT			
TP	21		S_2	V_2,C,E		
IN	175	V_2				
C_a	7.5			E,S_2 C		
A/C	0.41		St			
SD	1.7				C,E	

Table II. Continued.

Variable	Value	Ultra	Oligo	Meso	Eutro	Poly
			Eau Claire Lake, WI			
TP	89				C,V_2,E,S_2	
IN	500	V_2				
C_a	19.6				C,E,S_1	
A/C	0.23		St			
SD	0.8					C,E
			Willow Reservoir, WI			
TP	29				V_2,S_2,C,E	
IN	155	V_2				
C_a	9.2			E	S_2,C	
A/C	0.32		St			
SD	1.1				C,E	
			Glen Lake, NH			
TP	28			V_2	S_2,C,E	
IN	135	V_2				
C_a	3.8		E	S_1,C		
SD	1.7				C,E	
			Slatersville Reservoir, RI			
TP	32				S_2,V_2,C,E	
IN	205	V_2				
C_a	8.1		E	S_1	C	
SD	1.3				C,E	
			Lake Lamoille, VT			
TP	18			C,V_2,E,S_2		
IN	260	V_2				
C_a	3.5		E	S_1,C		
SD	1.5				C,E	
			Waterbury Reservoir, VT			
TP	7		C,E	V_2,S_2		
IN	230	V_2				
C_a	5.2		E	C,S_1		
SD	2.4		E	C		
			Lake Wissota, WI			
TP	43				C,E,V_2,S_2	
IN	240	V_2				
C_a	5.0		E	C,S_1		
SD	1.0				C,E	

[1] Total phosphorus (TP), inorganic nitrogen (IN), and chlorophyll-a (C_a) as µg l⁻¹; phytoplankton biovolume (PBv) as µl l⁻¹; araphid to centric diatom ratio (A/C); diatom index (DI); mean Secchi depth (SD) in m; phytoplankton production (PP) as g C m⁻² yr⁻¹; and phosphorus loading (PL) as g TP m⁻² yr⁻¹.

[2] C = Carlson (1977), E = USEPA National Eutrophication Survey (1974), N = Nygaard (1949), S_1 = Sakamoto (1966a), S_2 = Sakamoto (1966b), St = Stockner (1971), V_1 = Vollenweider (1968), V_2 = Vollenweider (1968) based on Thomas' data, V_3 = Vollenweider (1976), W = Wetzel (1975).

variation in its trophic state classification. Using the classification system of Vollenweider (1968), the trophic-state classification of Belton Reservoir ranges from ultraoligotrophic (based on inorganic nitrogen concentrations) to hypereutrophic (based on total phosphorus).

Trophic state classifications based on total phosphorus often predict a more eutrophic condition for reservoirs than is predicted by classifications based on algal production or algal taxa. For many of the impoundments listed in Table II, the major discrepancy in trophic-state classification is between classifications based on total phosphorus and those based on inorganic nitrogen, algal biomass (chlorophyll), or algal taxa (A/C, DI). For all impoundments in Table II (except Waterberry Reservoir, Vermont) for which a chlorophyll-based index and a total phosphorus-based index indicate different trophic states, the chlorophyll-based index places the impoundment in a less eutrophic classification. This relationship was consistent for all 33 of the Texas impoundments that differed in trophic-state classifications based on chlorophyll and total phosphorus, and for 25% of the 757 lakes and impoundments included in the USEPA NES (Hern et al., 1981). These results suggest that the trophic potential of these systems, as indicated by nutrient concentrations or nutrient loading rates, is not being realized; i.e., that it is not matched by an algal growth response.

3.2. Nutrient supply and trophic response in reservoirs

A basic assumption of trophic-state classifications derived from nutrient concentrations or loading rates is that the rate of supply of a limiting nutrient (phosphorus, in most cases) to the water body determines the nutrient levels available for primary production, and ultimately, the biological productivity of the ecosystem. However, the nutrient input to a reservoir may substantially exceed the actual nutrient availability for primary production.

For example, river-borne phosphorus entering a reservoir may be largely associated with suspended particles, either sorbed to suspended silts and clays or tightly bound in the clay lattice structure, and largely unavailable for phytoplankton production (Sonzogni et al., 1982). A large fraction of these suspended particles and their associated nutrients sediment from the water column in the up-lake portion of the reservoir. Some of these sediment-associated nutrients may be remobilized and released to the water column, but a significant fraction remains bound to deposited sediments and is not available for planktonic production (Chapra, 1980, Gloss et al., 1981). In thermally stratified systems, the inflowing river water (and river-borne dissolved nutrients) may plunge and traverse much of the reservoir as a subsurface density flow (Wunderlich, 1971), and thereby be similarly isolated from euphotic-zone phytoplankton. Density flows are common in reservoirs, particularly in hypolimnetic-discharge impoundments and in reservoirs in series which receive their inflow from the hypolimnetic discharge of a reservoir upstream (e.g., Paulson & Baker, 1981; Priscu et al., 1982; Elser & Kimmel, 1985).

Empirical models that relate algal abundance or trophic status to nutrient concentrations assume that phytoplankton growth and biomass is controlled by phosphorus or nitrogen availability. However, in reservoirs, it is common for abiotic factors unrelated to nutrient availability to be important determinants of algal production. For example, low light availability resulting from inorganic turbidity and/or 'optically deep' water columns (sensu Talling, 1961; where the mixing depth greatly exceeds the depth of the euphotic zone) commonly limits phytoplankton production in impoundments (Kimmel & Groeger, 1984; Kimmel et al., 1990). High concentrations of suspended sediments are often transported into reservoirs in river inflow and result in high levels of abiogenic turbidity, reduced water transparency, and limited light availability for primary production. Turbid (muddy) impoundments are common in regions of high erosivity (e.g., the midwestern US), where rivers carry large suspended sediment loads (Marzolf, 1984).

Although abiogenic turbidity is an obvious and common limiting factor in impoundments, we believe that short water-residence time is probably the primary factor explaining the difficulties encountered in applying nutrient-loading models and trophic-state classification schemes to reservoirs. Specifically, a short water-residence time (rapid flushing rate) places a temporal constraint on algal growth response to nutrient loading by controlling the time available for phytoplankton interaction with available nutrients and phytoplankton growth responses to those nutrients. Rapidly flushed impoundments are similar to continuous-flow chemostats in that algal standing crop in the system is controlled by flushing rate (i.e., by losses of biomass from the system) rather than by nutrient availability. The results of a number of studies (Brook & Woodward, 1956; Talling & Rzoska, 1967; Dickman, 1969; OECD, 1982; Pridmore & Mcbride, 1984; Soballe & Threlkeld, 1985; and Soballe & Kimmel, 1987) indicate that the direct flushing effects of water renewal on phytoplankton abundance are restricted to aquatic systems with water residence times < 60–100 d.

Additionally, short water-residence times invalidate common assumptions regarding trophic state and indicator taxa (Nygaard, 1949; Järnefelt, 1952; Stockner, 1971; Lowe, 1974). In reservoirs with very short water residence times or during periods of high flushing rates, the species-specific doubling time becomes a more important factor than nutrient availability in determining the proportions of various species in the community. Rapidly growing algae dominate the community, while slower growing species that are superior competitors for nutrients are washed out before they can establish large population sizes. For example, most of the trophic state indicators for Eau Clair Lake, Wisconsin suggest eutrophy, but the A/C ratio indicates oligotrophy. The annual average retention time of the reservoir is only 10 d (USEPA, 1975), which makes it unlikely that an indicator community would have sufficient time to become established.

Part of the trophic-state classification problem is that reservoirs, as a group, include such a diversity of environments that a single trophic-state classification scheme cannot be appropriately applied to all types of impoundments. Often within the same geographic region, reservoirs range from broad, deep, lake-like systems in which the advective influence of the impounded river is minimal (e.g., long-residence time, tributary storage impoundments) to narrow, shallow, run-of-the-river reservoirs that are physically dominated by advection (e.g., short-residence time, mainstem reservoirs. Cox (1984) compared the suitability of commonly used trophic-state classification methods for Tennessee Valley Authority (TVA) impoundments in the southeastern United States. She concluded that several of the classification methods readily characterized the tributary storage reservoirs, but none of the methods adequately characterized both tributary storage and mainstem TVA reservoirs. She concluded that there were basic differences in the environmental factors controlling trophic response to nutrient loading in the lake-like tributary storage impoundments and the river-like mainstem reservoirs (Table III).

Soballe & Kimmel (1987) recently conducted a large-scale, statistical comparison of factors influencing phytoplankton abundance (and trophic status) in rivers, lakes, and impoundments. Discriminant analysis with physical-chemical variables separated these three types of aquatic systems along a composite gradient that correlated with water residence time ($r = 0.7$), drainage area ($r = 0.7$), water depth ($r = 0.6$), stream flow ($r = 0.6$), and water clarity ($r = -0.4$). Natural lakes and rivers occupied opposite ends of this gradient and reservoirs occupied an intermediate position; however, the natural lake and impoundment groups overlapped extensively and did not form a useful dichotomy. Regression analysis showed algal abundance per unit phosphorus increasing in the sequence: rivers < impoundments < natural lakes, which paralleled intergroup differences in water residence times. However, algal abundance per unit phosphorus did not differ significantly between impoundments and lakes when systems with similar residence times were compared. Algal abundance per unit phosphorus was more variable for the reservoirs than for natural lakes because reservoirs, as a group, had shorter residence times than the lakes.

The same fundamental physical, chemical, and biotic processes operate in reservoirs and natural lakes; however, the magnitude and relative importance of these processes may differ considerably between and within system types. Clearly, differences between natural lakes and impoundments in trophic response are more strongly associated with factors such as water-residence time, abiogenic turbidity, water-column mixing, and the relative importance of advective flow, than with the mere presence or absence of a dam.

Table III. Comparison of trophic potential and trophic response characteristics of TVA tributary storage and mainstem reservoirs (modified from Placke, 1983).

Tributary storage	Mainstream
Trophic Potential	
Longer hydraulic retention time (average 138 days for 11 largest)	Shorter hydraulic retention time (average 10 days for all nine)
Deeper (av. mean depth = 18.1 m for 11 largest)	Shallower (av. mean depth = 6.6 m for all nine)
Thermal stratification strong	Not consistently thermally strong
Lower inorganic turbidity (av. Secchi 2.5 m for 11 largest)	Higher inorganic turbidity(av. Secchi 1.2 m for 4 surveyed)
Variable alkalinity (range 4.6 – 82.6 mg l^{-1} for 11 largest)	Relatively constant alkalinity (range 43.8–59.6 mg l^{-1} for 4 surveyed)
Lower unit phosphorus load, high Pretention (RP), low in-lake P conc. (average P Load = 4.0 g m^{-2} y^{-1}, RP = 0.48, P = 0.01 mg l^{-1} for 11 largest)	Higher unit phosphorus load, low retention, high in-lake P conc. (average P load = 31.3 g m^{-2} y^{-1}, RP = −0.03, P = 0.03 mg l^{-1} for 4 surveyed)
Trophic Response	
Chlorophyll production strongly correlated with P and N concentration	Chlorophyll production limited primarily by light and hydraulics rather than nutrients
Embayments kept free of macrophytes by drawdown	Significant algae and macrophyte problems in some embayments
Phytoplankton dominated by cyanobacteria (primarily *Microcystis*) during summer; algal abundance often high	Phytoplankton remain diversified during summer; algal abundance generally lower

3.3. Spatial and temporal heterogeneity in reservoirs

The above discussion assumes that adequate data are available for establishing valid classifications of reservoir trophic status. However, an additional layer of uncertainty is introduced by the spatially heterogeneous and temporally dynamic nature of river impoundments, which makes it much more difficult to acquire sufficient data for trophic-state classification in reservoirs than for most natural lakes.

The general assumption for many natural lakes, that spatial variation is related primarily to length of shoreline, depth, and wind-driven currents, is usually not applicable to reservoirs. Reservoirs characteristically exhibit longitudinal (uplake-to-downlake) gradients in turbidity, nutrient concentrations, mixing depth and euphotic depth, flushing rates, chlorophyll concentrations, phytoplankton productivity, fish standing stock, and other variables (Thornton *et al.*, 1981; Kennedy *et al.*, 1982; Kimmel & Groeger, 1984; Lind, 1984). Within a reservoir basin, the fertility of the mixed layer (e.g., in terms of phytoplankton productivity per m^3 or phytoplankton biomass per m^3) generally decreases from uplake to downlake as the basin volume increases and the advective nutrient supply decreases. Consequently, trophic state (as indicated by Secchi depth, nutrient concentrations, chlorophyll concentrations, etc.) usually shifts from more-eutrophic to more-oligotrophic conditions from uplake to downlake (e.g., Thornton *et al.*, 1981; Hannan *et al.*, 1981; Kennedy *et al.*, 1982; Groeger & Kimmel, 1984; Lind, 1984; Kimmel *et al.*, 1990).

The typical longitudinal zonation of environmental conditions and the uplake-to-downlake sequence of trophic states within individual reservoirs is illustrated generally in Fig. 1 and more specifically for three Texas reservoirs in Table IV. If the average annual chlorophyll concentration is used as the trophic state indicator, Belton Reservoir changes from eutrophic to oligo-mesotrophic from uplake to downlake along a 42 km longitudinal gradient. Stillhouse Hollow Reservoir changes from mesotrophic to oligotrophic along a 16 km gradient, and Waco

Table IV. Trophic-state classification (trophy) of up- and down- reservoir portions of three Texas reservoirs based on average annual chlorophyll-a concentrations and Secchi depth data: E = eutrophic, M = mesotrophic, O = oligotrophic.

Variable and region	Waco		Stillhouse Hollow		Belton	
	Trophy	Value	Trophy	Value	Trophy	Value
Chlorophyll-a (μg l^{-1})						
Up-reservoir	E	13.5	M	10.6	E	41.3
Down-reservoir	M	10.8	O	5.8	M	8.4
Secchi depth (m)						
Up-reservoir	E	0.6	E	0.9	E	0.5
Down-reservoir	E	0.8	M	3.0	M	2.9

Reservoir changes from eutrophic to mesotrophic along a gradient of only 6.4 km. A similar pattern of reduced trophic state from uplake to downlake is apparent in water transparency from the Secchi-depth data for these reservoirs (Table IV).

The longitudinal zonation illustrated in Fig. 1 is dependent on the advective influence of the impounded river and therefore it can be quite dynamic temporally. The riverine, transitional, and lacustrine zones within an impoundment expand and contract in response to watershed runoff events, inflow characteristics, and reservoir operations. For many impoundments, especially those in relatively dry climates, water movement through the impoundment and water renewal is seasonal and sporadic. Average water-residence times calculated from annual inflow data give little insight into the extreme variability in water residence time in many reservoirs. For example, the daily water-residence time of Normandy

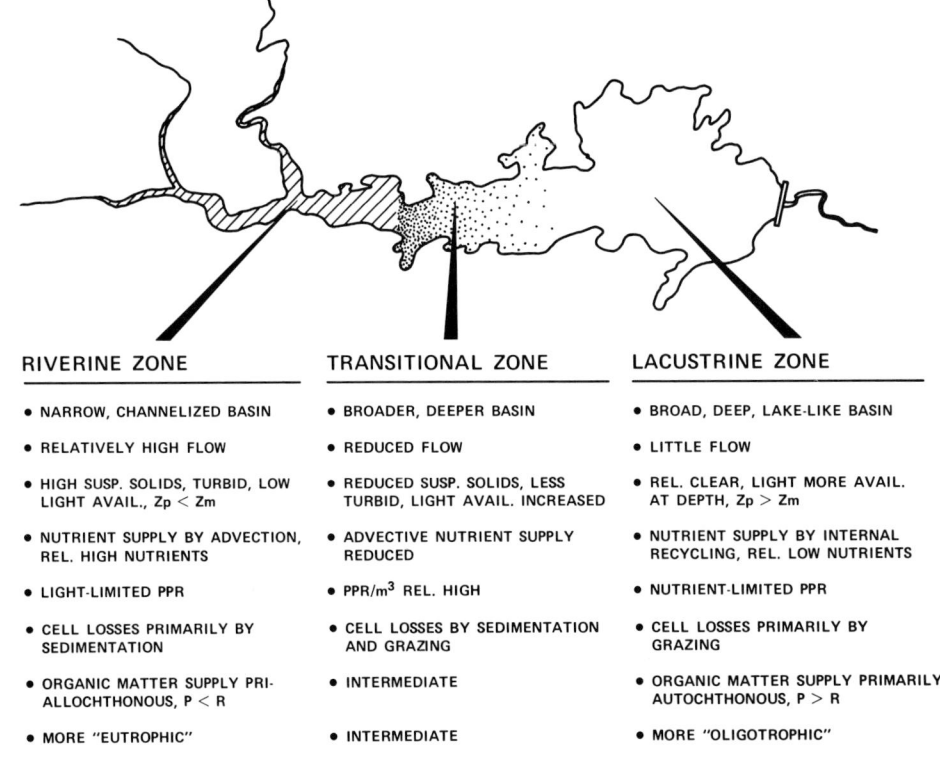

Fig. 1. Longitudinal zonation in the environmental factors controlling primary productivity, phytoplankton biomass, and trophic state within individual reservoirs (after Kimmel & Groeger, 1984).

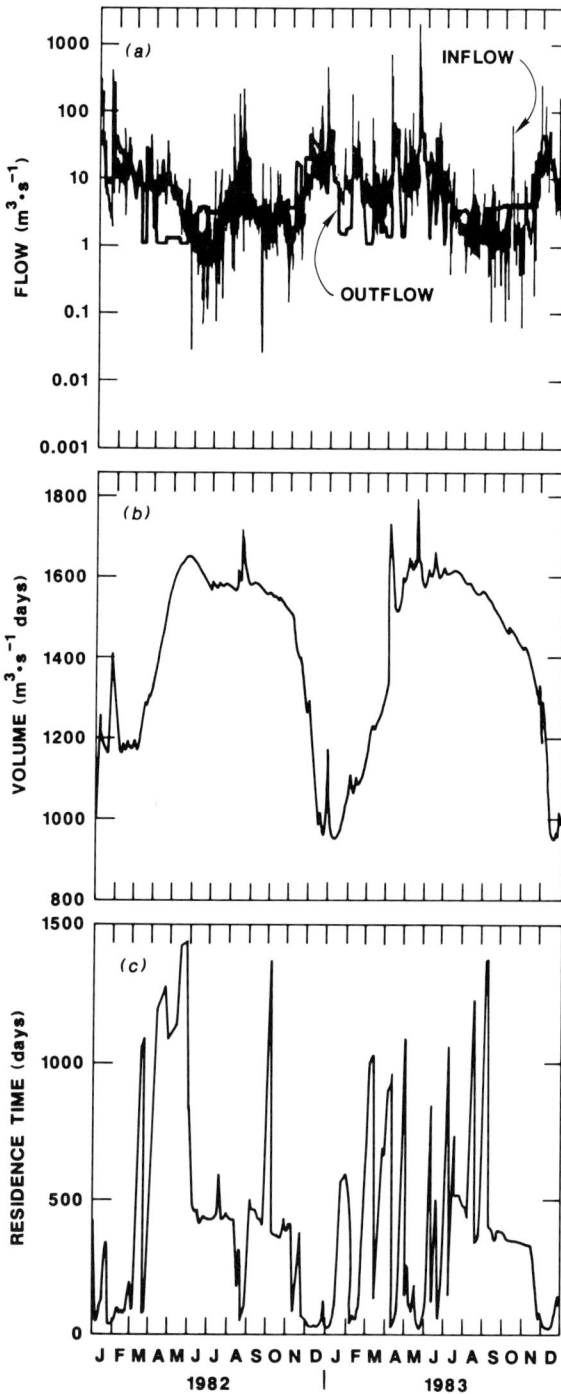

Fig. 2. A – Hydrologic inflow and outflow, B – water volume, and C – water-residence time for Normandy Reservoir, 1982–1983.

Reservoir, Tennessee (annual mean R = 170 d), ranges over two orders of magnitude (from 14 to 1400 days) and is quite variable during a single growing season (i.e., from early spring to late fall) (Fig. 2).

Similarly, water and nutrient movement through a reservoir can occur rapidly subsequent to a period of high precipitation and watershed runoff. Lind (1979) tracked phytoplankton blooms associated with nutrient-rich runoff water as they moved through Waco Reservoir, Texas (R = 152 d). His results showed that, during rainy periods, nutrient enriched runoff water traversed the 6.4 km length of the reservoir in only 2 to 4 weeks. Consequently, due to the short water-residence time, much of available nutrient supply entering reservoirs during storm events is probably converted to phytoplankton only once before being discharged downstream.

Because of the greater spatial and temporal variability in reservoirs, a more extensive sampling program is necessary for defining reservoir trophic status than is generally required for characterizing the trophic status of natural lakes. The common practice of obtaining samples and vertical profiles from only a single station in the deepest part of the lake is obviously inadequate for characterizing such spatially and temporally dynamic systems. Thornton *et al.* (1982) have provided a detailed discussion of reservoir sampling strategies, and Pickett & Harvey (1988) provide a recent example of the use of preliminary synoptic-survey data for designing a reservoir sampling program.

4. Conclusions and recommendations

We have discussed some basic problems that make trophic-state classification more difficult for reservoirs than for natural lakes. Reservoir limnologists may be cognizant of these problems; however, many water-resources managers, water-board and commission members, and the general public are not fully aware of such uncertainties. As a result, reservoirs are misclassified or variously classified, and consequently, reservoirs may be ineffectively or inappropriately managed.

For water resources managers and decision makers, it is essential to be aware of the basis of and the uncertainties in the trophic-state classifications being used to set priorities and make management decisions. Specifically, a comparison of the trophic-state classifications of a number of reservoirs can only be valid if those classifications are based on the same variables, are derived using the same method of estimation, and are used to compare systems in which the controlling factors are similar. For groups of reservoirs that include both river-like and lake-like systems, this may require using two different methods for estimating the trophic status of impounments having long versus short water-residence times [e.g., see Cox's (1984) conclusions for TVA impoundments].

What kind of management strategies should be employed with reservoirs that range from eutrophic to oligotrophic within the same basin? If nutrient availability is the primary limiting factor for primary production and if reservoir-wide decreases in productivity and trophic state are desired, then taking measures to reduce the nutrient loading from the reservoir watersheds will usually be effective. However, a trophic gradient will still persist in the impoundments as a result of uplake-to-downlake differences in the advected nutrient supply and in other potential limiting factors (e.g., turbidity, water-residence time).

The uplake-to-downlake gradient in trophic status in reservoirs strongly suggests that a corresponding zonation exists in the relative suitability of various portions of reservoirs for different uses (e.g., fishing, swimming and boating, water supply) (Siler *et al.*, 1986). Given the characteristic spatial heterogeneity of environmental conditions within impoundments, attempts to manage individual reservoirs as uniform bodies of water seem both naive and ineffective. We suggest that within-reservoir gradients in water quality and trophic status can be viewed as a positive attribute that provides a diversity of potential water uses (e.g., fishing, swimming and boating, water supply) in a single water body. Thus, the application of different management practices in various parts of a reservoir that are more suitable for certain water uses than for others (e.g., fishing uplake, swimming and water supply downlake) may comprise a more sensible and effective management approach for many reservoirs.

References

Brook, A. J. & W. B. Woodward, 1956. Some observations on the effect of water inflow and outflow on the plankton of small lakes. J. Animal Ecol. 25: 22–35.

Carlson, R. E., 1977. A trophic state index for lakes. Limnol. Oceanogr. 22: 361–369.

Chapra, S. C., 1980. Application of phosphorus loading models to river-run lakes and other incompletely mixed systems. In: Restoration of Lakes and Inland Water. EPA 440/5-81-010. US Environ. Prot. Agency, Washington, D.C.: 329–334.

Cox, J., 1984. Evaluating reservoir trophic status: the TVA approach. Lake Reserv. Manage. 2: 11–16.

Dickman, M., 1969. Some effects of lake renewal on phytoplankton productivity and species composition. Limnol. Oceanogr. 14: 660–666.

Elser, J. J. & B. L. Kimmel, 1985. Nutrient availability for phytoplankton production in a multiple-impoundment series, Can. J. Fish. Aquat. Sci. 42: 1359–1370.

Gloss, S. P., R. C. Reynolds, L. M. Mayer & D. E. Kidd, 1981. Reservoir influences on salinity and nutrient fluxes in the arid Colorado River Basin. In: H. G. Stefan (ed.), Proc. Symp. Surface Water Impoundments. Am. Soc. Civil Eng., New York, NY: 1618–1629.

Groeger, A. & B. Kimmel, 1984. Organic matter supply and processing in lakes and reservoirs. Lake Reserv. Manage. 2: 282–285.

Hannan, H. H., D. Barrows & D. C. Whitenberg, 1981. The trophic status of a deep-storage reservoir in Central Texas. In: H. G. Stefan (ed.), Proc. Symp. Surface Water Impoundments. Am. Soc. Civil Eng., New York, NY: 425–434.

Hern, S. C., V. W. Lambou, L. R. Williams & W. D. Taylor, 1981. Modifications of Models Predicting Trophic State of Lakes: Adjustment of models to account for biological manifestations of nutrients. Project Summary, EPA-600/S3-81-001.

Järnefelt, H., 1952. Plankton als indikator der Trophiegruppen der Seen. Suomal. Tiedeakat. Toim. (Annls Acad. Sci. Fenn.), ser A. IV Biol. 181: 1–20.

Kennedy, R. H., K. W. Thornton & R. C. Gunkel, 1982. The establishment of water quality gradients in reservoirs. Can. Water Res. J. 7: 71–87.

Kimmel, B. L. & A. W. Groeger, 1984. Factors controlling primary production in lakes and reservoirs: A perspective. Lake Reserv. Manage. 2: 277–281.

Kimmel, B. L., O. T. Lind & L. J. Paulson, 1990. Reservoir primary production. In: K. W. Thornton, B. L. Kimmel & F. E. Payne (eds.), Reservoir Limnology: An Ecological Continuum. John Wiley & Sons, New York, NY: 133–193.

Lind, O. T., 1979. Reservoir Eutrophication: Factors Governing Primary Production. Proj. Completion Report. Office Water Resources Tech. No. B-210-TEX.

Lind, O. T., 1984. Phytoplankton population patterns and trophic state relationships in an elongate reservoir. Verh. Internat. Verein. Limnol. 22: 1465–1469.

Lowe, R. L., 1974. Environmental requirements and pollution tolerance of freshwater diatoms. National Environmental

Research Center, Office of Research and Development, USEPA, Cincinnati, Ohio.

Marzolf, G. R., 1984. Reservoirs in the Great Plains of North America. In: F. B. Taub (ed.), Lake and Reservoir Ecosystems. Elsevier Sci. Publ., Amsterdam: 291–302.

Nigro, A. A., T. T. Terrell, L. G. Beckman & W. R. Persons, 1983. Assessment of the limnology and fisheries in Lake F.D. Roosevelt. 1982 Annual Report. US Bureau of Reclamation.

Nygaard, G., 1949. Hydrobiological Studies of Some Danish Ponds and Lakes II. [K. danske Vidensk. Selsk] Biol. Skr. 7, 293pp.

OECD, 1982. Eutrophication of Waters-Monitoring, Assessment, and Control. Organisation for Economic Co-operation and Development, Paris.

Paulson, L. J. & J. R. Baker, 1981. Nutrient interactions among reservoirs on the Colorado River. In: H. G. Stefan (ed.), Proc. Symposium on Surface Water Impoundments. Am. Soc. Civil Eng., New York, NY: 1647–1658.

Pickett, J. R. & R. M. Harvey, 1988. Water quality gradients in the Santee-Cooper Lakes, South Carolina. Lake Reserv. Manage. 4: 11–20.

Placke, J. F., 1983. Trophic status evaluation of TVA Reservoirs. Tennessee Valley Authority, Division of Air and Water Resources, Chattanooga, TN, 163p.

Pridmore, R. D. & G. B. McBride, 1984. Prediction of chlorophyll-a in impoundments of short hydraulic retention time. J. Environ. Manage. 19: 343–350.

Priscu, J. C., J. Verduin & J. E. Deacon, 1982. Primary productivity and nutrient balance in a lower Colorado River reservoir. Arch. Hydrobiol. 94: 1–23.

Sakamoto, M., 1966a. The chlorophyll amount in the euphotic zone in some Japanese lakes and its significance in the photosynthetic production of phytoplankton communities. Bot. Mag. (Tokyo) 79: 932–933.

Sakamoto, M., 1966b. Primary production by phytoplankton community in some Japanese lakes and its dependence on lake depth. Arch. Hydrobiol. 62: 1–28.

Siler, J. R., M. C. McInerny & W. J. Foris, 1986. Heterogeneity in fish parameters within a reservoir. In: G. E. Hall & M. I. Van den Auyle (eds.), Reservoir Fisheries Management: Strategies for the 80's. Allen Press, Lawrence, Kansas: 122–136.

Soballe, D. M. & B. L. Kimmel, 1987. A large-scale comparison of factors influencing phytoplankton abundance in rivers, lakes, and impoundments. Ecol. 68: 1943–1954.

Soballe, D. M. & S. T. Threlkeld, 1985. Advection, phytoplankton biomass, and nutrient transformations in a rapidly flushed impoundment. Arch. Hydrobiol. 105: 187–203.

Sonzogni, W. C., S. C. Chapra, D. E. Armstrong & T. J. Logan, 1982. Bioavailability of phosphorus inputs to lakes. J. Environ. Qual. 11: 555–563.

Stockner, J. G., 1971. Preliminary characterization of lakes in the experimental lakes area, northwestern Ontario, using diatom occurrences in sediments. J. Fish. Res. Board Can. 28: 265–275.

Talling, J. F., 1961. Photosynthesis under natural conditions. Ann. Rev. Plant Physiol. 12: 133–154.

Talling, J. F., & J. Rzoska, 1967. The development of plankton in relation to hydrological regime in the Blue Nile. J. Ecol. 55: 637–662.

Texas Department of Water Resources. The State of Texas Water Quality Inventory, 17th ed. Texas Department of Water Resources, Austin. (In Press)

Thornton, K. W., R. H. Kennedy, A. D. Magoun & G. E. Saul, 1982. Reservoir water quality sampling design. Water Res. Bull. 18: 471–48.

Thornton, K. W., R. H. Kennedy, J. H. Carroll, W. W. Walker, T. C. Gunkel & S. Ashby, 1981. Reservoir sedimentation and water quality – an heuristic model. In: G. H. Stefan (ed.), Proc. Symp. on Surface Water Impoundments. Am. Soc. Civil Eng., New York, NY: 654–661.

USEPA, 1974. An Approach to a Relative Trophic Index System for Classifying Lakes and Reservoirs. US Environmental Protection Agency, National Eutrophication Survey Working Paper No. 22.

USEPA, 1975. A Compendium of Lake and Reservoir Data Collected by the National Eutrophication Survey in the Northeast and North-Central United States. US Environmental Protection Agency, National Eutrophication Survey Working Paper No. 474.

Vollenweider, R. A., 1968. Scientific fundamentals of the eutrophication of lakes and flowing waters, with particular reference to phosphorus and nitrogen as factors in eutrophication. Tech. Rep. OECD Paris, DAS/CSI/58-27: 1–159.

Vollenweider, R. A., 1976. Advances in defining critical loading levels for phosphorus in lake eutrophication. Mem. Ist. Ital. Idrobiol. 33: 53–83.

Walker, W., 1984. Empirical prediction of chlorophyll in reservoirs. In: Lake and Reservoir Management. Proc. 3rd Annual Conf., North Amer. Lake Mgt. Soc., EPA-440/5/84-001: 292–297.

Wetzel, R. G., 1975. Limnology. Saunders. Co., Philadelphia, 744pp.

Wunderlich, W. O., 1971. The dynamics of density-stratified reservoirs. In: G. E. Hall (ed.), Reservoir Fisheries and Limnology. Am. Fish. Soc., Washington, D. C.: 219–231.

Chapter IV

Limnology of a subalpine pump-storage reservoir

B. Kiefer,[1] F. Schanz,[1] & D. Imboden[2]
[1] *Limnologische Station, Universität Zürich, Seestr. 187, CH–8802 Kilchberg, Switzerland;*
[2] *Eidgenössische Anstalt für Abwasserreinigung, Wasserversorgung und Gewässerschutz, CH–8600 Dübendorf, Switzerland*

Key words: reservoirs limnology, pump-storage, hydrodynamics, nutriens, mathematical models, simulation, plankton

Abstract

Wägitalersee is an oligotrophic pump-storage reservoir situated in subalpine Switzerland. An investigation carried out in 1983–1984 revealed that water input to the lake through precipitation and pumping occurred mainly between spring and autumn, whereas water outflow occurred mainly from October to March. Chlorophyll-a values ranged from 0.2 to 4.9 mg m^{-3} and biomass (wet weight) from 0.1 to 0.7 g m^{-3}. Phytoplankton growth was limited mainly by dissolved phosphorus. Diatom growth slowed down markedly in June due to silicate limitation at mean concentrations below 0.2 g m^{-3}. Further causes for phytoplankton losses are: coprecipitation with suspended material and zooplankton grazing. Primary productivity (PP), which varies from 3.8 to 107 mg C m^{-2} d^{-1}, is controlled by light climate, temperature and phosphorus availability. There is a close statistical relationship between PP and phosphorus input from pump water. No relationship could however be demonstrated between PP and the phosphorus input from inflowing rivers. Most of the year the areal photosynthetic efficiency lay around $2 \cdot 10^{-4}$ moles C mol^{-1} quanta but it increased in autumn to $10 \cdot 10^{-4}$ because of the changed physiological properties in the euphotic zone related to higher portions of blue-green algae.

The one-dimensional dynamical temperature simulation model DYRESM (Imberger, 1979) was extended by subroutines describing the intrusion of the pumped water and the nutrient supply to the euphotic layer by natural and artificial inflows. The model was calibrated using selected data. The simulated and measured temperatures were found to differ but little from one another. The surface temperatures differed by at most 3 °C (average difference 0.04 °C). The maximum difference in the rest of the depth profile was only 2 °C.

Pump-storage reservoirs act during spring and summer as very efficient energy traps. Furthermore pump-storage causes a strong recirculation of nutrients (phosphates) from the lake's depth to the euphotic layer. The mechanisms that lead to these phenomena have been quantified exactly using the extended model DYRESM.

1. Introduction

In the lower alpine and subalpine areas of Switzerland there exist a large number of small to medium-sized reservoirs which are utilized for the generation of hydro-power and which are also increasingly being used to store superfluous energy by pumping back water from lower-lying lakes and rivers. The construction of such pump-storage reservoirs and the introduction of pump-storage operation in already existing reservoirs require an environmental impact statement (Eidgenössisches Departement des Innern, 1986).

After a search of the relevant literature, Kiefer (1987) came to the conclusion that the present state of knowledge of the limnology of pump-storage operation is rudimentary. The following general statements can be made:

Reservoirs without pump-storage operation: The specific hydrological conditions pertaining to the reservoir influence stratification and mixing. The temperature gradient in summer is less pronounced in reservoirs than in natural lakes and mixing occurs more rapidly and lasts longer. The eddy diffusion is higher than in natural lakes. As a result of the underwater outflow, the reservoir acts as an 'energy trap'. Water temperatures are thus higher in reservoirs than in natural lakes.

The outflow of deep water results in nutrient depletion in the hypolimnion.

The annual primary production is often lower in reservoirs than in comparable natural lakes. The reasons for this are:
- a greater nutrient depletion in the epilimnion, since no exchange of epilimnetic water takes place in summer;
- the greater mixing depth of the epilimnion and the longer period of circulation;
- physical stress caused by strong variations in hydraulic loading and in the outflow and by variations in the light climate.

Reservoirs with pump-storage operation: According to Elster & Schmolinsky (1953) and Elster (1962), the intensive pump-storage operation in Schluchsee Reservoir (Federal Republic of Germany) causes a warming-up of the epilimnion and a reduction in the density gradient. The period of full circulation thus begins earlier and lasts longer. The same observations in a Scottish loch were made by Tippett (1978), who was of the opinion that the lengthened period of homothermy and the increased mixing depth during the period of stagnation led to a decrease in the primary production, although he did not verify this.

The question which led to the carrying out of this piece of research was formulated based on the contradiction between the requirements of the environmental impact statement (Eidgen. Depart. d. Innern, 1986) and the inadequate knowledge of the mechanisms determining the environmental conditions of the phytoplankton population: how can energy and mass exchange processes best be modelled in a pump-storage reservoir? The model should be capable of being used as a basis for predicting and interpreting the biological effects of hydrological measures implemented in reservoirs.

The example used in our investigations was the Swiss Reservoir Wägitalersee. Our quantitative description of energy and mass fluxes were based on version 5A of Imberger's (1979) Dynamic Reservoir Simulation Model (DYRESM).

2. Description of Wägitalersee

Geographical situation: Wägitalersee (Switzerland) is a subalpine pump-storage reservoir lying at 47° 06′ N 8° 54′ E. When completely filled, the water surface lies at 900 m a.s.l.

Hydrology: Table I contains the most important hydrological information. The main period of inflow is during the thaw between April and June. Five relatively large streams, very many small streams and two karst springs flow into the lake. In addition to these natural inlets, water is pumped into the reservoir from a storage basin. This basin is connected to the reservoir by means of a 3.5 km long underground pressure tunnel. Four groups of pumps, each group with a power rating of 4.4 MW, are built in to the main control area near the storage basin. The pumped inflow to the reservoir through the pressure tunnel occurs during periods of low electricity demand, i.e. during the night and at weekends. The water pumped up into the reservoir is comparatively strongly polluted with waste water, as untreated sewage from a small community in the area is conducted directly into the storage basin.

Fisheries and agriculture: Wägitalersee is well known for its abundance of trout, a the result of stocking with yearling rainbow trout. A healthy stock of perch living in the oxygen-rich deep water is able to maintain itself naturally. The reservoir is

Table I. Hydrological and morphological characteristics of Wägitalersee (see also Fig. 1).

Maximum volume: $147.4 \cdot 10^6$ m^3
Maximum volume utilizable for power production: $80 \cdot 10^6$ m^3
Maximum surface area (900 m a.s.l.): 4.17 km^2
Maximum depth (at the dam): 66 m (deepest point: 834 m a.s.l.)
Maximum length (900 m a.s.l.): 4.9 km
Maximum breadth (900 m a.s.l.): 1.2 km
Mean breadth (900 m a.s.l.): 840 m
Mean depth: 35.3 m (864.7 m a.s.l.)
Maximum depression of water surface (in winter): 20 m (880 m a.s.l.)
Mean residence time: 1.64 y
Location of mouth of pressure tunnel: 848.6 m a.s.l. (14.6 m above deepest point)
Natural catchment area: 42.7 km^2
Annual mean precipitation in catchment area: 2100 mm
Volume of storage basin: $285 \cdot 10^3$ m^3
Maximum pump capacity: 5 m^3 s^{-1}

fished intensively by hobby fishermen. The community in the reservoir's catchment area contains a stable population of 167, as well as 17 farms operating on an alpine farming basis. The sewage from the community is conducted through cesspits into the rivers that are flowing into the reservoir.

Trophic state: Wägitalersee has a mean total phosphorus concentration of 8 mg P m^{-3} and a nitrate nitrogen concentration of 300 mg N m^{-3}. The chlorophyll-a values range from 0.2 to 4.9 mg m^{-3} and the biomass (wet weight) from 0.1 to 0.7 g m^{-3}. We could not find any orthophosphate in the lake and the nitrate concentration was always stable. Wägitalersee can therefore be characterized as an oligotrophic water body (Thomas, 1952).

3. Influence of water management on biomass development and primary production

3.1. Materials and methods

Sampling: Sampling was carried out at position 3 (Fig. 1) from 21 December, 1982 to 12 July, 1984 at the following depths: 0.2, 0.5, 1.0, 2.5, 5.0, 7.5, 10.0, 12.5, 15.0, 20.0, 30.0 and 40.0 m (for sampling frequency, see Fig. 3). Primary production was measured between 21 December, 1982 and 16 March, 1984. Simultaneously to the production measurements, physical and chemical investigations of the inlets were also carried out. Continuous sampling equipment was used for the pump water analysis.

Phytoplankton: Counted according to Utermöhl (1958). We assumed a density of 1 g cm^{-3} in converting volume to wet weight.

Chlorophyll-a, b and phaeophytin-a: According to Schanz (1982).

Primary production: After Vollenweider (1969) with 4 µCurie of ^{14}C in 125 ml flasks.

Ortho P, soluble P, total P, NO_2—N, NH_4—N, dry matter: Using methods of EDI (1982).

NO_3—N: UV absorption at 275 nm after APHA (1976).

Water temperature: With thermometer TTM72, Züllig AG, Rheineck (CH).

Light (Photosynthetically Active Radiation, PAR, 400–700 nm): LI-185B with LI-192S, LICOR Inc., Nebraska (USA); daily input of PAR (1 June, 1983 to 31 March, 1984): LI-192S with

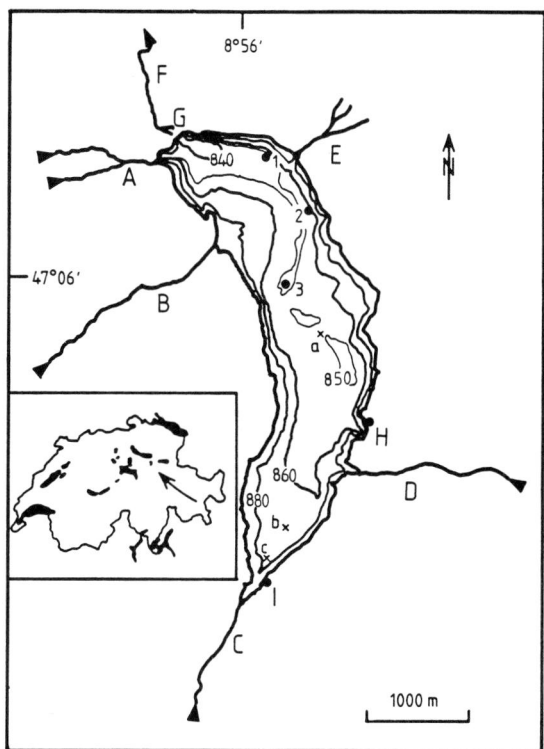

Fig. 1. Position of Wägitalersee in Switzerland (inset) and bathymetric map. Inlets: A = Schraehbach, B = Schlierenbach, C = Aberenbach, D = Ziggenbach, E = Innerthalbach, H = Fläschloch, I = Hundsloch; outlet: F = Wägitaler Aa; G = dam; 1 = mouth of the pressure tunnel (848.6 m a.s.l.), 2 = thermistor chain, 3 = biological and chemical sampling point.

printing integrator LI-550B. Missing light data were filled in by using the total radiation values of a nearby automatic measuring station (sensor solar 11C, CM6, Meteolabor, Wetzikon (CH)).

Hydrological data: From the Kraftwerk Wägital AG.

Statistics: Tests for normality: SAS (1982a). Data not normally distributed were transformed. Factor analysis with SAS (1982b) with standardization of the data matrix (Linder & Berchthold, 1982) and factor rotation (Flury & Riedwyl, 1983).

3.2. Water regime

Pump activities are normally frequent between spring and autumn (Fig. 2). In 1983 pumping was resumed in July after about 6 month of idleness due to pump maintenance. Daily maxima up to 0.5 · 10^6 m^3 d^{-1} and monthly means up to 0.12 · 10^6

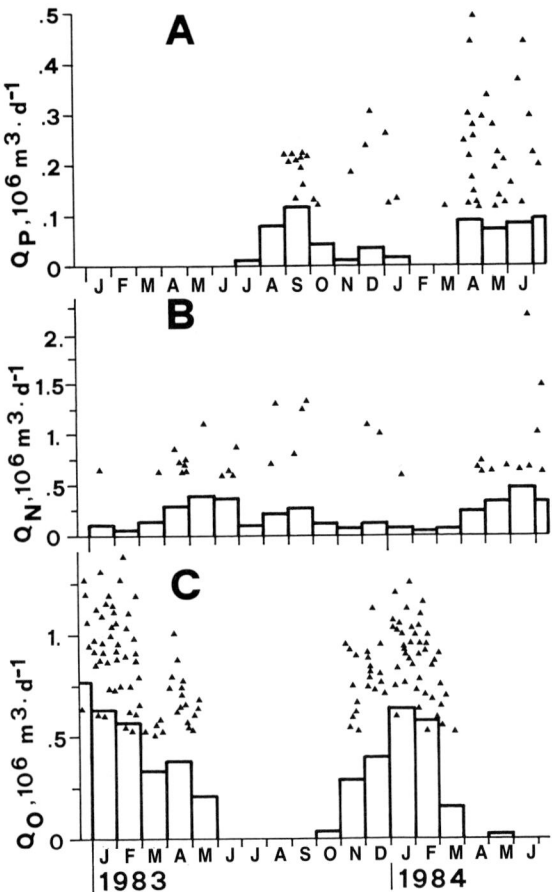

Fig. 2. Daily water input and output of Wägitalersee from 21 December 1982 to 12 July 1984. A: Volume of pump water (QP), B: Volume of natural inflows (QN), C: Volume of outflow for electric production (QO). Bars = monthly means; triangles = daily values, QP above $0.12 \cdot 10^6$ m^3 d^{-1}, QN above $0.6 \cdot 10^6$, QO above $0.5 \cdot 10^6$.

m^3 d^{-1} occurred (September, 1983). Natural inlets brought maximum input volumes into the reservoir during thunderstorms in June 1984 (up to $2.2 \cdot 10^6$ m^3 d^{-1}). Monthly means reach their maxima in spring during snow melting periods (about $0.4 \cdot 10^6$ m^3 d^{-1}). Water losses occur only by electricity production which is limited almost completely to the period October to May. Maximum output volumes occur between December and February (up to $1.35 \cdot 10^6$ m^3 d^{-1}). Monthly means reached $0.78 \cdot 10^6$ m^3 d^{-1} in December 1982 and $0.6 \cdot 10^6$ in January 1984.

Summarizing we can say that Wägitalersee is filled up during spring and summer months by natural inlets and pump activities. During summer the lake has almost no water loss. In contrast to this, great amounts of water are exported during winter months when the incoming water volume is smallest.

3.3. Biomass development and primary production

Chlorophyll-a (CHA) was used as a biomass parameter despite of several disadvantages (Gloschenko & Blanton, 1977). However a good relationship between CHA and the phytoplankton biomass as wet weight in Wägitalersee was found by Egloff (1986):

$$\text{CHA} = 18.0 + 1.37 \, \text{BM}; \quad R^2 = 0.738; \quad P < 0.0001 \tag{1}$$

where CHA from 0 to 30 m depth is in mg m^{-2} and BM, phytoplankton biomass, is in g m^{-2}. The chlorophyll values were minimal during winter months (around 20 mg m^{-2}; Fig. 3) and increased with increasing light intensity to reach a first maximum in May (1983: 40 mg m^{-2}) or June (1984: 58 mg m^{-2}). The phytoplankton population was dominated by the diatom *Stephanodiscus hantzschii* during this spring period (more than 50% of the total biomass). Further species of importance were: *Asterionella formosa, Rhodomonas minuta, Ceratium hirundinella* and *Dinobryon divergens*. The decrease of the phytoplankton biomass in June or July was caused partly by grazing by zooplankton (the density of which increased for example in 1984 from 1 to $5.5 \cdot 10^5$ individuals · m^{-2}) and partly by sedimentation of silicate depleted diatom cells (Egloff, 1986). A second phytoplankton maximum of up to 80 mg m^{-2} in the period from September to November 1983 was caused by diatoms (*Fragilaria crotonensis*), blue-greens (*Aphanothece clathrata, Gomphosphaeria naegeliana, Oscillatoria rubescens*) and dinoflagellates (*Ceratium hirundinella, Peridinium cinctum*).

The areal photosynthesis (Pz) attained maxima in May (77 mg C m^{-2} d^{-1}), July (108 mg C m^{-2} d^{-1}) and September (94 mg C m^{-2} d^{-1}). These observed maxima are typical for oligotrophic lakes (Schanz & Wälti, 1982). The intense daily production in May agrees well with the high insolation values on

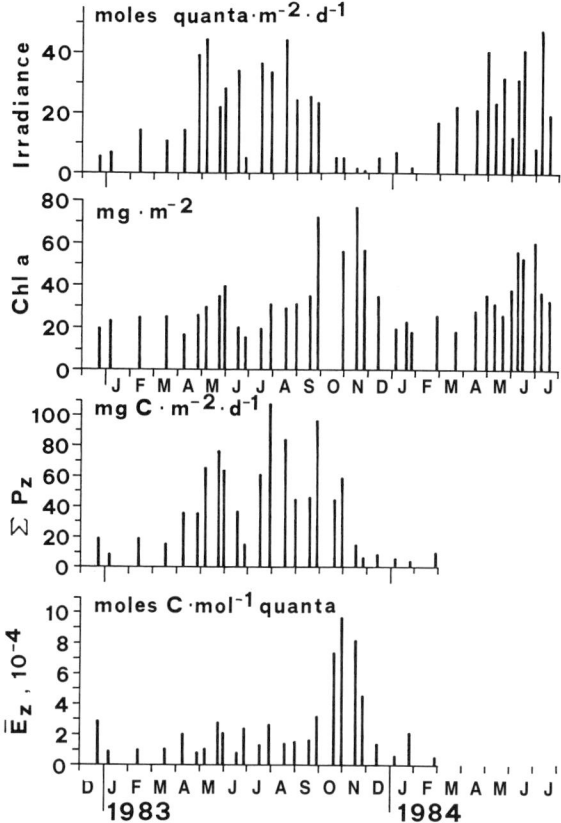

Fig. 3. Daily irradiance on the sampling date, areal chlorophyll-a (CHA), areal photosynthesis (P_z) and areal photosynthetic efficiency (E_z) in Wägitalersee. CHA, EP_z: integrals over the euphotic depth; E_z: calculated according to Dubinsky et al. 1984.

sunny sampling days (exceeding 40 moles quanta m^{-2} d^{-1}). Because of low zooplankton grazing activity the biomass increases rapidly as described by Egloff (1986). In autumn however, the conditions seem to be much more complicated: whereas most of the year the areal photosynthetic efficiency (E_z) lay around $2 \cdot 10^{-4}$ moles C mol^{-1} quanta, in October and November it went up to $10 \cdot 10^{-4}$. This points to a physiological change in the phytoplankton population as a result of alterations in species composition from spring (diatoms and flagellates) to autumn (diatoms and blue-greens). In addition to the improved nutrient conditions in Wägitalersee, caused by the considerable amounts of water pumped into the reservoir between July and September (Fig. 2) the more efficient light use at low light intensities is probably also responsible for the high biomass and production values in November, when the light intensity was only around 5 moles quanta m^{-2} d^{-1}.

The mean biomass during the spring period was considerably higher in 1984 than in 1983 (CHA sum under 1 m^2 surface area from 0 to 30 m depth, mean from 1 January to 12 July, 1983: 24 mg m^{-2}, 1984: 37). The sum of the light energy reaching Wägitalersee from 1 January to 12 July was almost the same in both years (1983: 3630 moles quanta m^{-2}; 1984: 3706). Assuming the same efficiency, the higher biomass in 1984 has to be a consequence of the increased nutrient load brought about by pump operation (Fig. 2).

3.4. Statistical analysis of hydrological, physical and biological data

Egloff's (1986) investigations in spring 1984 confirmed that the phytoplankton growth was phosphorus limited. Orthophosphate has never been detected in the lake water and total phosphorus concentrations were also very low (always below 16 mg m^{-3}). As discussed above, the results illustrated in Fig. 3 show the considerable influence of nutrient input on productivity and phytoplankton growth especially during autumn 1983 and spring 1984. However, the direct investigation of the phosphorus input from the two main sources (natural inlets, pump water) on biomass development or production was not possible because of several methodological problems in the determination of the phosphorus components available for algal use. We therefore made use of the indirect method of factor analysis (Table II). The situation can be very well interpreted on the basis of two factors alone, F1 and F2. Factor F1 is comprised mainly of parameters related to the natural inflows of the reservoir, and factor F2 of parameters related to the pump water inflow.

Factor 1: The coefficients of the nutrient parameters, of the suspended sediment and of the natural daily water inflow are 0.73 or more. The concentrations of the nutrients in the inflowing water is not exclusively influenced by the throughflow rate (Hynes, 1970) which explains the great difference of the calculated coefficients between 0.73 (I: NH_4—N) and 0.99 (I: NO_3—N). The high load of NO_3—N in the inlets always results in an increase of the NO_3—N concentration

Table II. Factor analysis with hydrological, physical and biological variables (period: Dec. 21, 1982 to March 26, 1984). R^2 = coefficient of determination of the variable and the factor solution; k = cumulative proportion of the total variance accounted for each factor. Tabulated values = correlation coefficients × 100. I = Natural inflow and P = Pump water, means of two consecutive days, L = Lake.

variables	F1	F2	F3	F4	R^2
I: NO_3—N	99	−1	8	1	0.98
I: Total phosphorus	97	18	5	−4	0.98
I: Soluble phosphorus	96	14	15	5	0.97
I: Ortho phosphorus	94	−8	23	−18	0.98
I: NO_2—N	90	27	19	−20	0.97
I: Suspended sediment	78	−40	28	−6	0.85
I: NH_4—N	73	16	−23	47	0.84
I: Water mass	98	5	11	−6	0.99
P: Total phosphorus	7	98	−8	−9	0.99
P: NO_3—N	4	97	−13	−6	0.96
P: Suspended sediment	4	97	−14	−11	0.97
P: P12*	23	88	7	6	0.84
P: P13*	4	97	−13	−6	0.96
P: Water flow rate	4	97	−13	−6	0.96
L: L15*	−32	81	18	−24	0.86
L: L16*	−2	−24	−29	91	0.99
L: L17*	25	24	21	−89	0.96
L: L18*	−12	−23	−64	67	0.95
L: L19*	31	0	92	−9	0.96
L: Primary production	8	64	15	−50	0.70
L: L21*	4	53	47	−31	0.77
L: L22*	33	27	80	−33	0.95
L: L23*	47	27	73	−20	0.89
L: L24*	22	−67	−20	42	0.73
L: Soluble phosphorus	−27	64	34	−8	0.60
L: Total phosphorus	31	55	24	−26	0.53
L: NH_4—N	11	−34	76	−28	0.80
L: L28*	2	38	−63	10	0.57
L: L29*	−46	−8	−19	−31	0.37
L: L30*	44	−29	−7	22	0.33
L: NO_3—N	86	0	29	−23	0.88
L: L32*	57	−34	−5	−49	0.70
L: NO_2—N	−59	58	4	0	0.69
L: transparency	−60	−49	−10	21	0.68
L: CHA	−67	15	−49	0	0.74
L: Chl b	−78	0	−40	29	0.88
k	0.435	0.741	0.901	0.963	

* variables of minor importance, presented in detail by Kiefer (1987)

in the uppermost water layers of the lake. This leads to the high coefficient of L: NO_3—N (0.86). Other parameters of the lake such as chlorophyll (L: CHA, −0.67; L: Chl b, −0.78), transparency (L: transparency, −0.60) and soluble phosphorus (L: Soluble P, −0.27) show negative correlation values. Total phosphorus, however, shows a small positive correlation (L: Total P, 0.31). These results can be interpreted as follows. A high total phosphorus input load of the natural inlets increases the total phosphorus concentration in the lake. The decrease of soluble phosphorus in the lake shows that no phosphorus was freed and became available for phytoplankton growth. On the contrary, the incoming particles seem to absorb considerable amounts of the soluble phosphorus present. A rapid sedimentation of the incoming particles is also capable of increasing the sedimentation of phytoplankton organisms (coprecipitation). As a consequence, a decrease in the chlorophyll concentrations can be observed.

Factor 2: Parameters of the pump water have high coefficients ranging from 0.81 (P: Water flow rate) to 0.98 (P: Total P). The following reservoir properties show smaller positive correlations: primary production (L: Primary Production, 0.64), soluble phosphorus (L: Soluble P, 0.64), total phosphorus (L: Total P, 0.55) and nitrite nitrogen (L: NO_2—N, 0.58). It is obvious that an increase in the pump water nutrient load results in increasing nutrient levels in the lake and a consequent increase in primary production. As mentioned in an earlier section, the pump water comes from the Rempenbecken, which receives the waste water of about 900 inhabitants. These additional nutrients have therefore a noticeable eutrophying effect on Wägitalersee. In contrast to the natural inlets there is no correlation between pump water volume and chlorophyll. We assume that low particle concentrations prevented coprecipitation.

3.5. Conclusions

In general the biomass development in reservoirs without pump facilities is much smaller than in similar natural water bodies where a deep-water discharge is missing (Kiefer, 1977). Main causes are:
(i) The export of nutrients accumulated in the hypolimnion and
(ii) The prolonged circulation periods, which reduce the mean light intensities for the phytoplankton population. If water is pumped from a lower lying water body into a pump-storage reservoir, severe eutrophication problems can result when the pump water is nutrient-rich.

Pump facilities should therefore not be installed in an existing hydroelectric power station without careful investigation of the nutrient concentrations in the reservoir on the one hand and in the pump water in the other. This is the only way of avoiding a change for the worse in the trophic state of the reservoir.

4. Quantification of vertical mass and energy fluxes using a dynamic model

4.1. DYRESM: a one-dimensional temperature model

4.1.1 General description of the model

The dynamics of the energy and mass exchange processes in Wägitalersee were investigated by means of a one-dimensional temperature model. The most commonly one-dimensional temperature models used today are those due to Ryan & Harlemann (1971) and Imberger (1979). Both models are based on the assumption that the physics of a lake is determined by two mechanisms which have a mutual influence on one another: viz. the heat flux and the flux of kinetic and potential energy. The exchange of heat at the lake surface results in a density gradient in the water column. As this density gradient increases, the potential energy content of the lake falls with respect to the potential energy content of a vertically homogeneous water body with the same total heat content. Working against this process are on the one hand convection in the upper water layers, which depends on heat loss from the surface, and on the other the entry of turbulent kinetic energy into the lake, which is dependent on wind and also on inflow and outflow. Potential and kinetic energy are always in a dynamic balance and together determine the mixing depth.

We decided on Imberger's (1979) model for the following reason: The 'Ryan model' divides the lake into horizontal layers, between which energy and mass exchange processes take place. The thickness of each layer and its position with respect to the lake bottom are constant (Eulerian coordinate system). Imberger's model DYRESM on the other hand is based on a Lagrangian view. The water body is divided into a variable number of horizontal layers, each of variable thickness. These alter in position, density and volume according to kinematic requirements. In DYRESM, mixing is modelled by the fusion of neighbouring layers. Inflow and outflow are modelled by increasing or decreasing the thickness of the relevant layer. The Imberger model is thus especially useful in cases where inflow and outflow do not lie at the same depth, giving rise to vertical advection. In addition, in DYRESM the layer thickness is a function of the density gradient: the steeper the density gradient, the thinner the layer.

Of central importance when modelling lake temperatures is the expression used in the calculation of the vertical eddy diffusion coefficient k_z. In a good many reservoirs, vertical mixing is determined in the main by the advection induced by the underwater outflow and the inflows (Octavio *et al.*, 1977). The vertical eddy diffusion coefficient can in such cases be set as small as the molecular diffusion coefficient or even neglected altogether. In deep natural lakes, and also in Wägitalersee, where vertical advection due to the outflow is only important during the winter months, the hypolimnion warms up more rapidly than one would expect based only on molecular diffusion and advection. Imboden (1980) has shown that k_z in the lake of Lucerne must be 100–200 times greater than the molecular diffusion coefficient. In addition, k_z is a function of stability (Brunt-Väisälä frequency) (Imberger, 1979; Imboden, 1980) and of the kinetic energy entering the lake as a result of wind action and inflow (Imberger, 1979). All the above-mentioned points are taken into account in DYRESM version 5A. The vertical eddy diffusion coefficient is considered on the one hand as a function of stability and on the other hand as a function of the kinetic energy entering the lake as a result of wind action and inflow. This approach is however still not completely satisfactory, since in general not enough research into eddy diffusion in lakes has been carried out; nevertheless results obtained in the examples cited by Imberger (1979) using this approach were found by him to be satisfactory. The model DYRESM was used to compute daily mean lake temperature profiles. The user-oriented modular structure of the program makes it easy to use and to modify. It is thus suitable for extending and elaborating into a nutrient and production model.

For this investigation, the basic algorithms of version 5A of the model describing the following mechanisms were adopted unchanged and are described in detail by Imberger (1979):
– energy exchange processes at the lake surface;
– the turbulent kinetic energy balance;
– vertical diffusion in the hypolimnion;
– depth of intrusion of the natural inflows;
– outflow dynamics.

4.1.2. Modifications to the model

All calculations were carried out on the Universtity of Zürich's IBM 3033 computer. Modifications to the standard program DYRESM5A were necessary in order to fit our specific needs: these modifications were written in FORTRAN IV.

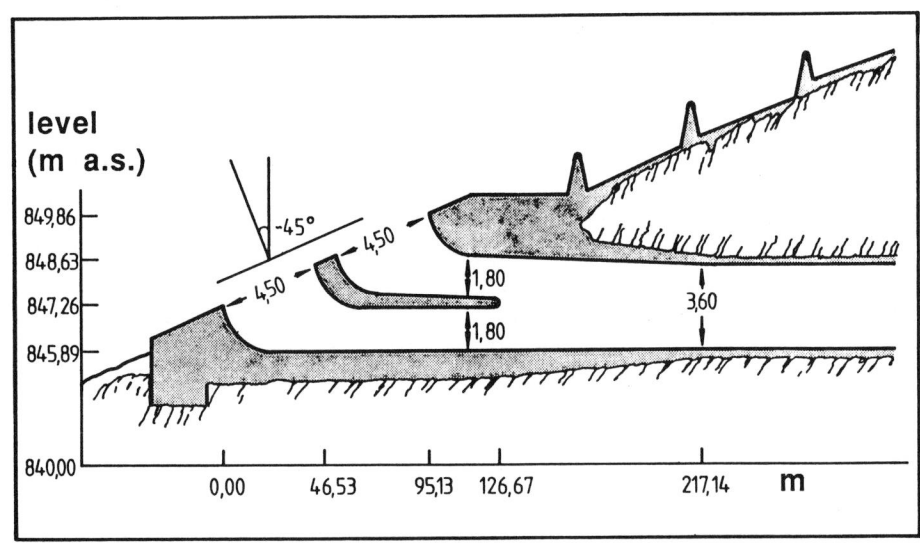

Fig. 4. Geometry of the mouth of the pressure tunnel. Water pumped out of the storage basin into the reservoir leaves the mouth of the pressure tunnel as two individual jets, which, although possessing high momentum, become completely mixed after at most 19 m.

4.1.2.1. The depth of intrusion of the pumped inflow. The calculation of the intrusion depth of an underwater pumped inflow is not taken into account in version 5A of DYRESM. This underwater pumped inflow however plays a decisive role in the determination of mass and energy exchange in Wägitalersee. It was thus necessary to compute the intrusion depth of the pump water as exactly as possible. Figure 4 shows that the mouth of the pressure tunnel through which the pump water enters the reservoir has a complex geometry. About 130 m before entering the reservoir, the stream of water is divided into two. The water enters the reservoir through two rectangular openings, each measuring 4.5 · 4.0 m, which are inclined at an angle of 45° to the perpendicular. None of the available dilution models can therefore be used in the calculation of the intrusion depth (N. Brooks pers. comm., 1985).

Estimation of pump water plume dilution: The aim of the following is to estimate the dilution (μ) of the pump water with lake water which occurs while the pump water plume is rising or falling to its final intrusion depth. All remarks in the following section are based on personal communications from N. Brooks. The openings through which the pump water enters the reservoir are almost square. It may however be assumed that they are circular with equivalent cross-sectional areas. The first question which arises is whether the water jet occupies the whole extent of the cross-section of the mouth, or whether water enters the tunnel from the lake also. Based on the definition of the Froude number, a critical diameter y_c can be defined below which no water from the lake enters the tunnel mouth. This is the case if the density difference between lake water and pump water is small enough or if the velocity of the pump water is large enough.

$$y_c = \sqrt[3]{(Q^2 \cdot p)/(D^2 \cdot \Delta p \cdot g)} \quad (2)$$

where:
y_c – critical diameter of a circular tunnel mouth [m]
Q – volume flux of pump water [m³ s⁻¹]
p – density of pump water [kg m⁻³]
Δp – density difference between lake water and pump water [kg m⁻³]
g – acceleration of gravity (9.81 m s⁻²)
D – effective diameter of the tunnel mouth [m]

Let us now analyze the situation by choosing the critical case with large p in summer (inflow temperature 15 °C, hypolimnetic temperature 5 °C). $Q = 11.1$ m³ s⁻¹ then yields a value of 9.7 m for y_c. The whole cross-section of the tunnel mouth is thus taken up with the pump water jet in summer.

The next question arises as to whether the pumped input has to be treated as a double or a single jet, because of the geometry of the opening (Fig. 4). The distance the water travels to turn from a turbulent jet into a plume or an impulse jet, is called the Zone of Flow Establishment (ZFE). The dilution within the ZFE can be estimated using the method of Fischer (1979). The dilution with the surrounding water is not very high and has a value of approximately 2. Assuming that a double jet is formed because of the geometry of the tunnel mouth, the length of the ZFE can be calculated as follows:

The cross-sectional area of one jet in the pressure tunnel is 5.1 m². In a typical situation $Q = 5.6$ m³ s⁻¹ and the exit velocity is 1.1 m s⁻¹. Thus the ZFE for the two jets is 16 m. If the jets are coupled to a single jet (cross-sectional area within the pressure tunnel = 10.2 m²), the ZFE is 23 m. If both jets fill the whole mouth of the tunnel ($A = 36$ m²) a single jet may be assumed and the ZFE then equals 42 m. In reality the value of ZFE ranges between 20 and 25 m as the jet never fills the whole diameter of the mouth of the pressure tunnel.

A further question arises as to the distance between the mouth of the pressure tunnel and the point where the jets are mixed. Taking the extreme double-jet example (cross-sectional area per jet within the pressure tunnel = 5.1 m²), the two jets would be fully mixed at a distance of 19 m. This means that outside the ZFE, a single jet is always found.

Finally, the relative influence of the buoyancy on the impulse of the jet has to be considered. If the jet has a lower density than the surrounding lake water, the buoyancy will make the jet convert to a plume. The distance required for this to occur is given by the following mathematical relationship:

$$l_m = M^{3/4} / B^{1/2} \quad (3)$$

$$M = 0.25 \, \pi D^2 W^2 \quad (4)$$

$$Q = 0.25 \, \pi D^2 W \quad (5)$$

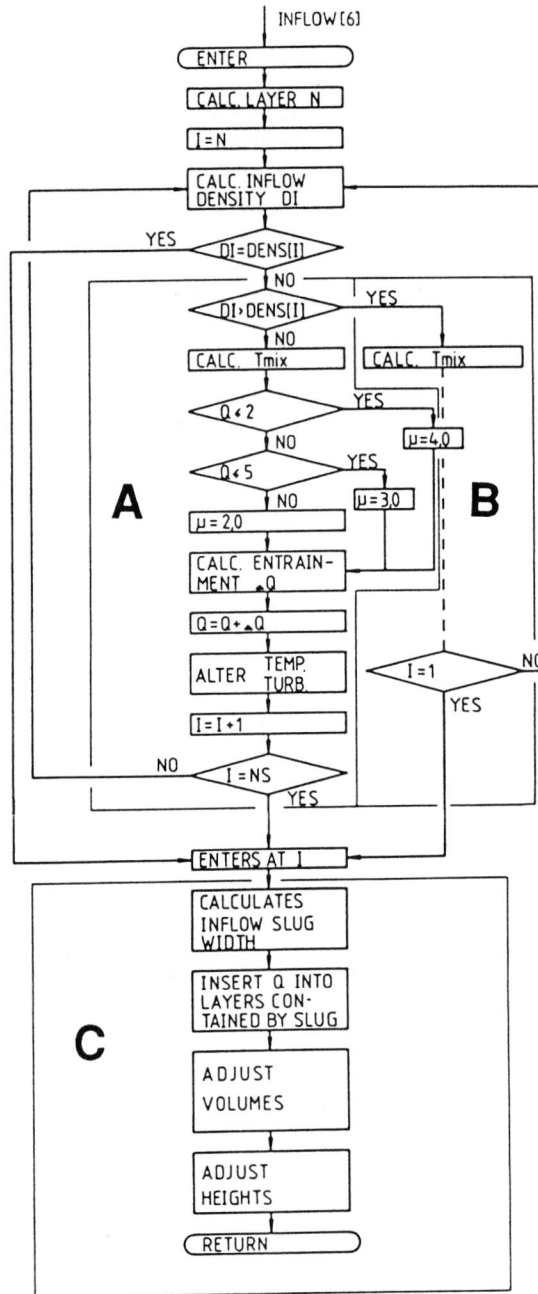

Fig. 5. Flow chart of subroutine PUFLOW in the modified version of model DYRESM5A. After the determination of the number of the layer in which the mouth of the pressure tunnel lies, the decision is made whether the density of the pumped inflow is greater than, equal to or less than that of the locally surrounding reservoir water. In logical blocks A and B the pump water is diluted with the mixture of reservoir water and inflow water. The temperature and concentration of suspended sediment in the water mixture correspond to the mean values in the water column directly above and below the mouth of the

where:
l_m – distance for the establishing of a plume (m)
B – buoyancy flux ($m^{3/2}\,s^{-3/2}$)
M – momentum flux ($m^4\,s^{-1}$)
Q – volume flux ($m^3\,s^{-1}$)
W – mean velocity at the mouth of the pressure tunnel ($m\,s^{-1}$)
D – cross-sectional distance (m)

Assuming $W = 1.1\,m\,s^{-1}$, $Q = 11.1\,m^3\,s^{-1}$ and $D = 3.6$ m the distance for the establishing of a plume (l_m) is 8.2 m. That means, the pumpwater jet is dominated by its impulse. Therefore it mixes only little with lake water.

The angle at which the pump water jet enters the reservoir can be taken to be 45°. Summarising, the following conclusions arise:
1. The 'zone of flow establishment' (Fischer, 1979) is 20–25m. At this distance from the tunnel mouth, the dilution factor $\mu = 2.0$.
2. The maximum vertical distance which the plume travels is 40m. This corresponds to a maximum resultant distance of 70m. At 70 m, μ can be taken to be 4.0.

The calculation of the intrusion depths of the pumped inflow was performed in a separate subroutine (PUFLOW). Its logical structure is illustrated and explained in Fig. 5.

4.1.2.2. Ice cover. The model was altered so that all energy exchange processes between the surface of the reservoir and the atmosphere ceased when the

pressure tunnel. Calculations showed the error in the intrusion depth resulting from this approximation to be negligibly small. The volume of water intruded was calculated as the product of the dilution factor and the pumpwater volume entering the reservoir. The difference between the volume of water intruded and the pumpwater volume (ΔQ) corresponds to the volume of reservoir water entrained into the pumpwater plume during the time taken for the plume to reach its final intrusion depth. Water from each layer through which the plume passes is mixed into the plume. The volume of reservoir water mixed into the plume is proportional to the thickness of the layer through which it is passing. The volume of each layer is reduced by the amount of water mixed into the plume during its passage. The dilution factor m is dependent on the volume flux. For large inflow rates, $Q > 5\,m^3\,s^{-1}$ (these are relevant to the trophic state of the reservoir), $\mu = 2.0$. The intrusion of the plume into its intrusion depths is performed in logical block C similarly to the intrusion of the natural inflows.

temperature of the two uppermost layers of the reservoir decreased to a value lower than 0 °C. The program breaks off the simulation computations 14 days after the point at which ice cover is considered to have begun.

4.1.2.3. Natural inflows. Four natural inflows are taken into account in version 5A of the program. We extended this number to include the five most important natural inflows.

4.1.2.4. Nutrient and suspended sediment loading. The daily load of nutrients and suspended sediment was computed per unit volume for each layer from the daily mean concentrations in the inflows and from the daily hydraulic load.

4.2. Input data to the model DYRESM

Imberger (1979) assumes that the computation of daily mean lake temperature profiles suffices for many purposes and that therefore daily mean meteorological and hydrological data also suffice as input data to the model. This situation could however be improved upon in the case of Wägitalersee by making use of hourly wind speeds and fortnightly or even weekly values of the vertical extinction of short-wave radiation instead of assuming a constant value.

The morphological and physical input data to the model are listed in Table III.

The influence of inorganic suspended sediment on water density: The suspended sediment concentrations in the Wägitalersee inflows undergo extreme variations according to the amount of water transported, and must therefore be taken into account when determining the water density (cf. Nydegger's (1967) observations on Brienzersee). The density of suspended sediment in the Wägitalersee inflows was measured and found to be 2.79 ± 0.02 g cm^{-3}. Since suspended sediment concentrations were only indirectly measured by means of the light transmission of the water at 420 nm, it was necessary to investigate the relationship between these two parameters. The suspended sediment concentration can be computed as follows (for a detailed description of the method, see Kiefer, 1987):

$$SS = -3.94 \cdot 10^{-2} \ln \sigma_{420} + 0.181 \quad (6)$$

where:
SS – suspended sediment concentration [kg m^{-3}]
σ_{420} – transmission at 420 nm (10 cm cuvette) [%]

The density of inflow water and lake water can be calculated from the water temperature and the inorganic suspended sediment concentration as follows:

$$p_W = p_{WT}(1 - SS/p_{SS}) + SS \quad (7)$$

where:
p_W – water density [kg m^{-3}]
p_{WT} – water density calculated from water temperature alone [kg m^{-3}]
SS – inorganic suspended sediment concentration [kg m^{-3}]
p_{SS} – density of inorganic suspended sediment [kg m^{-3}]

4.3. The calibration of the model DYRESM and the vertical mixing processes in the reservoir

4.3.1. Intrusion of the inflows
Natural inflows: The coefficient of friction c_D is the crucial parameter for the characterization of the mixing of inflowing water with lake water (Imberger, 1979). The determination of the dilution of the inflowing water must be carried out during periods of slight stratification, as the depth of intrusion is then very sensitive to slight density differences in the inflows. For calibration purposes, the intrusion depths during spring 1984 were determined and compared with the measured profiles of suspended sediment and total phosphorus. During periods of high inflow, the intrusion depths could be identified as horizontal layers of raised suspended sediment and total phosphorus concentrations. Periphyton from the inflowing streams were also to be found at these depths (Kiefer, 1987). For four of the inflowing streams, a value of 0.13 for c_D yielded a very good agreement between calculated intrusion depths and layers of suspended sediment and total phosphorus in the reservoir during the calibration period. In the case of the Schlierenbach, a stream which enters the lake with a very flat gradient, it was found that the inflowing water reached its final intrusion depth with practically no dilution. The dilution algorithm could thus in this case be

Table III. Model DYRESM5A: input data and values of model coefficients.

Coefficient of kinematic viscosity for water (10 °C): $1.14 \cdot 10^{-6}$ m² s⁻¹
Molecular diffusion coefficient for heat: $10 \cdot 10^{-8}$ m² s⁻¹
Annual mean proportion of photosynthetically active radiation entering at the lake surface (B_l): 0.94
Annual mean attenuation coefficient of photosynthetically active radiation (E): 0.41 W m⁻²
Height of overflow above deepest point: 66.0 m
Height of mouth of pressure tunnel above deepest point: 14.6 m
Maximum width of reservoir (A): 1200 m
Width of reservoir at height of mouth of pressure tunnel (B): 270 m
Maximum length of reservoir (C): 4900 m
Length of reservoir at height of mouth of pressure tunnel (D): 120 m
Width of overflow: 5.0 m
Width of mouth of pressure tunnel: 3.6 m (assuming that the jet is not totally filling the mouth of the pressure tunnel (see section 4.1.2.1.))
V-shaped cross-section of natural inflows: $\alpha_1 = 47°$; $\alpha_2 = 0.01°$ (This inflow enters the lake over a flat delta); $\alpha_3 = 84°$; $\alpha_4 = 74°$; $\alpha_5 = 65°$
Angle of natural inflows with respect to the horizontal: $\varrho_1 = 13°$; $\varrho_2 = 3.3°$; $\varrho_3 = 2.1°$; $\varrho_4 = 5.2°$; $\varrho_5 = 11.3°$
Half-angle of lake cross-section at mouth of pressure tunnel (γ): 87.5°
Angle of lake bottom with respect to horizontal at height of mouth of pressure tunnel (δ): 0.69°
Intrusion of inflows: $c_D = 0.13$
Wind-induced mixing in epilimnion: $\eta = 1.23$
Convective mixing: $c_K = 0.01$
Mixing by damping of internal waves: $c_S = 1.05$
Turbulent kinetic energy variations in the mixed layer: $c_T = 0.51$
Eddy diffusion: $k = 0.1$
Input of turbulent kinetic energy by pumped inflow: multiplied by a factor of 0.01 (see Kiefer, 1987).
Global radiation measurements were taken from a climatological station 17 km from the reservoir. According to Kiefer (1987), the values from this station are applicable to Wägitalersee when multiplied by a factor of 0.7 between 1 October and 31 March.
Period of sunshine: As the error in the relative length of the period of sunshine is dependent on the absorption of long-wave radiation only to the extent of 17%, measurements of this parameter from the above-mentioned climatological station can also be used.
Air temperature: Mean of the daily maximum and minimum.
The humidity is probably on the one hand the meteorological parameter with the greatest error in its determination and on the other the parameter with the greatest influence (sensitivity) on the calculation of the heat flux between atmosphere and lake. The absolute humidity was taken to be 80% of the humidity measured at the above-mentioned climatological station (for details see Kiefer, 1987).
Wind speeds: Arithmetic means of the hourly wind speeds measured 10 m above the lake surface. The wind speed has been measured on the lake surface 200 m west of the inlet of Innerthalbach.
Pumped flow volume: The volume of water pumped back into Wägitalersee from the storage basin was calculated indirectly from the energy used by the four pumps.
Flow volume of the natural inflows: The total daily natural inflow was calculated from the daily water balance of the hydro-electric station and the daily precipitation on to the surface of the reservoir. The method used to distribute the total daily inflow among the five main streams entering the lake is described by Kiefer (1987).
Turbine outflow: The turbine outflow was calculated indirectly from the energy production. The maximum outflow occurred mainly during peak demand times, i.e. in the morning between 06.00 and 08.00 and over the noon period between 11.00 and 13.00. On average over the year the greatest turbine outflow occurred between November and March.
Water temperature of the pumped inflow: Mean value of the twice-daily measured surface water temperature of the storage basin. The error in the daily mean water temperature associated with this method can be roughly corrected by subtracting 2 °C from the calculated mean.
Water temperatures of the natural inflows: Computed according to Kiefer (1987).
Nitrate, orthophosphate, total phosphorus and suspended sediment loads of the inflows: The model DYRESM5A required daily mean concentrations in each of the inflows in order to compute nutrient and suspended sediment intrusion depths. Daily mean concentrations were calculated according to Kiefer (1987). Kiefer (1987) discusses the quality of the input data in detail.

neglected. The values for c_D found by this procedure were confirmed by using them for the calculation of the intrusion depths in autumn 1983. The values are, with one exception, much greater than Imberger's (1979) values for the Collier River at Wellington Reservoir ($c_D = 0.05$). This is due to the fact that the streams dealt with in this paper are mountain streams flowing into the reservoir on beds strewn with large boulders. This causes an intensification of mixing with reservoir water, as these stream beds continue under the water surface of the reservoir.

Pumped inflow: The validity of the approach used to determine the dilution of the pump water during intrusion (cf. Section 3) was tested by comparing the calculated intrusion depth with suspended sediment and total phosphorus. No deviations were discerned during the whole simulation period.

4.3.2. Mixing processes in the body of the reservoir

The processes discussed here are described in detail by Imberger (1979). The individual terms responsible for mixing in the body of the reservoir were determined by comparing measured and calculated daily mean temperatures. The calibration periods extended from 16 March 1983 to 28 February 1984 and from 25 April 1984 to 12 July 1984. The parameters to be determined were fitted to the temperatures measured during one of the two calibration periods and verified by comparing with the temperatures measured during the other period.

The fact that temperatures were measured in only 12 depths in the reservoir sets a practical limit to the accuracy of determination of the calibration constants. Special importance was attached to the correct simulation of the temperatures in spring and autumn, as deviations of the computed from the measured temperature profiles at these times cause the greatest errors in the determination of the nutrient supply to the trophogenic layer.

Wind-induced mixing in the epilimnion: Determination of the parameter n, the efficiency of the conversion of wind energy into turbulent kinetic energy of the water body, must be carried out on days when stratification is slight and the mean wind strong. If measured lake temperatures are compared with daily mean wind speeds, it is obvious that no individual wind event causes a noticeable increase in the mixing depth of the epilimnion. The same is true of the simulated lake temperatures, if the value of n valid for the Wellington Reservoir (1.23) is chosen. This result was confirmed by Kiefer's (1987) statistical investigations. Although no statistical connection between wind speed and the depth of the mixed layer could be established, according to the simulated temperature profiles the upper 2–4 m of the reservoir should be almost always completely mixed in spring and summer. However almost no sign of an epilimnion is found in the measured temperature profiles. In addition to this, differing values of n were shown to have no effect on the depth of the mixed layer during temperature simulations. The mixing which is (wrongly) predicted by the model must therefore be caused by other factors.

The slight influence of the wind on mixing in the epilimnion can be explained as follows:
(a) Wägitalersee is well-protected from the wind by mountains surrounding the reservoir on three sides.
(b) The wind fetch is too small in Wägitalersee to allow the development of Langmuir cells.
(c) The destabilizing effect of the dominating thermal winds is compensated for by the simultaneous warming up of the surface water by solar radiation.

Kiefer (1987) discusses these observations by comparing them with measurements in other lakes.

Convective mixing: The coefficient c_K determines the efficiency of convective mixing. Imberger (1979) chose a value of 0.125 for this coefficient. As c_K has a strong influence on the mixing depth, it must be determined during periods of strong convective mixing, i.e. in autumn and in spring. The mixing efficiency obtained using Imberger's (1979) value of 0.125 is too large. With this value the reservoir circulates too long in spring, and the temperature of the hypolimnion after stratification sets in is on average 0.8 °C too high. Additionally, mixing down to the lake bottom in autumn occurs too early. These errors can be rectified by setting c_K to a value of 0.01. Convection has little influence on the mixing depth of the epilimnion in summer, since during 24 hrs the surface water hardly undergoes any cooling. With a value of 0.01 for c_K instead of 0.125, the mixing depth of the epilimnion in summer is reduced by merely a few decimeters.

Mixing by the damping of internal waves: The parameter c_s, which quantifies the efficiency of the production of vertical mixing by the turbulence engendered by shear stress at the thermocline, affects the temperature gradient in the thermocline by influencing mixing in the epilimnion. As c_s

increases, the mixing depth increases and the thermocline becomes less pronounced. This mechanism is especially important in spring at the onset of stratification. The effect on the steepness of the temperature gradient in the thermocline causes changes in the magnitude of the eddy diffusion coefficient. The reason why c_s must be altered in order to simulate the thermocline correctly will be dealt with in the section on the determination of the vertical eddy diffusion coefficient. Imberger's (1979) value of c_s (0.2) is much too small. The value 1.05 used by us allows a good simulation of the steepness of the thermocline and thus of the energy input into the hypolimnion. The reason why the efficiency c_s can be greater than 1 will also be explained in the section on the calibration of the vertical eddy diffusion coefficient.

The mixing depth of the epilimnion is scarcely affected by an increase in c_s. An observation made by Hirata & Muraoka (1983) confirms this result: he found that seiches had hardly any effect on the depth of the mixed layer of a Japanese lake 94 m deep, 6.5 km long and 1.9 km broad during the summer stagnation period. During the onset of stratification, small changes in c_s cause large changes in the mixing depth and in the steepness of the temperature gradient. Thus the determination of c_s to three significant figures was possible.

Vertical mixing by eddy diffusion: Our attempts to determine an optimal value for the calibration constant of the eddy diffusion coefficient k showed that the expression chosen by Imberger (1979) is inadequate. In this expression the energy input from the wind and that from the inflows are given equal weight. As Kiefer (1987) has shown, this is not valid in the case of Wägitalersee. An algorithm taking into account differing calibration parameters for the turbulent kinetic energy input from wind and inflows would yield better results. This would however necessitate a much more intensive study of the relevant mixing processes than was possible within the framework of our project, and was therefore not attempted by us. The problem thus had to be tackled in a more roundabout way. The turbulent kinetic energy input due to shear at the thermocline is proportional to the square of the current speed difference (ΔU) between two horizontal layers. Further, according to Imberger (1979) ΔU is proportional to the friction velocity $u^* = C_{10}^{0.5} u_{10}$, where C_{10} is the wind stress coefficient and u_{10} is the wind speed 10 m above the lake surface [m s^{-1}]. Since the thermocline in Wägitalersee is not very pronounced, this approach results in positive values of ΔU for all neighbouring layers of differing density during seiches. The input of turbulent kinetic energy into the hypolimnion can therefore be adjusted by choosing a suitable value for the coefficient c_s.

Theoretically, this approach may be called into question; in practice however, it allows the differing efficiencies of wind and inflows to be taken into account in computing the turbulent kinetic energy input. The best agreement between computed and measured values are obtained by setting $k = 0.1$ and $c_s = 1.05$. The eddy diffusion coefficients k_z thus lay between 10^{-6} m^2 s^{-1} in the

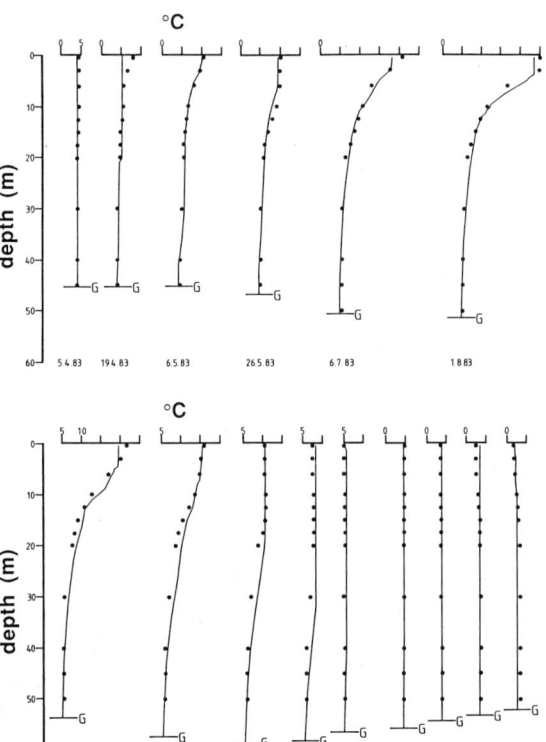

Fig. 6. Typical calculated (line) and measured (dotted) temperature profiles. G = reservoir bottom. Reservoir temperatures computed with the extended version of model DYRESM differ only slightly from measured values. Surface temperatures differ by at most 3 °C (mean difference: 0.04 °C). The maximum deviation in the deeper layers of the reservoir is only 2 °C. It may therefore be assumed that the model performs a realistic simulation of the vertical energy and mass exchange processes in the reservoir.

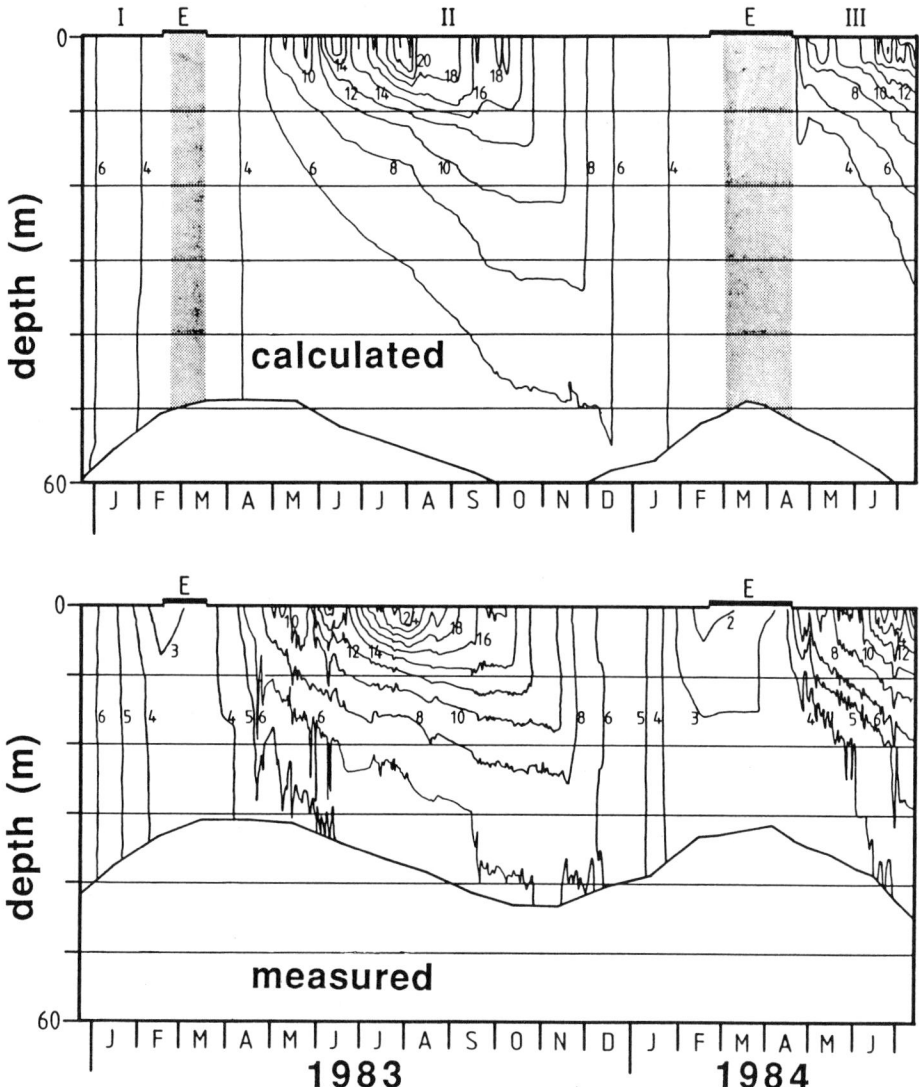

Fig. 7. Isotherms of calculated and measured temperatures in Wägitalersee. Starting values for the three simulation periods (I, II, III): measured temperatures on 21 December 1982, 16 March 1983 and 25 April 1984. Dotted area: temperatures not calculated. *E* denotes periods of ice cover. A comparison of the isotherms in the two figures reveals strong short-term fluctuations in the measured temperatures in spring and summer, especially in the deep water. These fluctuations are a result of the drifting of surface water due to wind action. Spatial inhomogeneity cannot be avoided, as the temperature profile was measured at one point only in the reservoir. The differing depths of the reservoir bottom in the two figures is a result of the fact that the temperature measurements, in contrast to the model computations, were not carried out at the deepest point in the reservoir.

thermocline and 10^{-4} m² s⁻¹ in depths with small temperature gradients.

Results of the calibration procedure: The model calculations are divided into three periods:

(i) 21 December 1982–28 February 1983;
(ii) 16 March 1983–28 February 1984;
(iii) 25 April 1984–12 July 1984.

The simulations within each of these periods are based on the temperature profile measured on the first day of each period. This approach was made necessary by the fact that the validity of the temperature model ceases with the onset of ice cover. The temperature simulation results are presented in Figs. 6 and 7. They are based on the information contained in Table II. The computed and measured temperature profiles illustrated in Fig. 6 show surprising agreement: the deviation between the two never exceeds 3 °C at the surface and 2 °C below 5 m. The results of the simulation are especially good in spring and autumn, when stratification can alter considerably in a very short time.

Imberger (1979) mentions that the deviation between measured and calculated temperatures in Wellington Reservoir never exceed 1 °C. The reasons for the greater deviations in the case of Wägitalersee are as follows: Wägitalersee is filled in summer, during stratification. The water level of Wellington Reservoir rises only during the period of complete turnover, which simplifies the energy and mass exchange processes considerably.

Imberger (1979) had daily temperature measurements from the inflows at his disposal, whereas we estimated them using a statistical approach. The DYRESM5A model was developed especially for the Wellington Reservoir. Many of the mixing algorithms have not been tested sufficiently on other lakes and are in part not generally valid (e.g. the method used in the computation of the eddy diffusion coefficients).

4.4. Energy and mass exchange processes in the reservoir

4.4.1. Heat content

The heat content of a lake is determined by three energy flux processes:
1. Radiation and heat exchange at the lake surface;
2. Advective energy fluxes due to inflows and outflows;
3. Internal heat transport by means of currents and turbulence.

Increases in the heat content due to biological, chemical and physical processes such as friction, heat of solution and biochemical reactions are negligible. Similarly, heat exchange processes with the lake bottom can also be neglected. The influence of precipitation falling directly on to the lake surface can be considered together with the inflows.

The individual heat exchange processes are described in detail by Kiefer (1987). The most important results concerning the heat balance in the year 1983 are illustrated in Fig. 8. Wägitalersee

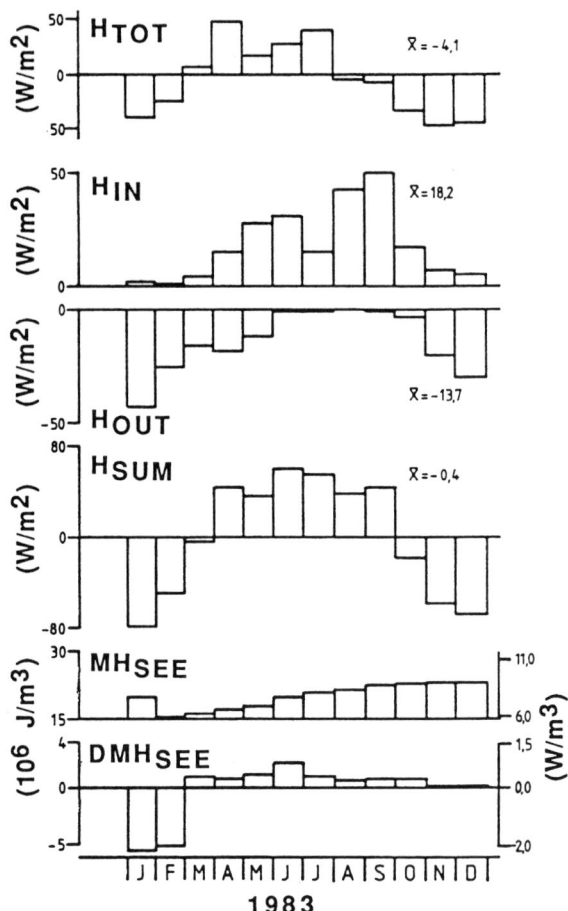

Fig. 8. The energy balance of Wägitalersee in 1983.
H_{TOT} = monthly mean sum of the energy fluxes at the surface;
H_{IN} = monthly mean advective energy flux of all inflows;
H_{OUT} = monthly mean energy flux of the outflow;
H_{SUM} = monthly mean of all energy fluxes;
MH_{SEE} = volume specific energy content of the reservoir (relative to 0 °C);
DMH_{SEE} = monthly changes in MH_{SEE};
$\bar{\chi}$ = mean energy flux.
This figure makes clear the increase in heat content of the reservoir during spring and summer caused by pump-storage operation.

can be seen not only as a reservoir of water, but also of energy. The reservoir acts as an energy trap in spring and summer, since the surface water remains relatively cold due to the influence of the inflows on the upper water layers and the energetic mixing of surface water with colder deep water (a result of the advective heat transport caused by pump operation). The mean lake temperature lags the equilibrium temperature (Kuhn, 1977) by an amount dependent on the volume and temperature of the water inflow. This mechanism is reversed in autumn and winter, when the surface temperature of the reservoir is much warmer than the equilibrium temperature. The heat 'trapped' in the reservoir is lost on the one hand by the heat flux through the lake surface (long wave radiation; evaporative and convective heat loss) and on the other hand by advective heat transport due to the outflow through the turbines. The turbine outflow is of decisive importance in this context as a result of the fact that the volume reduction brought about in autumn by the massive outflow of deep water means that less water remains in the reservoir to be cooled off. During the period of inverse stratification in winter, the temperature of the water exported from the lake via the deep-water outlet is greater than the mean lake temperature; this results in a further reduction of the mean lake temperature. The great heat retention efficiency of Wägitalersee (4.1 W m^{-2}) becomes clear if we compare it with the value of the Lake of Zürich (-5.3 W m^{-2}) obtained by Kuhn (1977). The mean residence times of the water in these two lakes are comparable (Wägitalersee: 1.6 y; Lake of Zürich: 1.3 y).

4.4.2. The water mass balance and intrusion depth of the inflows

The water mass balance: According to Kuhn (1977), precipitation and evaporation from the surface of lakes in the lowlands and subalpine region of Switzerland can be considered to be approximately equal. The water balance of a lake can thus be described by the following equation:

$$dV/dt = \Sigma_i Q_{in} - \Sigma_j Q_{out} \qquad (8)$$

where:
V — lake volume [m^3]
t — time [s]
Q_{in}, Q_{out} — inflow, outflow [m^3 s^{-1}]

Annual variations in the inflows is characteristic of streams in the subalpine and lower alpine regions: extremely low flow volumes during winter ($< 150 \cdot 10^3$ m^3 d^{-1}) and annual maxima during the period of thaw (ca. $400 \cdot 10^3$ m^3 d^{-1}). A period of relatively low flow is often observed in summer (e.g. ca. $100 \cdot 10^3$ m^3 d^{-1} in July 1983), succeeded by periods of higher flow as a result of frequent thunderstorms (e.g. ca. $250 \cdot 10^3$ m^3 d^{-1} in September 1983). Flow rates decrease markedly after October, since precipitation in higher areas begins to fall as snow. Extremely high values can be observed in spring (e.g. $2300 \cdot 10^3$ m^3 d^{-1} in June 1984) and in late summer (e.g. $1300 \cdot 10^3$ m^3 d^{-1} in September 1983).

Corresponding to the wishes of the owners of the hydro-electric station, economic grounds alone determine pump operation times. Water is therefore pumped out of the storage basin into the reservoir especially at night and at weekends. Pumped inflow to the reservoir is at its greatest when the inflow to the storage basin is high. Pumped inflow and natural inflow thus tend to attain high values simultaneously. Pump maintenance in 1983 resulted in normal pump operation being delayed until the middle of July. Mean monthly pumped inflow was at a maximum in September 1983 (ca. $100 \cdot 10^3$ m^3 d^{-1}) and at a minimum in February 1984 (practically no pumped inflow); a high pumped inflow also occurred during the thaw (ca. $90 \cdot 10^3$ m^3 d^{-1}). Peak rates occurred during several weekends in spring 1984 – up to ca. $460 \cdot 10^3$ m^3 d^{-1} were registered during this time. The daily outflow is also determined by the owners of the hydro-electric station. In the course of the day water is allowed to flow out of the reservoir during several hours in the morning, over the noon period and in the afternoon. In the course of a year, the energy demand (and thus the outflow) tends to be greatest especially between October and March to April. As a result of maintenance work on the dam in early 1983, the water level in the reservoir had to be kept as low as possible up till the end of May: this resulted in the outflow of up to $1000 \cdot 10^3$ m^3 d^{-1} in April and May 1983. From June to the middle of October 1983, almost no water was allowed to flow out. Following this, the outflow increased up until January 1984 ($650 \cdot 10^3$ m^3 d^{-1}), only to sink again to zero in April 1984.

Inflow intrusion: The intrusion of the inflows into the reservoir can be characterized as follows: the pump water intrudes at a depth of 9–15 m in spring and summer. Towards autumn the intrusion depth undergoes a rapid increase. Between the middle of October and the beginning of May the pump water flows into the lower 20 m of the

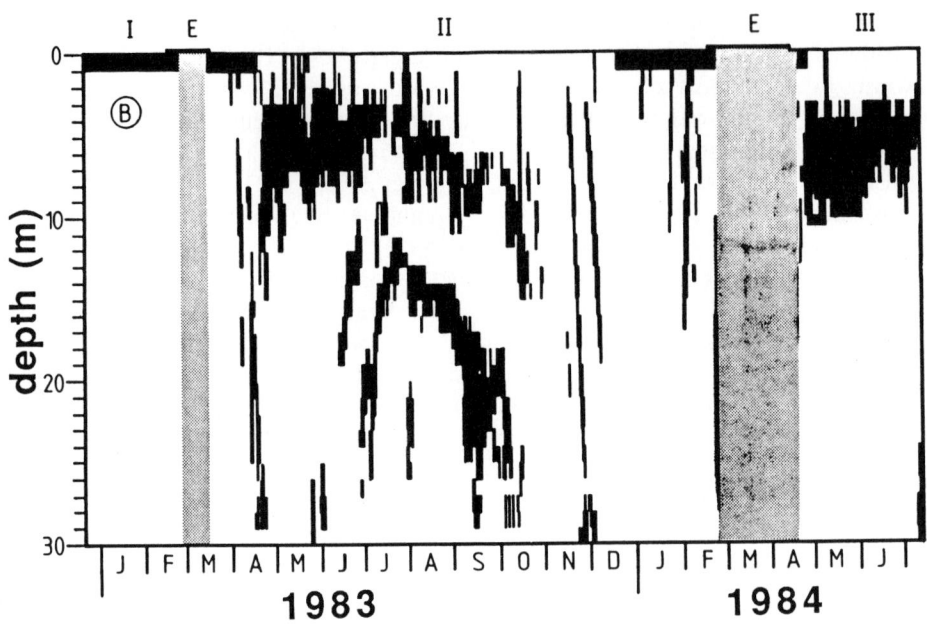

Fig. 9. Depths of intrusion of the natural inflows of Wägitalersee calculated for the uppermost 30 m of the water body. The dotted areas mark periods when the intrusion depths were not calculated. Ice cover is denoted by *E* and the three simulation periods by I, II and III.

reservoir. The intrusion depth of the pumped inflow in spring and summer is not sensitive to small temperature variations. During the onset of stratification, the intrusion depth is to a large extent controlled by the concentration of suspended sediment.

The intrusion depths of the natural inflows are affected by the morphometry of the individual stream beds under the water surface of the reservoir (cf. Fig. 9). The water of the Schlierenbach, which has a flat delta and a level stream bed, mixes very little with lake water. Its intrusion depth corresponds to the original density of its water and is thus vary sensitive to temperature variations. The other four inflows possess steep, rough stream beds. Mixing with lake water is therefore extensive, with the result that they tend to intrude at a shallower depth than the Schlierenbach, and the intrusion depths are insensitive to small temperature variations of the inflowing water. The intrusion depths lay between 3 and 15 m during spring and summer. The intrusion of the natural inflows almost always occurs at a shallower depth than that of the pumped inflow. In autumn the intrusion depths increase slowly; in spring they increase faster than that of the pumped inflow.

Since the pump water intrudes into the deep water of the reservoir in late autumn and winter, it is likely that it soon leaves the reservoir again on account of the large outflow volume through the turbines at this time; this is also true during turnover. This mechanism is however without great importance for the natural inflows. During a short period at the end of November and the beginning of December intrusion occurs into the deep water of the reservoir, but with a very small inflow volume. During a short period in spring they also intrude into depths down to as far as the lake bottom.

4.5. The influence of mass exchange processes on the primary production and biomass

Nutrient recycling in the pump-storage reservoir: Kiefer (1987) was able to demonstrate a close correlation between primary production and the flow of pump water. His statistical results can be explained in terms of the mixing processes taking place in the reservoir. In addition to the above-mentioned 'heat trap' effect, these mixing processes apparently also result in nutrient recycling. The intrusion of pump water into the upper layers of the reservoir in spring and summer inputs

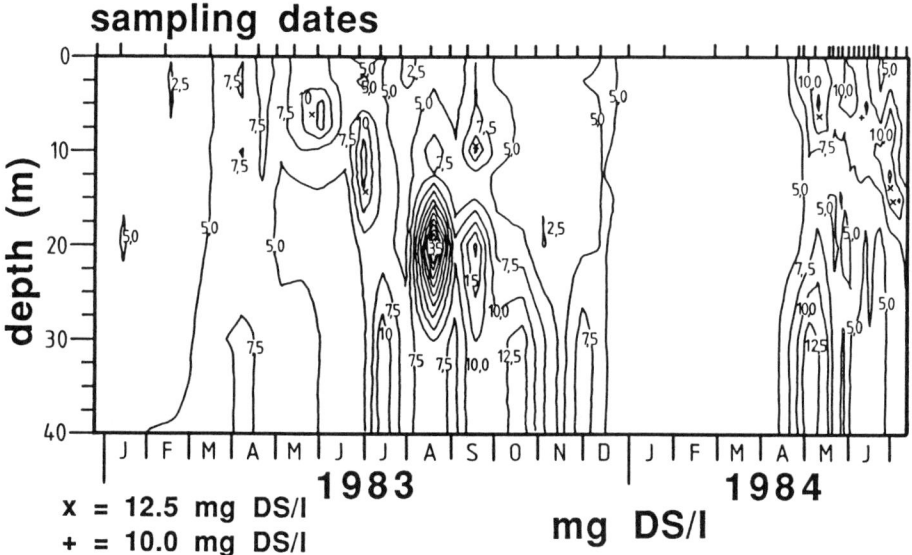

Fig. 10. Distribution of suspended sediment in Wägitalersee. 'Dates of sampling' refers to turbidity measurements. The extreme fluctuations in allochthonous inorganic suspended sediment in the reservoir is due on the one hand to the strongly varying suspended sediment loads in the inflows and on the other to rapid sedimentation in the reservoir itself. The hydrological event which caused the large concentration maximum in August 1983 also brought about a strong decrease in the phytoplankton biomass as a result of co-precipitation. DS = Dry matter; estimation according to Kiefer (1987)

allochthonous nutrients into the trophogenic zone. At the same time, mixing with hypolimnetic water also results in the upward transport of autochthonous nutrients from the relatively nutrient-rich hypolimnion into the trophogenic zone.

The reduction of primary production and biomass caused by precipitation: Although the natural inflows intrude into the euphotic zone in summer and their nutrient input can thus not be neglected – the natural inflows bring on average 8 kg P d^{-1} into the reservoir – Kiefer's (1987) statistical investigations showed no positive correlation between primary production and biomass (measured as chlorophyll-a) on the one hand and nutrient loading on the other. The negative correlation with suspended sediment load was however very pronounced. This phenomenon can be explained in terms of the sedimentation of allochthonous suspended sediment.

Figure 10 illustrates the extreme fluctuations in suspended sediment concentrations in the reservoir caused by the natural inflows. Hakanson & Jansson's (1983) one-box model yields a value of 2.6 g SS m^{-2} d^{-1} for the sedimentation rate. The sedimentation rate undergoes extreme variations during the course of a year. Between October 1983 and the onset of the ice break-up in April 1984 and also between December 1982 and March 1983 the observed values are relatively small (< 2.5 g SS m^{-2} d^{-1}). With the onset of the spring thaw these values increase sharply. In April 1983 the sedimentation rate (5.8 g SS m^{-2} d^{-1}) is about the same as the rate of suspended sediment input. The same is true of May 1984 with an observed sedimentation rate of 5.0 g SS m^{-2} d^{-1}. Between April and October 1983 the sedimentation rate decreased from 5.8 g SS m^{-2} d^{-1} to 0.5 g SS m^2 d^{-1}. In 1984 on the other hand, it remained at a constant level of 5.0 g SS m^{-2} d^{-1} in May and June and rose to over 7.5 g SS m^{-2} d^{-1} in July. This was due to pump-storage operation. In 1983 the suspended sediment load decreased at a steady rate, whereas between April and June 1984 an increase in the suspended sediment load occurred as a result of an increase in the volume of pump water. The explanation for the strong increase in sedimentation at the beginning of July 1984 lies in an inflow event which was responsible for bringing 37.5 g SS m^{-2} d^{-1} on average into the reservoir within a period of three days.

The negative effect of high suspended sediment concentrations on the phytoplankton can be explained by the efficient co-precipitation of phytoplankton brought about by the rapid sedimentation of the suspended sediment. Co-precipitation of phytoplankton with allochthonous suspended sediment from the pumped inflow could not however be shown. The reason for this lies in the fact that a large proportion of the suspended sediment brought into the storage basin by the inflows undergoes rapid sedimentation in the storage basin itself and is not pumped into the reservoir.

4.6. The effects of various different scenarios of pump-storage operation on the phytoplankton nutrient supply

In order to obtain a clearer picture of the effect of varying degrees of pumped inflow and outflow on the reservoir, the following two extreme scenarios were modelled:
1. Scenario 'ZERO': No water whatever is pumped back into the reservoir from the storage basin. The outflow is reduced by the amount of pumped inflow eliminated. The remaining parameters of the model correspond to the conditions prevailing during the period of investigation.
2. Scenario 'MAX': The maximum possible pumped inflow ($258 \cdot 10^3$ m^3 d^{-1}) is pumped back into the reservoir during the night and allowed to flow through the turbines into the storage reservoir during peak demand periods. All remaining parameters correspond, as before, to the conditions prevailing during the period of investigation. As a result of the intensive pump-storage operation, the water temperature in the storage basin will probably decrease: this effect is neglected. A suspended sediment load of 0.039 kg SS · m^{-3} is assumed in the pump water; this corresponds to the mean suspended sediment concentration measured in the storage basin.

The beginning of each simulation period was chosen to coincide with the beginning of the corresponding period during which the natural temperature conditions had been modelled. Each simulation began with a measured temperature profile. The results obtained for the reservoir temperatures simulated according to both scenarios mentioned above are presented in Fig. 11.

Water temperatures: Summarizing, both scenarios emphasise the basic mechanisms already brought to light in the preceding sections. Pump-storage operation causes a rise in the heat content of the reservoir primarily as the result of a temperature increase in the hypolimnion. The reservoir becomes less stable and mixing down to the lake bottom in autumn occurs earlier. The reservoir freezes over later because of its greater heat content. The temperature increase of the deep water occurs mainly as a result of the advective heat input due to the pumped inflow. However, the decrease in the stability of the reservoir also results in an increase in vertical eddy diffusion.

Inflow intrusion: In both extreme scenarios, the intrusion depths of the natural inflows during spring and summer remain practically the same as under the conditions actually pertaining. The reason for this is that in all scenarios, the natural inflows mix with the surface water of the reservoir, the temperature of which is independent of the scenario chosen. In autumn, the intrusion of the natural inflows into the deep water of the reservoir occurs almost two weeks earlier in scenario 'ZERO' than in scenario 'MAX', only to rise to the surface again in the third week of November, five weeks earlier than in the case of maximum pumped inflow. The intrusion depth of the pump water during spring and summer lies in general about 5 m deeper in scenario 'MAX' than in the existing situation. In autumn, the intrusion of the pumped inflow into greater depths occurs more rapidly than in the existing situation; in spring on the other hand the intrusion of the pumped inflow into the euphotic zone does not occur until 14 days later than in the existing situation.

The nutrient supply to the euphotic zone: The nutrient supply to the euphotic zone is calculated with the intrusion of the inflows assuming the nutrients to be conservative tracer. This is allowed if we only discuss the intrusion of the nutrients and not their fate after the intrusion.

In scenario 'ZERO', the nutrient supply to the phototrophic layer in spring and summer is lower than in the existing situation, since the deep-water recirculation caused by pumped inflow is missing and eddy diffusion is lower. In autumn however, the conditions are more favourable for primary production, since the phytoplankton light climate is better as a result of the lateness of the onset of

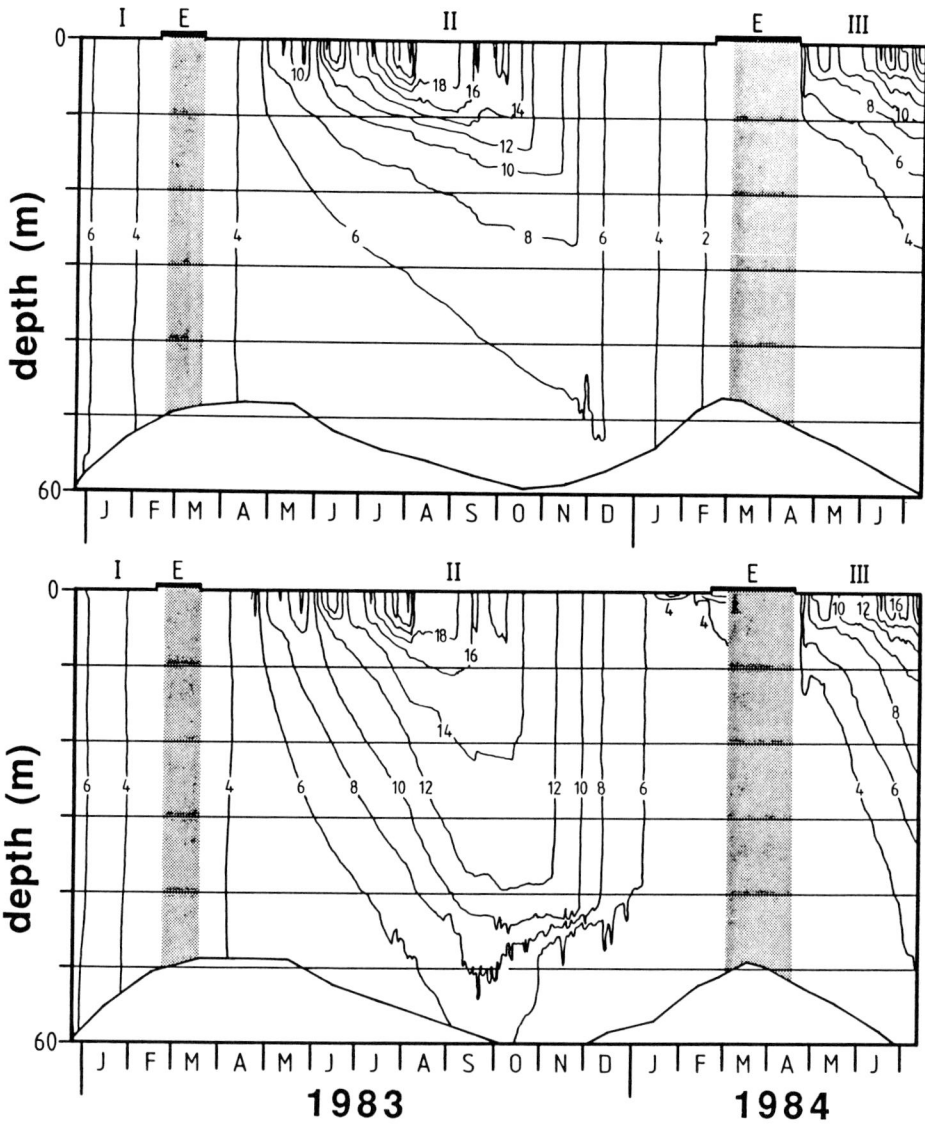

Fig. 11. Simulated temperatures of Lake Wägital for the two scenarios 'ZERO' and 'MAX'. Starting values for the three periods (I,II,III) see Fig. 7. Dotted area: temperatures not calculated. *E* denotes periods of ice cover. The two scenarios show that pump-storage operation increases the heat content of the lake. The lake becomes less stable, in harvest it mixes earlier down to the bottom and the beginning of ice cover is later. The heating of the hypolimnic water results from the advective heat flux caused by the intrusion of the pump water and from the lower stability of the lake that increases the eddy diffusion.

autumn turnover. The fact that the reservoir cools off strongly in winter should on the other hand bring about a production decrease.

In scenario 'MAX', the nutrient supply to the trophogenic layer is increased on the one hand by the nutrient content of the pump water and of the deep water carried upwards in the pump water plume, and on the other by a strong increase in eddy diffusion, which brings about an additional transport of nutrients from the depths of the reservoir to the surface. Pump-storage operation thus results in the reservoir becoming a very efficient nutrient recirculation system in spring and summer. The effects of pump-storage operation could be particularly drastic in a eutrophic lake where phosphate from the sediments can re-enter solution. The simultaneous removal of deep water from the lake complicates the situation however, so that only by modelling can such a case be dealt with quantitatively. In scenario 'MAX' the reservoir is mixed down to 40 m in late autumn, which probably results in a lower primary production than that measured by Kiefer (1987) for the same period.

References

APHA, American Public Health Association, 1976. Standard Methods for the Examination of Water and Wastewater. APHA-AWWA-WPCF, Washington, 1193 pp.

Dubinsky, Z., T. Berman & F. Schanz, 1984. Field experiments for in situ measurement of photosynthetic efficiency and quantum yield. J. Plankton Res. 6: 339–349.

Egloff, J., 1986. Die Frühjahrsentwicklung der Phytoplanktonpopulation im Wägitaler See. Diplomarbeit Universität Zürich, Zürich, 85 pp.

Eidgenössisches Departement des Innern (EDI), 1982. Richtlinien für die Untersuchung von Abwasser und Oberflächenwasser. EDMZ, Bern.

Eidgenössisches Departement des Innern (EDI), 1986. Entwurf Verordnung über die Umweltverträglichkeitsprüfung. Eidgenössische Drucksachen- und Materialienzentrale, Bern, 27 pp.

Elster, H. J., 1962. Untersuchungen über die Rolle des Rheinwassers im Schluchsee und den Zwischenstaubecken der Schluchseewerk AG 1951–1954. Arch.Hydrobiol., Suppl. 15: 430–455.

Elster, H. J. & F. Schmolinsky, 1953. Morphometrie, Klimatologie und Hydrographie der Seen des südlichen Schwarzwaldes II. Arch. Hydrobiol., Suppl. 20: 375–441.

Fischer, H. B., 1979. Mixing in Inland and Coastal Waters. Academic Press, New York, 483 pp.

Flury, B. & H. Riedwil, 1983. Angewandte Multivariate Statistik. G. Fischer Verlag, Stuttgart, 180 pp.

Gloschenko, W. A. & J. O. Blanton, 1977. Short term variability of Chlorophyll-a concentration in Lake Ontario. Hydrobiologia 53: 203–212.

Hakanson, L. & M. Jansson, 1983. Principles of Lake Sedimentology. Springer Verlag, Berlin, 205 pp.

Hirata, T. & K. Muraoka, 1983. Internal seiche and wave in Lake Chuzenji. Verh. Int. Verein. Limnol. 21: 410–422.

Hynes, H. B. N., 1970. The Ecology of Running Waters. Liverpool University Press, Liverpool, 555 pp.

Imberger, J., 1979. Mixing in reservoirs. In: H. B. Fischer (ed.), Mixing in Inland and Coastal Waters. Academic Press, New York, 483 pp.

Imboden, D. M., 1980. The impact of pumped storage operation on the vertical temperature structure in a deep lake. A mathematical model. Proc. of the Clemson Workshop on environmental impacts of pumped storage hydroelectric operations: 125–147.

Isler, J., 1968. Fachtagung: Schwingungen in Wasserkraftzentralen, und 75. Hauptversammlung des Schweizerischen Wasserwirtschaftsverbandes. Wasser, Energie, Luft 78,11/12: 313–319.

Kiefer, B., 1987. Untersuchungen zum Einfluss des Wasserregimes eines voralpinen Pumpspeicher Sees (Wägitaler See) auf die Nährstoffversorgung der Phytoplanktonpopulation. Diss. Universität Zürich, 217 pp.

Kuhn, W., 1977. Berechnung der Temperatur und Verdunstung alpiner Seen auf klimatologischer und thermodynamischer Grundlage. Arbeitsber. Schweiz. Meteor. Zentralanstalt 70: 1–79.

Kuhn, W., 1978. Aus Wärmehaushalt und Klimadaten berechnete Verdunstung des Zürichsees. Vierteljahrsschr. Naturforsch. Ges. Zürich 123: 261–283.

Linder, A. & W. Berchthold, 1982. Statistische Methoden II. Birkhäuser Verlag, Basel, 295 pp.

Martin, D.B., 1978. Comparative limnology of a deep-discharge reservoir and a surface-discharge lake on the Madison River. Freshwat. Biol. 8: 33–42.

Nydegger, P., 1967. Untersuchungen über den Feinstofftransport in Flüssen und Seen, über Entstehung von Trübehorizonten und zuflussbedingten Strömungen im Brienzersee und einigen Vergleichsseen. Beitr. Geol. Schweiz. Hydrol. 16: 1–92.

Octavio, H. K. A., G. H. Jirka & D. R. F. Harlemann, 1977. Vertical heat transport mechanisms in lakes and reservoirs. MIT, Dep. of Civil Eng., Rep. 227, 66 pp.

Ryan, P. J., & D. R. F. Harlemann, 1971. Prediction of the annual cycle of temperature changes in stratified lake or reservoir; mathematical model and user's manual. MIT, Dep. Civ. Eng., Rep. 137, 112 pp.

SAS, 1982a. User's Guide: Basics. SAS Institute, Cary, 789 pp.

SAS, 1982b. User's Guide: Statistics. SAS Institute, Cary, 478 pp.

Schanz, F., 1982. A fluorometric method for determining chlorophyll-a and phaeophytin-a concentrations. Arch. Hydrobiol., Beih. Ergebn. Limnol. 16: 91–100.

Schanz, F. & K. Wälti, 1982. Primary productivity in freshwater environments. In: A. Mitsui & C. C. Black (eds.), CRC Handbook of Biosolar Resources. Vol. 1, 2. CRC Press, Boca Raton: 389–394.

Stroud, R. H., 1973. Influence of reservoir discharge location on the water quality, biology and sport fisheries of reservoirs and tail-waters. In: W. C. Ackermann, F. G. White, E. B. Worthington & J. L. Ivens (eds.), Man-Made Lakes. Their Problems and Environmental Effects. Amer. Geophys. Union, Washington: 315–340.

Thomas, E., 1952. Empirische und experimentelle Untersuchungen zur Kenntnis der Minimumstoffe in 46 Seen der Schweiz und angrenzender Gebiete. Monatsbull. Schweiz. Ver. Gas- u. Wasserfachm. 20: 1–15.

Tippett, R., 1978. Effect of pumped storage hydroelectric scheme on the stratification and ecology of a Scottish loch. Verh. Int. Verein. Limnol. 20: 2697–2700.

Utermöhl, H., 1958. Zur Vervollkommnung der quantitativen Phyto-plankton Methodik. Mitt. Internat. Verein. Limnol. 9: 1–38.

Vollenweider, R. A., 1969. A Manual on the Methods for Measuring Primary Production in Aquatic Environments. International Biological Program Handbook 12. Blackwell, Oxford, 213 pp.

Chapter V

A hierarchy of mathematical models: towards understanding the physical processes in reservoirs

B. Henderson-Sellers
The University of New South Wales, School of Information Systems, P.O. Box 1, Kensington, N.S.W. 2033, Australia

Key words: reservoirs, physical limnology, hydrodynamics, mathematical models

Abstract

Both for day-to-day reservoir management and for longer term planning exercises, mathematical models can provide a useful tool to supplement insights gained from both field and laboratory observations. In reservoir water quality studies, there is an increasing availability of models of a wide range of complexity. Discrete models, such as assessments of trophic state, may now be complemented by a range of simulation models, often implemented in terms of a computer algorithm. These continuous models may be characterised by the number of spatial dimensions represented or by the number of physical, chemical and biological variables included. In the context of reservoir management, it is perhaps the selection, by the reservoir manager, of the model most appropriate to a specific management goal that is one of his greatest concerns.

1. Introduction

Good engineering management of reservoir water quality may be facilitated by the availability of good and proven mathematical models (Reckhow & Chapra, 1983; Henderson-Sellers, 1984). These can be used as part of the design procedure, for example, to assist in decisions relating to the implementation of destratification devices and location of inlets and outlets (e.g. Burns & Powling, 1981; Smalls & Petrie, 1983; Henderson-Sellers & Markland, 1987); as well as in day-to-day predictions of the quality characteristics of the water so that appropriate treatment can be assessed or so that in-lake management can be undertaken in order to minimise treatment costs. For river regulation reservoirs, models will similarly predict outlet water characteristics (e.g. temperature, dissolved oxygen) so that the impact on downstream fisheries, for example, may be assessed (Graham & Willer, 1982).

2. Modelling concepts

In attempting to simulate reservoir water quality it is imperative that a multidisciplinary, integrated (systems science) approach be adopted. Not only is it vital to represent the chemistry and biology but it is also important that these complex processes be amalgamated with a representation of reservoir physics in such a way that feedbacks between biota and physical characteristics are possible. This amalgamated approach may, in respect of reservoirs, be termed 'engineering limnology' (Henderson-Sellers, 1984). Furthermore, any such integrated study programme must include not only numerical modelling, but also field observational programmes and laboratory studies. All three aspects must play an equal, and complementary, part.

Modelling studies may serve not only as an aid to understanding the nature and behaviour of reservoirs, but also as planning and management tools. As such, scientific detail may be sacrificed in the short term to gain the advantage of an engineering tool to solve real problems. The most appropriate model is determined by many factors, from the evaluation of the current state-of-the-art models (often within a predetermined computational budget) to an appreciation of the implicit temporal and spatial scales which any model builder must have encapsulated within the numerical algorithm, and which, unfortunately, are not always 'visible' to the reservoir manager (as model user). Indeed it is not always the case that the most complex model (for example, CE-QUAL-R1 (Environmental Laboratory, 1982)) will provide the greatest advances in understanding. For example, Straškraba & Šerá (1989) present an empirical model of temperature stratification for Klíčava Reservoir, Czechoslovakia, in which the difference between the surface and deep water temperature permits the authors to gain insight into the effect of different retention times on the strength and stability of thermal stratification.

In water quality modelling at the present time, there is a wide range of available models. The commonly used and relatively simplistic trophic state assessments are in fact a simple example of a *discrete* model. Assessments of the trophic state as oligotrophic, mesotrophic, eutrophic etc. have demonstrated that it is not possible to assign a single discrete category to most lakes (e.g. Hannan *et al*, 1981; Lind & Terrell, 1989). In any such simple model, where there is, in effect, a discontinuity in the total range of attainable values, the 'grey' range associated with the borderline is not recognised within the classification scheme. Furthermore, with respect to trophic state classification, these problems were recognised in the OECD (1982) report, in which categories with 100% certainty of assignment were superceded by a two-part trophic state assessment of probabilities plus trophic state category. However, despite this, as well as other attempts to smooth over these problems in terms of a compound trophic state index (e.g. Carlson, 1977; Porcella *et al*., 1980), it is still clear that there is a limited usefulness to such simple approaches.

At the same time, it is becoming increasingly feasible to provide reservoir managers with software packages of *continuous*, simulation models which encapsulate higher temporal and spatial resolution, yet which require little additional time or user education in comprehension and application in decision-making. A wide range of complexity exists in model formulation for continuous models. This can be exemplified most simply, perhaps, in terms of the dimension of the model. Many of the more biologically and chemically orientated eutrophication models use the concept of the continuously stirred reactor (CSR). This model, which is essentially a zero-dimensional model, assumes that all state variables are homogeneously distributed throughout the water body. Consequently there is no possibility of understanding phenomena that are heterogeneous with depth or which occur on seasonal or diel timescales; yet the models are very useful for longer timescale, year-to-year changes of, for example, phosphorus (e.g. Chapra, 1977).

One-dimensional models in which variations in the vertical direction are represented (or sometimes in one-dimensional riverine reservoir models, the longitudinal variability of some depth and width averaged parameter of interest) are most represented in the literature. They have been used most extensively in both physical (e.g. stratification) as well as chemical (e.g. dissolved oxygen) models; although in ecosystem and eutrophication models (which attempt to describe the biota in more detail), vertical variability is seldom well resolved – at best only two or three layers in the vertical may be utilised.

Two-dimensional models are used to describe lateral and longitudinal variations in variable values but averaged over the depth. These may be useful for run-of-the-river reservoirs (e.g. Edinger & Buchak, 1983) which have

(i) a shallow depth and hence are unlikely to stratify and
(ii) a strong throughflow, which tends to ensure the water is well-mixed in the vertical direction.

Three-dimensional models have been developed more recently, largely in hydrodynamical investigations of reservoir currents (e.g. Paul & Lick, 1981; Strub & Powell, 1986). However, it is important to note that although the state-of-the-art in physical/thermodynamical models (Henderson-Sellers & Markland, 1987) centres on 1-D models and in physical/dynamical models centres on 3-D models, there is an urgent need to fuse these apparently disparate approaches into fully three-dimensional representations of thermodynamics + dynamics + biochemical processes.

One further difference between the different modelling approaches presented can be identified in terms of the division of the variables in the model between state variables (which are the prognostic dependent variables within the model) and the driving (or forcing variables) which are prespecified either from data or some sort of prior knowledge. For example, if it is intended ultimately to model biological interactions, some representation of the epilimnetic depth is needed. Should this be calculated (with either a dynamic thermocline or mixed layer model) or is it possible to prespecify this depth – for example as a linear function of time of year over the summer stratification period? The former approach appears to be more adaptable to unusual circumstances; yet for a typical year's weather and for a reservoir with a long time series of observations and managerial expertise, the latter approach may be as (or even more) successfully applied to the specific study area. However, it is most unlikely that such a model could be transferred to other locales (e.g. Henderson-Sellers & Reckhow, 1989).

3. Reservoir thermodynamics

The important role of heat transfer within a water body can be divided between the surface energy budget (including shortwave radiative penetration) and the sub-surface turbulent mixing.

3.1. Surface energy forcing

Almost all energy exchanges at the surface occur within the top few millimetres (McAlister & Macleish, 1969) and can thus be considered as occurring at the air-water interface; the exception being that a portion of the shortwave radiation penetrates this surface layer. It is important to determine fairly accurately the total heat budget of the lake, as well as the feedbacks associated with the surface energy fluxes. The net available energy, γ_N can be given as the sum of the non-reflected incoming shortwave radiation, γ_0, the non-reflected incoming atmospheric longwave radiation, γ_{ri}, the outgoing longwave radiation, γ_{ro}, the evaporative energy flux, γ_e and the convective (or sensible) heat loss, γ_c (In this calculation the heat flux associated with precipitation is negligible because there is no phase change). Thus:

$$\gamma_N = \gamma_0 + \gamma_{ri} - \gamma_{ro} - \gamma_e - \gamma_c \qquad (1)$$

The portion of the shortwave radiation which penetrates (about 60%) can have a marked effect not only on the subsurface heat budget but also upon the potential biotic niches. The differentiation, between wavelength regions, of this fraction of solar radiation has been examined by e.g. Field & Effler (1983), Kirk (1983), Effler et al., (1984).

3.2. Stratification modelling

There are basically two approaches to modelling thermal stratification: eddy diffusion models which *predict* the existence of the mixed layer; and integral or turbulent kinetic energy (TKE) models which *assume, a priori*, the existence of such an homogeneous mixed layer. The latter are computationally cheaper, but may not be as able to represent biological parameters, such as diel vertical migration of blue-green algae (Reynolds & Walsby, 1975). These model types can be exemplified by

(i) the mixed layer model DYRESM (e.g. Spigel & Imberger, 1980) and
(ii) the eddy diffusion model *EDD1* (Eddy Diffusion Dimension 1) (Henderson-Sellers, 1988).

These two types of models have been developed in parallel for some years and both have been shown to be more than adequate for simulating a wide range of lakes and reservoirs. Clearly a direct comparison of these models, already undertaken analytically (Henderson-Sellers & Davies, 1989), needs to be repeated numerically using common data sets (cf. comparable oceanic testing of Martin, 1985).

Existing models have often been designed for a single application and hence include tuning to one particular data set (e.g. Johnson & Ford, 1981). To be of wider applicability, it is vital to include a conceptual framework for such a model. One technique available is to utilise observational data sets from a wide range of sites (cf. preliminary results given by Henderson-Sellers and Reckhow, 1989) so that each parameter may be assessed for its significance and also for its sensitivity to perturbations which may result from deficiencies in either forcing data or in the model parametrisations. For example, the impact of flood or drought events is difficult to assess from a UK data base as a result of the rarity of such events; errors in cloud cover are likely to be less important in the UK than in the subtropics where the absolute magnitude of the incident radiation is considerably higher so that a small error in cloud cover estimation can result in a significant error in calculation of the surface heat flux which is the major energy source for the reservoir energy budget (McGuffie & Henderson-Sellers, 1988).

The research areas in reservoir water quality modelling which require immediate attention appear to be

1. the integration of CSR and process models for biological and chemical variables with higher resolution (temporal and spatial) physico-chemical (largely thermodynamic and hydro-dynamic) 1, 2 and 3 dimensional models;
2. the need to formulate better diffusion parametrisations for throughflows (especially for low retention time reservoirs and pumped storage reservoirs). At present mass conservation is included but, in general, the mixing feedbacks between an underflow or interflow and the otherwise lentic water body itself are excluded.

As a final thought, mathematical modellers should be at least cogniscent of (and hopefully will utilise) new mathematical theories, concepts and tools. In reservoir modelling there has been little discussion of the potential use of *catastrophe theory* (but cf. Renguet & Dubois, 1981) and the concept of *chaos* (Holden, 1986). These tools are now available to the modeller – how can they best be used to improve both our understanding and ability to simulate reservoir ecosystems?

4. Conclusions

Both for day-to-day reservoir management and for longer term planning exercises, mathematical models can provide a useful tool to supplement insights gained from both field and laboratory observations. Models of a wide range of complexity are becoming increasingly available for reservoir water quality studies and it is the selection, by the reservoir manager, of the model most appropriate to a specific management goal that is perhaps one of his greatest problems.

References

Burns, F. L. & I. J. Powling (eds.), 1981. Destratification of Lakes and Reservoirs to Improve Water Quality. Australian Government Publishing Service, Canberra, 915 pp.

Carlson, R. E., 1977. A trophic state index for lakes. Limnol. Oceanogr. 22: 361–369.

Chapra, S. C., 1977. Total phosphorus model for the Great Lakes. Procs. ASCE J. Env. Eng. Div. 103 (EE2): 147–161.

Edinger, J. E. & E. M. Buchak, 1983. Developments in LARM2: a longitudinal-vertical, time-varying hydrodynamic reservoir model, Technical Report E-83-1, US Army Engineer Waterways Experiment Station, Vicksburg, Mississippi.

Effler, S. W., M. C. Wodka & S. D. Field, 1984. Scattering and absorption of light in Onondoga Lake, Procs. ASCE J. Env. Eng. Div. 110: 1134–1145.

Environmental Laboratory, 1982, CE-QUAL-R1: A Numerical One-Dimensional Model of Reservoir Water Quality, User's Manual, Instruction Report E-82-1, US Army Corps of Engineers, Waterways Experiment Station, CE, Vicksburg, Mississippi.

Field, S. D. & S. W. Effler, 1983. Light-productivity model for Onondaga Lake, N.Y. Proc. ASCE J. Env. Eng. Div. 109: 830–844.

Graham, D. S. & D. C. Willer, 1982. Design of retrofit hydro plant for water quality. In: P. E. Smith (ed.), Applying Research to Hydraulic Practice, Am. Soc. Civil Eng., New York: 65–75.

Hannan, H. H., D. Barrows & D. C. Whitenberg, 1981. The trophic status of a deep-storage reservoir in central Texas. In: H. G. Stefan (ed.), Procs. of the Symposium on Surface Water Impoundments, Vol. I. Am. Soc. Civil Eng., New York: 425–434.

Henderson-Sellers, B., 1984. Engineering Limnology. Pitman, London, 356 pp.

Henderson-Sellers, B., 1986. Calculating the surface energy balance for lake and reservoir modelling: a review. Rev. Geophys. 24: 625–649.

Henderson-Sellers, B., 1988. Sensitivity of thermal stratification models to changing boundary conditions. Applied Mathematical Modelling 12: 31–43.

Henderson-Sellers, B. & A. M. Davies, 1989. Thermal stratification modelling for oceans and lakes. Annual Rev. Numerical Fluid Dynamics and Heat Transfer 2: 86–156.

Henderson-Sellers, B. & H. R. Markland, 1987. Decaying Lakes. The Origins and Control of Cultural Eutrophication. Wiley, Chichester, 254 pp.

Henderson-Sellers, B. & K. H. Reckhow, 1989. Application of a lake thermal stratification model to various climatic regimes. Arch. Hydrobiol. Beih., Ergebn. Limnol. 33: 71-78.

Holden, A. (ed.), 1986. Chaos. Princeton University Press, Princeton, 324 pp.

Johnson, L. S. & D. E. Ford, 1981. Verification of a one-dimensional reservoir thermal model, presented at ASCE 1981, Convention and Exposition, St.Louis, Missouri, October 1981.

Kirk, J. T. O., 1983. Light and Photosynthesis in Aquatic Ecosystems. Cambridge University Press, Cambridge, 401 pp.

Lind, O. T. & T. T. Terrell, 1989. Do lake-based trophic classifications apply to impoundments? Arch. Hydrobiol. Beih., Ergebn. Limnol. 33: 647.

McAlister, E. D. & W. Macleish, 1969. Heat transfer in the top millimeter of the ocean. J. Geophys. Res. 74: 3408–3414.

McGuffie, K. & B. Henderson-Sellers, 1988. Accuracy of oceanic cloud amounts for surface flux determination. Proc. Conf. Remote Sensing of the Atmosphere and Oceans (Canberra, Australia, 16–24 February 1988).

Martin, P. J., 1985. Simulation of the mixed layer at OWS November and Papa with several models. J. Geophys. Res. 90: 903–916.

OECD, 1982. Eutrophication of Waters. Monitoring, Assessment and Control, OECD, Paris.

Paul, J. F. & W. J. Lick, 1981. A numerical model for three-dimensional, variable-density hydrodynamic flows, USEPA Report, Washington, DC.

Porcella, D. B., S. A. Peterson & D. P. Larsen, 1980. Index to evaluate lake restoration. Proc. ASCE, J. Env. Eng. Div. 106 (EE6): 1151–1169.

Reckhow, K. H. & S. C. Chapra, 1983. Engineering Approaches for Lake Management. 2 vols, Ann Arbor Sci., Ann Arbor, Michigan.

Renguet, E. & D. M. Dubois, 1981. Approche stochastique de la theorie des catastrophes. In: D. M. Dubois (ed.), Progress in Ecological Engineering and Management by Mathematical Modelling. Editions Cebedoc sprl, Liege, Belgium: 49–86.

Reynolds, C. S. & A. E. Walsby, 1975. Water blooms. Biol. Rev. 50: 437–481.

Straškraba, M. & Z. Šerá, 1989. An empirical model of temperature stratification for Klíčava Reservoir, Czechoslovakia. Arch. Hydrobiol. Beih. Ergebn. Limnol. 33: 69.

Smalls, I. C. & L. G. Petrie, 1983. Low cost destratification in small upland reservoirs. Procs. Tenth Federal Convention, Australian Water and Resources Council.

Spigel, R. H. & J. Imberger, 1980. The classification of mixed-layer dynamics in lakes of small to medium size. J. Phys. Oceanogr. 10: 1104–1121.

Strub, P. T. & T. M. Powell, 1986. Wind-driven transport in stratified closed basins; direct versus residual circulations. J. Geophys. Res. 91: 8497–8508.

Chapter VI

Modelling of physical, chemical and biological processes in Polish lakes and reservoirs

J. Uchmański,[1] W. Szeligiewicz,[1] & M. Loga[2]
[1] *Institute of Ecology, Dziekanòw Leŝny, 05092 Lomianki, Poland;* [2] *Institute of Environmental Engineering, Warsaw Technical University, ul. Nowowiejska 20, 00653 Warsaw, Poland*

Key words: reservoirs, lakes, limnology, plankton, eutrophication, simulation, nutrients, mathematical models

Abstract

Three models of phosphorus cycling in a lake or a reservoir are presented. All models are calibrated for data collected from Lake Głębokie, Jorzec and Inulec (Masurian Lakeland, North-Eastern Poland). The one-layer model of phosphorus cycling in the epilimnion is presented. Also two-layer model of phosphorus cycling in the epilimnion and hypolimnion is described. The multi-layer model presented consists of two linked submodels: thermal and biological one. Results of numerical experiments concerning the influence of changes of phosphorus load to the water body on the values of biological and chemical variables are discussed. Also a model of exchange of phosphorus between bottom sediments and near-bottom water is presented. A method of parameter estimation is described, which uses Fourier series or polynomial approximation of experimental points for limiting the number of estimated parameters, and supports the process of parameter estimation by sensitivity analysis.

1. Introduction

The most general form of the model describing phosphorus cycling in a lake or reservoir ecosystem is given by the following equation:

$$V \, dP/dt = I - S - O \qquad (1)$$

where P is total phosphorus concentration, V is volume of water body, I is input of total phosphorus, O is output of total phosphorus and S is rate of sedimentation and both biological and chemical changes of phosphorus.

Various processes leading to the input (I), output (O) and various changes (S) give different forms of Eq. (1). Predicting concentration of total phosphorus in the water body in relation to the phosphorus load was used by Uchmański & Szeligiewicz (1988) to estimate parameters of Eq. (1) for lakes located in northern Poland.

Eq. (1) is a very general description of matter balance in a water body. It describes a lake or a reservoir as a black box. For practical purposes it may be enough. However for ecological studies it is necessary to treat a lake or reservoir as vertically stratified (at least two water layers should be distinguished in deep water body: the epilimnion and hypolimnion) and horizontally heterogeneous. Various trophic groups should be included into ecological analysis. Also bottom of the lake or reservoir is an important part of the ecosystem.

Therefore the Eq. (1) should be separately applied to each physical layer in the water body and to each biological and chemical variable included in the model.

This paper describes an attempt at such ecologically more precise description of physical, chemical and biological processes and relations between them taking place in a lake and reservoir ecosystems.

All the models presented here are based on data collected during long-term studies of the Lakes Inulec, Głębokie and Jorzec situated in the watershed of the River Jorka in the Masurian Lakeland, Poland (Ejsmont-Karabin *et al.*, 1983; Godlewska-Lipowa, 1983; Hillbricht-Ilkowska,

1983; Hillbricht-Ilkowska & Ławacz, 1983; Penczak *et al.*, 1985; Planter *et al.*, 1983; Spodniewska, 1983; Węgleńska *et al.*, 1983; Węgleńska *et al.*, 1987; Woroniecka-de Wachter, 1983). The basic characteristics of the lakes are presented in Table I.

Table I. Limnological characteristics of Lakes Jorzec, Inulec and Głębokie in the Mazurian Lakeland, northeastern Poland (according to Bajkiewicz-Grabowska, 1985).

	LAKE		
	Głębokie	Inulec	Jorzec
volume (m³)	5.31×10^6	7.61×10^6	2.20×10^6
area (ha)	46.0	161.0	41.0
maximum depth (m)	4.3	10.1	11.6
average depth (m)	11.8	4.6	5.5
retention time (year)	0.8–1.5	1.7–3.2	0.2–0.4

2. One-layer model: phosphorus cycling in the epilimnion

2.1. The model

Let us assume that we can describe the ecosystem using one-layer model. Such description is appropriate in the case of a shallow lake or when the model is confined to one layer only in a stratified water body, for instance, the epilimnion. This layer can be treated as being physically homogeneous. The changes of temperature within the layer as in the whole water body can be described by simple polynomial or trigonometric functions.

The one-layer model of phosphorus cycling used in our investigations describes time-related changes in the following biotic and abiotic variables: concentration of phosphate-phosphorus P—PO_4, phytoplankton, non-predatory zooplankton, predatory zooplankton, bacteria, and detritus. The values of all these variables are expressed in μg P l^{-1}. Moreover, the model describes temporal changes occurring in the water temperature, incident light intensity and the distribution of light intensity with depth. A diagram of the relationships between variables is shown in Fig. 1. Equations of the model are given in Table II and driving functions in Table III. A detailed description of the model is presented in Uchmański (1988).

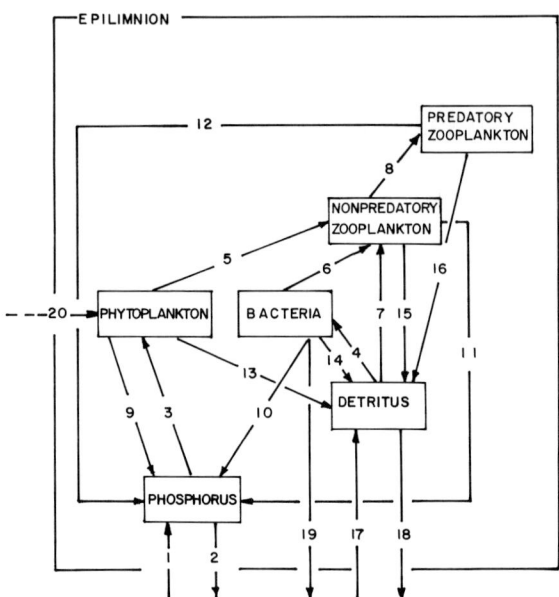

Fig. 1. A diagram of the one-layer model of phosphorus cycling in the epilimnion. Symbols for the processes described: 1 – loading with dissolved phosphorus from external sources and loading with phosphorus from deeper part of lake, 2 – sedimentation of phosphorus, 3 – utilization of phosphorus by phytoplankton, 4 – utilization of detritus by bacteria, 5 – grazing on phytoplankton by non-predatory zooplankton, 6 – grazing on bacteria by non-predatory zooplankton, 7 – grazing on detritus by non-predatory zooplankton, 8 – predation of predatory zooplankton on non-predatory one, 9 – excretion of phosphorus by phytoplankton, 10 – excretion of phosphorus by bacteria, 11 – excretion of phosphorus by non-predatory zooplankton, 12 – excretion of phosphorus by predatory zooplankton, 13 – phytoplankton mortality, 14 – bacterial mortality, 15 – mortality of non-predatory zooplankton, 16 – mortality of predatory zooplankton, 17 – loading of detritus with food supplied for fish, 18 – detritus sedimentation, 19 – sedimentation of bacteria associated with detritus, 20 – symbol for solar energy received by epilimnion. This figure does not show that many of these processes depend on temperature.

The sources of phosphorus to the model system comprise the loading from the sediments or the deeper layers of water body P_r and the external loading P_e and excretion of the two trophic types of zooplankton $Q_{Z_{np}}$ and Q_{Z_p}. Bacteria are also involved in the transformation of detritus into phosphorus Q_b. Non-predatory zooplankton consume phytoplankton, detritus and bacteria. Predatory zooplankton feeds on non-predatory zooplankton. Detritus consists of undigested residues of the food of the two trophic types of

Table II. Equations of the one-layer model.

dZ_p/dt	$= G^{max}{}_{Z_p} F^1{}_Z(T) F^2{}_{Z_p}(Z_{np}) A_{Z_p} Z_p - Q_{Z_p} - M_{Z_p}$
dZ_{np}/dt	$= G^{max}{}_{Z_{np}} F^1{}_Z(T) F^2{}_{Z_{np}}(F,B,D) A_{Z_{np}} Z_{np} - G_{Z_p} Z_p - Q_{Z_{np}} - M_{Z_{np}}$
dF/dt	$= G^{max}{}_f F^1{}_f(T) F^2{}_f(I) F^3{}_f(P) F - G^{max}{}_{Z_{np}} F^1{}_Z(T) c_1 f^1{}_{Z_{np}}(F) Z_{np} - Q_f - M_f$
dB/dt	$= G^{max}{}_b F^1{}_f(T) F^2{}_b(D) B - G^{max}{}_{Z_{np}} F^1{}_Z(T) f^2{}_{Z_{np}}(B) Z_{np} - M_b - S_b - Q_b$
dD/dt	$= M_f + M_b + M_{Z_{np}} + M_{Z_p} + (1-A_{Z_{np}}) G_{Z_{np}} Z_{np} + (1-A_{Z_p}) G_{Z_p} Z_p + P_{fish} - G^{max}{}_{Z_{np}} F^1{}_Z(T) f^3{}_{Z_{np}}(D) Z_{np} - G_b B - S_{det}$
dP/dt	$= P_e + P_r + Q_f + Q_{Z_p} + Q_{Z_{np}} + Q_b - G_f F - S_p$

G_{Z_p}	$= G^{max}{}_{Z_p} F^1{}_Z(T) F^2{}_{Z_p}(Z_{np}),\ G_{Z_{np}} = G^{max}{}_{Z_{np}} F^1{}_Z(T) F^2{}_{Z_{np}}(F,B,D) A_{Z_{np}}$
G_f	$= G^{max}{}_f F^1{}_f(T) F^2{}_f(I) F^3{}_f(P),\ G_b = G^{max}{}_b F^1{}_f(T) F^2{}_b(D)$
$F^1{}_Z(T)$	$= \exp(-v_z(T^{opt}{}_Z - T)^2),\ F^1{}_f(T) = \exp(-v_f(T^{opt}{}_f - T)^2)$
$F^2{}_f(I)$	$= (I/I_{opt})\exp(1 - I/I_{opt}),\ F^3{}_f(P) = P/(K_f + P),\ F^2{}_b(D) = D/(gB + D)$
$F^2{}_{Z_p}(Z_{np})$	$= 1-\exp(-K_p Z_{np})$
$F^2{}_{Z_{np}}(F,B,D)$	$= c_1 f^1{}_{Z_{np}}(F) + c_2 f^2{}_{Z_{np}}(B) + c_3 f^3{}_{Z_{np}}(D)$
$f^1{}_{Z_{np}}(F)$	$= 1-\exp(-K^1{}_n F),\ f^2{}_{Z_{np}}(B) = 1-\exp(-K^2{}_n B),\ f^3{}_{Z_{np}}(D) = 1-\exp(-K^3{}_n D)$
Q_{Z_p}	$= q_{Z_p} F^1{}_Z(T) Z_p,\ Q_{Z_{np}} = q_{Z_{np}} F^1{}_Z(T) Z_{np},\ Q_f = q_f F^1{}_f(T) F,\ Q_b = q_b F^1{}_f(T) B$
M_{Z_p}	$= m_{Z_p} Z_p,\ M_{Z_{np}} = m_{Z_{np}} Z_{np},\ M_f = m_f F,\ M_b = m_b B$
S_{det}	$= s_{det} D,\ S_b = s_{det} mB,\ S_p = s_p P$

where: P – phosphorus, F – phytoplankton, B – bacteria, D – detritus, Z_{np} – non-predatory zooplankton, Z_p – predatory zooplankton, I – light intensity, T – temperature.

Table III. Driving functions of the one-layer model.

temperature in epilimnion:

$$T = a_{epi} - b_{epi} \cos(2\pi(t - c_{epi})/365)$$

light intensity:

$$I = (d - g \cos(2\pi(t - h)/365)) e^{-nz}$$

phosphorus loads:

P_e, P_r = polynomials of t, P_{fish} = constant

where t is number of day in the year, z – depth, n – vertical light extinction coefficient.

zooplankton ($A_{Z_{np}}$ and A_{Z_p} – coefficients of assimilation efficiency of the non-predatory and predatory zooplankton, respectively). Concentration of detritus is treated as to increase as a result of supplying food for farming fish (P_{fish}) and due to the mortality of all living components of the system (M_f, $M_{Z_{np}}$, M_{Z_p}, M_b). Matter is lost from the model system due to sedimentation of detritus particles (S_{det}) and the associated bacteria (S_{bdet}), and also by physical transportation of phosphorus caused by water movements (S_p). Primary production depends in the model on light intensity (I). Moreover, almost all the processes occurring in the lake depend on temperature (functions $F^1{}_f$ and $F^1{}_z$).

2.2. Results of simulation

The model was calibrated to the data from the lake Głębokie. Table IV shows the parameters of the model, which were obtained by iteration solving of the set of equations from Table II. The values of the variables used in the model are mean values for the upper six meters of the lake depth in order to simulate the process of mixing in the epilimnion and to approximate the output of the model to empirical data, which were measured at these depths. These solutions are shown graphically in

Table IV. Estimated values of the one-layer model parameters.

Phosphorus:			Predatory zooplankton		
$s_p = 0.1$ d^{-1}			$G^{max}{}_{Z_p}$	= 0.65	d^{-1}
			K_p	= 0.04	(μg P·l^{-1})$^{-1}$
			A_{Z_p}	= 0.6	
			q_{Z_p}	= 0.04	d^{-1}
			m_{Z_p}	= 0.01	d^{-1}

Phytoplankton:					
$G^{max}{}_f$	= 1.3	d^{-1}	Bacteria:		
K	= 8.0	μg P l^{-1}	$G^{max}{}_b$	= 2.5	d^{-1}
I_{opt}	= 1464.4	J cm^{-2} d^{-1}	g	= 2.0	
$T^{opt}{}_f$	= 16	°C	q_b	= 0.005	d^{-1}
v_f	= 0.004	(°C)$^{-2}$	m_b	= 0.05	d^{-1}
q_f	= 0.001	d^{-1}	m	= 0.03	
m_f	= 0.15	d^{-1}			

Non-predatory zooplankton:			Detritus:		
			s_{det}	= 0.3	d^{-1}
$G^{max}{}_{Z_{np}}$	= 1.3	d^{-1}			
c_1	= 0.6				
c_2	= 0.3				
c_3	= 0.1				
$K^1{}_n$	= 0.05	(μg P l^{-1})$^{-1}$			
$K^2{}_n$	= 0.01	(μg P l^{-1})$^{-1}$			
$K^3{}_n$	= 0.2	(μg P l^{-1})$^{-1}$			
$T^{opt}{}_Z$	= 20	°C			
v_Z	= 0.007	(°C)			
$A_{Z_{np}}$	= 0.5				
$q_{Z_{np}}$	= 0.03	d^{-1}			
$m_{Z_{np}}$	= 0.01 d^{-1} for $t < 136$ and $t > 290$				
	= 0.05 d^{-1} for $136 \leq t \leq 290$				

Fig. 2. In Fig. 3A the simulated values of phosphorus concentrations are compared with the empirical values obtained from Lake Głębokie in 1976, and show a good agreement.

2.3. Stability analysis

The set of model equations in Table II was solved for a situation when all the time-dependent parameters and other functions of time occurring in the model were constant. In particular, all sources of epilimnion loading with phosphorus were made constant.

The solutions of the model equations with such constant parameters show that, for a certain neighbourhood of the Table IV parameter values, this model has a singular point in the phase space, which is of the type of a stable focus. After initial oscillations which are damped down in time, all the variables will attain equilibrium values (Fig. 3B). With initial conditions as occurred in the lake in mid-March of 1976, equilibrium conditions were reached after about three months. But because the real lake values of the model parameters are continuously changing, the model variables are 'pursuing' the permanently escaping equilibrium values, and in addition, changes resulting from interactions between ecosystem components will be superimposed on this process.

2.4. Numerical experiments with the model

2.4.1. The consequences of stopping fish farming

The set of model equations in Table II was solved at $P_{fish} = 0$. The amount of phosphorus P_{fish} supplied from May to October enriched the pool of detritus. Thus, when the food supplying for fish is stopped, the concentration of detritus is reduced along with the concentration of bacteria feeding on detritus. The effect of this treatment on the other model variables is small, but they also are reduced a little.

2.4.2. Changes in lake loading with phosphorus

The most important source of phosphorus in the epilimnion of Lake Glebokie is the loading from deeper layers. It is estimated that the epilimnion of Lake Glebokie receives up to several hundred kilograms of phosphorus per month. The maximum is observed in spring.

In one group of experiments with the model, the level of P loading to the lake from deeper layers was multiplied by a certain constant (e.g. 0.5) but the timing of the loading remained unchanged. Figure 3C shows that the consequence of the reduced loading from the deeper layers was a flattening of all the curves; all the model variables were reduced and the smaller maxima disappeared. When the P loading from deeper layer was totally

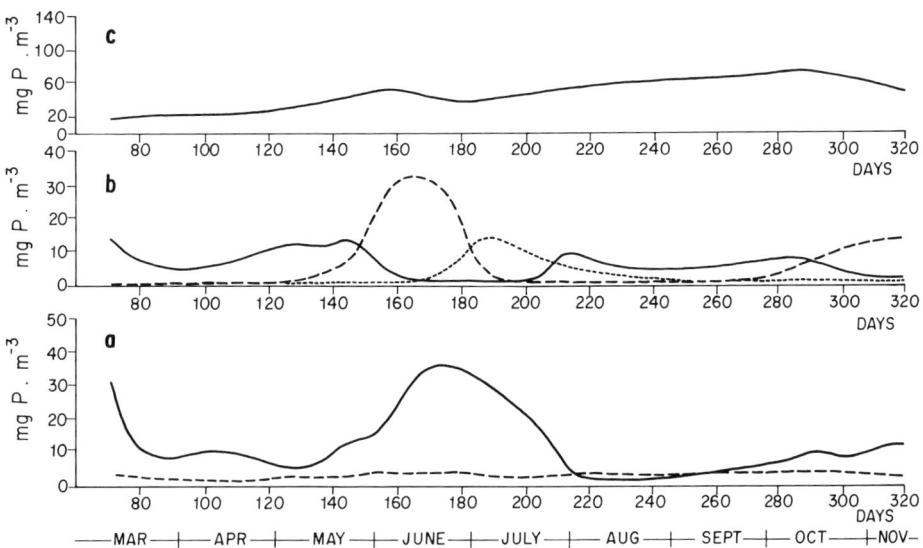

Fig. 2. Results of simulation of the one-layer model with parameters estimated for Lake Głębokie. a – Solid line represents the concentration of phosphorus, and dashed line is the concentration of detritus. b – Solid line concentration of phytoplankton, dashed line concentration of non-predatory zooplankton, dash-dot line concentration of predatory zooplankton, c – Graph of the concentration of bacteria.

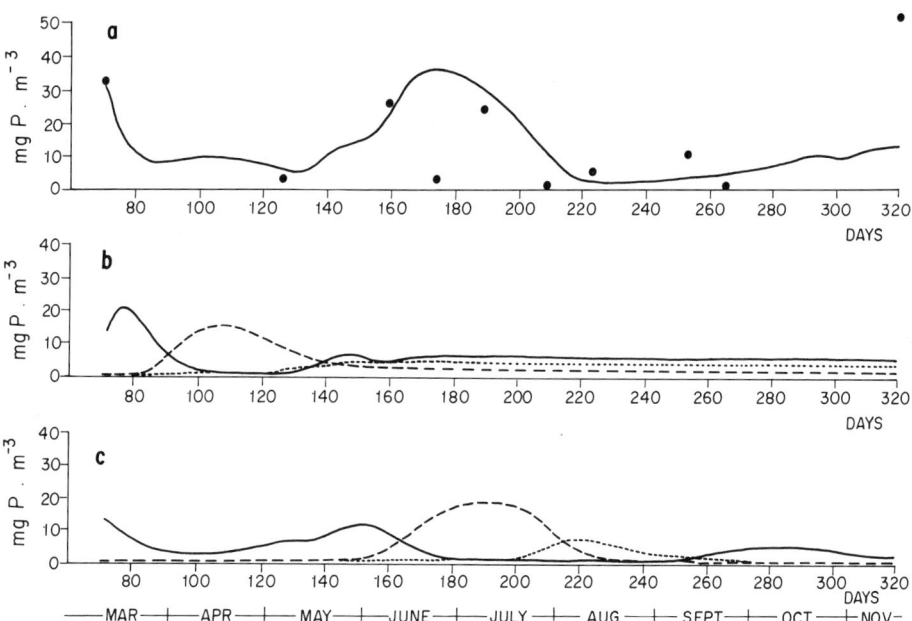

Fig. 3. a – One-layer model. Comparison of the model simulations with real measurements of the concentration of dissolved phosphorus taken from Lake Głębokie. Solid line – results of simulation, dots results of measurements. b – Analysis of the stability of the one-layer model of the epilimnion of Lake Głębokie. Solution of model equations was carried out with constant values of all the parameters and functions of the time. Solid line – concentration of phytoplankton, dashed line – concentration of non-predatory zooplankton, dash-dot line – concentration of predatory zooplankton. c – Results of simulations, when the external loading with phosphorus is reduced by half each day. Compare with Fig. 2b.

eliminated, the values of all the variables except for bacterial concentration became very low. The increase in the concentration of bacteria during May is related to the onset of P loading the lake with food for fish. In the same way, increasing P loading from deeper layers accounts for an initial increase of the values of all the variables. For P loadings increased by a factor of 1.5 times, the model variables became negative, for which there is no biological interpretation. This implies that the structure of the model should be changed so that the solutions for such an increased loading from deeper layers can be biologically interpreted.

Another numerical experiment was carried out which simulated an instantaneous loading of the lake with a large amount of phosphorus. It was found that the model was more sensitive to this type of loading if it occurred in the first half of the season and especially during the spring-summer peak of phosphorus concentration. In the second half of the season, instantaneous P-loading had little effect.

Another model simulation was carried out with constant monthly P loadings from deeper layers. If this load ranged between 100 and 200 kg P $\cdot$ (lake month)$^{-1}$ which is approximately the mean loading level to the lake during the first part of the season, no important changes (in variable values) were observed in the first half of the season but in the second half there were substantial increases in loadings, which then stimulated a second cycle of oscillations of all the variables.

2.4.3. Changes in temperature

All variables are temperature dependent, including zooplankton excretion. In the model, at a low concentration of non-predatory zooplankton, a temperature increase of 4 °C produces a relatively lower increase in excretion than in growth rate. For this reason, zooplankton appear earlier and markedly reduces the concentration of phytoplankton in spring. However, with larger values of the model variables, increasing temperature produces a relatively higher increase in phosphorus excretion. As a result, all peaks become flat and the dynamics of the model become less variable. Due to the increased excretion, only the concentration of dissolved phosphorus increased, which results in an earlier appearance of the summer phytoplankton peak.

If temperature is lowered, a greater decrease in phosphorus excretion is observed. Then the maxima of all biotic variables are increasing, whereas the concentration of dissolved phosphorus is decreasing.

2.5. General remarks

The epilimnion of Lake Glebokie consists of two largely independent subsystems. One consists of the microbial loop based upon bacteria and the other the phytoplankton-based foodwebs. Any exchange of matter between these subsystems operates via the detritus and dissolved phosphorus. Stopping fish farming loads of the lake with detritus, changes bacterial concentration but has little effect on other variables. Conversely, changes in the P loading of lake has a strong effect on phytoplankton and both types of zooplankton, but only relatively small changes in the concentration of bacteria.

The annual distribution and mean level of P loading from the deeper water layers have a major influence on the values of the model variables. Especially important is the spring peak of the loading of phosphorus from deeper layers.

3. Two-layer model: phosphorus cycling in the epilimnion and hypolimnion

3.1. The model

For a lake or reservoir which is thermally stratified the vertical heterogeneity can be described as two layers. The whole system consists now of two subsystems: the epilimnion and hypolimnion. The model is represented respectively by two submodels which work as a cascade. The modelling approach adopted here for describing of biological and chemical processes is a simplified version of the earlier one-layer model of the phosphorus cycling in the epilimnion (Table V and VI).

The state variables in the model for the epilimnion subsystem (Fig. 4) are: concentrations of phosphate – phosphorus, phytoplankton and zooplankton concentrations. In the hypolimnion subsystem there are: concentration of P—PO$_4$, zooplankton and bacteria.

In the submodel for the epilimnion, phosphorus

Table V. Equations of the two-layer model.

EPILIMNION:

$$dZ/dt = G^{max}_z F^1_z(T) F^2_z(F) A_z Z - Q_z - M_z$$

$$dF/dt = G^{max}_f F^1_f(T) F^2_f(I) F^3_f(P) F - G^{max}_z F^1_z(T) F^2_z(F) Z - M_f$$

$$dP/dt = I_1 + M_f + Q_z + M_z + (1-A_z) G^{max}_z F^1_z(T) F^2_z(F) Z - G^{max}_f F^1_f(T) F^2_f(I) F^3_f(P) F - S_p$$

HYPOLIMNION:

$$dZ/dt = G^{max}_z F^1_z(T) F^3_z(B) A_z Z - Q_z - M_z$$

$$dB/dt = G^{max}_b F^1_b(T) F^2_b(P) B - G^{max}_z F^1_z(T) F^3_z(B) Z - Q_b - M_b$$

$$dP/dt = I_2 + Q_b + M_b + Q_z + M_z + (1-A_z) G^{max}_z F^1_z(T) F^3_z(B) Z - G^{max}_b F^1_b(T) F^2_b(P) B - S_p$$

$F^1_z(T) = \exp(-v_z (T^{opt}_z - T))^2$, $\quad F^1_f(T) = \exp(-v_f (T^{opt}_f - T))^2$,

$F^1_b(T) = \exp(-v_b (T^{opt}_b - T))^2$

$F^2_f(I) = (I/I_{opt}) \exp(1-I/I_{opt})$, $\quad F^3_f(P) = P/(K_f + P)$,

$F^2_b(P) = P/(gB + P)$, $\quad F^2_z(F) = 1 - \exp(-K_z F)$,

$F^3_z(B) = 1 - \exp(-K_z B)$

$Q_z = q_z F^1_z(T) Z$, $\quad Q_b = q_b F^1_b(T) B$,

$S_p = s_p P$, $M_p = m_f F$, $M_b = m_b B$, $M_z = m_z Z$

where: P – concentration of phosphorus, F – concentration of phytoplankton, B – concentration of bacteria, Z – concentration of zooplankton, T – temperature, I — light intensity.

Table VI. Driving functions of two-layer model.

temperature in epilimnion:

$T = a_{epi} - b_{epi} \cos(2\pi(t - c_{epi})/365)$

temperature in hypolimnion:

$T = a_{hyp} - b_{hyp} \cos(2\pi(t - c_{hyp})/365)$

light intensity:

$I = d - g \cos(2\pi(t - c)/365) e^{-nz}$

phosphorus loads:

I_1, I_2 = polynomials of t

where t is number of day in the year, z – depth, n – light extinction coefficient.

is utilised by the phytoplankton. Phytoplankton serves as the food for zooplankton. During the process of excretion, zooplankton release phosphorus (Q_z). As the model does not take directly into account the existence of detritus, it is assumed that the amount of the phosphorus which is quantitatively equivalent to the decaying cells of phytoplankton (M_f) and zooplankton (M_z) enriches the phosphorus pool. It is also assumed that part of organic matter not digested by zooplankton (A_z) goes to the pool of phosphorus. These two assumptions can be explained by the fast process of biochemical destruction of the organic matter. The term of the phosphorus sedimentation (S_p) is also introduced.

In the hypolimnion subsystem, phytoplankton is substituted by bacteria and the process of phosphorus excretion (Q_b) by bacteria is incorporated into the hypolimnion model.

Epilimnetic depth, epilimnetic and hypolimnetic temperature and intensity of solar radiation were approximated by the Fourier series or polynomials.

3.2. Phosphorus inputs

The evaluation of phosphorus inputs to both layers of the lake was based on the monthly phosphorus budgets produced by Ławacz (1985), who also assumed a two layered lake. External inputs to the epilimnion came from point sources (mainly streams), diffuse sources from the watershed and

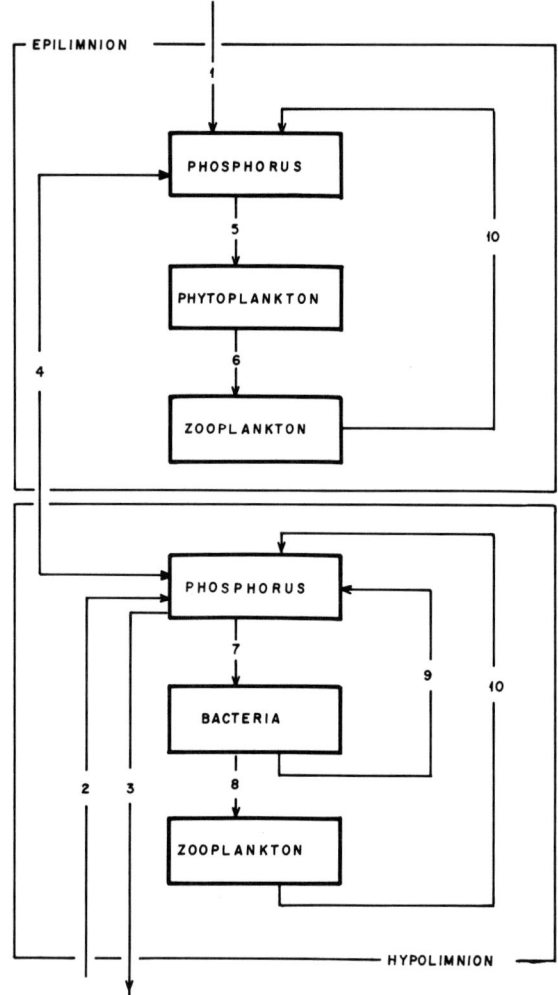

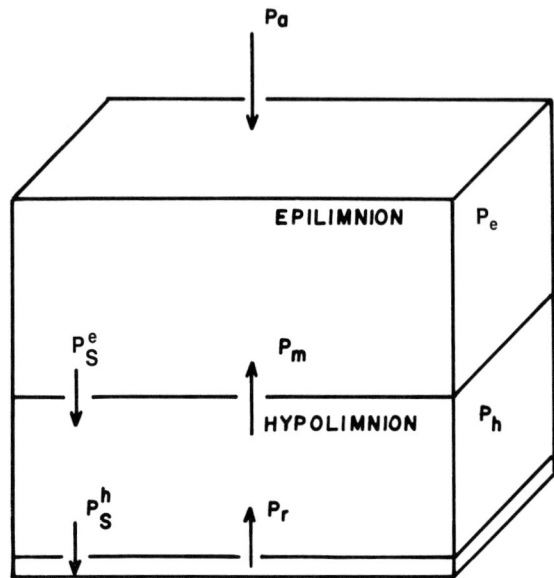

Fig. 5. An explanation of the calculation of the rate of phosphorus exchange between the hypolimnion and epilimnion. P_e, P_h, – amount of total phosphorus in the epilimnion and hypolimnion respectively, P_a – rate of external loading of the lake with phosphorus, P_r – rate of internal loading with phosphorus, P_s^e, P_s^h – rates of sedimentation of phosphorus from the epilimnion and hypolimnion respectively, P_m – rate of exchange of phosphorus between the epilimnion and hypolimnion. The calculations were based on monthly budgets of total phosphorus in both layers (Ławacz, 1985). All rates are per month. P_m and P_s^h were unknown. They were calculated from the following set of equations: $P_e(i+1) = P_e(i) + P_a(i) + P_m(i) - P_s^e(i)$ and $P_h(i+1) = P_h(i) + P_s^e(i) - P_m(i) - P_s^h(i)$, where $i+1$ and i denote numbers of subsequent months.

Fig. 4. A diagram of the two-layer model of phosphorus cycling in the epilimnion and hypolimnion. Symbols for the processes described in the model: 1 – loading of the lake with phosphorus from external sources, 2 – loading of the lake with phosphorus from internal sources, 3 – sedimentation of phosphorus from the hypolimnion, 4 – phosphorus exchange between the epilimnion and hypolimnion including sedimentation of phosphorus from the epilimnion, 5 – utilization of phosphorus by phytoplankton, 6 – grazing on phytoplankton by zooplankton, 7 – utilization of phosphorus by bacteria, 8 – grazing on bacteria by zooplankton, 9 – excretion of phosphorus by bacteria, 10 – excretion of phosphorus by zooplankton.

atmospheric sources. The phosphorus pool of the hypolimnion is enriched largely by internal loading. An important transfer of phosphorus is the exchange between these two layers. This consists of the nutrient coming from both the hypolimnion into the epilimnion and by sedimentation from the epilimnion.

The components of monthly phosphorus budgets are shown in Fig. 5, as well as the equations for evaluating the unknown amount of the phosphorus exchange between the layers and the value of the sedimenting phosphorus from the hypolimnion.

In order to obtain continuous in time functions of phosphorus input, approximations of the monthly values were carried out using Fourier series.

3.3. Results of simulations

The model was calibrated to measurements originating from Lake Jorzec, whose morphometric and hydraulic features are listed in Table I.

Table VII. Two-layer model: parameters of the epilimnion model.

Phytoplankton:
- G^{max}_f = *1.0* d^{-1}
- K_f = *20.0* $(\mu g\ P\ l^{-1})^{-1}$
- I_{opt} = *1464.4* $J\ cm^{-1}\ d^{-1}$
- v_f = *0.0025* $(°C)^{-2}$
- T^{opt}_f = *16.0* °C
- m_f = *0.03* d^{-1}

Zooplankton:
- G^{max}_z = *1.05* d^{-1}
- K_z = *0.027* $(\mu g\ P\ l^{-1})^{-1}$
- v_z = *0.0045* $(°C)^{-2}$
- T^{opt}_z = *18.0* °C
- A_z = *0.5*
- q_z = *0.02* d^{-1}
- m_z = 0.01 d^{-1} for $t < 150$
  0.04 d^{-1} for $150 \leq t \leq 250$
  0.01 d^{-1} for $250 < t$

Phosphorus:
- s_p = *0.35* d^{-1}

Table VIII. Two-layer model: parameters of the hypolimnion model.

bacteria:
- G^{max}_b = *2.1* d^{-1}
- g = *8.0*
- v_b = *0.004* $(°C)^{-2}$
- T^{opt}_b = *8.0* °C
- m_b = *0.1* d^{-1}
- q_b = *0.1* d^{-1}

zooplankton:
- G^{max}_z = *1.0* d^{-1}
- K_z = *0.05* $(\mu g\ P\ l^{-1})^{-1}$
- v_z = *0.009* $(°C)^{-2}$
- T^{opt}_z = *12.0* °C
- A_z = *0.5*
- q_z = *0.05* d^{-1}
- m_z = *0.07* d^{-1}

phosphorus:
- s_p = *0.25* d^{-1}

The set of parameters used in the simulations is shown in Tables VII and VIII. The results of simulations of all state variables for the epilimnion and hypolimnion are shown in Figs. 6 and 7.

Qualitatively, these simulations confirm the existence in the epilimnion of two phytoplankton blooms and two periods of enhanced zooplankton growth. Compared with events in the lake, it seems that the second simulated phytoplankton maximum appeared later in the model. The precise timing of the first bloom is difficult to evaluate because of a gap in measurements. The simulated appearance of the maximum concentrations of zooplankton in the epilimnion are also shifted in time compared with those observed in the lake.

The results of simulation of zooplankton in the

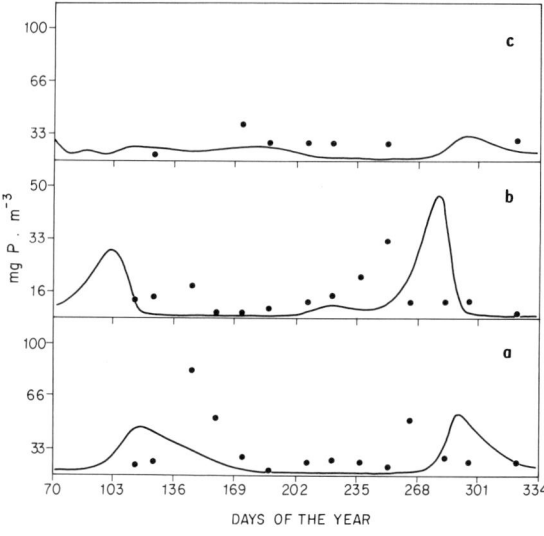

Fig. 6. Results of simulation of the two-layer model with parameters estimated for concentrations of phosphorus, phytoplankton and zooplankton in the epilimnion of the lake Jorzec. Solid lines – results of simulations, dots – measured values. a – concentration of zooplankton, b – concentration of phytoplankton, c – concentration of phosphorus.

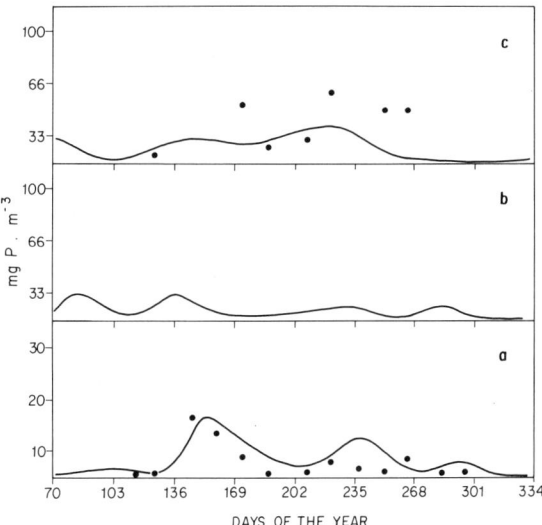

Fig. 7. Results of simulations of the two-layer model with parameters estimated for concentrations of phosphorus, bacteria and zooplankton in the hypolimnion of the Lake Jorzec. Solid lines – results of simulations, dots – measured values. a – concentration of zooplankton, b – concentration of bacteria, c – concentration of phosphorus.

hypolimnion shows much better agreement with measurements, than for the epilimnion. Both the timing of appearance and the values of maximum concentrations of zooplankton are in agreement with the field data. Concentration of phosphorus in the deeper water layers appear to be underestimated. Quantitative evaluation of the simulation results for bacteria is impossible, as there is lack of field data.

It is worth-while to compare the values of some parameters which appear in the two subsystems of the model and relations between them (see Table VII and VIII). The biggest maximal growth rate is given by bacteria. The values of maximal growth rates of zooplankton are similar in both water layers but mortality and sedimentation coefficient are bigger in hypelimnion than the epilimnion ones. It was assumed that from June to September the epilimentic zooplankton was characterized by a higher mortality than at other times of year due to increased fish predation. The larger value of the zooplankton mortality coefficient for the hypelimnion might be caused by the fact that some part of zooplankton in the lower layer happened to be the species which migrated downwards because of predatory pressure in the upper layers.

The results obtained from the hypolimnion submodel especially the phosphorus concentrations demonstrate that the more complicated equations for description of phosphorus or even the separate submodel of phosphorus exchange between sediments and near-bottom water is necessary.

4. Multi-layer model of phosphorus cycling

Stratification has been included into the ecological models of lakes or reservoirs usually by modelling separately the biochemical processes for layers corresponding to the epilimnion and hypolimnion (see Section 3) and by introducing a simple approximation of mass and heat exchange between them. More precise description of the vertical structure of the water body can be attained by subdivision of water basin into many layers parallel to the surface of the water and by using one dimensional partial differential equations to describe the changes of state variables at each point along the vertical axis.

This kind of one-dimensional model, has been published recently in the literature (e.g. Markowski & Harleman, 1973; Chen & Orlob, 1975; Imboden & Gächter, 1978; Kinnunen et al., 1982). They are capable of more adequate simulation of some of the phenomena in stratified lakes or reservoirs, especially in relation to vertical transport processes.

The construction of these models is based on the very popular one-dimensional thermal model describing the vertical temperature structure of lakes and reservoirs. For an excellent review of these models, the reader can consult Henderson-Sellers, 1984). The main assumption of both thermal and ecological models of this type is that differences in water quality can be observed only along the vertical axis of water body.

4.1. General description of the model

The model presented here is a deterministic one-dimensional simulation model of phosphorus cycling in lakes or reservoirs. It consists of two linked submodels: one, thermal and the other, biological. The thermal model describes vertical temperature profiles and the mixing depth. The biological model calculates the vertical concentration profiles for phosphate phosphorus, phytoplankton and zooplankton. The input of the biological model like transport of phosphorus depends on the output from the thermal model. Any influence of the output of the biological model on the thermal model is ignored.

4.2. The thermal submodel

The thermal model was based on the MIT model of Ryan & Harleman (1971) and on the wind mixing algorithm of Stefan & Ford (1975). Heat exchange at the surface of a lake due to evaporation, conduction and radiation was calculated according to Jurak's (1976) relationships developed for Polish conditions. After distribution of heat in the water body due to vertical advection, diffusion, internal absorption of solar radiation, inflow and outflow of heat with river or watercourses is considered at each time step, and intermediate temperature profiles are developed (Fig. 8). The final current profile is formed by mixing some upper layers due to wind and convective cooling.

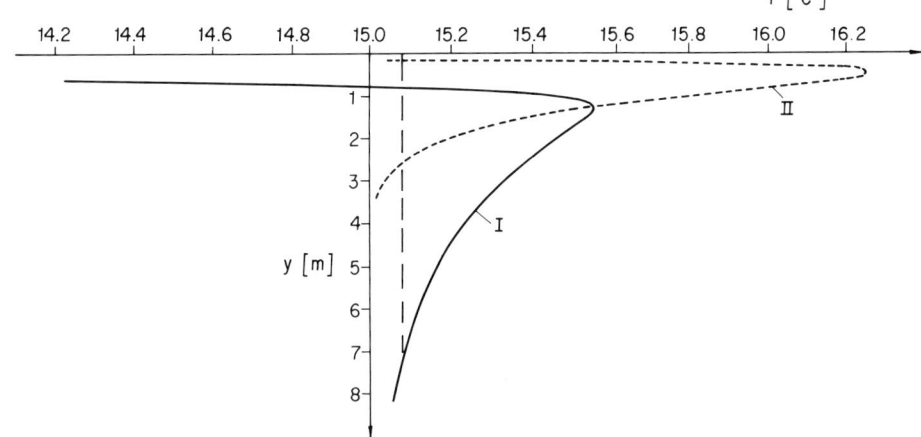

Fig. 8. An example of intermediate temperature profiles predicted by the model before applying mixing mechanisms.
I: $n = 1.7$ m^{-1}, II: $n = 0.34$ m^{-1}.

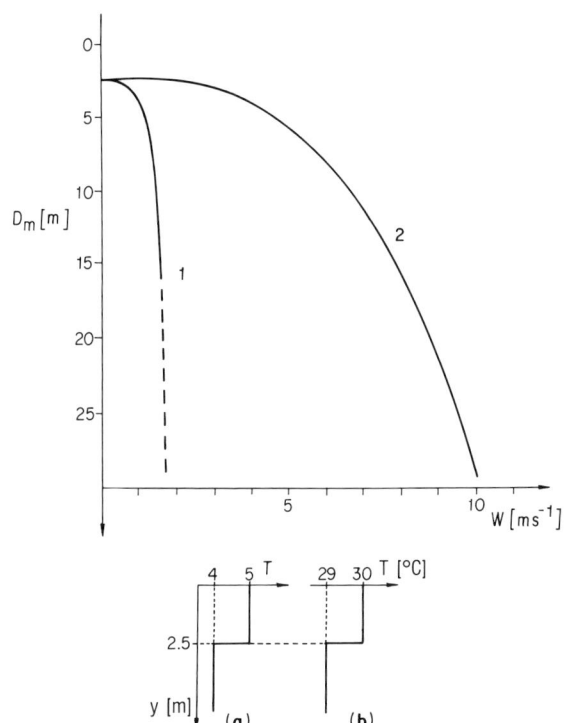

Fig. 9. Full mixing depth (D_m) versus wind speed (W) according to the wind mixing algorithm calculated for the two different simple temperature profiles and consequence to thermal-density gradients. Line 1 and 2 correspond to calculations with temperature-density profiles a and b, respectively.

These processes form fully mixed surface layers approximating to the epilimnion. The wind mixing algorithm is based on the balance of energy from wind generated water currents and energy necessary for mixing some upper layers. The latter energy depends on the mass differences and the distance between these layers. The layers can be mixed if the wind energy is high enough. An example of the relationship between the full mixing depth and wind speed according to the wind mixing algorithm is shown in the Fig. 9.

4.3. The biological model

The state variables of the biological model are concentrations of phosphate-phosphorus, phytoplankton and zooplankton, expressed in phosphorus units per unit volume. The phosphorus fluxes between these constituents are described according to the model AQUAMOD (Straškraba & Gnauck, 1983) (see Fig. 10). The first step involves calculation of phosphorus, phytoplankton and zooplankton concentrations at various depths from the equations of mass balance (Table IX). These consist of both the fluxes of these constituents between water layers and the sources due to biochemical reactions within each layer (Fig. 11). It is then assumed that this material is redistributed due to full mixing in the uppermost layers of the lake. The depth of full mixing has already been calculated in the thermal model.

Table IX. General equations of the multi-layer biological model.

$$\frac{\delta P}{\delta t} = \frac{E}{A} \frac{\delta}{\delta y} \left(A \left(\frac{\delta P}{\delta y} \right) \right) - \frac{1}{A} \frac{\delta}{\delta y} (AVP) + S_{P_h} + S_P$$

$$\frac{\delta F}{\delta t} = \frac{E}{A} \frac{\delta}{\delta y} \left(A \left(\frac{\delta F}{\delta y} \right) \right) - \frac{1}{A} \frac{\delta}{\delta y} (A(V+W)F) + S_{F_h} + S_F$$

$$\frac{\delta Z}{\delta t} = \frac{E}{A} \frac{\delta}{\delta y} \left(A \left(\frac{\delta Z}{\delta y} \right) \right) - \frac{1}{A} \frac{\delta}{\delta y} (AVZ) + S_{Z_h} + S_Z$$

where:

P, F and Z	– concentration of phospohorus, phytoplankton and zooplankton
t	– time
y	– vertical space coordinate
V	– speed of vertical advection
W	– speed of phytoplankton sedimentation
A	– horizontal cross sectional area
E	– eddy diffusion coefficient
S_P, S_F and S_Z	– rate of change of P, F and Z due to biological and chemical reactions
S_{P_h}, S_{F_h} and S_{Z_h}	– change of P, F and Z due to lateral inflows and outflows of phosphorus, phytoplankton and zooplankton with river or due to phosphorus release from bottom sediments

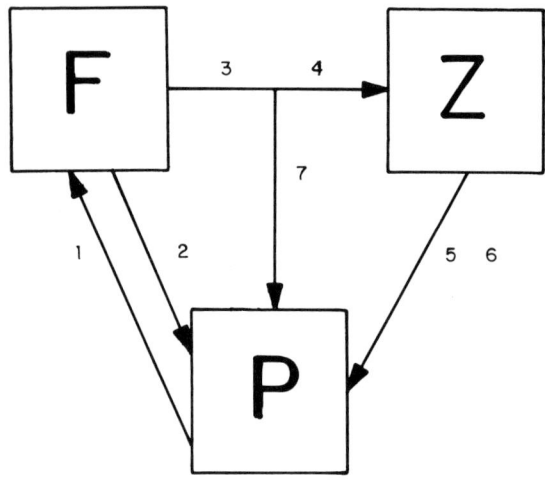

Fig. 10. A schematic representation of mass fluxes between phytoplankton (F), zooplankton (Z) and phosphorus (P), according to the model AQUAMOD (Straškraba & Gnauck, 1983). 1 – phosphorus assimilation by phytoplankton, 2 – phosphorus release by phytoplankton, 3 – consumption of phytoplankton by zooplankton, 4 – assimilation of zooplankton, 5 – excretion of zooplankton, 6 – mortality of zooplankton, 7 – phosphorus release from undigested residues of phytoplankton. See Table X for equations.

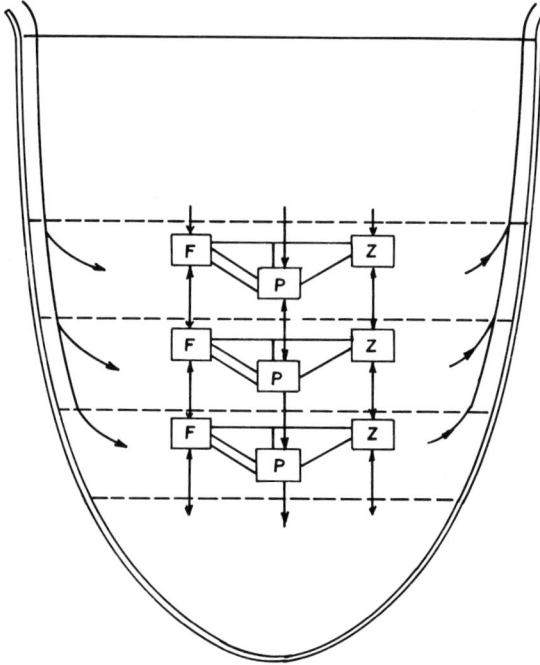

Fig. 11. A schematic representation of fluxes of the mass inside and between layers. Lateral inflow and outflow of the mass due to river flow through is also denoted.

These calculation steps finally result in developing of an instantaneous profile of the state variables.

In order to illustrate the main features of the multi-layer model some simulations were carried out with field data from the dimictic, eutrophic Lake Głębokie (see Table I). The equations of the model were solved with a computational step of the order of 0.1 – 1.0 day. Meteorological data for each day were used. The lake basin was divided into the layers of 1 m thickness.

4.4. Results of simulations

Simulations for temperature are shown in Fig. 12 and for surface concentrations of phosphorus (P), phytoplankton (F) and zooplankton (Z) in Fig. 13. The agreement between calculated and measured concentrations of phytoplankton is rather poor, probably because not all factors affecting phytoplankton have been incorporated into the model.

As can be seen in Fig. 13 and Fig. 14, the changes of P in the surface water of the lake probably coincide with changes of the full mixing depth as predicted by the thermal model and with incident solar radiation. Increasing the depth of full mixing results in the entrainment of P-rich layers but worsens the average light conditions. Therefore, P increases in the surface layer in such a situation. Decreasing the depth of full mixing may result in a decrease of P due to faster assimilation of phosphorus in better light conditions. Moreover, the minimum measured and simulated concentrations of P in the surface water correspond to maximum solar radiation and minimum depth of full mixed layer, respectively, as it is shown in Fig. 14. The reason for this is weak light limitation of assimilation of phosphorus by phytoplankton in such conditions in the surface layers of the lake. It may not be the rule because primary production in this model depends on the other agents: temperature, and phosphorus and phytoplankton concentrations.

4.5. Numerical experiments

During numerical experiments with the model it was assumed that temperature did not change in these calculations.

An increased phosphorus load into the epilimnion of the lake results mainly in increasing the concentration of P and Z in the surface water, whereas the mean phytoplankton level remained relatively unchanged but the amplitude of temporary changes of P, F and Z increased. The phytoplankton in the model is little sensitive to the changes of P.

A flushing of the epilimnion due to river trough flow assumed in the calculations result mainly in

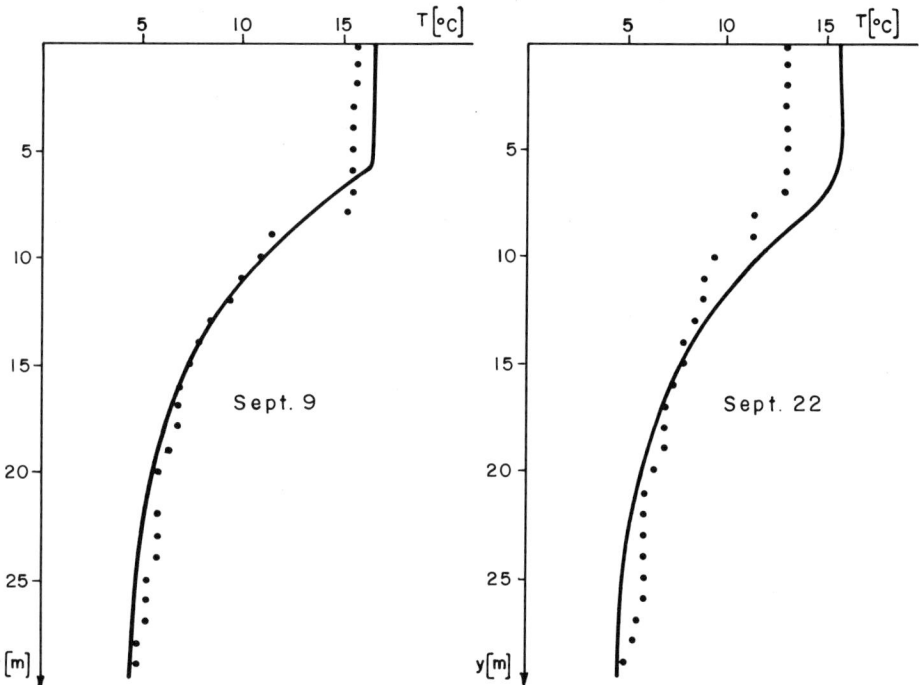

Fig. 12. Examples of a comparison of values of temperature observed in the Lake Głębokie (black dots) and predicted from thermal model (solid lines).

112

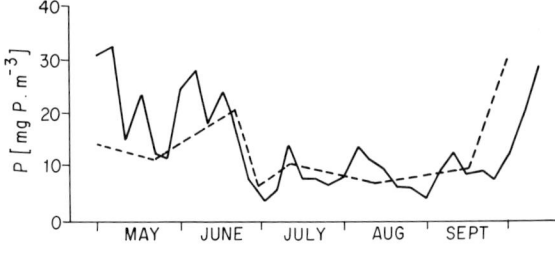
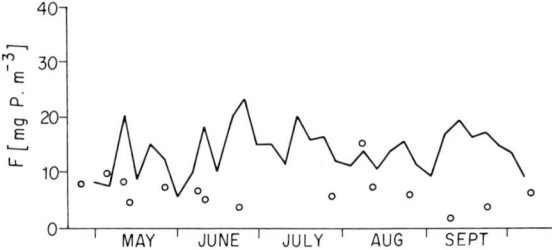
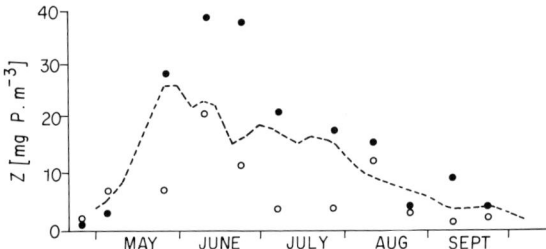

Fig. 13. A comparison of measured and predicted values of phosphorus (upper graph), phytoplankton (middle graph) and zooplankton (lower graph) concentrations for the surface water of the Lake Głębokie. Phosphorus: solid line – model simulations, broken line – measurements. Phytoplankton: solid line – model simulations, circles – measurements. Zooplankton: broken line – model simulations, black and white circles – measurements for non- predatory and predatory zooplankton, respectively.

decreasing of P and Z and decreasing of amplitude of changes of P, F and Z.

Increasing of speed of phytoplankton sedimentation results in lower value of P, F and Z in the surface water with zooplankton becoming almost extinct due to insufficient availability of food. Moreover, more distinct phytoplankton maxima may occur below the upper mixed layers.

If phytoplankton in the model is more adapted to low light intensity by decreasing the parameter of I_k (see Table X), or if light attenuation in the water is decreased then P, F and Z also decrease in the surface water (Fig. 15). It seems to be a paradox because P should be higher at lower F, as P loading

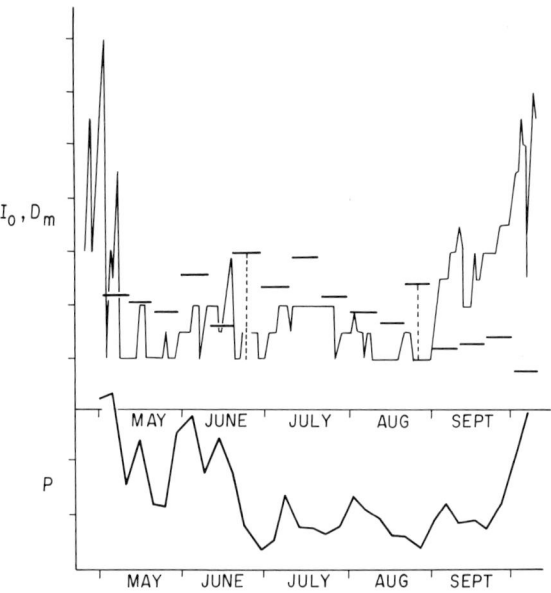

Fig. 14. The time course of the predicted concentrations of phosphorus P (lower graph solid line) (values are the same as in the Fig. 13) coincide mainly with changes of full mixing depth D_m (upper graph – thin line) and with incident solar radiation I_0 (upper graph – solid horizontal sections). Vertical broken lines in the upper graph indicate the period of time when high radiation and small depth of full mixing occur simultaneously.

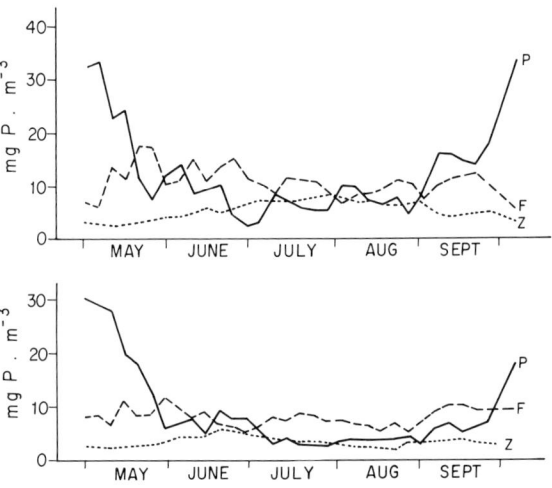

Fig. 15. An effect of 'introducing' phytoplankton adapted to lower solar radiation into the lake. P – phosphorus concentration, F – phytoplankton concentration and Z – zooplankton concentration in the surface layer of the lake. Upper graph-simulation of phytoplankton adapted to higher solar radiation than lower graph.

Table X. Equations describing biochemical processes in the multi-layer model.

$$S_F = \overset{(1)}{G} - \overset{(2)}{r_F T F} - \overset{(3)}{g F Z}$$

$$S_z = \overset{(4)}{g F Z e} \frac{k_F}{k_F + F} - \overset{(5)}{(m + r_Z) Z}\overset{(6)}{}$$

$$S_p = -\overset{(1)}{G} + \overset{(2)}{r_F T F} = \overset{(7)}{g F Z} (1 - \frac{e k_F}{k_F + F})\overset{(5)}{} + \overset{(6)}{(m + r_Z) Z}$$

where:

$G = G_{max}(T, F) \; f(I) \; f(P) \; \text{fotop} \; F$
$G_{max}(T, F) = \alpha \exp 0.09 \, T F^{-0.52}$

$$f(I) = \frac{2(I/2I_K)}{1 + (I/2I_K)^2}, \quad I = \frac{I_o \exp(-n_Z)}{\text{fotop}}$$

$$f(P) = \frac{P}{k_p + P}, \quad \text{fotop} = 12 - 4\cos(2\pi t/360)$$

1, 2, 3, 4, 5, 6, 7 – flux rates of mass between phosphorus, phytoplankton and zooplankton pools due to processes defined in Fig. 10
T – temperature of the water at the depth y
I – mean solar radiation at the depth y
I_o – daily solar radiation reaching a surface
fotop – photoperiod
For other symbols see Table XI.

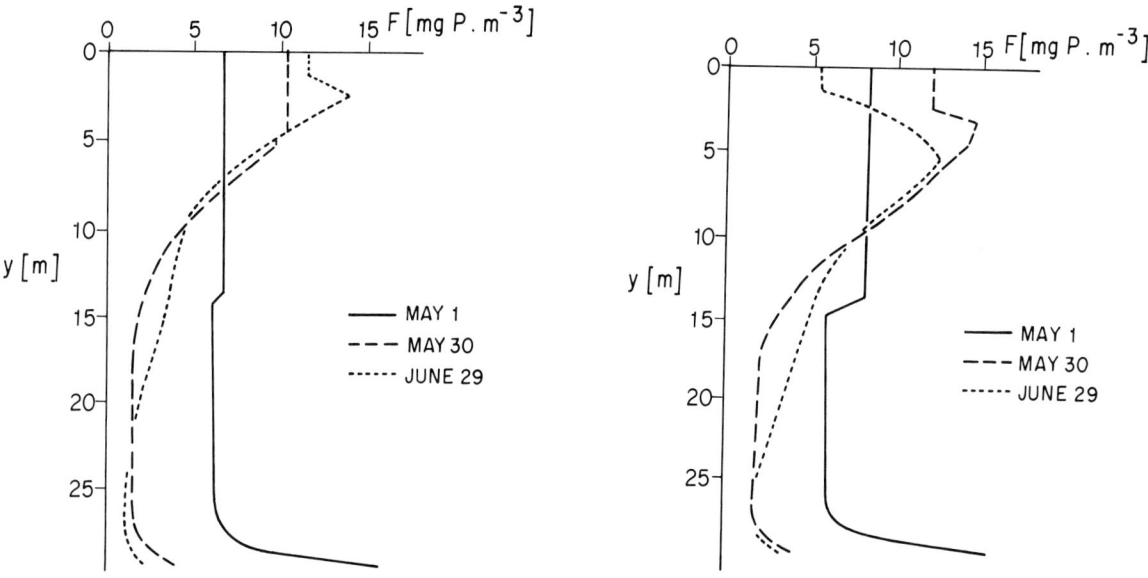

Fig. 16. Vertical profiles of phytoplankton concentrations calculated for phytoplankton adapted to relatively higher (left graph) and lower (right graph) solar radiation. One can see more distinct phytoplankton maxima in the right graph below the mixed layers.

Table XI. The parameters of the biological multi-layer model.

Symbol	Value	Unit	Meaning	Source
α	0.5*	h^{-1}	rate of primary production	AQUAMOD
r_F	0.005	$°C^{-1} d^{-1}$	excretion coef. of phytopl.	AQUAMOD
r_Z	0.01	d^{-1}	respiration coef. of zooplankton	AQUAMOD
e	0.6		assimilation coef. of zoopl.	AQUAMOD
g	0.0165	$m^3 (mg P)^{-1} d^{-1}$	grazing coef. of zoopl.	Assumption
m	0.07*	d^{-1}	mortality of zooplankton	Assumption
k_F	60	$mg P m^3$	saturat. constant of phytopl.	AQUAMOD
k_Z	10	$mg P m^3$	saturat. constant of zoopl.	AQUAMOD
I_k	2.4	$g cm^{-2} h^{-1}$	param. of photosynth. model	Assumption
W	0.3*	$m d^{-1}$	sediment. of phytoplankton	Assumption
E	0.5	$m^2 d^{-1}$	diffusion coefficient	Assumption

* These values were used for the numerical experiments only. For a comparison of simulated with observed values of the state variables it was assumed that: m increased in this calculations from the value 0.01 [d^{-1}] at 15th April to the value 0.12 [d^{-1}] at 15th May and then it remained constant, $W = 0.2 + 0.03\ T$ [$m\ d^{-1}$], $\alpha = 0.1$ [h^{-1}].

into the lake was the same as before. An explanation of the point can be drawn from Fig. 16. A peak of phytoplankton can be seen in the metalimnion due to optimal conditions for primary production in this layer and due to photoinhibition in the epilimnion. The higher phytoplankton concentrations near the lake bottom result from the model assumption that sedimenting phytoplankton is not trapped by the bottom sediments. Higher phosphorus assimilation result in lower P in the metalimnion and in lower diffusional phosphorus fluxes from the metalimnion to the epilimnion which are proportional to the difference in phosphorus concentration between these layers. The high phytoplankton concentrations support high zooplankton concentrations in the metalimnion.

If bottom sediments do not release phosphorus, then the P-levels predicted by the model are much lower in the hypolimnion. The phosphorus gradient within the metalimnion remains almost the same. This is because phosphorus regeneration by phytoplankton and zooplankton takes place in this region.

4.6. Final remarks

The description of biological processes in the multilayer model has been simplified here, and many hypothetical assumptions have been made which could cause deviations of simulated from measured values. For example, primary production, food consumption, sedimentation rate, regeneration of phosphorus etc. are not directly dependent on any particular location along the vertical axis of the lake. It is obviously not true as these processes may by influenced by local environmental conditions. For instance, both selfshading effects and dampening down of turbulence within the metalimnion are observed. Different organisms possess ecological and physiological adaptations or planktonic communities from different depths have different species composition. Similarly, the parameters of phytoplankton for the layers under the epilimnion ought to change because phytoplankton in the model plays a role of detritus in this part of the lake. Moreover, in the case of the Lake Głębokie the primary production is probably limited by both phosphorus and nitrogen (Woroniecka-de Wachter, 1983) and not by phosphorus only. In addition, bacteria play an important role in the cycling of these elements in the lake (see Section 3).

Nevertheless, the model might simulate the direction of changes of such a simplified biotic environment under influence of the changes in the temperature structure of a lake and under processes of vertical transport of the mass. By means of such a model it is possible to investigate the phenomenon of inhibition of mass flux by the metalimnion and other processes connected with the vertical structure of a lake, such as the influence of the intensity of vertical turbulence on the level of biological production and on the loss processes from the epilimnion (Szeligiewicz, 1989).

5. Model of exchange of phosphorus between sediments and near-bottom water

5.1. The model

The block diagram in Fig. 17 illustrates the model of phosphorus exchange between sediments and near-bottom water. A detailed description of the model is presented in Mitraszewski & Uchmański (1988). The following variables are described by the model: phosphorus in near-bottom water, phosphorus in interstitial water, exchangeable phosphorus in the active layer of sediments and non-exchangeable phosphorus in the active layer of sediments.

The values of these variables are results of balance between rates of following processes: sedimentation of phosphorus, transfers of sedimenting phosphorus into exchangeable phosphorus, transfer of exchangeable phosphorus into interstitial phosphorus, diffusion of phosphorus from interstitial water into near-bottom water and sorption and desorption of phosphorus in sediments. The influence of organisms living in sediments is also included into the model by means of rates of phosphorus exchange between interstitial water and near-bottom water due to the activity of bottom fauna. The model equations are presented in Table XII.

Concentration of dissolved oxygen in the bottom parts of the lake is one of the main factors influencing exchange of phosphorus between sediments and near-bottom water. In eutrophic lakes release of phosphorus from sediments predominate under anaerobic conditions. The opposite direction of exchange is observed under aerobic conditions.

According to Kamp-Nielsen (1974) under anaerobic conditions phosphorus diffuses from interstitial into near-bottom water at a rate proportional to the difference in concentration of phosphorus between these compartments. The diffusion coefficient depends on temperature. In aerobic conditions, the process of phosphorus sorption predominates. The rate of this process depends on temperature and is described by the logarithmic function of phosphorus concentration in interstitial water. The rate of change of exchangeable phosphorus into interstitial one is proportional to the amount of exchangeable phosphorus with a proportional constant dependent on temperature (Jørgensen et al., 1975).

The phosphorus exchange between interstitial and near-bottom water due to activity of bottom fauna is connected with their consumption of oxygen: the lower the concentration of oxygen, the more intensive the ventilation movements of organism's body, 'pumping' water rich in oxygen. As a byproduct the phosphorus is transported from interstitial to near-bottom water. The activity of two groups of organisms are discussed in the model: chironomids and tubificids.

The following assumptions were made concerning the rate of phosphorus exchange due to benthic animals: it is proportional to the biomass of living organisms; in chironomids, it is proportional to the concentration of phosphorus in interstitial water; and, in tubificids, it is proportional to the difference between concentration of phosphorus in interstitial and near-bottom water; it is related by a Michaelis-Menten relationship to the oxygen concentration with the half-saturation constant equal to respiration rate of an individual.

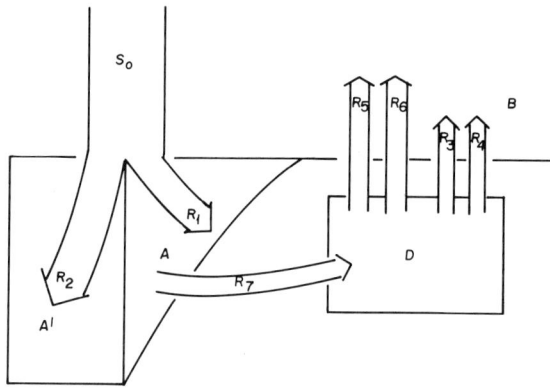

Fig. 17. A diagram of the model of phosphorus exchange between sediments and near-bottom water. Horizontal line indicates the border between sediments and near-bottom water. A – exchangeable phosphorus in the active layer of sediments, A' – non-exchangeable phosphorus in the active layer of sediments, D – phosphorus in the interstitial water, B – phosphorus in the near-bottom water, S_0 – sedimenting phosphorus, R_1 – transfer of sedimenting phosphorus into exchangeable phosphorus, R_2 – transfer of sedimenting phosphorus into non-exchangeable phosphorus, R_3 – release of phosphorus due to activity of Chironomidae, R_4 – release of phosphorus due to activity of Tubificidae, R_5 – sorption or desorption of phosphorus in the sediments, R_6 – diffusion of phosphorus into near-bottom water, R_7 – transfer of exchangeable phosphorus into interstitial phosphorus.

Table XII. Equations of the model describing exchange of phosphorus between sediments and near-bottom water.

anaerobic conditions:
$dA/dt = R_1 - R_7,$ $dD/dt = R_7 - R_5 - R_6 - R_3 - R_4,$
$dB/dt = R_5 + R_6 + R_3 + R_4$

aerobic conditions:
$dA/dt = R_1 - R_7,$ $dD/dt = R_7 - R_5 - R_3 - R_4,$
$dB/dt = R_5 + R_3 + R_4$

$R_1 = a S_0$
$R_3 = \mu_1 T N_1 P_1 (r_1 w_1^{r_2} / (r_1 w_1^{r_2} + O_2))$
$R_4 = \mu_2 T N_2 (P_1 - P_2)(r_1 w_2^{r_2} / (r_1 w_2^{r_2} + O_2))$
$R_5 = D'_0 e^{\beta/T} (d_1 \ln P_2 + d_2),$ $R_6 = D_0 e^{\beta/T} (P_1 - P_2),$
$R_7 = k A Q^{T-20}$

A – amount of exchangeable phosphorus in active layer of sediments in experimental core, D – amount of phosphorus in interstitial water in experimental core, B – amount of phosphorus in near-bottom water in experimental core, P_1 – concentration of phosphorus in interstitial water, P_2 – concentration of phosphorus in near-bottom water; rates: S_0 – sedimentation of phosphorus, R_1 – changing of sedimenting phosphorus into exchangeable phosphorus, R_3 – release of phosphorus into near-bottom water due to activity of Chironomidae, R_4 – release of phosphorus into near-bottom water due to activity of Tubificidae, R_5 – sorption or desorption of phosphorus in sediments, R_6 – diffusion of phosphorus into near-bottom water, R_7 – changing of exchangeable phosphorus into interstitial phosphorus; T – temperature, w_1, w_2 – individual body weights of Chironomidae and Tubificidae, N_1, N_2 – number of individuals of Chironomidae and Tubificidae in the experimental core, O_2 – oxygen concentration, a, d_1, d_2, D'_0, k, α, β, μ_1, μ_2, r_1, r_2, Q – constants.

A constant rate of phosphorus sedimentation is assumed in the model.

5.2. Results of simulation

The model was calibrated for data collected during laboratory experiments with cores of bottom sediments taken from Lake Głębokie and Inulec (see Table I) (Planter & Wiśniewski, 1985; Wiśniewski & Planter, 1987). The concentrations of phosphorus in near-bottom and interstitial water after six days of incubation of cores was measured.

The concentration of phosphorus in near-bottom water as resulting from the simulation was compared with the experimental values. An example of the results from some simulations is presented in Fig. 18. The relative differences between simulated and empirical values are not higher then 30%, most often approximately 10%. The highest discrepancy was observed when exchange of phosphorus between sediments and near-bottom water was simulated for unrealistic conditions.

5.3. Sensitivity analysis

It is possible to estimate the influence of bottom fauna on exchange of phosphorus between sediments and near-bottom water by means of sensitivity analysis. This involves solving the model equations by substituting various values of chironomids and tubificids biomass. Such an

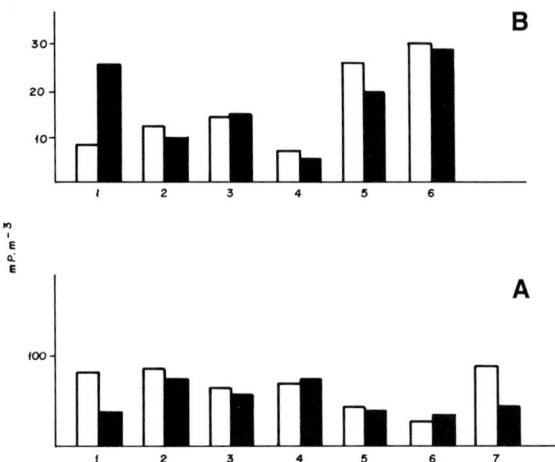

Fig. 18. Results of model simulation (solid bars) as compared with results of experiments (open bars) for the model of phosphorus exchange between sediments and near-bottom water. Parameters were estimated for data from the Lake Inulec during autumnal conditions. On the vertical axis is concentration of phosphorus in the near-bottom water, on horizontal axis the number of experimental cores. A – anaerobic conditions, B – aerobic conditions.

exercise shows that the phosphorus concentration in near-bottom water was 30% less with zero chironomid biomass than the concentration of P with realistic levels of chironomid biomass. A similar numerical experiment conducted with tubificid biomass gave a 10% reduction in the phosphorus concentration in the near-bottom water.

In all simulations, a greater sensitivity of the model was observed for anaerobic than aerobic conditions. The sensitivity was also greater for the model which had been calibrated for a lake with small variation of temperature and oxygen conditions near bottom.

6. Parameter estimation in modelling of lake and reservoir ecosystems

In the case when values of the model parameters cannot be obtained from field measurements or laboratory experiments, the only possibility left is to estimate their value. Two kinds of procedures are usually applied when parameters have to be estimated.

The first one is called the 'trial and error' method. Using this method, one tries to isolate a small group of parameters, usually those which play the key role in the model, and to obtain in several simulation runs, the proper values of these parameters, assuming that the others are known.

In the case of so-called formal procedures, it is necessary to define an objective function which permits a comparison to be made between values of simulated state variables and of a real system represented by the measurements. The least square objective function is the most commonly used. The task of the parameter estimation in this case is to search for the values of the parameters which minimize the value of the objective function. Application of these formal methods from the computational point of view is rather hopeless.

Therefore, these traditional methods do not offer an efficient tool for estimating parameters of ecological models. A new method was developed by Loga (1988). It allows to increase the set of the simultaneously estimated parameters.

It was proposed to substitute solutions of some of the model equations by the Fourier series or polynomial approximations of the field data and to estimate parameters of other equations. The values of the parameters obtained with some of the state equations excluded should be equally good after merging these equations.

This estimating procedure can be supported by the results of sensitivity analysis. Let us define the sensitivity coefficient as the ratio of nominal value of the state variable and state variable value for the parameter changed by 1%. During estimation, the sensitivity coefficient was calculated and appropriate curves were plotted for the parameters which appeared in the recently merged equation. Subsequent simulations were performed with parameter values modified according to sensitivity analysis results. Such procedure enables us to determine also parameters which show the greatest influence on the simulation results.

References

Bajkiewicz-Grabowska, E., 1985. Factors affecting nutrient budget in lakes of the r. Jorka watershed (Masurian Lakeland, Poland) I. Geographical description, hydrographic components and man's impact. Ekol. pol. 33: 173–200.

Chen, C. W. & G. T. Orlob, 1975. Ecologic simulation for aquatic environments. In: B. C. Patten (ed.), System Analysis and Simulation in Ecology. Vol. 3, Academic Press, New York: 475–588.

Ejsmont-Karabin, J., L. Bownik-Dylińska, & W. Godlewska-Lipowa, 1983. Biotic structure and processes in the lake system of r. Jorka (Masurian Lakeland, Poland) VII. Phosphorus and nitrogen regeneration by zooplankton as the mechanism of the nutrient supply for bacteria and phytoplankton. Ekol. pol. 31: 719–746.

Godlewska-Lipowa, W., 1983. Biotic structure and processes in the lake system of r. Jorka (Masurian Lakeland, Poland) V. Biomass and production of bacterioplankton. Ekol. pol. 31: 667–678.

Henderson-Sellers, B., 1984. Engineering Limnology. Pitman, Boston, London, Melbourne, 335 pp.

Hillbricht-Ilkowska, A., 1983. Biotic structure and processes in the lake system of r. Jorka (Masurian Lakeland, Poland) XII. Productivity, structure and dynamics of lake biota (a synthesis of research). Ekol. pol. 31: 801–834.

Hillbricht-Ilkowska, A. & W. Ławacz, 1983. Biotic structure and processes in the lake system of r. Jorka (Masurian Lakeland, Poland) I. Land impact, loading and dynamics of nutrients. Ekol. pol. 31: 539–585.

Imboden, D. M. & R. Gächter, 1978. A dynamic lake model for trophic state prediction. Ecol. Modelling 4: 77–98.

Jørgensen, S. E., L. Kamp-Nielsen & O. S. Jacobsen, 1975. A submodel for anaerobic mud-water exchange of phosphate. Ecol. Modelling 1: 133–146.

Jurak, D., 1976. Heat exchange coefficients for cooling lakes. J. Hydrol. Sci. 3: 123–133.

Kamp-Nielsen, L., 1974. Mud-water exchange of phosphate and other ions in undisturbed sediments cores and factors affecting the exchange rates. Arch. Hydrobiol. 73: 218–237.

Kinnunen, K., B. Nyholm, J. Niemi, T. Frisk, T. Kyla-Harakka & T. Kauranne, 1982. Water quality modeling in finnish water bodies. Publ. Water Res. Inst. 46, Vesihallitus National Board of Waters, Finland, Helsinki, 99 pp.

Loga, M., 1988. Methods for improving the efficiency of calibration of a model of phosphorus cycling in a lake ecosystem. Ekol. pol. 36: 387–406.

Ławacz, W., 1985. Factor affecting nutrient budget in lakes of r. Jorka watershed (Masurian Lakeland, Poland) XI. Nutrient budget with special consideration to phosphorus retention. Ekol. pol. 33: 357–381.

Markowski, M. & D. R. F. Harleman, 1973. Prediction of water quality in stratified reservoirs. J. Hydraulics Div., ASCE 99, HY 5: 729–745.

Mitraszewski, P. & J. Uchmański, 1988. A numerical model of phosphorus exchange between sediments and the near-bottom water in a lake. Ekol. pol. 36: 317–346.

Penczak, T., M. Molińska, W. Galicka & A. Prejs, 1985. Biotic structure and processes in the lake system of r. Jorka (Masurian Lakeland, Poland) VII. Input and removal of nutrients with fish. Ekol. pol. 33: 301–309.

Planter, M., W. Ławacz & A. Tatur, 1983. Biotic structure and processes in the lake system of r. Jorka (Masurian Lakeland, Poland) II. Physical and chemical properties of water and sediments. Ekol. pol. 31: 587–611.

Planter, M. & R. J. Wiśniewski, 1985. Factor affecting nutrient budget in lakes of the r. Jorka watershed (Masurian Lakeland, Poland) IX. The exchange of phosphorus between sediments and water. Ekol. pol. 33: 329–344.

Ryan, P. J. & D. R. F. Harleman, 1971. Temperature Prediction in Stratified Water: Mathematical Model User's Manual. Environmental Protection Agency, Water Pollution Control Research Series, 16130 DJH 01/71, 125 pp.

Spodniewska, I., 1983. Biotic structure and processes in the lake system of r. Jorka watershed (Masurian Lakeland, Poland) IV. Structure and biomass of phytoplankton. Ekol. pol. 31: 635–665.

Stefan, H. & D. E. Ford, 1975. Temperature dynamics in dimictic lakes. J. Hydraulics Div. ASCE, 101, HY 1: 97–114.

Straškraba, M. & A. Gnauck, 1983. Aquatische Ökosysteme, Modelierung und Simulation. VEB Gustav Fisher Verlag, Jena, 279 pp.

Szeligiewicz, W., 1989. Modeling of organic particle flux through the metalimnion in lakes. Arch. Hydrobiol. Beih. Ergebn. Limnol. 33: 169–177.

Uchmański, J., 1989. Numerical experiments with a simulation model of phosphorus cycling in the epilimnion of eutrophic Lake Głębokie. Arch. Hydrobiol. Beih. Ergebn. Limnol. 33: 147–156.

Uchmański, J. & W. Szeligiewicz, 1988. Empirical models for predicting water quality, as applied to data on lakes of Poland. Ekol. pol. 36: 285–316.

Węgleńska, T., L. Bownik-Dylińska & J. Ejsmont-Karabin, 1983. Biotic structure and processes in the lake system of r. Jorka watershed (Masurian Lakeland, Poland) VI. Structure and dynamics of zooplankton. Ekol. pol. 31: 679–717.

Węgleńska, T., L. Bownik-Dylińska, J. Ejsmont-Karabin & I. Spodniewska, 1987. Plankton structure and dynamics, phosphorus and nitrogen regeneration by zooplankton in Lake Głębokie polluted by aquaculture. Ekol. pol. 35: 173–208.

Wiśniewski, R. J. & M. Planter, 1987. Phosphate exchange between sediments and the near-bottom water in relationship to oxygen conditions in a lake used for intensive trout cage culture. Ekol. pol. 35: 219–236.

Woroniecka-de Wachter, U., 1983. Biotic structure and processes in the lake system of r. Jorka watershed (Masurian Lakeland, Poland) III. Production and photosynthesis efficiencies of phytoplankton. Ekol. pol. 31: 613–633.

Chapter VII

Sedimentation and mineralization of seston in a eutrophic reservoir, with a tentative sedimentation model

J.A. Galvez[1,2] & F. X. Niell[1]
[1] *Departamento de Ecología, Facultad de Ciencias, Universidad de Málaga, Campus de Teatinos, 29071-Málaga, Spain;* [2] *Present address: Area de Ecología, Departamento de Biología, Facultad de Ciencias del Mar, Iniversidad de Cádiz, Campus Río S. Pedro, 11510-Puerto Real (Cádiz), Spain*

Key words: reservoirs, limnology, sedimentation, seston, mathematical models

Abstract

Seston vertical flux was estimated by using sediment traps in a eutrophic stratified reservoir in southern Spain (Marbella, Málaga). At the same time, phytoplankton flux was estimated in the laboratory. Phytoplankton epilimnetic flux represents a high percentage of seston vertical flux. However, in hypolimnion the percentage of phytoplankton decreases, and the role of detritus becomes more important. During sedimentation, mineralization of organic matter is carrying on, and nutrient regeneration in the euphotic zone enhances primary production. Regeneration at the thermocline level is related to the presence of a deep chlorophyll maximum. A functional relationship between primary production and sedimentation produces evident regularities when different freshwater environments are compared.

1. Introduction

In aquatic environments, circulation processes aside, the density of suspended particles determines their sinking. However, most of the planktonic organisms without swimming capacity have a slightly larger density than the water, which accounts for their sinking (Hutchinson, 1967; Smayda, 1970; Margalef, 1974). This process is very important for phytoplankton cells, because they can be taken out of the trophogenic zone and transported to tropholitic depths where the cells can take advantage of high nutrient concentration (Sournia, 1981). Sedimentation is a general process which affects not only live organisms and their capacity for maintenance in the water but it also concerns all suspended particles: organic and inorganic, detrital or not. Therefore, sedimentation is related to the global processes in the aquatic ecosystem:
– As a transference of energy, as matter, from the surface layers to deeper zones. Primary production and allochthonous inputs balance this matter flow.
– As regeneration of nutrients *in situ* during sedimentation, allowing new and successive cycles of production.
– As, in specific conditions, a nutrient source (especially soluble reactive phosphorus) from the settled matter on the bottom, which can contribute in an important way to the enhancement of eutrophication.
– As factor of decrease in the storage capacity of the basin. Settling matter which reaches the bottom, the amount of which is a result of the previous processes, can reduce it. Usually, this phenomenon is not important, but sometimes, especially in water storage reservoirs with high erosion rates of their sides can be a big problem.

The aim of this work is to assess the seston vertical flux at different levels of the water column of a stratified water reservoir and quantify the importance of regeneration and mineralization during the sedimentation process. At the same time this work attempts to assess the contribution of phytoplankton in the seston vertical flux and the relationship between primary production and sedimentation from the results obtained through other aquatic environments.

2. Methods

La Concepción is a eutrophic reservoir situated in the south of Spain (Marbella, Málaga), with an area of 2.41 km², a total capacity of 61 · 10⁶ m³ and a maximum depth of 68 m.

From 1985 to 1986, during stratification, sampling was carried out at a station close to the dam (Fig. 1).

Seston settling flux was estimated by using duplicate sediment traps, placed at 2, 6, 10, 15, 25 m and near the bottom. Sediment traps were cylindrical (Bloesch & Burns, 1980; Blomqvist & Håkanson, 1981) with an aspect ratio (height : diameter) of 3 (Gardner, 1980; Blomqvist & Kofoed, 1981), which is sufficient because of the low water turbulence during the stratified period.

Seston flux has been calculated from dry weight of the matter collected in the traps. Chlorophyll flux has been estimated from the acetonic extractions of an aliquot of the sample collected in traps, according to Talling & Driver (1963) calculations.

Close to the traps situated in the photic zone, several slides were placed to estimate organic matter due to fouling and, in this way, to correct the values of seston trapped at each depth.

Other variables measured in water were temperature and oxygen, with a YSI Mod. 57 monitor, chlorophyll-a (Talling & Driver, 1963) and phytoplankton biovolume by using an Image Semiautomatic Analyzer IBAS-1.

2.1. Laboratory estimations of phytoplankton vertical flux

The experiment was carried out with the water collected at the end of April, 1986, at 0, 2, 6, 10, 15, 25 and 40 m. Light and temperature conditions measured *in situ* were reproduced in the laboratory.

Sedimentation of cells was estimated according to Burns & Rosa (1980) and Rathke *et al.* (1981), in cylinders of 20 cm high by 2.5 cm diameter, and a transferable chamber as bottom. At 2, 4 and 96 hours after the beginning of the experiment, each water column was carefully transferred to a new chamber and cell numbers in chambers were counted by using the Utermöhl (1958) method. The process of sedimentation was undisturbed by these manipulations as evidenced in Fig. 2 for curves of two species: *Fragilaria crotonensis* and *Tetraedron minimum*. The initial concentration of cells was obtained by the addition of densities estimated from the different counts.

Net transport (Nt) at time t is calculated by Eq. (1):

$$Nt = Vb \, (Ct - Co) \qquad (1)$$

where:
Vb = volume of transferable chamber
Ct = density at time t
Co = initial density

Phytoplankton flux (F) could be calculated by:

$$F = Nt \, / \, a \, t \qquad (2)$$

where:
a = cross sectional area of the column

Phytoplankton mean sinking velocity (V) is given by

$$V = F \, / \, Co \quad \text{(Hargrave \& Burns, 1979)} \qquad (3)$$

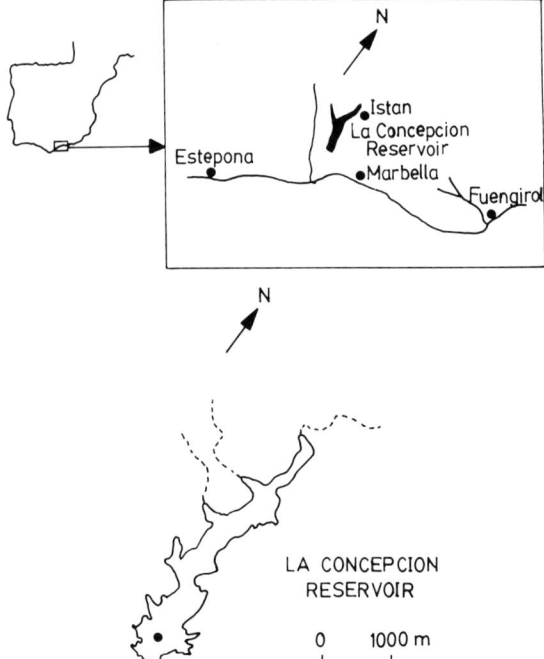

Fig. 1. Location of sampling zone.

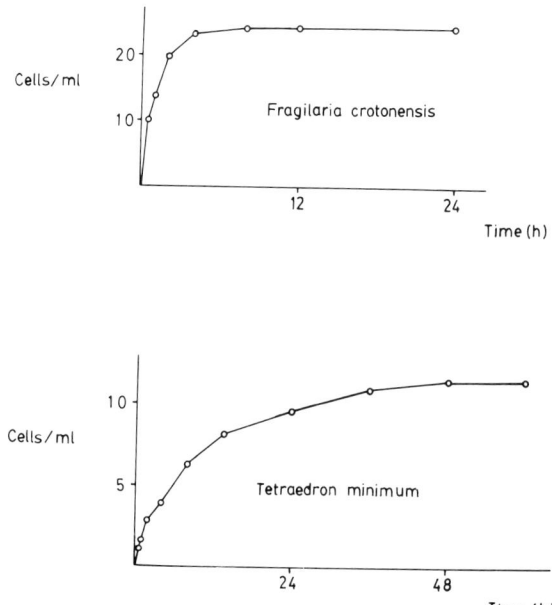

Fig. 2. Sinking curves of *Fragilaria crotonensis* and *Tetraedron minimum* in settling chambers.

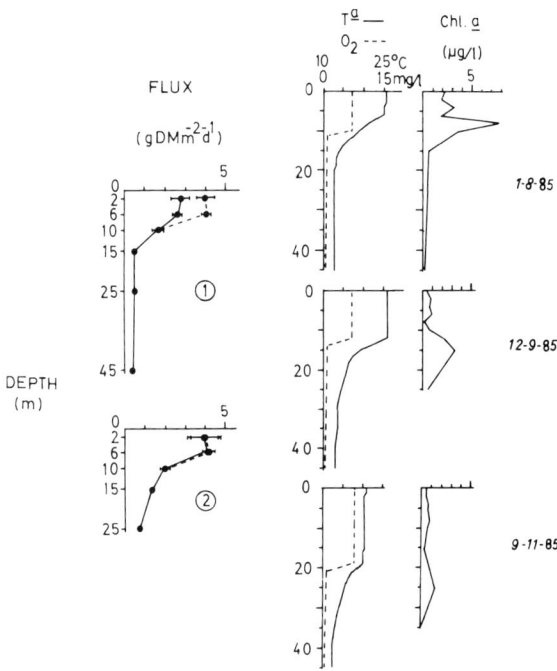

Fig. 3. Seston dry matter (DM) flux in 1985. Dotted lines show seston flux without correcting attached growth. Temperature, oxygen and chlorophyll-a are registered in sampling dates.

The used values are those estimated at two hours of sedimentation (linear phase) given the asymptotic process in the enclosed columns.

3. Results

3.1. Seston settling flux

During stratification of 1985, dry matter (DM) settling flux ranged between 3 and 4 g m^{-2} d^{-1} in the photic zone and decreased to 0.4–0.6 g m^{-2} d^{-1} in the hypolimnion (Fig. 3). Thus, about 15% of seston reached the bottom. During this period, chlorophyll-a concentration was low (0–8 µg l^{-1}). Maximal decrease in flux was detected between 6 and 15 m, at the thermocline level. Development of attached matter is important in initial periods, but irrelevant during the succession.

In 1986 the study started when stratification was initiated (Fig. 4). At that time, chlorophyll-a concentration was very high (1–22 µg l^{-1}), and the fluxes measured in the photic zone were more than 10 g DM m^{-2} d^{-1}, decreasing linearly with depth. During the stratification period the chlorophyll-a concentration as well as the dry matter flux decreased, showing the same pattern as the previous year. Chlorophyll flux showed a pattern close to the seston flux, although in the deep stations it was almost zero. The percentage of chlorophyll-a on dry matter in seston flux decreased, from 0.9–0.7 ‰ in the photic zone, to 0.1–0 ‰ in hypolimnion. During the stratified period of 1986, the percentage of seston which reached the bottom was approximately 10% of epilimnion flux.

Slopes of maximal gradient (dry matter flux decrease per depth) showed different values. In 1985 the slope was 0.25 and increased to 0.55 (absolute value). In 1986, changes were from 0.55 to 0.70 and increased during stratification.

3.2. Mineralization at the thermocline

The term mineralization will be used here in the sense of particulate sestonic matter disappearing between two chosen depths. It has been estimated as a percentage of difference of fluxes between 6 m and 15 m (Table I).

Before the thermocline was established (case 3), mineralization at this level was relatively low

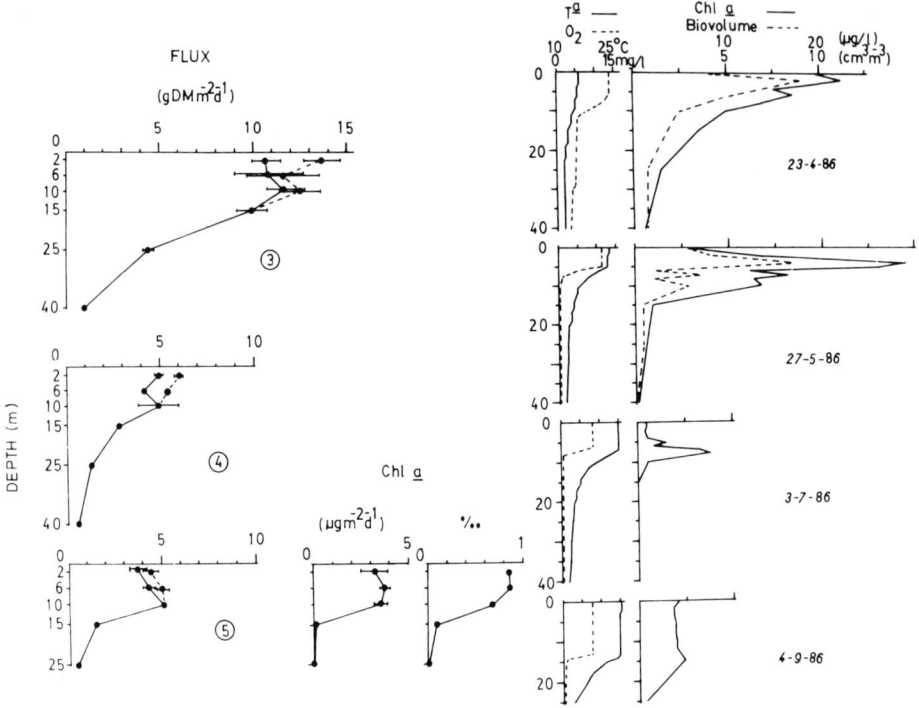

Fig. 4. Seston dry matter (DM) flux in 1986. Dotted lines show seston flux without correcting attached growth. Temperature, oxygen, chlorophyll-a and phytoplankton biovolume are registered in sampling dates. Chlorophyll-a flux and parts per thousand (‰) of chlorophyll-a in seston flux were estimated in last case (5).

(8.3%). However, in stratified water, mineralization of seston reached 80%.

The epilimnetic nutrient cycle shows, in a quantitative way, inorganic carbon and nutrients dissolved by mineralization which are available for assimilation by primary producers (Ohle, 1984). During 1986, there was an increase in it parallel to the stratification process.

Table I. Percentage of mineralization between 6 m (F6) and 15 m (F15) and epilimnetic nutrient cycle in different periods (see Figs. 3 and 4).

		% Mineralization between 6 and 15 m. (F6–F15) · 100/F6	Epilimnetic nutrient cycle: 100/(100-% mineralization)
1	1985	79.6	4.9
2	1985	67.5	3.1
3	1986	8.3	1.1
4	1986	32.1	1.5
5	1986	65.5	2.9

3.3. Phytoplankton vertical flux

Mean vertical flux is a function of phytoplankton biomass and mean sinking velocity. At 2 and 6 m both parameters show high values (Fig. 5). With depth, the phytoplankton biovolume and mean sinking velocity decrease, causing flux reduction. Mean sinking velocity depends, to a great extent and in a simple way, on the mean cell volume of the phytoplankton community, according to Stokes law. However, on the surface and at 2 m, in spite of the high mean cell volume, sinking velocity is not very high; this could be due to species such as *Peridinium cinctum* with a high biovolume (40,000 μm^3), which controls its vertical position by swimming (Berman & Rhode, 1971).

Phytoplanktonic organisms in the deep hypolimnion are smaller than those growing in the photic zone (1,000 $\mu m^3 \cdot$ cell^{-1} in epilimnion and 125 $\mu m^3 \cdot$ cell^{-1} near bottom). This difference is related to the specific composition.

Ratios of seston flux to phytoplankton flux are shown in Table II.

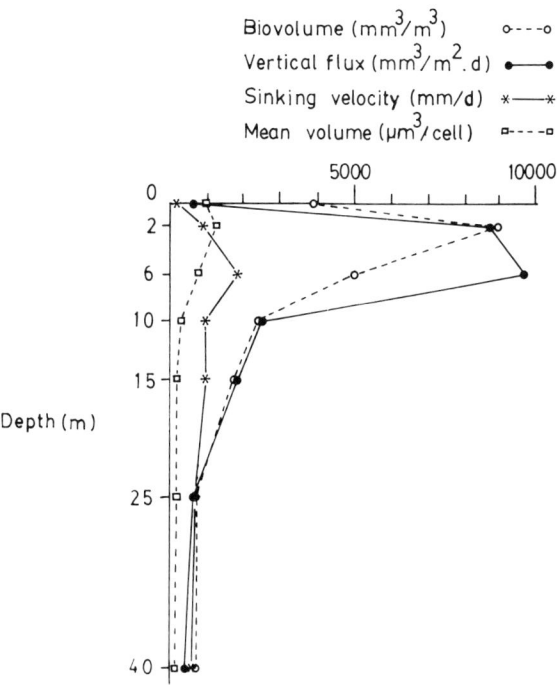

Fig. 5. Phytoplankton total biovolume, mean cell volume, vertical flux and sinking velocity estimated in laboratory.

In epilimnion, phytoplankton flux is important in relation to seston flux. However, in hypolimnion, the importance of detritus in seston flux is higher than phytoplankton.

Table II. Ratios of seston flux (Fs) to phytoplankton flux (Fp) in different depths. H = Phytoplankton dry weight to wet weight (WW) conversion factor = 3 (Mullin et al., 1966; Nalewajko, 1966; Strathmann, 1967; Margalef, 1974).

Depth (m)	Fs (g DM m^{-2} d^{-1})/ Fp (g WW m^{-2} d^{-1}) H
2	1.21
6	1.11
10	4.54
15	5.12
25	6.23
40	2.31

4. Discussion

Seston vertical flux in reservoirs, and in aquatic environments in general, is not merely an accumulation of matter in the benthic system but it also has a great importance in the planktonic one. In the stratified La Concepción Reservoir, the matter which reaches the bottom is only a small percentage of the epilimnion seston flux. These relationships were also mentioned by other authors; Kimmel & Goldman (1977) found in Castle Lake (California) that the 27% of POC reached the sediment surface. In Lake Wingra (Wisconsin) it was estimated that the 30% of the settling organic matter was involved in the long term accumulation of bottom deposits (Gasith, 1976). In epilimnion there occurs a greater reduction in seston vertical flux, suggesting an epilimnetic mineralization of organic matter, meaning regeneration of nutrients which will be taken up by phytoplankton (regenerated production) as opposed to allochthonous nutrients (new production). In Horw Bay (Lake Lucerne) and in Lake Rotsee (Switzerland), only 14 and 30%, respectively, of the measured annual primary production could be entrapped in the sediment traps. This suggests a high mineralization of organic carbon in the euphotic zone, and that 55–85% of the phosphorus needed for primary production can be regenerated by mineralization (Bloesch et al., 1977). In Lake Frederiksborg (Denmark) gross primary production amounted to 560 g C m^{-2} y^{-1}, and sedimentation to 154 g C m^{-2} y^{-1}, the latter being 28% of the former; the remaining 72% was mineralized in the water mainly by algae and bacteria (Andersen & Jacobsen, 1979). Hargrave (1972) estimated in Lake Esrom (Denmark) that 68% of the production in the euphotic zone was consumed in the water column. In an experiment in artificial tanks, Golterman (1972) demonstrated the possibility of a rapid bacterial recycling of nitrogen and other nutrients and suggested that the rate of mineralization is the main factor controlling algal primary production in lakes. This statement is reinforced with the above cited results in different freshwater environments.

In La Concepción Reservoir, phytoplankton vertical flux is quantitatively important in the photic zone, reaching between 30 and 40% of the total seston flux. The decrease in the hypolimnion shows the mineralization and fragmentation of phytoplankton cells during sedimentation. In Horw Bay and Rotsee, 77–79% of the total phytoplankton was decomposed in the epilimnion of both lakes before it could sink into the hypolimnion (Bloesch, 1974). Fragmentation contributes to a greater concentration of detritus, as the smaller particles, that reduce the mean sinking velocity of the particulate matter. Although sometimes detritus can aggregate; e.g. in Lake Tahoe (California), Paerl (1973) found that some groups of microorganisms are responsible for detrital aggregation. Detritus are quantitatively very important in seston vertical flux, especially in hypolimnion. In Lake Frederiksborg, particulate detritus was the main contributor to sedimentation (Andersen & Jacobsen, 1979).

In La Concepción Reservoir, mineralization increases at the thermocline level, where the gradient of density is responsible for a decrease in the sinking velocity of the particles. Often, this zone coincides with the lower level of the photic zone. Some phytoplankton species are well adapted to low light intensities and proliferate due to the availability of nutrients in this regeneration zone. In this way, in La Concepción Reservoir, *Ceratium hirundinella* is the main species of a deep chlorophyll maximum (DCM). DCMs are common in reservoirs (Gálvez, 1986; Niell *et al.*, 1987, Gálvez *et al.*, 1988) in lakes (Bowers, 1980; Priscu & Goldman, 1983; Abbott *et al.*, 1984) as well as marine environments (Cullen & Eppley, 1981; Cullen *et al.*, 1983; Takahashi & Hori, 1984), although there are differences in chlorophyll concentration, phytoplankton species and mechanisms responsible for its formation.

In reservoirs of arid and semiarid regions, the inorganic matter which settle may be quantitatively more important than organic matter, due to erosive processes which transport inorganic material to the water. Thus, flux may be higher and refractory material for bacterial mineralization too. In this way, matter which reaches the bottom is greater, decreasing the storage capacity of the reservoirs. Shalash (1982) estimated the life span of the dead storage capacity in High Aswan Dam (Egypt) at about 360 years, and Duck & McManus (1989) stated that some reservoirs have become totally filled in with sediment due to increasing runoff and sediment erosion.

Different studies carried out in freshwater and marine environments show a relationship between primary production and settled matter. But functions which relate both variables have only been elaborated for oceans. Suess (1980) summarized the results of sediment traps in the world's oceans and described a model that assumed that the downward flux of organic carbon to be directly proportional to the rate of surface primary production with the proportionality constant being an inverse linear function of depth. Betzer *et al.* (1984) assume that downward flux of particulate organic carbon varies in direct proportion to the quotient of surface primary production raised to the power of 1.4 and depth raised to the power of 0.63. In continental waters the data are dispersed and a model was never developed. Sedimentation and primary production values compiled in different continental aquatic environments (Gasith, 1976; Lastein, 1976; Bloesch *et al.*, 1977; Kimmel & Goldman, 1977; Andersen & Jacobsen, 1979; Premazzi & Marengo, 1982; Lastein, 1983; Varela, 1983; Eadie *et al.*, 1984) allow us to establish a linear relationship between both variables (Fig. 6). This model is similar to Suess (1980) for oceans, but it is only an approximation and more data and work is needed. Dystrophic lakes (Sobrado and Wingra) present characteristics of high amounts of macrophytes relative their volume. This biomass and the inputs of allochthonous material are very refractory to mineralization. For these reasons these lakes must be discounted from the general model. In conclusion, sediment fluxes are bigger than those expected for their primary production. In La Concepción Reservoir, when chlorophyll-a concentration is high, seston vertical flux is higher than in periods with low chlorophyll-a amount. Thus, one can see a seasonality function between primary production and sedimentation. When considering the level below the photic zone, La Concepción Reservoir presents a sedimentary flux larger than the above referred to lakes. A higher primary production also could be expected. If dry weight is transformed to organic carbon (Hargrave, 1985), it is possible to estimate, as the model predicts, primary production around 600–700 g C m^{-2} y^{-1}. These values supported the assessment that waters in southern European reservoirs are highly productive.

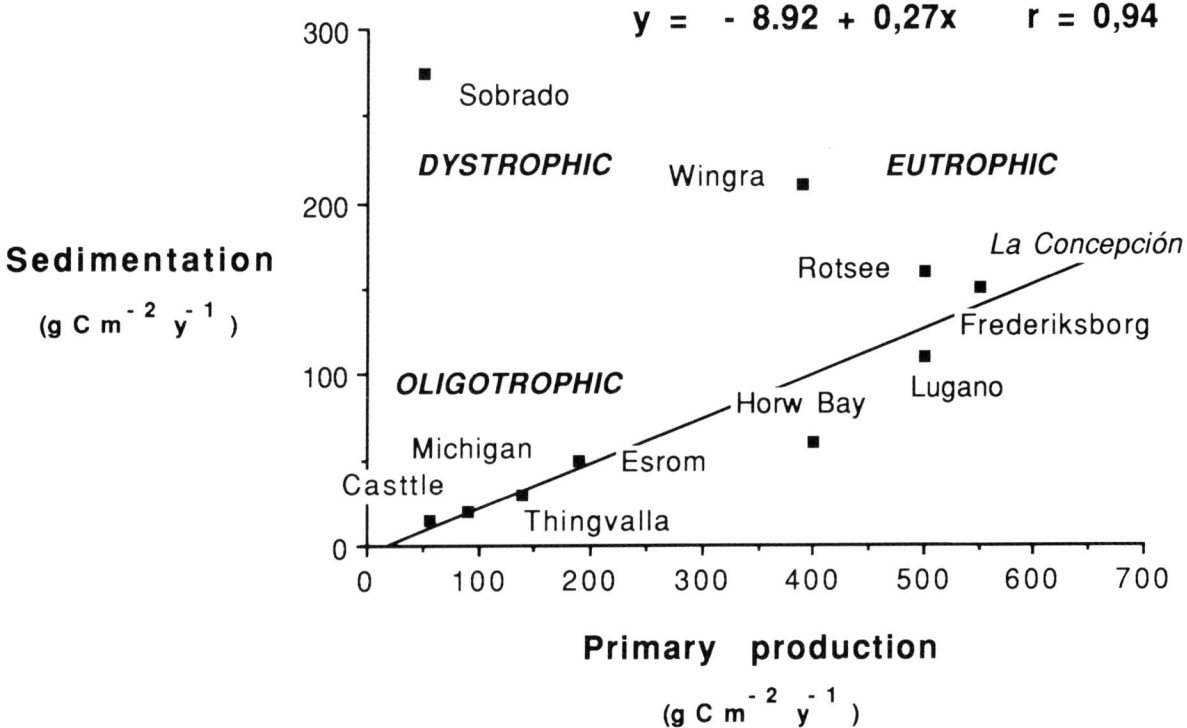

Fig. 6. Relationship between primary production and sedimentation from different freshwater environments. Data taken from Gasith (1976), Lastein (1976), Bloesch *et al.* (1977), Kimmel & Goldman (1977), Andersen & Jacobsen (1979), Premazzi & Marengo (1982), Lastein (1983), Varela (1983) and Eadie *et al.* (1984). Primary production expected for La Concepcion Reservoir according to the linear function.

Sedimentary flux is a function of primary production except for dystrophic lakes. Sinking particulate matter undergoes mineralization, which means, in the photic zone, nutrient regeneration. This enhances primary production, especially at the thermocline where a DCM sometimes is formed. Phytoplankton flux is quantitatively important in the epilimnion, decreasing in the hypolimnion where detritus present a higher percentage in the flux. Mineralization and fragmentation are responsible factors for decreasing the hypolimnetic flux.

Acknowledgements

We are grateful to Dr. J. Bloesch for his suggestions and help which have improved this paper. This work was supported by a F.P.I. grant from Junta de Andalucia and C.A.I.C.Y.T. plan No. 543.

References

Abbott, M. R., K. L. Denman, T. M. Powell, P. J. Richerson, R. C. Richards & C. R. Goldman, 1984. Mixing and the dynamics of the deep chlorophyll maximum in Lake Tahoe. Limnol. Oceanogr. 29: 862–878.

Andersen, J. M. & O. S. Jacobsen, 1979. Production and decomposition of organic matter in eutrophic Frederiksborg Slotssö, Denmark. Arch. Hydrobiol. 85: 511–542.

Berman, T. & W. Rodhe, 1971. Distribution and migration of *Peridinium* in Lake Kinneret. Mitt. Int Verein. Limnol. 15: 266–276.

Betzer, P. R., W. J. Showers, E. A. Laws, C. D. Winn, G. R. Di Tullis & P. M. Kroopnick, 1984. Primary productivity and particle fluxes on a transect of the equator at 153° W in the Pacific Ocean. Deep Sea Res. 31: 1–11.

Bloesch, J., 1974. Sedimentation und Phosphorhaushalt im Vierwaldstättersee (Horwer Bucht) und im Rotsee. Schweiz. Z. Hydrol. 36: 71–186.

Bloesch, J. & N. M. Burns, 1980. A critical review of sediment trap technique. Schweiz. Z. Hydrol. 42: 15–55.

Bloesch, J., P. Stadelmann & H. Bührer, 1977. Primary production, mineralization and sedimentation in the euphotic zone of two Swiss lakes. Limnol. Oceanogr. 22: 511–526.

Blomqvist, S. & L. Håkanson, 1981. A review on sediment traps in aquatic environments. Arch. Hydrobiol. 91: 101–132.

Blomqvist, S. & C. Kofoed, 1981. Sediment trapping – a subaquatic in situ experiment. Limnol. Oceanogr. 26: 585–590.

Bowers, J. A., 1980. Effects of thermocline displacement upon surface chlorophyll maxima in Lake Michigan. J. Great Lakes Res. 6: 367–370.

Burns, N. M. & F. Rosa, 1980. *In situ* measurement of the settling velocity of organic carbon particles on 10 species of phytoplankton. Limnol. Oceanogr. 25: 855–864.

Cullen, J. J. & R. W. Eppley, 1981. Chlorophyll maximum layers of the Southern California Bight and possible mechanisms of their formation and maintenance. Oceanol. Acta 4: 23–32.

Cullen, J. J., E. Stewart, E. Renger, R. W. Eppley & C. D. Winant, 1983. Vertical motion of the thermocline, nitracline and chlorophyll maximum layers in relation to currents on the Southern California Shelf. J. Mar. Res. 41: 239–269.

Duck, R. W. & J. McManus, 1989. Variations in reservoir sedimentation in Scotland in response to land use changes. Arch. Hydrobiol. Beih. Ergebn. Limnol. 33: 19–26.

Eadie, B. J., R. L. Chambers, W. S. Gardner & G. L. Bell, 1984. Sediment trap studies in Lake Michigan: resuspension and chemical fluxes in the Southern Basin. J. Great Lakes Res. 10: 307–321.

Gálvez, J. A., 1986. Heterogeneidades verticales pigmentarias en el embalse eutrófico de La Concepción (Málaga). Tesis de Licenciatura. Universidad de Málaga, 122 pp.

Gálvez, J. A., F. X. Niell & J. Lucena, 1988. Description and mechanism of formation of a deep chlorophyll maximum due to *Ceratium hirundinella* (O. F. Müller) Bergh. Arch. Hydrobiol. 112: 143–155.

Gardner, W. D., 1980. Field assessment of sediment traps. J. Mar. Res. 38: 41–52.

Gasith, A., 1976. Seston dynamics and tripton sedimentation in the pelagic zone of a shallow eutrophic lake. Hydrobiologia 51: 225–231.

Golterman, H. L., 1972. The role of phytoplankton in detritus formation. Mem. Ist. Ital. Idrobiol. 29(suppl.): 89–103.

Hargrave, B. T., 1972. A comparison of sediment oxygen uptake, hypolimnetic oxygen deficit and primary production in Lake Esrom, Denmark. Verh. Int. Verein. Limnol. 18: 134–139

Hargrave, B. T., 1985. Particle sedimentation in the ocean. Ecol. Model. 30: 229–246.

Hargrave, B. T. & N. M. Burns, 1979. Assessment of sediment trap collection efficiency. Limnol. Oceanogr. 24: 1124–1136.

Hutchinson, G. E., 1967. A Treatise On Limnology, vol 2. Wiley & Sons, New York, 1115 pp.

Kimmel, B. L. & C. R. Goldman, 1977. Production, sedimentation, and accumulation of particulate carbon and nitrogen in a sheltered subalpine lake. In: H. L. Golterman (ed.), Interactions between Sediment and Freshwater, Proc. Int. Symp. Amsterdam: 148–155.

Lastein, E., 1976. Recent sedimentation and resuspension of organic matter in eutrophic Lake Esrom, Denmark. Oikos 27: 44–49.

Lastein, E., 1983. Decomposition and sedimentation processes in oligotrophic, subarctic Lake Thingvalla, Iceland. Oikos 40: 103–112.

Margalef, R., 1974. Ecologia. Ed. Omega. Barcelona, 955 pp.

Mullin, M. M., P. S. Sloan & R. W. Eppley, 1966. Relationship between carbon content, cell volume and area in phytoplankton. Limnol. Oceanogr. 11: 307–311.

Nalewajko, C., 1966. Dry weight, ash, and volume data for some freshwater planktonic algae. J. Fish. Res. Bd. Can. 23: 1285–1288.

Niell, F. X., J. A. Gálvez, F. López-Figueroa & P. Algarra, 1987. Usos no fotosintéticos de la luz en plantas acuáticas. Actas del IV Congreso Español de Limnologia: 13–23.

Ohle, W., 1984. Measurement and comparative values of the Short Circuit Metabolism (SCM) of lakes by POC relationship of primary production of phytoplankton and settling matter. Arch. Hydrobiol. 19: 163–174.

Paerl, H. W., 1973. Detritus in Lake Tahoe: structural modification by attached microflora. Science 180: 496–498.

Premazzi, G. & G. Marengo, 1982. Sedimentation rates in a Swiss Italian lake measured with sediment traps. Hydrobiologia 92: 603–610.

Priscu, J. C. & C. R. Goldman, 1983. Seasonal dynamics of the deep-chlorphyll maximum in Castle Lake, California. Can. J. Fish. Aquat. Sci. 40: 208–214.

Rathke, D. E., J. Bloesch, N. M. Burns & F. Rosa, 1981. Settling fluxes in Lake Erie (Canada) measured by traps and settling chambers. Verh. int. Verein. Limnol. 21: 383–388.

Shalash, S., 1982. Effects of sedimentation on the storage capacity of the High Aswan Dam Reservoir. Hydrobiologia 92: 623–635.

Smayda, T. J., 1970. The suspension and sinking of phytoplankton in the sea. Oceanogr. Mar. Biol. Ann. Rev. 8: 353–414.

Sournia, A., 1981. Morphological bases of competition and succession. In: T. Platt (ed.), Physiological bases of phytoplankton ecology. Can. Bull. Fish. Aquat. Sc. 210: 339–346.

Strathmann, R. R., 1967. Estimating the organic carbon content of phytoplankton from cell volume or plasma volume. Limnol. Oceanogr. 12: 411–418.

Suess, E., 1980. Particulate organic carbon flux in the oceans surface productivity and oxygen utilization. Nature 228: 260–263.

Takahashi, M. & T. Hori, 1984. Abundance of picophytoplankton in the subsurface chlorophyll maximum layer in subtropical and tropical waters. Mar. Biol. 79: 177–186.

Talling, J. F. & D. Driver, 1963. Some problems in the estimation of chlorophyll-a in phytoplankton. Proc. Conf. Prim. Prod. Meas. Mar. and Fresh. Hawaii, 1961: 142–146.

Utermöhl, H., 1958. Zur Vervollkommung der quantitativen Phytoplankton Methodik. Mitt. int. Verein. Limnol. 9: 1–38.

Varela, M., 1983. Estructura y producción estacional en un sistema lacustre de dimensiones reducidas y amplio desarrollo de la vegetación macrofítica litoral. Competencia de macrófitos y fitoplancton. Tesis Doctoral. Universidad de Santiago, 355 pp.

Chapter VIII

Impacts of growth factors on competitive ability of blue-green algae analyzed with whole-lake simulation

O. Varis
Laboratory of Hydrology and Water Resources Management, Helsinki University of Technology, Rakentajanaukio 4 A, SF–02150 Espoo, Finland

Key words: reservoirs, lakes limnology, blue-greens, mathematical models, simulation, nutrients, competition

Abstract

The competitive ability of N-fixing blue-green algae (*Aphanizomenon flos-aquae*) was studied in Lake Kuortaneenjärvi, Finland. A 15-state dynamic lumped parameter model simulating N, P, and algae was used. The observed and modeled behavior of the lake was extrapolated using the model within a much wider range of input values for N and P loads, water temperature and irradiance, than was available at the model identification. For each input variable and run, a fixed amount of perturbation was introduced into the whole input array. An increase in the P load improved and an increase in the N load weakened the competitive ability of blue-greens. This was mainly due to the changes in the N:P ratio. Higher water temperatures greatly improved the facilities of cyanobacteria, but the competition was not sensitive to irradiance. The different roles of physiological and ecological factors appear clearly in the outcome. The simulation results coincide well with the existing ecological knowledge. This suggests that the assumptions used in the model identification do not involve major errors.

1. Introduction

The problems caused by lake eutrophication involve biotic responses to various inputs into the ecosystem. In this study, the inputs considered are the P and N loads on the lake, water temperature and total irradiance, and the problems caused by them are blooms of blue-green algae in late summer.

This study has two aims, elucidation of the ecology of eutrophicated short retention time lakes and assessment of the credibility and validity of the modeling approach used. In the study of the lake ecology, the emphasis is on factors affecting the competitive ability of N-fixing blue-green algae, mostly *Aphanizomenon flos-aquae* (L.) Ralfs, in a case in which N limitation grows in importance as the season advances.

From the modeling point of view, we are faced with a very common situation. A large-scale problem has to be analyzed with inadequate quantitative data for reliable statistical analysis or a guarantee of the observability of the system. But the large amount of information already available facilitates the approach to the problem. In the present study, a simulation model constructed by Varis (1984) on the basis of existing data (by induction) and knowledge (by deduction) was used and analyzed. So, the behavior of the model was validated, not against measured data, since the model is not purely data based abstraction, and the amount of data was strongly limited, but against causal knowledge from lake ecology.

2. Lake Kuortaneenjärvi

Lake Kuortaneenjärvi (62° 45′ N, 23° 30′ E), Finland, has a catchment of 1280 km². As the district has a relatively dense population and is poor in lakes, the lake is an important source of recreation. Being slightly eutrophicated, however, Lake Kuortaneenjärvi suffers from blue-green algal blooms. Municipal waste waters are

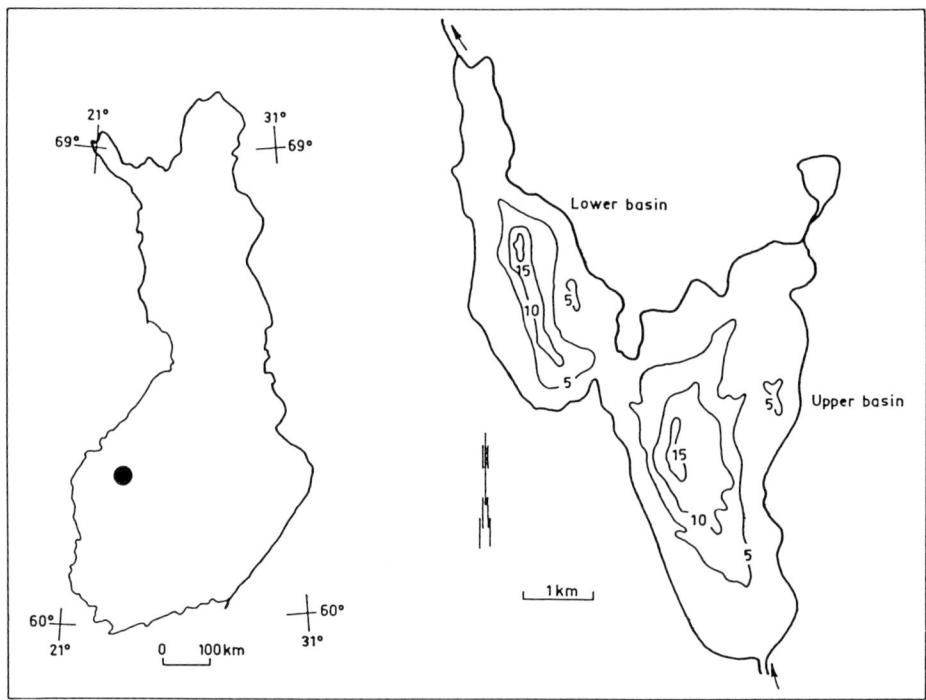

Fig. 1. Location and bathymetric map of Lake Kuortaneenjärvi.

responsible for 5 to 10% of the total nutrient load, but in summer the proportion rises to 30%. The main part of the load comes from diffuse sources. The lake consists of two main basins (Fig. 1). The results of this study are for the upper, southern basin. For a detailed description of the lake, the reader is referred to Stenmark (1982).

The water quality of the lake has been monitored since 1961, with the annual number of observations below 10. In 1980, an intensive programme with 23 sampling dates was carried out. On some dates, only chlorophyll-a (CHA), dissolved oxygen and water temperature were measured. Some water quality characteristics are presented in Table I. Water temperature and irradiance observations are presented in Fig. 2, since they are among the input variables that were perturbed. Algal counts were carried out 10 times for both basins during the ice-free period. The total biomasses and the percentages of different algal groups in the upper basin are given in Table II.

The water balance calculations were based on daily observations of the water level and precipitation. Some hydrological and morphological data on the lake are presented in Table III. Both the water flow and the nutrient load (Fig. 3) fluctuate greatly. Flooding, due to melting snow in spring and to a low evaporation-to-precipitation ratio in fall, clearly dominates the annual balances. During flooding, the theoretical retention time of the whole lake may reduce to one week.

From its hydrological and limnological properties, the lake apparently has very much in common with reservoirs. Short retention time, high non-algal turbidity and heterogeneity in time

Table I. Some water quality characteristics from the upper basin of Lake Kuortaneenjärvi at 1 m depth during the ice-free period (5 May – 31 October) in 1980.

Variable	Min.	Median	Max.	Unit
CHA	52	158	260	$\mu g\ l^{-1}$
Total N	499	660	953	$\mu g\ l^{-1}$
NO_3—N	4	74	318	$\mu g\ l^{-1}$
NH_4—N	18	40	73	$\mu g\ l^{-1}$
Total P	40	64	77	$\mu g\ l^{-1}$
PO_4—P	8	13	40	$\mu g\ l^{-1}$
Total N:P	8.5	10	24	(in weight)
DIN:DIP	3.3	5.1	27	(in weight)
pH	6.5	6.8	7.3	–
Color	90	140	240	Pt scale

Table II. Percentual amounts of algae groups from total planktonic algae biomass in the Southern basin of Lake Kuortaneenjärvi in summer 1980.

Date	20.5	4.6	16.6	1.7	14.7	28.7	11.8	26.8	8.8	22.8
Chlorophyta	12	0.04	0.5	2.4	2.0	1.0	0.3	1.7	0.5	5.0
Euglenophyta	0.2	0.06	0.004	0.1	0.7	0.08	0.3	0.2	0.1	0.2
Chrysophyta	41	80	66	58	33	47	72	46	12	40
Pyrrophyta	46	20	33	38	48	28	5.0	21	48	24
Cyanobacteria										
A. flos-aquae	–	–	–	0.8	12	21	19	24	37	26
Others	0.4	0.08	0.6	1.1	3.9	3.0	2.7	6.7	1.9	4.4
Total (mg m^{-3})	560	2800	2700	4200	2100	3200	2700	1400	2600	950

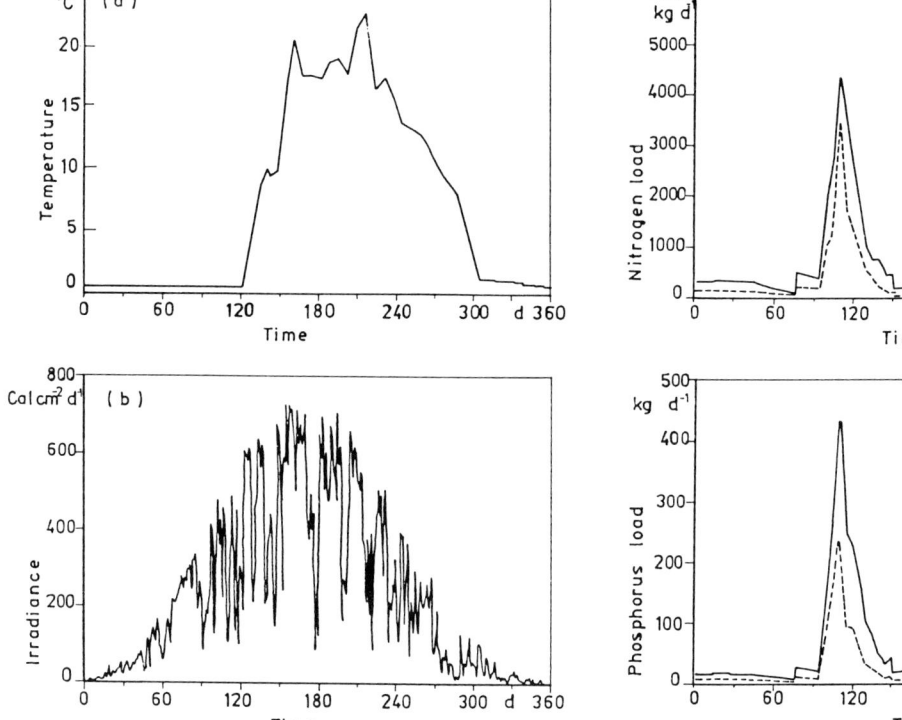

Fig. 2. Observed water temperature at 1 m depth (a) and irradiance from Jyväskylä airport (b) in 1980.

Fig. 3. Nutrient inflow, (a) phosphorus, (b) nitrogen, in 1980 (according to Stenmark, 1982). Solid line denotes total load, broken line dissolved inorganic load.

and space have often been mentioned being features in which reservoirs differ most essentially from natural lakes. Lind *et al.* (1992) point to the difficulty of applying 'lake derived classifications' of water quality to reservoirs due to short retention time and the other features mentioned above. These features are very common in Finnish lakes, many of which are geologically essentially younger than most of the temperate zone lakes, on studies of which the limnological research and classifications have greatly been focused. In the district of Southern Ostrobothnia, Western Finland, in the watersheds of the two major rivers, Kyrönjoki and Lapuanjoki, there exists 35 natural lakes and six impoundments. 18 of them are being regulated (Anonymus, 1977). All of them have a short retention time and a high water color value (Table IV). Their mass balances are very much

Table III. Hydromorphological characteristics of Lake Kuortaneenjärvi.

	Whole lake	Southern basin	Northern basin
Mean depth (m)	3.7	4.0	3.2
Max. depth (m)	16	16	16
Retention time (d)	69	49	20
Surface area (km²)			
HW 1980	18.0	11.9	6.1
MW 1980	16.4	10.9	5.5
NW 1980	15.6	10.5	5.1
Volume (10^6 m³)			
HW 1980	71.5	50.0	21.5
MW 1980	59.3	42.1	17.2
NW 1980	50.0	36.0	14.0

Table IV. Some characteristics of 35 natural lakes and 6 reservoirs in the basins of Rivers Kyrönjoki and Lapuanjoki, Western Finland. Data from the National Board of Waters (Anonymus, 1977).

Variable	Unit	Min.	Median	Max.
Surface area	km²	1.0	2.0	16.4
Volume	10^6 m³	0.5	3.7	60.0
Retention time	years	0.01	0.28	1.71
Water color	mg Pt l^{-1}	60	200	320

Table V. State and input variables of the model (Varis, 1984). Unit for state variables is µg l^{-1}.

State variables	Input variables
Dissolved inorganic phosphorus	Water volume
Dissolved inorganic nitrogen	Organic phosphorus load
Biomass of cyanobacteria	Inorganic phosphorus load
Cyanobacteria phosphorus	Organic nitrogen load
Cyanobacteria nitrogen	Inorganic nitrogen load
Phytoplankton biomass	Outflow
Phytoplankton phosphorus	Mean water temperature
Phytoplankton nitrogen	Water temperature at 1 m depth
Allochtonous detrital phosphorus	Wind direction
Autochtonous detrital phosphorus	Wind speed
Detrital nitrogen	Total irradiance
Organic phosphorus in sediment	
Inorganic phosphorus in sediment	
Organic nitrogen in sediment	
Inorganic nitrogen in sediment	

dominated by floods in spring and in fall and most of them are spatially heterogeneous.

3. The model

The dynamic simulation model used consisted of 15 state variables, i.e. ordinary differential equations. It was very nonlinear with respect to states and parameters and had a lumped parameter structure. The state and input variables are given in Table V. The annual dynamics of P and N were simulated with the constraint of mass conservation. The two main basins of the lake were described with identical models. As the stratification of the lake is only temporary (Stenmark, 1982), the basins were assumed to be continuously mixed in the model. A constant time-step of 24 hours was used. The simulation period, one year, was divided into 360 time-steps. A detailed description of the model equations and parameters is presented by Varis (1984, 1988), but an overview of the structure of the descriptions directly related to the algal growth mechanisms is provided also here.

Both the algal fractions considered, cyanobacteria and the (other) phytoplankton, were described with three state variables: biomass, intracellular P and intracellular N. This two-step growth control scheme was slightly modified from the equations of Nyholm (1976). The state equations for biomass for the two algal fractions had the form:

$$dX_t/dt = (\mu_{x,t} - L_{x,t}) X_t \quad (1)$$

where X_t is biomass at time t ($t = 1, 360$), $L_{x,t}$ is a sum of loss terms consisting of distinct biological and hydrological losses, and $\mu_{x,t}$ is the growth rate. The intracellular nutrient concentration equations were the following:

$$dY_{x,t}/dt = \mu_{y,t} X_t - L_{x,t} Y_{x,t} \quad (2)$$

where $Y_{x,t}$ is the intracellular nutrient concentration and $\mu_{y,t}$ is the nutrient uptake rate specific to the algal group and the nutrient. Growth and uptake rates were obtained by multiplying the corresponding parameters $\alpha_{x/y}$, specific to each state, by all the control factor values $F_{f,t,x/y}$ affecting the process. x/y denotes x or y:

$$\mu_{x/y, t} = \alpha_{x/y} \, \pi_f \, F_{f, t, x/y} \quad (3)$$

where f is the factor (N, P, temperature or irradiance). The equations for the controlling factors are not presented in detail in this paper. For the most essential differences in parameter values between the two algal groups, the reader is referred to Varis (1984; 1988), or Table II by Kettunen (1992).

Due to the dark water (see Table I), a steep temperature gradient forms in the upper water layers of the lake on sunny days (Stenmark, 1982). Therefore, the temperature at 1 m depth was taken as a driving variable of algal growth instead of the mean water temperature of the whole water column, which controlled other temperature dependent processes. All the temperature control descriptions were of van't Hoff form.

The model contains also a rough description for interactions between the water mass and sediment. The descriptions are based mostly on experience reported by Kettunen & Stenmark (1982). The asymmetrical form of the lake was taken into account in the description for resuspension in terms of a cosine function dependent on wind direction and speed.

The model identification was carried out manually using the data for 1980. The submodels were first calibrated separately and then iterated to fulfil the mass continuum. This modular, decoupling approach greatly reduces the degree of freedom in identification because at each step only a few parameters are to be estimated. This procedure also helps to diminish the drawbacks of uncertainty and unreliability discussed widely by e.g. Ackerman *et al.* (1974), Young (1977; 1983) and Beck (1983), which are unavoidably connected with the modeling approach used.

The results and the identification procedure are described in detail by Varis (1984). A crucial point, which should be mentioned in this context also, is the importance of the ratio between the dissolved inorganic nutrients N and P, DIN:DIP. The good adequacy gained, with respect to the ratio, in model calibration (Fig. 4) facilitated the use of the model in an analysis of the ecological competition between N-fixers and other algae. In many studies (e.g. Hrbáček, 1964; Istvánovics *et al.*, 1986), the N:P ratio has been found to play an important role in controlling the formation of an *Aphanizomenon flos-aquae* bloom.

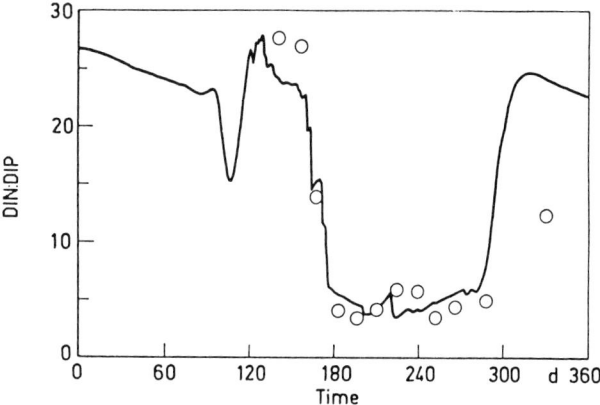

Fig. 4. Observations and calibration results for the ratio of inorganic nitrogen to phosphorus (———— = simulation, o = observation).

4. Principle of the analysis

The dynamic growth factor analysis consists of two parts. First, the relative physiologically limiting roles of P, N, water temperature and irradiance were simulated. Then, the impacts of the factors on competition between blue-greens and the other phytoplankton were studied.

The relative physiologically limiting roles were calculated according to the formula:

$$R_{f, t, \text{phyt}} = (1 - F_{f, t, \text{phyt}} / FOPT_{f, \text{phyt}}) / \sum_{f=1}^{N} (1 - F_{f, t, \text{phyt}} / FOPT_{f, \text{phyt}}), \quad (4)$$

where $R_{f, t, \text{phyt}}$ denotes the proportion of limitation of factor f, $FOPT_{f, \text{phyt}}$ is the 'optimal' value of the factor f for the growth of phytoplankton and N is the number of limiting factors taken into account. Cyanobacteria were excluded from the first analysis, since the intensity of N-fixation depends on the N compounds available as well as on irradiance and temperature (Ward & Wetzel, 1980).

The impacts of the factors on competition were studied in terms of perturbation analysis. Feasible ranges around the nominal (1980) input values were chosen for each of the four input variable sets considered, i.e.

(i) inorganic and organic P load,
(ii) inorganic and organic N load,
(iii) water temperature (1 m depth and mean) and
(iv) irradiance.

For each run, the same quantity of perturbation was introduced into the input values of each of the 360 days. The range chosen was divided into 9 equal intervals. So, 10 runs were required for each input variable studied.

For the P and N loads, the range chosen was from 40% to 220% of the nominal values, i.e. the simulations were made with the loading multiplied by factors ranging from 0.4 to 2.2, with intervals of 20% of the respective input value for each day. The range was approximated to cover the great natural fluctuations in the flow and load regimes.

Water temperature was shifted within a range of 65% to 105% of the nominal values. The maximum observed value for water temperature at 1 m depth in 1980 was 22.9 °C and the mean for June to August was about 18 °C. Now the maximum ranged between 15.0 °C and 24.0 °C and the mean between 12 °C and 19 °C. The values are realistic as compared with observations from other years from Lake Kuortaneenjärvi and with a water temperature frequency analysis from 18 Finnish lakes by Kuusisto (1981).

For total irradiance, application of a linear range analogous to those described above would not have been reasonable. As the first step, a function describing maximal daily values of irradiance $R_{max,t}$ throughout a year was formulated:

$$R_{max,t} = [(0.52 + 0.48 \sin(t - 82)]^{1.2} \cdot 31.5 \cdot 10^6 \text{ [J m}^{-2} \text{ d}^{-1}] \quad (5)$$

Daily minima $R_{min,t}$ were approximated to be 10% of the maximal values. Thereafter, the daily values R_t of 1980 were perturbed and scaled between minimum and maximum functions:

$$R'_t = [R_t - R_{min,t}]/[R_{max,t} - R_{min,t}] \quad (6)$$

The scaled values R'_t had an even distribution between [0, 1]. The shifted irradiance values $R_{p,t}$ were obtained with the formula:

$$\beta = R'_t [1 + a \sin(R'_t \cdot 180)] \times [R_{max,t} - R_{min,t}] + R_{min,t} \quad (7)$$

$$R_{p,t} = \begin{cases} 1, & \text{if } \beta > 1 \\ \beta, & \text{if } 0 < \beta < 1 \\ 0, & \text{if } \beta < 0 \end{cases}$$

the parameter a receiving 10 values within a range [0, 1] with equal intervals. Now the annual mean of scaled irradiance ranged from 0.3 to 0.7.

5. Results

According to the simulation results, N dominated over P as the nutrient limiting phytoplankton growth for almost the whole summer (Fig. 5). Fig. 6 suggests that the importance of nutrients started to grow in April and increased till August. First the main factor was the increase in irradiance reaching the water. After the snow melted and the ice broke

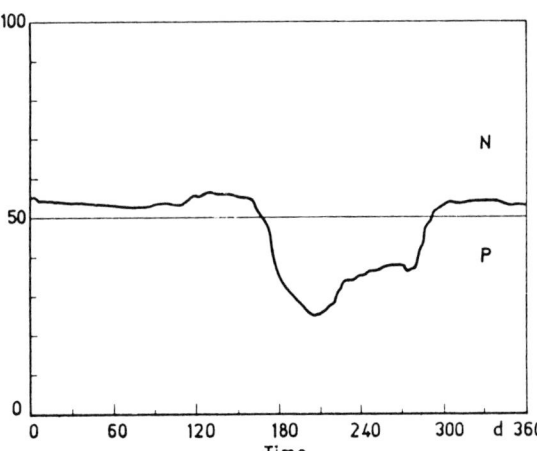

Fig. 5. Simulated relative importance of dissolved inorganic N and P to phytoplankton growth limitation (Cyanobacteria excluded).

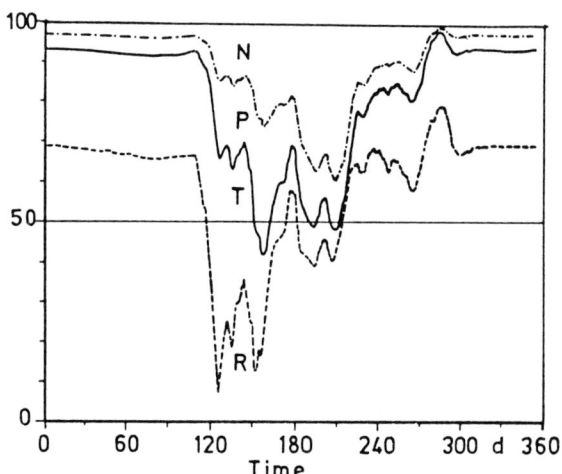

Fig. 6. Simulated relative importance of dissolved inorganic N and P, water temperature (*T*) and irradiance (*R*) to phytoplankton growth limitation (Cyanobacteria excluded).

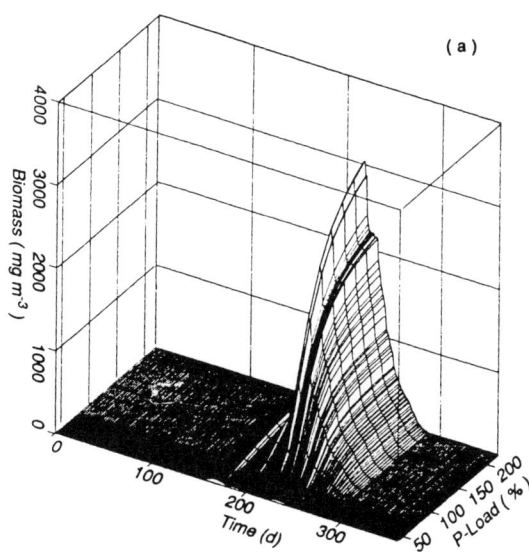 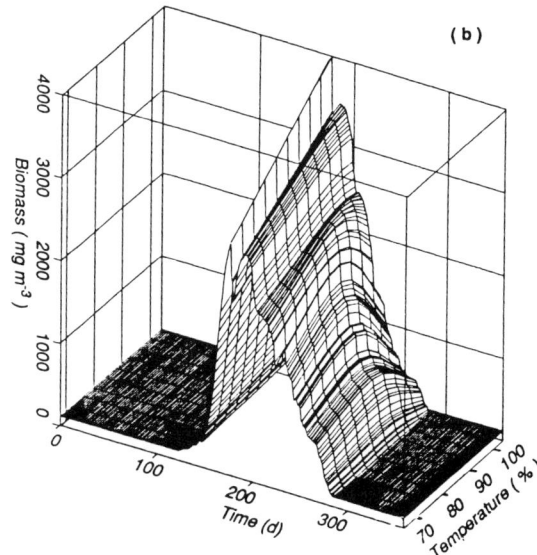

Fig. 7. Simulated fresh biomass of (a) Cyanobacteria and (b) other phytoplankton plotted against phosphorus load and time (from the beginning of the year). The 1980 load is used as the nominal (100%) situation.

up, the importance of the low water temperature gradually decreased. Affecting the extinction value in the growth equations of algae (see Varis, 1984), the wind-induced resuspension caused variations in the relative importance of irradiance as a growth-limiting factor.

An increase in the P load appeared to favor cyanobacteria in competition (Fig. 7). N-fixing blue-greens were able to compete successfully with the other algae in late summer, as N limited the algal growth and plenty of energy was available. The results show no changes in the timing of the maximum biomass of cyanobacteria.

The phytoplankton benefited remarkably from an increase in the N load (Fig. 8). This was most evident in late summer, with an excessive increase

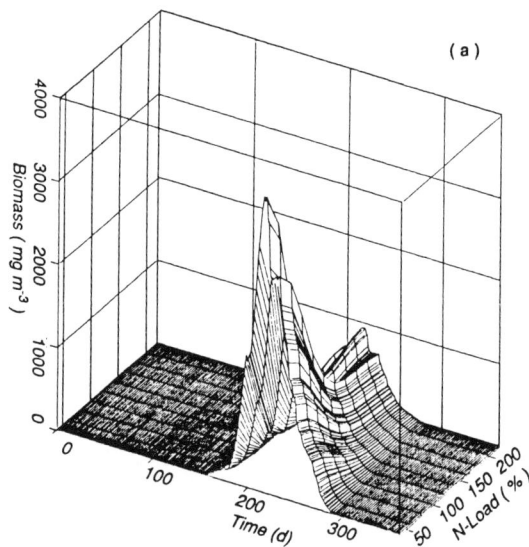 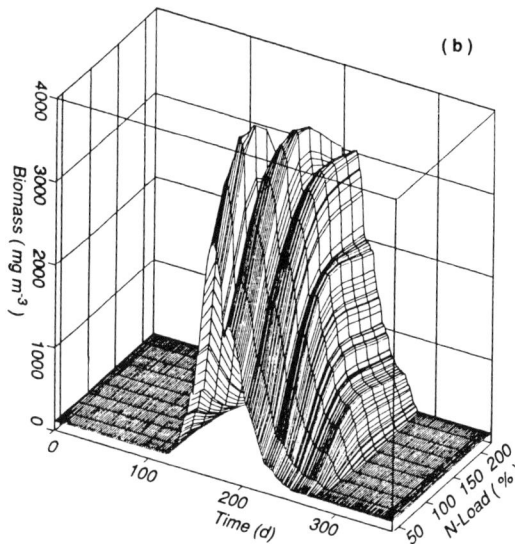

Fig. 8. Simulated fresh biomass of (a) Cyanobacteria and (b) other phytoplankton plotted against nitrogen load and time (from the beginning of the year). The 1980 load is used as the nominal (100%) situation.

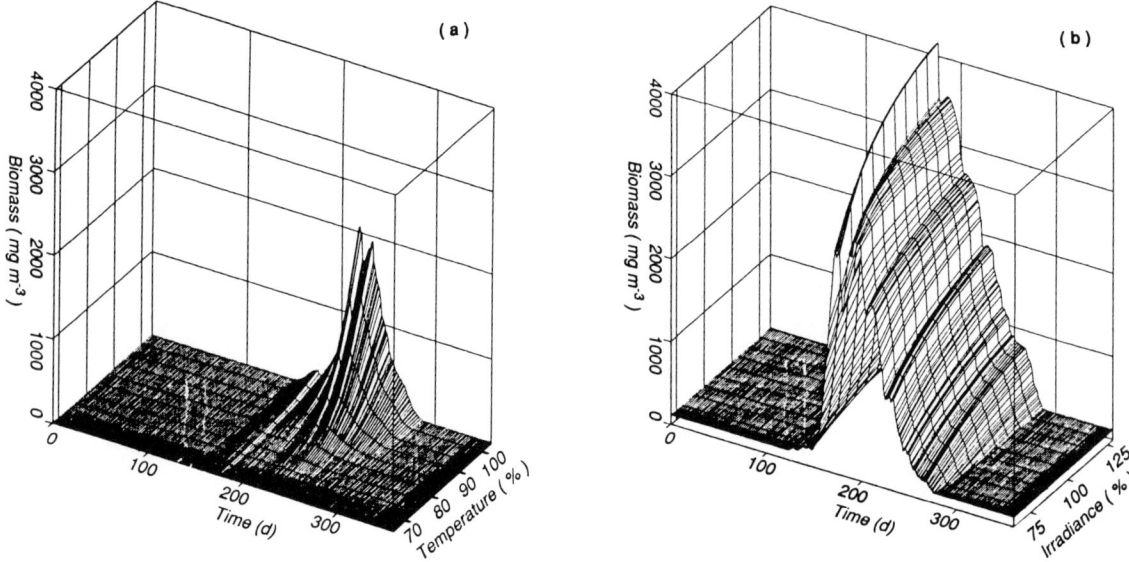

Fig. 9. Simulated fresh biomass of (a) Cyanobacteria and (b) other phytoplankton plotted against water temperature and time (from the beginning of the year). The 1980 load is used as the nominal (100%) situation.

in biomass and a delay in the biomass peak. The competitive ability of the cyanobacteria was diminished by a decrease in the importance of N as a limiting nutrient and also directly by the great phytoplankton biomass.

The competition was very sensitive to changes in water temperature (Fig. 9). Cyanobacteria benefited greatly from higher temperatures. The impact of temperature perturbation closely resembles that of the P load.

The shifts in irradiance did not cause notable changes in the competition (Fig. 10). The cyanobacteria peak took place slightly later, but no alterations were evident in the level of the peak. An

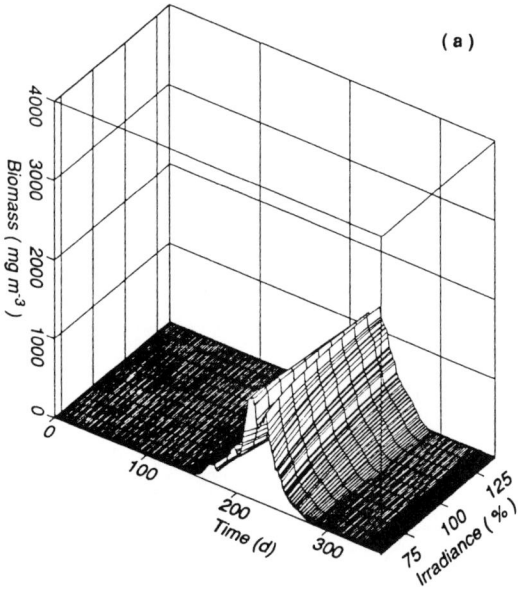
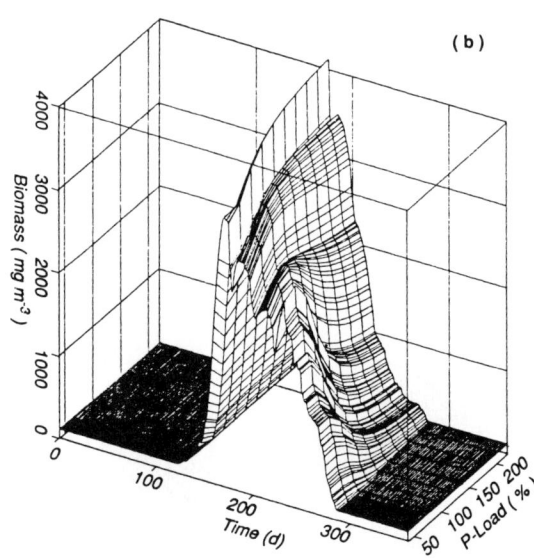

Fig. 10. Simulated fresh biomass of (a) Cyanobacteria and (b) other phytoplankton plotted against irradiance and time (from the beginning of the year). The 1980 load is used as the nominal (100%) situation.

increase in irradiance caused a slight collinear response in phytoplankton biomass. This effect was greater at low than at high irradiance.

6. Discussion and conclusions

According to the results in Figs. 5 and 6, N was more important than P as a nutrient limiting phytoplankton growth during summer. This result accords well with Fig. 4 and the studies on N:P ratios referred to below.

Redfield (1958) determined a critical N:P ratio of 15 by atoms (about 7 by weight) for nutrient limitation of algal growth. Chiaudani & Vighi (1974) and Forsberg et al. (1978) suggested that N limits the growth if the DIN:DIP ratio decreases below 5, and P will be limiting if the ratio is above 10 or 12. The critical ratio varies widely, however, among different species (Tilman, 1977; Rhee, 1978; Rhee & Gotham, 1980).

When water temperature and irradiance were also taken into consideration (Fig. 6), the limiting factor controversy was seen in a new light. Since these factors are usually more difficult to control than nutrient loads, their relative importance has not often been properly emphasized in whole-lake studies. The competition between blue-greens and other algae responded very strongly to changes in nutrient loads (Figs. 7 and 8). An increased P load favored cyanobacteria, while an increased N load favored other algae. This behavior is much due to changes in the N:P ratio rather than solely to changes in the absolute concentrations of the nutrients. To put it more precisely, a high P load compared to the N load caused scarcity of N earlier in the growth season and so promoted earlier formation of favorable competitive conditions for cyanobacteria.

Many recent studies (e.g. Schindler, 1977; Niemi, 1979; Flett et al., 1980; Rhee & Gotham, 1980; Kanninen et al., 1982; Smith, 1983; Sakshaug & Olsen, 1986; Sommer et al., 1986; Varis, 1991; 1993) have also emphasized the great importance of the N:P ratio to the competitive ability of N-fixing blue-green algae. It has even been proposed that removal of N from sewage effluents could be optimized or that waters could be fertilized with N to prevent blooms of cyanobacteria (Ripl et al., 1979; Leonardson & Ripl, 1980).

Although the relative importance of temperature as a limiting factor was notably smaller than that of irradiance, except in May and early June, the algae, especially the blue-greens, responded very much more strongly to changes in temperature. However, the algal periodicity may vary in its sensitivity to the input variables with the time of the year. During the model identification process and in a temporal input perturbation studies with the same model, it appeared that the competition (model output) was very sensitive to perturbations of model inputs (Varis, 1988) and outflow (Varis, 1989) in spring.

In view of the behavior of the model, the following hypothetical scheme can be suggested, even if all the processes mentioned are not explicitly included in the model: Warming of the water in spring induces development of a large algal biomass. The algae consume the dissolved inorganic N in the lake fairly rapidly, and high water temperatures and oxygen depletion contribute to decrease the N concentration. The earlier the decrease in the N:P ratio starts to favor cyanobacteria, the more time they have to form blooms before temperature and irradiance become more important limiting factors than nutrients.

A number of studies suggest blue-green algae to be favored by warm water. Hammer (1964) reported that large *Aphanizomenon flos-aquae* biomasses seldom occurred in temperatures below 20 °C. In algal assays, Tilman et al. (1986) found that an increase in temperature caused a remarkable improvement in the competitive ability of blue-greens. However, no clear temperature preference was found for *Anabaena* in comparison to several other genera in results reported by Straškraba & Gnauck (1985).

The results are subject to errors and uncertainties from several sources. The modeler of a large-scale environmental system is faced with the problem of model aggregation; from the real system with infinite dimensions, he must select the items essential to the problem studied. The data base available also constitutes a constraint for modeling. As regards the present model, the data available did not allow consideration of many features of the lake which could be important to the problem studied, and naturally only a part of the competition mechanisms are considered. Hrbáček (1964) and Lynch (1980) pointed out, that zooplankton and fish have an important role in controlling *A. flos-aquae* blooms. Algal periodicity is also affected by other plants and bacteria. The model used cannot take into account possible changes in dominant algal species. Zevenboom & Mur (1980) suggested, for instance, that *A. flos-aquae* dominates over *Oscillatoria agardhii* when nitrates are scarce. Also the role of dissolved inorganic carbon may be important to the competition (see e.g. Lange, 1967; 1970; Kuenzel, 1969; King, 1970; Shapiro, 1973). Also several other inorganic and organic substances are apparently noteworthy in this context.

The ecosystem and the model studied here are both decidedly nonlinear. Extrapolation of the model behavior to scenarios outside the range of model identification is a source of uncertainty for the results in such a case.

Due to the various sources of uncertainty and error, the outcome of the model may not be considered to be very quantitative. However, the very logical behavior and its close agreement with ecological *a priori* knowledge point to the conclusion that the assumptions and simplifications made in the model identification do not involve major errors. The model is thus very applicable in elucidative studies of the functioning of the ecosystem of the lake. The degree of deductive validation is not at the same level as in classical statistical analysis, but perhaps good enough for management purposes. Besides the application of temporal input sensitivities (Varis, 1988) mentioned above, it has already been applied by Kettunen (1992) at investigations on the information gained from lake monitoring.

Acknowledgements

I am grateful to Mr. J. Kettunen for innumerous discussions rich in ideas. Comments of M. Straškraba on the manuscript are acknowledged. Prof. P. Vakkilainen contributed plenty of encouragement during the study, prof. H. Seppänen gave valuable comments on the ecological questions and Mr. M. Stenmark and Mr. M. Karonen presented many interesting ideas. Mrs. A. Damsröm revised the English. I am glad to address my thanks to them all. The study was financed by Maa- ja Vesitekniikan Tuki ry association.

References

Ackerman, B. A., S. R. Ackerman, J. W. Sawyer & D. W. Henderson, 1974. The Uncertain Search for Environmental Quality. The Free Press, New York, 386 pp.

Anonymus, 1978. A general plan for the water use in Southern Ostrobothnia (in Finnish). Natl. Bd. Waters, Finland, Rep. 140.

Beck M.B., 1983. Uncertainty, system identification, and the prediction of water quality. In: M. B. Beck & G. Van Straten (eds.), Uncertainty and Forecasting of Water Quality. Springer-Verlag, Berlin: 3–68.

Chiaudani, G. & M. Vighi, 1974. The N:P ratio and tests with *Selenastrum* to predict eutrophication in lakes. Water Research 8: 1063–1069.

Flett, R. J., D. W. Schindler, R. D. Hamilton & N. E. R. Campbell, 1980. Nitrogen fixation in Canadian precambrian shield lakes. Can. J. Fish. Aquat. Sci. 37: 488–493.

Forsberg, C., S-O. Ryding, A. Claesson & A. Forsberg, 1978. Water chemical analysis or algal assay? – Sewage effluent and polluted lake water studies. Mitt. internat. Verein. Limnol. 21: 352–363.

Hammer, V. T., 1964. The succession of "bloom" species of blue-green algae and some causal factors. Verh. internat. Verein. Limnol. 15: 829–836.

Hrbáček, J., 1964. Contribution to the ecology of water-bloom forming blue-green algae *Aphanizomenon flos-aquae* and *Microcystis aeruginosa*. Verh. internat. Verein. Limnol. 15: 837–846.

Istvánovics, V., L. Vörös, S. Herodek, L.G. Tóth, & I. Tátrai, 1986. Changes in phosphorus and nitrogen concentration and of phytoplankton in enriched lake enclosures. Limnol. Oceanogr. 31: 798–811.

Kanninen, J., L. Kauppi & E-R. Yrjänä, 1982. The role of nitrogen as a growth limiting factor in the eutrophic Lake Vesijärvi, Southern Finland. Hydrobiologia 86: 81–85.

Kettunen, J., 1993. Design of limnological observations for detecting processes in lakes and reservoirs. In: M. Straškraba, J.G. Tundisi & A. Duncan (eds.), Comparative Reservoir Limnology and Water Quality Management. Kluwer Academic Publishers, Dordrecht: 139–146.

Kettunen, J. & M. Stenmark, 1982. Wind induced resuspension and phosphorus exchange between resuspended sediment and lake water. In: I. Bergström, J. Kettunen & M. Stenmark (eds.), Physical, Chemical and Biological Dynamics in Sediment. Div. Water Eng. Helsinki Univ. Technol. Report 26.

King, D. L., 1970. The role of carbon in eutrophication. J. Water Poll. Control Fed. 42: 2035–2051.

Kuenzel, L. E., 1969. Bacteria, carbon dioxide and algal blooms. J. Water Poll. Control Fed. 41: 1737–1747.

Kuusisto, E., 1981. Temperatures of Finnish water courses in 1961–1975 (in Finnish, with English summary). Publ. Water Research Inst. 44, 40 pp.

Lange, W., 1967. Effect of carbohydrates on the symbiotic growth of planktonic blue-green algae with bacteria. Nature 215: 1277–1278.

Lange, W., 1970. Cyanophyta-bacteria systems: Effects of added carbon compounds or phosphate on algal growth at low nutrient concentrations. J. Phycol. 6: 230–234.

Leonardson, L. & W. Ripl, 1980. Control of undesirable algae and introduction of algal successions in hypertrophic lake ecosystems. In: J. Barica & L-R. Mur (eds.), Hypertrophic Ecosystems. Junk, The Hague: 57–65.

Lind, O. T., T. T. Terrell & B. L. Kimmel, 1993. Problems in reservoir trophic-state classification and implications for reservoir management. In: M. Straškraba, J. G. Tundisi & A. Duncan (eds.), Comparative Reservoir Limnology and Water Quality Management. Kluwer Academic Publishers, Dordrecht: 57–67.

Lynch, M., 1980. *Aphanizomenon* blooms: Alternate control and cultivation by *Daphnia pulex*. In: W. C. Kerfot (ed.), Evolution and Ecology of Zooplankton Communities. Univ. Press New England, Hanover, N.H.: 299–304.

Niemi, Å., 1979. Blue-green algal blooms and N:P ratio in the Baltic Sea. Acta Bot. Fenn. 110: 57–61.

Nyholm, N., 1976. A mathematical model for the growth of phytoplankton. International Symposium Experimental Use Algal Cultures in Limnol. Sondefjord, Norway, 31 pp.

Redfield, A. C., 1958. The biological control of chemical factors in the environment. Amer. Sci. 46: 205–221.

Rhee, G-Y., 1978. Effects of N:P atomic ratios and nitrate limitation on algal growth, cell decomposition and nitrate uptake. Limnol. Oceanogr. 23: 10–25.

Rhee, G-Y. & I. J. Gotham, 1980. Optimum N:P ratios and coexistence of planktonic algae. J. Phycol. 16: 486–489.

Ripl, W., L. Leonardson, G. Lindemark, G. Andersson & G. Cronberg, 1979. Optimization of treatment plant / recipient system (in Swedish). Vatten 35: 96–103.

Sakshaug, E. & Y. Olsen, 1986. Nutrient status of phytoplankton biomass in Norwegian waters and algal strategies for nutrient competition. Can. J. Fish. Aquat. Sci. 43: 389–396.

Schindler, D. W., 1977. Evolution of phosphorus limitation in lakes. Science 195: 260–262.

Shapiro, J., 1973. Blue-green algae: Why do they become dominant. Science 179: 382–384.

Smith, V. H., 1983. Low nitrogen to phosphorus ratio favor dominance by blue-green algae. Science 221: 669–671.

Sommer, U., Z. M. Gliwicz, W. Lampert & A. Duncan, 1986. The PEG model of seasonal succession of planktonic events in fresh waters. Arch. Hydrobiol. 106: 433–471.

Stenmark, M., 1982. Nutrient Cycle in Lake Kuortaneenjärvi (in Finnish, with English summary). M.Sc. Thesis. Div. Water Eng., Helsinki Univ. Technol., Espoo, Finland, 74 pp.

Straškraba, M. & A. Gnauck, 1985. Freshwater Ecosystems. Modelling and Simulation. Elsevier, Amsterdam, 309 pp.

Tilman, D., 1977. Resource competition between planktonic algae: an experimental and theoretical approach. Ecology 58: 338–348.

Tilman, D., R. Kiesling, R. Steiner, S. S. Kilham & F. A. Johnson, 1986. Green, bluegreen and diatom algae: Taxonomic differences in competitive ability for phosphorus, silicon and nitrogen. Arch. Hydrobiol. 106: 473–485.

Varis, O., 1984. Water quality model for Lake Kuortaneenjärvi, a polyhumic Finnish lake. Aqua Fenn. 14: 179–187.

Varis, O., 1988. Temporal sensitivity of *Aphanizomenon flos-aquae* bloom formation – A whole-lake simulation study with input perturbations. Ecol. Modelling 43: 137–153.

Varis, O., 1989. Simulated impacts of flow regulation on blue-green algae in a short retention time lake. Arch. Hydrobiol. Beih. Ergebn. Limnol. 33: 181–189.

Varis, O., O. Malve & J. Kettunen, 1988. Blue-Green Algae in Lake Enjärvi – Modeling and analysis of periodicity. In: A. Marani (ed.), Advances in Environmental Modelling. Elsevier, Amsterdam: 507–524.

Varis, O., 1991. A canonical approach to diagnostic and predictive modelling of phytoplankton communities. Arch. Hydrobiol. 122: 147–166.

Varis, O., 1993. Cyanobacteria dynamics in a restored Finnish lake: a long-term simulation study. Hydrobiologia, in press.

Ward, A. K. & R. G. Wetzel, 1980. Interactions of light and nitrogen source among planktonic blue-green algae. Arch. Hydrobiol. 90: 1–25.

Young, P. C., 1977. A general theory of modeling for badly defined systems. In: G. C. Vansteenkiste (ed.), Modeling, Identification and Control in Environmental Systems. North-Holland / American Elsevier, New York: 103–135.

Young, P. C., 1983. The validity and credibility of models for badly defined systems. In: M. B. Beck & G. van Straten (eds.), Uncertainty and Forecasting of Water Quality. Springer-Verlag, Berlin: 69–98.

Zevenboom, W. & L-R. Mur, 1980. N_2-fixing cyanobacteria: Why they do not become dominant in Dutch, hypertrophic lakes. In: J. Barica & L-R. Mur (eds.) Hypertrophic Ecosystems. Junk, The Hague: 123–130.

Chapter IX

Design of limnological observations for detecting processes in lakes and reservoirs

J. Kettunen
Laboratory of Hydrology and Water Resources Management, Helsinki University of Technology, Rakentajanaukio 4 A, SF-02150 Espoo, Finland

Key words: reservoirs, lakes, limnology, sampling, mathematical models, water quality

Abstract

Ways of optimizing water quality sampling were studied in Lake Kuortaneenjärvi, Finland. The aim was to design a measurement programme supported by prior knowledge of the lake. The behaviour of the lake was represented by a nonlinear model developed earlier. Timing of sampling for algal counts and observations of inorganic P and N were designed to maximize the estimation accuracy of the model parameters. The design was constructed using the Simplex search procedure. Optimal sampling times coincided with the extreme values of the parameter sensitivity of the model output. When the number of observations was kept low, the algorithm converged towards a unique solution, even when the search started with rather crude initial guesses. An increased number of observations decreased the uniqueness. This was considered to be due to multi-collinearity of the postulated parameter values and correlated measurements. The optimality and robustness of the solution is discussed. It is concluded that a comprehensive sensitivity analysis should precede implementation of the design.

1. Introduction

Water quality models are widely applied in lake management and limnological research. Sophisticated simulation languages and parameter estimation packages were introduced during the 1970s and 1980s to aid modellers in data processing. Unique conclusions about the modelling process and the uncertainty inherent in it have been reported (e.g. Orlob, 1983; Beck & Van Straten, 1983). However, comprehensive calibration and validation of models is, in most cases, impossible because of a lack of field data. A large number of models have not been validated at all. It is thus clear that methodological effort should be directed towards more appropriate design of observations.

This study was motivated by the idea of using the modelling approach to guide the observation design. The idea is not new. Its roots go back to the times of the great pioneer of experimental design, R. Fisher. From his and later work a unique theory of linear experiments was developed. Box & Lucas (1959) extended the linear theory into nonlinear process studies. They illustrated the theoretical results with a simple model for chemical reactors equivalent to the Streeter & Phelps (1925) BOD-DO model, well known in water quality research. Nonlinear theory has been further developed by various scientists including Draper & Hunter (1966; 1967) and Atkinson & Hunter (1968).

In this study, experimental design methods are applied to outline the optimal timing of water quality observations in Lake Kuortaneenjärvi, Western Finland. The aim was to establish a sampling programme with the purpose of re-calibrating the most characteristic rate and temperature dependence parameters of cyanobacteria and other groups of algae in the model simulating the lake.

2. Notation and the optimality criterion

The system-experiment model is assumed to be represented in the observation interval $[t_0, T]$ by the structure (1)–(4)

$$\mathbf{x}'(t, \mathbf{p}) = \mathbf{f}[\mathbf{x}(t, \mathbf{p}), \mathbf{u}(t), t; \mathbf{p}], \mathbf{x}_0 = \mathbf{x}(t_0, \mathbf{p}) \quad (1)$$
$$\mathbf{y}(t, \mathbf{p}) = \mathbf{g}[\mathbf{x}(t, \mathbf{p}); \mathbf{p}] \quad (2)$$
$$\mathbf{z}(t_k, \mathbf{p}) = \mathbf{y}(t_k, \mathbf{p}) + \mathbf{e}(t_k), k = 1, 2, \ldots N \quad (3)$$
$$\mathbf{h}[\mathbf{x}(t, \mathbf{p}), \mathbf{u}(t), \mathbf{p}] \geq 0 \quad (4)$$

where $\mathbf{x} \in \mathbf{R}^n$ is the state vector of water quality and ' denotes the time derivative; $\mathbf{u} \in \mathbf{R}^r$ is the input vector; $\mathbf{y} \in \mathbf{R}^L$ is the output vector of the model. The dimension of $\mathbf{y}$ is not necessarily the same as the dimension of $\mathbf{x}$. $\mathbf{f}$ is a nonlinear vector valued function which defines the postulated structure of the model dynamics parameterized by a parameter vector $\mathbf{p} \in \mathbf{R}^p$; $\mathbf{g}$ is a nonlinear vector valued function which describes the known measurement process; $\mathbf{h}$ represents all the auxiliary, mainly differential and algebraic equality or inequality constraints known *a priori* relating to $\mathbf{x}$, $\mathbf{u}$ and $\mathbf{p}$; $\mathbf{z}(t_k, \mathbf{p}) \in \mathbf{R}^L$ is a vector valued discrete time measurement vector at measurement times t_k. $\mathbf{z}$ relates the noisy free model output $\mathbf{y}(t_k, \mathbf{p})$ to the measurement errors $\mathbf{e}(t_k)$, which are assumed to be white, Gaussian noise with zero mean and known variance $\sigma^2(t_k)$. N is the number of discrete samples in time.

The task of the observation design is to choose sampling times t_k that guarantee the most accurate parameter estimates of the simulation model. To solve the problem, it is assumed that the model given below is structurally correct and that the mean population of the parameter vector is known *a priori*. The parameter vector is assumed to be normally distributed, having the mean value $\mathbf{p}$ and variance-covariance $\mathbf{V}_0(\mathbf{p})$.

Determination of an optimal sampling schedule can be tackled using the Cramer-Rao theorem (e.g. Goodwin & Payne, 1977). According to this, the covariance matrix $\mathbf{V}(\mathbf{p})$ for all unbiased *a posteriori* parameter estimates has the lower bound given by the inverse of the Fisher information matrix $\mathbf{M}(\mathbf{p})$:

$$\mathbf{V}(\mathbf{p}) \geq \mathbf{M}^{-1}(\mathbf{p}) \quad (5)$$

$\mathbf{V}(\mathbf{p})$ is a function of the number of samples and analyses for each measurement vector $\mathbf{z}$ and of the timing of sampling. The uncertainty of the prior parameter values also affects $\mathbf{V}(\mathbf{p})$. Matrix $\mathbf{V}$ is given by the equation:

$$\mathbf{V}(\mathbf{p}) = (\mathbf{S}^T \mathbf{P}^{-1} \mathbf{S} + \mathbf{V}_0^{-1}(\mathbf{p}))^{-1} \quad (6)$$

where $\mathbf{P}$ denotes the variance-covariance matrix of measurement error, and $\mathbf{S}$ is determined by the following equation:

$$\mathbf{S} = \begin{matrix} \delta y_1(t_1)/\delta p_1 & \cdots & \delta y_1(t_1)/\delta p_p \\ \delta y_2(t_1)/\delta p_1 & \cdots & \delta y_2(t_1)/\delta p_p \\ \vdots & & \vdots \\ \delta y_L(t_1)/\delta p_1 & \cdots & \delta y_L(t_1)/\delta p_p \\ \vdots & & \vdots \\ \delta y_{L-1}(t_N)/\delta p_1 & \cdots & \delta y_{L-1}(t_N)/\delta p_p \\ \delta y_L(t_N)/\delta p_1 & \cdots & \delta y_L(t_N)/\delta p_p \end{matrix} \quad (7)$$

where the partial derivatives are evaluated at the known (*a priori*) nominal values of the parameter vector.

A proper choice of measurements will minimize the uncertainty of *a posteriori* parameter estimates. Thus the uncertainty will approach the theoretical lower bound $\mathbf{M}^{-1}$ given by the Eq. (5).

The optimal sampling strategy $t_1^*, t_2^*, \ldots, t_N^*$ for parameter estimation can be obtained by maximizing det $(\mathbf{V}^{-1}(\mathbf{p}))$ (Bard, 1974). In the literature on experimental design, the maximization criterion is called D-optimality (e.g. St. John & Draper, 1975).

3. The Lake Kuortaneenjärvi model

The prior field data used in this work were collected from Lake Kuortaneenjärvi, Western Finland, in 1980. Twenty-three sets of samples were taken for water chemical analyses, temperature and chlorophyll-a. Ten algal counts were made. The water budget of the lake was constructed according to daily observations of inflows, outflows and precipitation. Daily radiation and wind velocity records were obtained from nearby meteorological stations. The lake and the data have been described in more detail by Stenmark (1982).

Varis (1984a,b) constructed a model simulating the main transformations of phosphorus and nitrogen, and separately, the biomasses of nitrogen-fixing cyanobacteria and other planktonic algae in the lake. The model state variables x are given in Table I. The structure of the model is illustrated in Fig. 1.

Table I. State variables of the model used in the observation design.

Symbol	Meaning	Unit	Remarks
DIP	Dissolved inorganic phosphorus	mg m^{-3}	
DIN	Dissolved inorganic nitrogen	mg m^{-3}	
C	Biomass of cyanobacteria	mg m^{-3}	Constraint of simulation C $\geq$ 5 mg m^{-3}
F	Biomass of phytoplankton other than C	mg m^{-3}	Constraint F $\geq$ 50 mg m^{-3}
CN*	Nitrogen in cyanobacteria cells	mg m^{-3}	
CP*	Phosphorus in cyanobacteria cells	mg m^{-3}	
FN*	Nitrogen in phytoplankton cells	mg m^{-3}	
FP*	Phosphorus in phytoplankton cells	mg m^{-3}	
ALP	Alloctonous detrital phosphorus	mg m^{-3}	
AUP	Autochtonous detrital phosphorus	mg m^{-3}	
ND	Detrital nitrogen	mg m^{-3}	
POSED	Organic phosphorus in sediment	mg m^{-3}	1
PSSED	Inorganic phosphorus in sediment	mg m^{-3}	1
NOSED	Organic nitrogen in sediment	mg m^{-3}	1
NSSED	Inorganic nitrogen in sediment	mg m^{-3}	1

* Nutrient fractions of cyanobacteria and other phytoplankton are dynamic state variables (For details, see Varis, 1984b).
^{1} – per volume of lake water.

The model for Lake Kuortaneenjärvi has 33 parameters. In his analysis of average sensitivities (e.g. Jørgensen & Mejer, 1978), Varis (1984a) concluded that, of the model output **y**, the concentrations of inorganic nutrients and phytoplankton and cyanobacteria biomasses are extremely sensitive to errors in the growth and respiration parameters and the temperature dependencies of algal growth. For this reason it was found necessary to design further measurements in order to re-evaluate them.

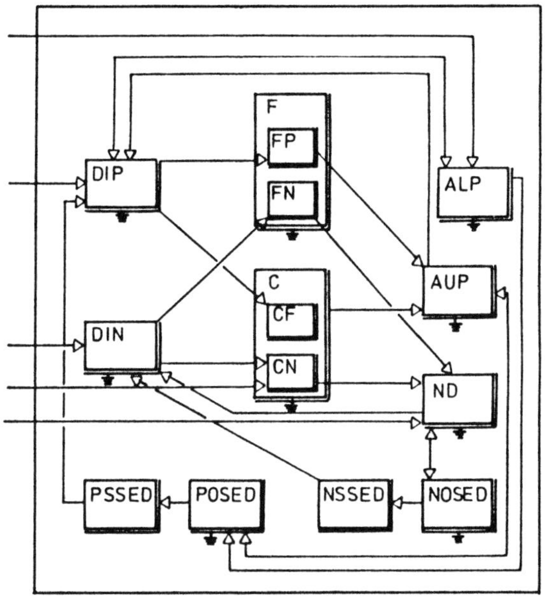

4. A priori sensitivity analysis of parameters and construction of the optimal design

The observation design procedure started with an *a priori* sensitivity analysis of the model parameters. The parameters included in the analysis and their nominal values are given in Table II. The output variables analyzed were soluble inorganic phosphorus (DIP) and nitrogen (DIN), and biomasses of phytoplankton (F) and cyanobacteria (C). Simulated model output using nominal para-

Fig. 1. Flow diagram of the model of Lake Kuortaneenjärvi (the meaning of state variables is given in Table I.)

Table II. Parameters in the sensitivity analysis.

Symbol	Meaning	Nominal value	Unit
p_1	Max. growth rate of F	0.65	d^{-1}
p_2	Respiration rate of F	0.16	d^{-1}
p_3	Temperature coeff. of F. (growth in van 't Hoff's eq.)	1.02	–
p_4	Max. growth rate of C	0.60	d^{-1}
p_5	Respiration rate of C	0.185	d^{-1}
p_6	Temperature coeff. of C (growth in van 't Hoff's eq.)	1.09	–

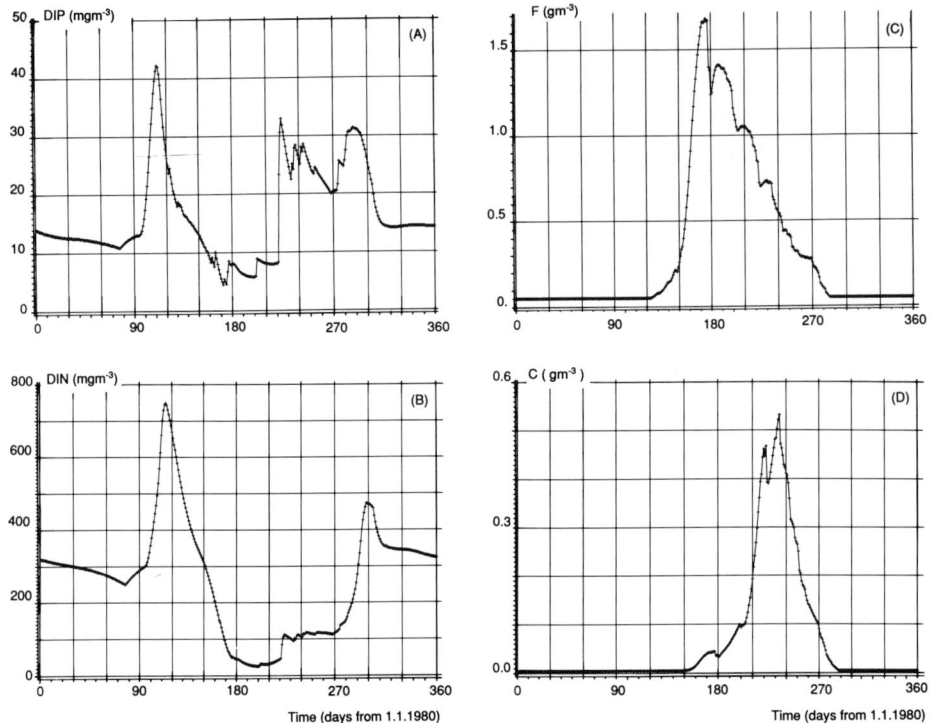

Fig. 2. Simulated values of soluble inorganic phosphorus (A), soluble inorganic nitrogen (B), algal biomass (C) and biomass of cyanobacteria (D) in Lake Kuortaneenjärvi, 1980.

meter values is shown in Fig. 2. The computational results correspond well to the values observed in the lake in 1980 (Varis, 1984b; 1992).

Optimal designs were studied in two different cases. First, the optimal timing of algal counts was considered. Counts of nitrogen-fixing cyanobacteria and other planktonic algae were regarded as independent components of the measurement vector. Second, the observation vector was considered to comprise of DIP and DIN measurements and algal counts.

Nelder & Mead's (1965) direct search Simplex algorithm was used to maximize det ($\mathbf{V}^{-1}(\mathbf{p})$). This has been documented at Fortran-code level by O'Neill (1971), Chambers & Ertel (1974) and Benyon (1976). The elements of the sensitivity matrix $\mathbf{S}$ (Eq. 7), denoted by $s(t_k)_{ij}$ were substituted by first-order difference approximations according to the formula:

$$s(t_k)_{ij} \approx \{ y_i(t_k, p_j + \Delta p_j) - y_i(t_k, p_j) \} / \Delta p_j$$
$$i = 1, \ldots L, j = 1, \ldots p, k = 1, \ldots, N \qquad (8)$$

The perturbation of the parameters Δp_j was 1% of the nominal value. To avoid computational problems, all the components of the sensitivity matrix were scaled by dividing them by the maximum values of the respective output variables and multiplying them by the respective prior nominal parameter values. As an exception to this, algal and cyanobacterial biomasses were scaled by the maximum value of the total algal biomass of the lake. It was assumed that all the components of the scaled observation vector had equal variances. Furthermore, the parameters and observations were assumed to be non-correlated. The assumptions allowed the optimality criterion to be reduced to the form:

$$\max J = \det (\mathbf{V}_0^{-1} + \mathbf{S}^T \mathbf{P}^{-1} \mathbf{S}) \approx$$
$$\det (\lambda \mathbf{I} + \mathbf{S}^T \mathbf{I} \mathbf{S}) = \det (\lambda \mathbf{I} + \mathbf{S}^T \mathbf{S}) \qquad (9)$$

where $\mathbf{I}$ denotes the identity matrix. In this study prior parameter values were assumed to be accurate, and the value $\lambda = 0$ was included in the criterion.

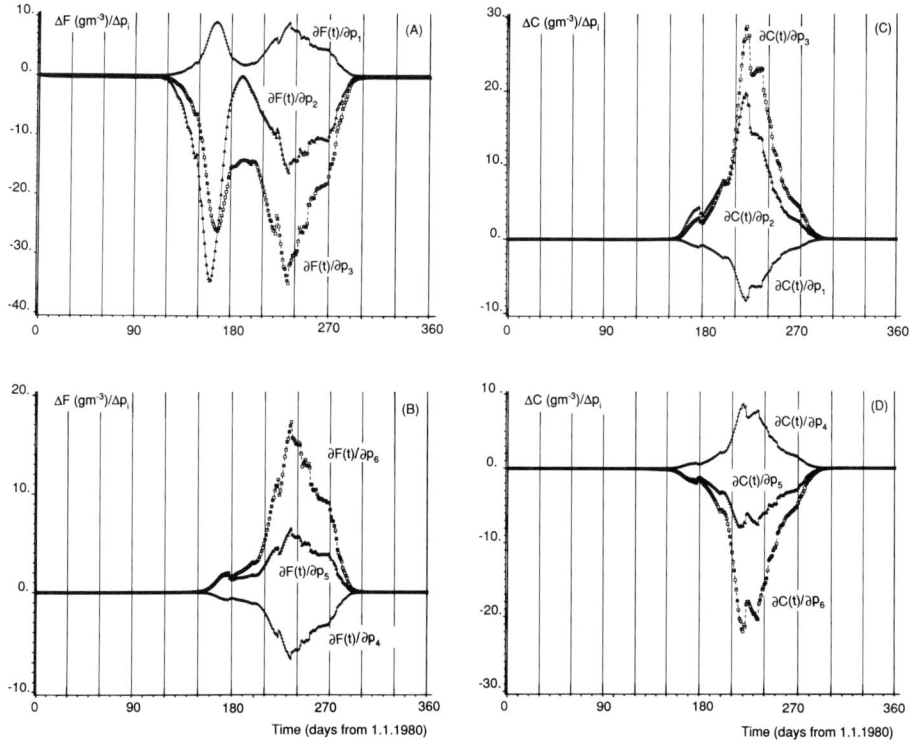

Fig. 3. (A)–(B) Sensitivities of F output to parameters p_1–p_3 and p_4–p_6, and (C)–(D) sensitivities of C output to parameters p_1–p_3 and p_4–p_6. Unscaled values of the model of Lake Kuortaneenjärvi, 1980. Time axis indicates the number of days from the beginning of the year.

5. Optimal timing of samples for algal counts

The unscaled sensitivities of phytoplankton and cyanophyta output to the parameters studied are shown in Fig. 3 A-D. Physically these values imply that unit error in the parameter value will lead to an error in output that is indicated by the respective sensitivity value. However, it should be kept in mind that the results are valid only in close proximity to the nominal solution.

One can see that the F predicted by the model is extremely sensitive to the parameters p_1–p_3 between days 155–175 and 210–270. The maximum sensitivity of the cyanobacteria to all the parameters falls during the period 200–250. Thus, the sensitivity analysis would suggests that the most intensive sampling of algal counts should be carried out during the 20-day period beginning on 5th June (day No. 155) and the 60–70-day period beginning on 20th July (day No. 200).

Some 27 Simplex runs were carried out to establish the optimal strategy for algal sampling. Algorithm was used to choose 6 to 10 observation points between days 90 and 300. The initial guesses for optimization were the uniform sampling schedules in time. For example, one of the runs had the initial guess that optimal sampling would begin at the end of March (day No. 90) and last till the end of October (day No. 300), with a sampling frequency of three weeks. However, owing to the results of the sensitivity analysis, most of the initial guesses were initiated approximately on day 135 or later, and uniform sampling was considered to continue until day 270 or earlier. Figure 4 summarizes of optimized timings.

The main finding was that the optimal times for sampling occurred during the summer period (dates 140–260). This is not surprising. The most natural time to measure growth and respiration rates and their temperature dependencies is when the respective processes take place in the ecosystem. However, the result indicated that the algorithm is capable of detecting the optimal periods.

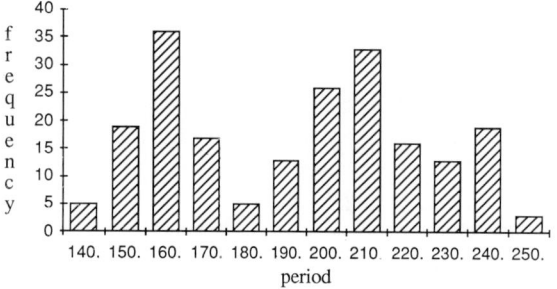

Fig. 4. Results of 27 Simplex runs to optimize algal count sampling. Frequencies of occurrence of optimal sampling times in 10-day periods. Figures of abscissa-axis indicate the first day of the period.

Each of the 27 runs included an optimal date between days 160 and 170. Few of the runs located two or more sampling times during that period. Another period in every optimum was between days 210 and 220. The periods 150–160, 200–210 and 240–250 occurred in two-thirds of the runs and occasionally there were several samplings during those periods.

Several trials were conducted in an effort to include more than 15 sampling points in the sampling programme. They usually ended up with either replicating the most frequent sampling times given by previous analysis or suggesting uniform sampling. The latter was always the case when optimization of more than 50 sampling points was attempted.

As an overall result, it was concluded that the optimal periods for algal counts are those indicated by the 27 optimization runs. Intensively sampling should be carried out during them.

6. Optimal timing of samples for nutrient analyses and algal counts

The next phase of the study was to increase the dimension of the observation vector from two to four. Simultaneous measurements of DIP, DIN, F and C were considered. First the sensitivities of DIP and DIN were analyzed. The results are shown in Fig. 5.

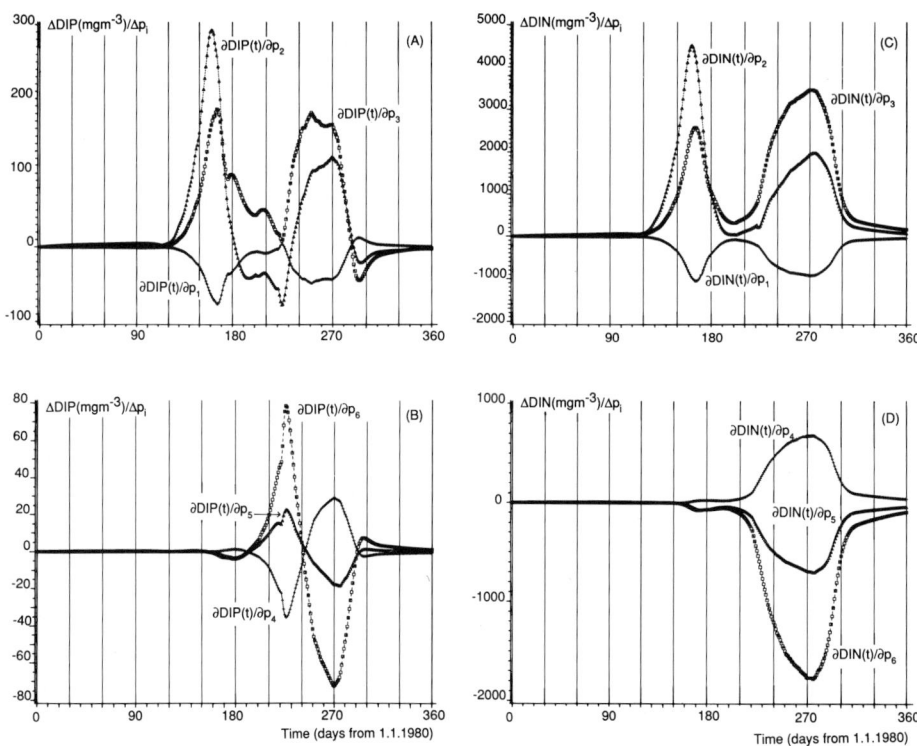

Fig. 5. Sensitivities of DIP output (A)–(B) and DIN output (C)–(D) to parameters p_1–p_3 and p_4–p_6. Unscaled values of the model for Lake Kuortaneenjärvi, 1980. Time axis indicates the number of days from the beginning of the year.

The sensitivity of the output of inorganic soluble phosphorus, DIP, has two main maxima to parameters p_1–p_3. These occur during periods 150–180 and 230–280. DIP is most sensitive to parameters p_4–p_6 between periods 210–230 and 260–280. The output of soluble inorganic nitrogen is sensitive to any of the parameters during periods 160–175 and 250–290. As shown above, the sensitivity analysis suggests that the most intensive sampling should be carried out in June and during the 80-day period beginning at the end of July.

Six optimization runs were performed to find the optimal sampling times. The runs are summarized in Fig. 6. The analysis resulted in a unique solution. The best way of using sampling resources to detect the parameters of growth and respiration rates and their temperature dependencies would be to collect the samples during the following six periods, 160–170, 180–190, 200–210, 210–220, 230–240 and 240–250.

Several optimization runs were performed to choose more than 15 sampling times in the programme. The results indicated that if uniform sampling was given as the initial guess, Simplex was not able appreciable to improve the goal function values and usually ended up with the initial guess. If suitable replication of the optimal sampling times found above was given as the initial guess, the goal function values were considerably higher than with uniform sampling schedules. The conclusion is, that the sampling should be concentrated on those times.

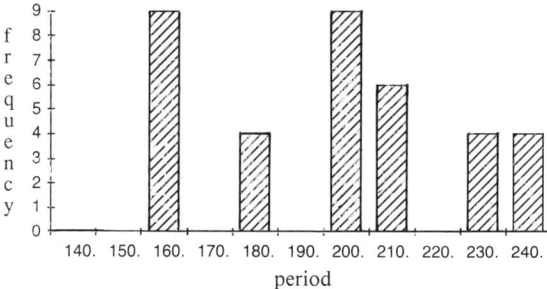

Fig. 6. Results of 6 Simplex runs to optimize sampling for algal counts and DIP and DIN analyses. Frequencies of occurrence of optimal sampling times in 10-day periods. Figures of abscissa-axis indicate the first day of the period.

7. Discussion and conclusions

The generally accepted result is that the maximum number of D-optimal samples equals the number of parameters to be estimated. If extra sampling resources are available, they should be used for suitable replications of a set of observations whose number equals p. According to Box (1970), this result is supported not by any explicative theory, but by the evidence of the practical experience. The present study also supports it. The slight non-uniqueness in timing the samples for algal counts is attributed to the multicollinearity of the model parameters and observations that was not given adequate consideration in this study. It could equally be a result of the properties of the search algorithm, which was not analyzed in detail in this study. On the other hand, Evans (1979) discussed the performance of Simplex in the linear modelling context and concluded that the results achieved with it surpass those of sequential algorithms usually applied in the experimental design.

The Simplex algorithm has several advantages for observation designs, the main one being its simplicity. Another is the ability to fix several sampling points simultaneously. In the sampling design, Simplex may not be the most effective if the initial guesses are poor. This is because it may try the same choices of sampling instances several times.

The D-criterion of optimal designs is restricted by the assumption that the structure of the model chosen is correct and that the magnitude of the parameters is known in advance. If not so, the optimality of the designs is of minor value. However, even in such cases it is recommended that the criterion be used. It will, namely, help the researcher to falsify his hypothesis of the model structure or at least the parameter values, which is usually not the case with the non-designed experiments. Furthermore, there are no restrictions to formulating several structures of the model *a priori* and constructing a combination of designs to test them.

Two possible applications of the method presented above are apparent. On the one hand it will guide the design of observations *a priori*. When there is a large number of parameters optimal timing of sampling cannot be judged from the sensitivity plots. The method actually assembles the information in a very compact form thus facilitating appropriate designs. On the other hand, it can guide the modeller to choose the optimal periods of historical data to be used in judging the inadequacies of the model structure.

Application of the method led to the recommendation that sampling should be concentrated on a few intensively studied periods. A similar conclusion was reached by Jørgensen (1980). Before the programme is implemented, it should be studied further in the light of different hypotheses and using more appropriate statistical analysis of the model. Also the robustness of the solution should be considered.

Climatic conditions cannot be controlled in the 'observation experiment'. Therefore the sampling schedule should be flexible. This can be achieved by adding an updating procedure to guide the real time realization of the schedule.

There may be arguments why traditional procedures are preferable to modelling oriented sampling. One objection to the latter is that uniform sampling even in time, when used for estimating averages and detecting trends in data, requires impossibly dense sampling programs as pointed out by Somlyody et al. (1986).

Acknowledgements

This study was carried out at the Laboratory of Hydrology and Water Resources Engineering, Helsinki University of Technology. I thank Prof. P. Vakkilainen for encouraging me throughout the work. The comments of M. Straškraba and his colleagues are gratefully acknowledged. I would also like to thank Mr. O. Varis and Mr. H. Sirviö for their cooperation and the inspiring ideas they contributed to my study. Financial support from the Maa- ja vesitekniikan tuki ry association is gratefully acknowledged.

References

Atkinson, A. C. & W. G. Hunter, 1968. The design of experiments for parameter estimation. Technometrics 10: 271–289.
Bard, Y., 1974. Nonlinear Parameter Estimation. Academic Press, New York, 341 pp.
Beck, M. B. & G. Van Straten (eds.), 1983. Uncertainty and Forecasting of Water Quality. Springer-Verlag, Berlin, 358 pp.
Benyon, P. R., 1976. Function minimization using a Simplex procedure. Applied Statistics 25: 97.
Box, G. E. P. & H. L. Lucas, 1959: Design of experiments in nonlinear situations. Biometrika 46: 77–99.
Box, M. J., 1970. Some experiences with a nonlinear experimental design criterion. Technometrics 12: 569–589.
Chambers, J. M. & J. E. Ertel, 1974. Function maximization using a Simplex procedure. A remark on algorithm AS 47. Applied Statistics 23: 250–251.
Draper, N. R. & W. G. Hunter, 1966. Design of experiments for parameter estimation in multiresponse situations. Biometrika 53: 525–533.
Draper, N. R. & W. G. Hunter, 1967. The use of prior distributions in the design of experiments for parameter estimation in non-linear situations, multiresponse case. Biometrika 54: 662–665.
Evans, J. W., 1979. Computer augmentation of experimental designs to maximize $|X^TX|$. Technometrics 21: 321–330.
Goodwin, C. & R. L. Payne, 1977. Dynamic System Identification: Experiment Design and Data Analysis. Academic Press, New York, 261 pp.
Jørgensen, S. E., 1980. Lake Management. Pergamon Press, Oxford, 167 pp.
Jørgensen, S. E. & H. Mejer, 1978. Examination of a lake model. Ecol. Modelling 4: 253–279.
Nelder, J. A. & R. Mead, 1965. A Simplex method for function minimization. The Computer Journal 7: 308–313.
O'Neill, R., 1971. Function minimization using a Simplex procedure. Algorithm AS 47. Applied Statistics 20: 338–345.
Orlob, G. T. (ed.), 1983. Mathematical Modeling of Water Quality: Streams, Lakes and Reservoirs. International Series on Applied Systems Analysis 12. IIASA. Wiley & Sons, Chichester, 518 pp.
Somlyody, L., J. Pinter, L. Koncsos, I. Hanacsek & I. Juhasz, 1986. Estimating averages and detecting trends in water quality data. Monitoring to Detect Changes in Water Quality Series. (Proceedings of the Budapest Symposium, July 1986). IAHS Publ. No. 157: 61–69.
St. John, R. C. & N. R. Draper, 1975. D-optimality for regression designs: A review. Technometrics 17: 15–23.
Stenmark, M., 1982. Nutrient Cycle in Lake Kuortaneenjärvi. (In Finnish with English summary.) M.Sc. Thesis, Division of Water Engineering, Helsinki University of Technology, Helsinki, 74 pp.
Streeter, H. W. & E. B. Phelps, 1925. A study of the pollution of the Ohio River. III Public Health Bulletin, U.S Public Health Service. 141, 75 pp.
Varis, O., 1984a. A water quality model for Lake Kuortaneenjärvi. (In Finnish with English summary) M.Sc. Thesis. Division of Water Engineering, Helsinki University of Technology, Helsinki, 62 pp.
Varis, O., 1984b. Water quality model for Lake Kuortaneenjärvi, a polyhumic Finnish lake. Aqua Fennica 14: 179–187.
Varis, O., 1993. Impacts of growth factors on competitive ability of blue-green algae analyzed with whole-lake simulation. In: M. Straškraba, J. G. Tundisi & A. Duncan (eds.), Comparative Reservoir Limnology and Water Quality Management. Kluwer Academic Publishers, Dordrecht: 127–137.

Chapter X

Remote sensing estimation of total chlorophyll pigment distribution in Barra Bonita Reservoir, Brazil

E. M. L. M. Novo,[1] C. Z. F. Braga[1] & J. G. Tundisi[2]
[1] *National Institute for Space Research, Caixa Postal 515, São José dos Campos, São Paulo-12201-970, Brazil;* [2] *USP/EESC/Centre for Water Resources and Applied Ecology, School of Engineering at São Carlos, Saõ Carlos, São Paulo-13560, Brazil*

Key words: reservoirs, chlorophyll, remote sensing, tropics

Abstract

This paper reports preliminary results of a research project carried on the Barra Bonita Reservoir to assess how the current remote sensing technology can provide useful information for limnological studies. Water samples collected simultaneously to the Landsat-5 overpass were used to calibrate Thematic Mapper data and to generate an empirical model for estimating the total surface chlorophyll pigment distribution in the reservoir.

1. Introduction

The management of inland water is based largely on empirical relations between forcing functions and responses of biological communities. Those models, however, were mostly based on a limited subsampling of inland water, and large areas of freshwater comprising a diversity of natural ecosystems usually are not adequately sampled.

Remote sensing technology can be applied to water studies so as to make those subsamples more representative of the variety of freshwater ecosystems. Water color data have been collected since 1968 (Clark & Ewing, 1974) as a remote indicator of the water biological properties. Since 1978 a special sensor (Coastal Zone Color Scanner) has been used to assess mesoscale processes in the ocean. Recent work is mainly focused on the estimation of water column primary productivity using pigment concentration derived from satellite data (Kirk, 1983). Inland waters, however, have several distinctive characteristics which requires special attention if remote sensing is to succeed (NASA, 1987). Current remote sensing technology has had modest success in the examination of inland waters, mainly because of wide range of optical conditions occurring in space and time in these ecosystems. A large amount of research is still needed to make remote sensing applications on inland water management operational.

This paper reports preliminary results of a research project oriented to assess the information content of TM/Landsat data as far as water quality parameters are concerned.

2. Theoretical background

Water remote sensing studies have involved two main directions. The first research line (McCluney, 1976; Philpot & Klemas, 1981; Bricaud & Sathyendranath, 1981; Witte *et al.*, 1982) is oriented to assess what are the water components affecting water spectra. The second research line is oriented to the use of orbital remote sensing data to estimate water parameters (Welby *et al.*, 1981; Nielsen *et al.*, 1983; Schiebe *et al.*, 1984; Braga, 1988).

Water components affecting water spectra were classified into four groups according to laboratory researches performed by Bricaud & Sathyendranath (1981): live phytoplankton; biogenic detritus matter associated to phytoplankton; terrigenous matter and suspended sediments; and dissolved organic matter. Spectral features of these components were measured in laboratory and parameters (concentration, cell diameter, etc) affecting water spectra were identified.

Experimental results showed that absorption efficiency per unit of chlorophyll concentration varied widely from one water body to the other. Figures 1 and 2 illustrate some of those laboratory experiments.

For spherical cells experimental results sug-

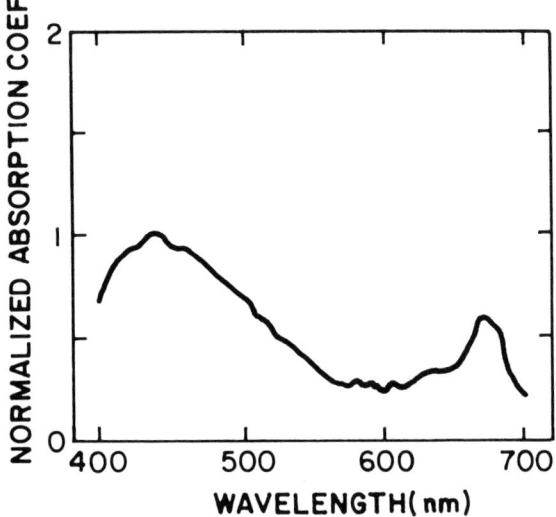

Fig. 1. Absorption curve (normalized at 440 nm) of phytoplankton and covarying detrital matter. Obtained in the study of Prieur & Sathyendranath according to Bricaud & Sathyendranath (1981).

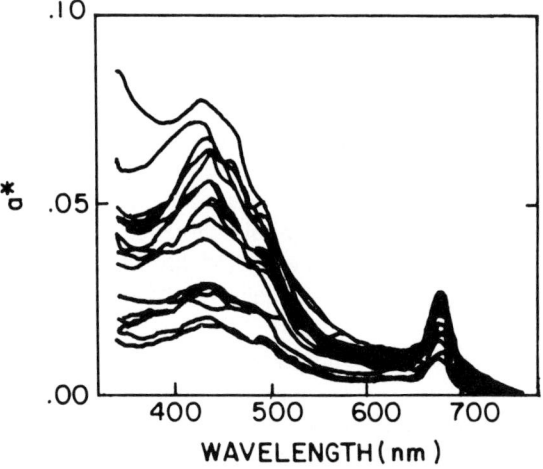

Fig. 2. Spectral values of specific absorption, a* (expressed in m^{-1} (mg (CHA + pheo-a) m^{-3})$^{-}$) determined on intact cells during exponential growth, for different algal species in batch cultures. From Bricaud & Sathyendranath (1981).

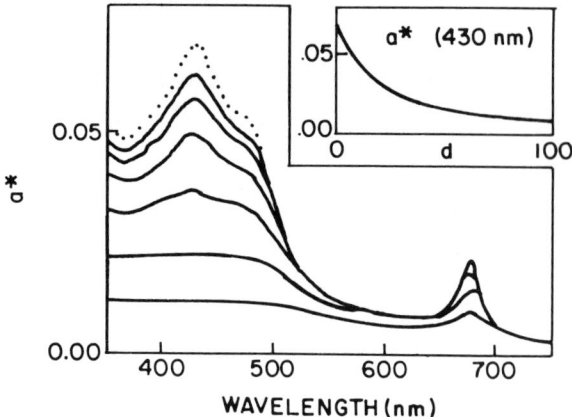

Fig. 3. Change in spectral values of the specific absorption coefficient a* (in m^{-1} (mg CHA m^{-3})$^{-1}$) for algal cells containing the same material but differing size (diameter, d, in μm). The dotted curve represents the spectral absorption values of this material (arbitrarily chosen) as dispersed in solution (d → 0). The continuous curves, from the upper to the lower one, correspond respectively to the diameters 2, 4, 8, 16, 32 and 64 μm. Insert: variations with the cell diameter of the specific absolute value of absorption at 430 nm. From Bricaud & Sathyendranath (1981).

gested a decrease in the absorption coefficient as the cell size increases (Fig. 3). A comprehensive review on this subject can be found in Braga (1988).

MSS/Landsat data were applied to detect algal blooms of the water surface. MSS band 6 (red/near-infrared wavelengths) was found to be more sensitive to algal blooms (Nielsen *et al.*, 1983). Similar results were found (Schiebe *et al.*, 1984) using TM/Landsat data. According to them, the best waveband for estimating chlorophyll concentration was TM3 (red wavelengths).

3. Study area

Barra Bonita Reservoir is located in the Tiete River, at 22° 29′ S and 48° 34′ W. Tiete drainage Basin has 32,330 km^2 and the reservoir open water is around 324 km^2. The reservoir catchment basin encompasses the most populated and urbanized area in Brazil. Sugar cane plantation is the main agricultural activity. More information on the area can be found in Calijuri (1988).

4. Methodology

Two data sets were used in this study: a) digital data from TM/Landsat bands 1,2,3,4 (path 220; row 76) referring to July, 17th 1988; and b) water quality parameters determined *in situ* and in laboratory according to procedures defined in Golterman *et al.* (1978). Table I presents the spectral range of each TM band used in this study. The following parameters were determined for each of the eight water sampling stations: Secchi depth; surface temperature; light penetration; total suspended solids (organic and inorganic); total chlorophyll pigment concentration; and nutrients.

TM digital data were processed according to the methodology suggested by Godoy & Novo (1989). An average of 16 pixels was obtained for each sample area to derive water reflectance values in the four TM bands used.

Water sample variables and reflectance data were submitted to linear correlation analysis. Stepwise multiple regression was applied to derive a model to estimate chlorophyll concentration from combinations of different wavebands. The model was then implemented into the digital image processing system to classify the chlorophyll concentration in the entire reservoir.

Table I. Spectral range of utilized TM/Landsat-5 bands.

Band	Spectral range (nm)
1	(450–520)
2	(520–600)
3	(630–690)
4	(760–900)

5. Results and discussion

5.1. Correlation analysis

Correlation coefficients between optically active water variables and remotely sensed reflectance are presented in Table II.

Only chlorophyll concentration presented significant linear correlation with the remotely sensed reflectance in the visible bands at the 5% significance level. The correlation coefficients were inverse for the whole visible spectrum, what was

Table II. Correlation coefficients between water sample variables and remotely sensed reflectance ($n = 30$).

	CHA	TSS	SECCHI
B1	−0.59	−0.29	−0.19
B2	−0.77	−0.48	−0.05
B3	−0.74	−0.51	−0.07
B4	−0.30	−0.21	−0.12

B1 = average reflectance of band 1.
B2 = average reflectance of band 2.
B3 = average reflectance of band 3.
B4 = average reflectance of band 4.
CHA = Total chlorophyll pigment concentration.
TSS = Total suspended solids.
SECCHI = Secchi depth.

excepted only for bands 1 and 3, which correspond to the chlorophyll absorption bands.

For band 2 the inverse correlation was not expected, since an increase in chlorophyll concentration tends to shift the maximum water reflectance towards the green (Clark & Ewing, 1974). Figure 4, however, suggests that for certain types of aquatic systems the correlation can be inverse.

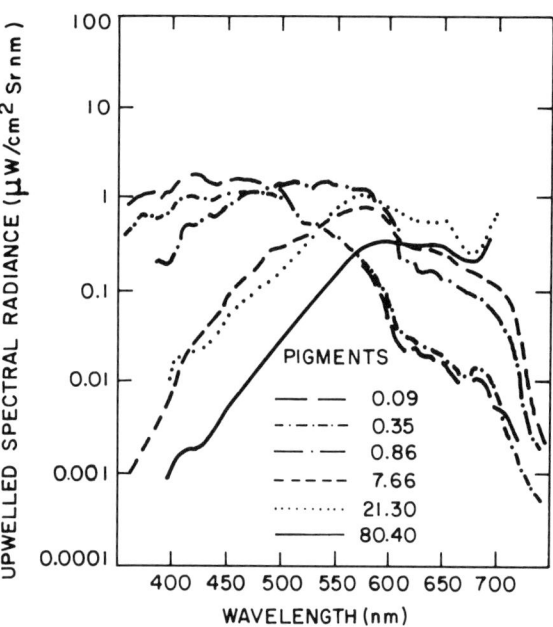

Fig. 4. Upwelled spectral radiance as a function of chlorophyll pigment concentration for various ocean waters. From NASA (1987).

The 80 µg l⁻¹ chlorophyll concentration curve presents the lowest radiance value in the blue region (430 nm), that is in accordance with the theory. It would be also expected a very high radiance in the green region (550 nm) for this curve, too, which did not happen. The 0.86 µg l⁻¹ curve also presented an anomalous behaviour in the green region but not in the blue one. In the green region the radiance does not follow a pattern either increasing or decreasing as the concentration does. In the red region the radiance value increases as the chlorophyll concentration increases. Considering that all the water samples used to build the figure presented very low TSS concentration, departure from theory can be explained by differences in phytoplankton composition. Literature shows the influence of the concentration range on the spectral behavior of the optically active substances (Kirk, 1983).

Figure 5 shows the effect of changes in turbidity and chlorophyll concentration on the water spectral reflectance. For the red region, considering the chlorophyll concentration, it would be expected the highest absorption coefficient for curve c, followed by a, b and d, respectively. If turbidity is assessed, however, one can conclude that this variable is controlling water spectral response. The absorption coefficient increases with increasing turbidity for the whole visible spectrum, except for the blue region (400–440 nm).

Another important aspect to explain the inverse correlation coefficient is then the presence of other water components affecting the turbidity. Those components can overcome the effect of chlorophyll absorption on the water spectra.

The effect of the attenuation coefficient over the water column actually sensed by TM system was also assessed by ratioing the chlorophyll and TSS concentrations to the Secchi depth (Braga, 1988). Results in Table III show a decrease in the correlation between chlorophyll and remotely sensed reflectance when ratioed data is used and an increase for TSS case. These differences between ratioed and raw data can indicate the chlorophyll and TSS distributions in the water column. Chlorophyll tends to be concentrated in the euphotic remotely sensed depth, while TSS is varying within the whole water column.

Table III. Correlation coefficient between normalized water variable and remotely sensed reflectance.

	CLSECCHI	TSECCHI
B1	−0.362	−0.311
B2	−0.624	−0.547
B3	−0.603	−0.566
B4	−0.221	−0.289

where:
CLSECCHI = Normalized total chlorophyll pigment concentration.
TSECCHI = Normalized total suspended solids.

5.2. The chlorophyll model

Based upon the correlation analysis one could identify chlorophyll as the major measured parameter to control water reflectance. By inverting the physical model (Curran & Hay, 1986), chlorophyll concentration could be estimated by using remotely sensed reflectance as independent variables to run a stepwise regression algorithm. The resulting model is described in Table IV.

As one can observe the model for estimating chlorophyll concentration included band TM4 which originally presented the lowest correlation.

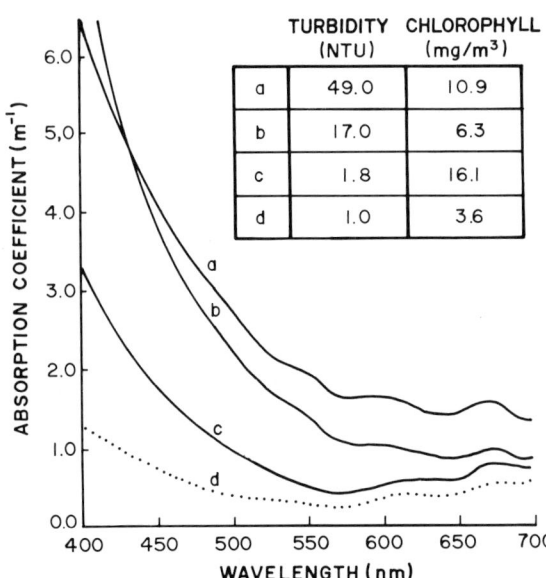

Fig. 5. Total absorption spectra of various natural waters in southeastern Australia. The values of turbidity (NTU) and total chlorophyll content (mg m⁻³) for each curve are listed beside. Adapted from Kirk (1983).

Table IV. Stepwise selection for chlorophyll.

Variables in the model	Regression coefficient	Adjusted R squared	Standard error of estimate
B2	−0.431		
B4	1.324	0.861	0.296
Const	8.294		

The model accounts for the colinearity among the independent variables (Wonnacott & Wonnacott, 1977).

The model accounts for 86% of the chlorophyll variation producing a standard error of 0.296 μg l^{-1}. In the range of variation in the chlorophyll concentration (3.96–6.15 μg l^{-1}) it represents a 6% error in relation to the average.

The model was then implemented into the digital processing system producing the spatial distribution of the chlorophyll concentration in the Barra Bonita Reservoir (Fig. 6).

Four concentration classes were defined in order to fit the Kratzer & Brezonik (1981) Trophic Index Classification. According to their classification, oligotrophic aquatic systems present a Trophic State Index ranging from 21 to 40, which corresponds to chlorophyll concentrations up to 2.6 μg l^{-1}; mesotrophic systems correspond to concentration around 6.4 μg l^{-1} and eutrophic systems to concentration over that value.

Using chlorophyll distribution as an indicator of the reservoir trophic state one can observe that during the satellite overpass most of the water presented mesotrophic conditions with patches of eutrophic water concentrated in the Piracicaba River. Based on this information areas of pollutant discharges can be better identified and controlled by the environmental protection agencies.

6. Conclusion

These preliminary results were applied to optimize data collection of the ongoing research, since it is a long term project carried out through scientific cooperation between the National Institute for Space Research and the Centre for Water Resources and Applied Ecology (CRHEA), Brazil. The chlorophyll distribution map is in agreement with the existing results obtained through conventional limnological studies performed by CRHEA during the last 10 years, giving to them the missing synoptic view.

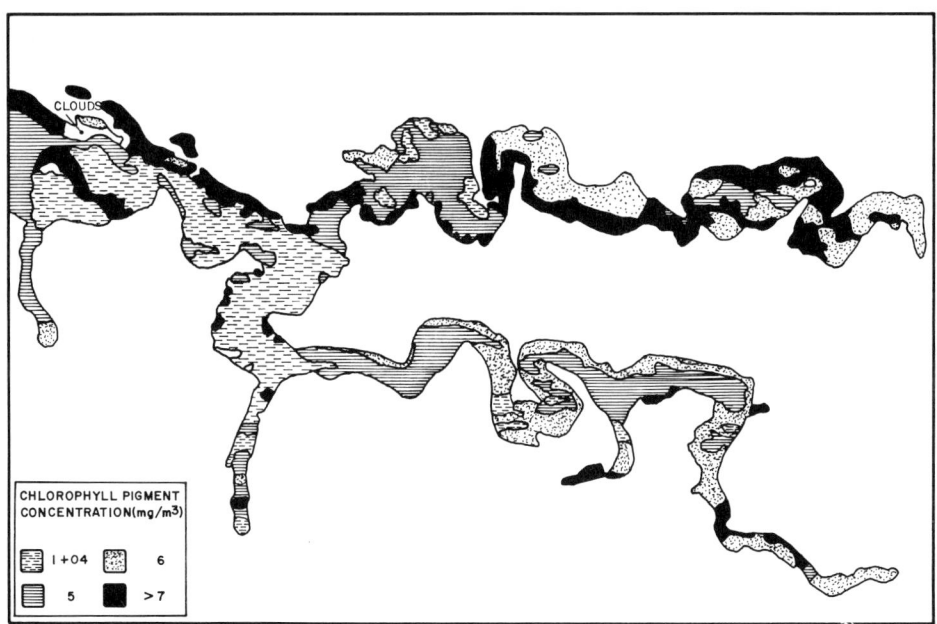

Fig. 6. Chlorophyll pigment distribution in Barra Bonita Reservoir, July 1988.

References

Braga, C. Z. F., 1988. Utilização de imagens dos satélites LANDSAT-5 e NOAA-9 na identificação de parâmetros físico-químicos da água na Baia da Guanabara. Dissertação de Mestrado em Sensoriamento Remoto e Aplicaçes, São José dos Campos, INPE, (INPE-4761-tdl/349).

Bricaud, A. & S. Sathyendranath, 1981. Spectral signatures of substances responsible for the change in ocean colour. Signatures Spectrales d'Objects en Teledetéction, Avignon, 8--11 sep.

Calijuri, M. C., 1988. Respostas fisioecológicas da comunidade fitoplanctônica e fatores ecológicos em ecossistemas com diferentes estágios de eutrofização. Doctorate Thesis, University of S. Paulo, School of Engineering, 293 pp.

Clark, G. L., & G. C. Ewing, 1974. Remote spectroscopy of the sea for biological production studies. In: N. G. Jerlov & E. S. Nielsen (eds.), Optical Aspects of Oceanography. Academic Press, New York: 389–413.

Curran, P. J. & M. Hay, 1986. The importance of measurement error for certain procedures in remote sensing at optical wavelengths. Photogrammetric Engineering and Remote Sensing 52: 229–241.

Godoy, M. Jr. & E. M. L. M. Novo, 1989. Processamento de dados TM/LANDSAT no monitoramento de águas interiores. São José dos Campos, INPE. (INPE- 4976-PRE/1533).

Golterman, H. L., R. S. Clymo & M. A. M. Ohnstad, 1978. Methods for Physical & Chemical Analysis of Fresh Waters. IBP Handbook No 8, Blackwell Scientific Publications, Oxford.

Kirk, J. T. D., 1983. Light and Photosynthesis in Aquatic Ecosystems. Cambridge University Press, Cambridge, 401 pp.

Kratzer, C. R. & P. L. Brezonik, 1981. A Carlson type trophic state index for nitrogen in Florida lakes. Water Resources Bulletin 17: 713–714.

McCluney, W. R., 1976. Remote measurement of water colour. Remote Sensing of Environment 5: 3–33.

NASA – National Aeronautics and Space Administration 1987. High Resolution Imaging Spectrometer. In: Science Opportunities for the 1990's, Vol. llc: 14.

Nielsen, A., P. Hansen, A. Malmgreen & V. Hansen, 1983. Introductory studies of natural contamination and man-made pollution in danish waters. In: Earsel Esa Symposium on Remote Sensing Applications for Environmental Studies, Brussels, 26–29 Apr. 1983, Proceedings. Paris, ESA: 203–210.

Philpot, W. & V. Klemas, 1981. Remote sensing of coastal pollutants using multispectral data. In: Annual William T. Pecora Memorial Symposium on Remote Sensing, 5. Satellite Hydrology. Proceedings, Sioux Falls, SD, Jun. 10–15, 1979. Minneapolis, Minnesota: 543–549.

Schiebe, F. R., J. C. Ritchie & G. O. Boatwright, 1984. A first evaluation of LANDSAT/TM data to monitor suspended sediments in lakes. In: NASA Goddard Space Flight Center LANDSAT-4 Science Investigations Summary, Proceedings of the LANDSAT-4 Early Results Symposium, Feb. 22–24 and the LANDSAT-4 Science Characterization Workshop, held at Grenbelt, MD, Dec. 6, 1983. Washington, DC, Vol.1: 141.

Welby, C. W., A. M. Whiterspoon & R. E. Holman, 1981. Trophic state determination for shallow coastal lakes from LANDSAT imagery. In Annual William T. Pecora Mmemorial Symposium on Remote Sensing, 5. Satellite hydrology, Proceedings. Sioux Falls, SD, Jun. 10–15, 1979. Minneapolis, Minnesota: 674–680.

Witte, W. G., C. H. Whitlock, R. C. Harris, J. W. Usry, L. R. Poole, W. M. Houghton, W. D. Morris & E. Gurganus, 1982. Influence of dissolved organic materials on turbid water optical properties and remote sensing reflectance. Journal of Geophysical Research 87: 441–446.

Wonnacott, T. W. & R. J. Wonnacott, 1977. Introductory Statistics. Wiley & Sons, New York.

Chapter XI

Succession of fish communities in reservoirs of Central and Eastern Europe

J. Kubečka
Hydrobiological Institute, Czechoslovak Academy of Sciences, Na sádkách 7, 370 05 České Budějovice, Czechoslovakia; Present address: Department of Biology, Royal Holloway and Bedford New College, (University of London), Egham, TW20 0EX, Surrey, England

Key words: fish community, reservoirs, Europe, succession, *Perca, Rutilus, Abramis*, spawning, competition

Abstract

The fish communities of 84 Central and Eastern European reservoirs were sub-divided according to their species compositions into six fish faunal types that are identical with the successional stages of reservoir ichthyocenoses. The six types are:

(1) Briefly existing fish faunas in which riverine species (especially salmonids) predominate. Found in 4% of reservoirs.
(2) Faunas characteristic of the reservoir initial filling period and with extraordinarily high (15–70%) percentage occurrence of northern pike. Found in 6% of reservoirs.
(3) Faunas in which perch (*Perca fluviatilis*) is predominant usually with one particular year class strongly represented (a 'cycling' population). Found in 9% of reservoirs.
(4) A transient fish fauna which is dominated by perch plus cyprinid fish. Found in 8% of reservoirs, with 20–50% of the fauna being perch and the rest represented by the predominant cyprinid species of type 5.
(5) A fauna dominated by cyprinids, usually by *Rutilus rutilus, Abramis brama* and/or *Blicca bjoerkna* together with a non-cycling perch population of less than 20%. This faunal type is the most frequently occurring one in Central and Eastern European reservoirs (61% of cases).
(6) The remaining reservoirs (12%) contain fish faunas that are dominated by coregonids or *Clupeonella* or *Carassius* or *Ctenopharyngodon* or *Hypophthalmichthys* or *Aristichthys* or *Cyprinus carpio* or *Pelecus cultratus*.

The characteristics and validity of these fish faunal types are discussed.

1. Introduction

The filling of reservoirs can be considered a catastrophic event for the impounded riverine fish community. Most fish species of Central and Eastern Europe are of riverine origin (Fernando & Holčík, 1989), but even these species may have different adaptive abilities in newly formed lentic environment. The composition of the fish fauna of the dammed river is then the first of the factors influencing the final fish composition.

Abiotic conditions in reservoirs, such as temperature regime, storage time, pH, oxygen conditions, presence of toxicants, water level fluctuations, are fundamentally important to the ichthyofauna and may be responsible for the absence of some fish species. According to Fernando & Holčík (1991), there is probably some reservoir size threshold above which riverine fishes cannot colonize the pelagial; this emphasises the importance of volume. Here we assume that the main groups of fish recorded are generally able to survive and satisfy their ecological requirements in the range of reservoirs studied.

The development of a stable fish stock in reservoirs has to be considered as one biological component of their limnological succession (Holčík *et al.*, 1989). Some patterns of change in the fish communities have been recorded in northern temperate lakes according to their degree of trophy – (Colby *et al.*, 1972, 1987; Kitchell *et al.*, 1977; Leach *et al.*, 1987; Prejs, 1978), which form a similar sequence as can be seen during the ageing of reservoirs. For example, Vostradovský *et al.* (1989) describe three possible stages of reservoir development: the first stage is dominated by salmonids, the second by perch and the third by cyprinids. All three stages can appear when the reservoir is built on a trout stream or is stocked by salmonids from the beginning and it is productive enough to attain a cyprinid fish fauna. Alternatively, the reservoir may be a cyprinid water right from its origin.

In lakes, eutrophication seems to be the main process determining the succession from a salmonid to a cyprinid water body, but other factors are influential, especially in reservoirs, such as dispersive, stochastic (according to Henderson, 1985) and anthropogenic factors. Normal development of the ichthyofauna, as described by Vostradovský *et al.* (1989), may be rejuvenated or 'enjuvented' (Christie *et al.*, 1987) by many events. Eutrophication can be also accelerated considerably by the activity of the fish stock itself (Hrbáček, 1962; Hrbáček *et al.*, 1986; Opuszynski, 1987; Persson *et al.*, 1988 etc.). All these facts make uncertain the idea that fish stock composition is determined solely by level of trophy.

When evaluating the development of fish in reservoirs, five groups of processes must be taken into account:
1. the time-course of reservoir productivity after filling (Poddubnyi, 1971; Reshetnikov *et al.*, 1982; Holčík *et al.*, 1989 etc.);
2. nutrient enrichment as a catchment area process;
3. development of more complex biotic interactions in reservoir (feeding, competition, predation);
4. the hydrological regime;
5. the management of the reservoir.

It was shown that the first two groups directly influence the abundance and biomass of the fish stocks, but only indirectly govern their species composition, which, in turn, is strongly influenced by the group 3, the biological interactions (Evans *et al.*, 1987). All the factors listed above also affect the rate of development of the reservoir's fish stock.

In reservoirs, there may be a good ecological agreement or some level of disagreement between the environmental requirements of the fish present and the nutrient and productivity level of the

Table I. Common and scientific names of the main fish species.

Salmonidae:

Brown trout *Salmo trutta* (L. 1758)
Rainbow trout *Oncorhynchus mykiss* Walbaum, 1855

Esocidae:

Pike *Esox lucius* (L. 1758)

Cyprinidae:

Roach *Rutilus rutilus* (L. 1758)
Common bream *Abramis brama* (L. 1758)
Rudd *Scardinius erythrophthalmus* (L. 1758)
Chub *Leuciscus cephalus* (L. 1758)
Dace *Leuciscus leuciscus* (L. 1758)
Bleak *Alburnus alburnus* (L. 1758)
White bream *Blicca bjoerkna* (L. 1758)
Asp *Aspius aspius* (L. 1758)
Tench *Tinca tinca* (L. 1758)
Leucaspius delineatus (Heckel, 1843)
Pelecus cultratus (L. 1758)
Grass carp *Ctenopharyngodon idella* (Valenc. 1884)
White carp *Hypophthalmichthys molitrix* (Valenc. 1884)
Bighead *Aristichthys nobilis* (Richardson, 1844)
Carp *Cyprinus carpio* (L. 1758)

Siluridae:

European catfish *Silurus glanis* (L. 1758)

Gadidae:

Burbot *Lota lota* (L. 1758)

Percidae:

Eurasian perch *Perca fluviatilis* (L. 1758)
Ruffe *Gymnocephalus cernua* (L. 1758)
Pikeperch *Stizostedion lucioperca* (L. 1758)

waterbody. A mature fish community is achieved only when the species composition attains a steady state. The degree of ecological disagreement or discordance then can indicate the direction and rate of change in the fish community.

The main purpose of the present paper is to establish a set of good examples of the species composition of fish communities in the man-made reservoirs of Central and Eastern Europe. These examples are then classified according to the presence of dominant species and discussed from the view of their maturity.

2. Data sources and methods

The sources of information for this paper come largely from published papers which, for some reservoirs, has been up-dated by unpublished data from named fish biologists. The percentage species composition of reservoir fish stocks has been obtained as follows:

1. There is information from 26 reservoirs of central European countries (Czechoslovakia, Germany and Poland; in some of them we have more than one observation of fish community composition) which gives the percentage composition of adult fish stocks (Table II). In only 9 reservoirs is either the fish stock density (from population censuses) or the fish stock biomass known. In 6 reservoirs, fish counts were carried out after the reservoirs were emptied; in 1 reservoir, after a poisoning; in 13 reservoirs, by means of night seining; in 7 reservoirs, by non-quantitative gill-netting using a variety of meshes. Species composition based on numerical abundance was available in 27 cases, and based on biomass from 9 cases.
2. Data on the species composition of the fry in reservoirs from Central and Eastern Europe comes largely from shore seining and is listed in Table II.
3. Data on the species composition of commercial catches comes from one source (Isayev & Karpova, 1989) and lists 30 reservoirs from European USSR.

Geographical locations of the Central and Eastern European reservoirs are given in Figs. 1 and 2.

The species composition of reservoir fish faunas were classified according to the six types defined in

Table II. Reservoirs and sources of published and unpublished data.

1. Adult stocks from Central European Reservoirs (each number indicates an example of fish composition)

Czechoslovakia

1	Bystřička (Václavík, 1956)
2	Morávka (Lojkásek, pers.comm.)
3	Šance (Lojkásek & Kubečka, pers.comm.)
4,5	Klíčava (Holčík & Pivnička, 1972; Pivnička, 1982)
6–8	Římov (Vostradovský *et al.*, 1990; Kubečka, 1990)
9	Podhora (Rozmajzlová *et al.*, 1986)
10,11	Slapy (Hanel *et al.*, 1983; Hanel, 1988)
12–14	Lipno (Vostradovský, 1964; Vostradovský *et al.*, 1986; Kubečka *et al.*, pers.comm.)
15	Lučina (Vostradovský *et al.*, 1983)
16	Záskalská (Švátora, 1989)

Table II. Continued.

17	Husinec (Kubečka, pers.comm.)
18	Stanovice (Křížek, 1987)
19	Luhačovice (Anonymus, 1956)
20	Orlík (Závěta, 1990)
21	Jordán (Kubečka & Böhm, 1991)
22	Želivka (Vostradovský *et al.*, 1988)
23	Mušov (Lusk, 1984)
24	Jesenice (Vostradovský, 1964)
25	Orava (Holčík, 1966)
26	Seč (Dohelský, 1956)
27	Fryšták (Václavík, 1956)

Poland

28	Malta (Mastynski, 1984)
29	Roznow (Jelonek & Amirowicz, 1987a)
30	Goluchow (Mastynski, 1984)
31,32	Goczalkowice (Starmach, 1986; Jelonek & Amirowicz, 1987b)

Germany

33	Bautzen (Schultz, 1988)

2. Commercial adult catches from 30 European Russian Reservoirs

(Isayev & Karpova, 1989)

3. Fish fry assemblages from Central and Eastern European Reservoirs

Czechoslovakia

1,2	Římov (Vostradovský *et al.*, 1983; Kubečka, 1990)
3	Kníničská (Kubečka, 1984)
4	Klíčava (Kubečka, 1984)
6	Slapy (Hanel, 1988)
7	Lipno (Vostradovský, 1965)

USSR

8	Volgogradskoye (Mirotvortsev, 1983)
9	Uglichshoye (Bergelson & Boitsov, 1981)
10	Rybinskoye (Konobeeva *et al.*, 1980)
11	Kamskoye (Pushkina, 1969)
12	Kuybyshevskoye (Vasyanin, 1958)
13	Kievskoye (Yerko, 1975)
14	Kegumskoye (Gaumiga, 1968)
15,16	Gorkovskoye (Galkin, 1965; Lesnikova & Kharitonova, 1979)
17	Ivankovskoye (Boitsov, 1975)
18,19	Kremencukskoye (Volkov, 1969; Volkov *et al.*, 1978)
20	Mozhayskoye (Spanovskaya *et al.*, 1980)

Poland

21	Sulejow (Kalinowski, 1989)

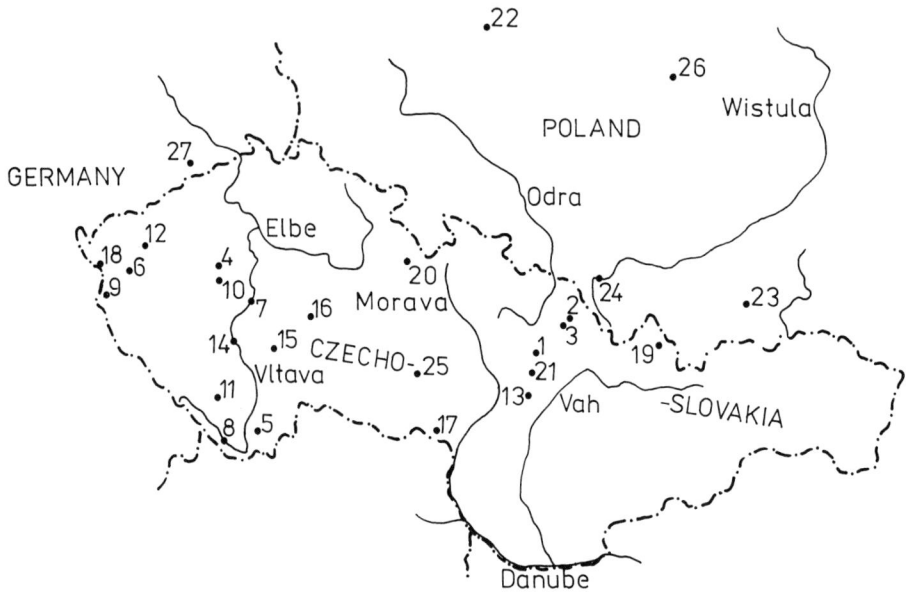

Fig. 1. Geographical location of Central European reservoirs: 1 – Bystřička, 2 – Morávka, 3 – Šance, 4 – Klíčava, 5 – Římov, 6 – Podhora, 7 – Slapy, 8 – Lipno, 9 – Lučina, 10 – Záskalská, 11 – Husinec, 12 – Stanovice, 13 – Luhačovice, 14 – Orlík, 15 – Jordán, 16 – Želivka, 17 – Musov, 18 – Jesenice, 19 – Orava, 20 – Seč, 21 – Fryšták, 22 – Goluchow, Malta Reservoir is probably near. 23 – Roznow, 24 – Goczalkowice, 25 – Kničiská, 26 – Sulejow, 27 – Bautzen.

the results section. For the most frequently occurring types such as perch-dominated (type 3), cyprinid-dominated (type 5) and transient (type 4), there was calculated for each species its mean percentage occurrence (with standard deviation), the number of reservoirs in which the species occurred and in which it was represented at various levels of percentage frequency ($> 5\%$, $> 20\%$).

Common or abbreviated names of fish species are used in the text and their full scientific names are listed in Table I.

3. Results and discussion

Six fish faunal types could be defined amongst the published and unpublished studies on 84 reservoirs from Central and East Europe. There were:

Type 1: Faunas with riverine (especially salmonid) which existed only briefly.

Type 2: Faunas characteristic of the reservoir filling period which contained extraordinarily high representation of the northern pike; in 6% of the reservoirs.

Type 3: Faunas in which perch is dominant; in 9% of the reservoirs.

Type 4: A transient fish fauna with perch plus cyprinid fish; in 8% of the reservoirs.

Type 5: Faunas in which cyprinids were dominant: in 61% of the reservoirs.

Type 6: The remaining reservoirs (12%) in which a variety of different fish species were dominant.

These are dealt in detail below.

3.1. Fish stock dominated by river species

Fish stocks dominated by river species seem to be a scarce phenomenon in Central and Eastern European reservoirs. River species seem to be dominant immediately after the filling of reservoirs and before a more stable fish community can develop. Where a reservoir is built in the non-salmonid section of the river, the species as *Leuciscus leuciscus, L. cephalus, Barbus, Chondrostoma, Phoxinus, Lampetra* become important (Holčík, 1966; Vostradovský, 1968; Starmach, 1986). The phase of the dominance of these species seems to be very short. Theoretically salmonids should be able to form a stable community, but the development of the fish stock of the Štrbské Pleso Lake shows that a shift from a

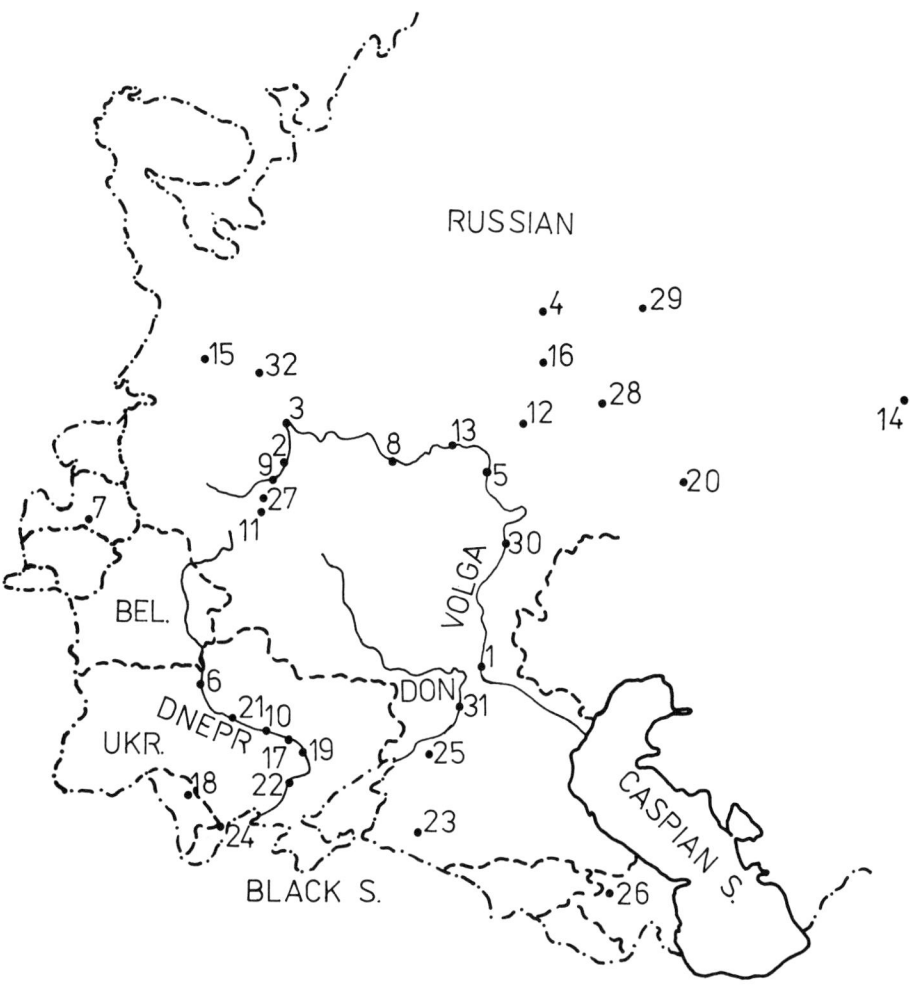

Fig. 2. Geographical location of Eastern European reservoirs: 1 – Volgogradskoye, 2 – Uglichkoye, 3 – Rybinskoye, 4 – Kamskoye, 5 – Kuybyshevskoye, 6 – Kievskoye, 7 – Kegumskoye, 8 – Gorkovskoye, 9 – Ivankovskoye, 10 – Kremenchugskoye, 11 – Mozhayskoye, 12 – Nizhnekamskoye, 13 – Tscheboksarskoye, 14 – Bukhtarminskoye, 15 – Verkhnesvirskoye, 16 – Votkinskoye, 17 – Dneprodzherdzhinskoye, 18 – Dubossarskoye, 19 – Zaporozhskoye, 20 – Iriklinskoye, 21 – Kanevskoye, 22 – Kakhovskoye, 23 – Krasnodarskoye, 24 – Kuchirganskoye, 25 – Manychskie Reservoirs, 26 – Mingechaurskoye, 27 – Ozerinskoye, 28 – Pavlovskoye, 29 – Reftinskie Reservoirs, 30 – Saratovskoye, 31 – Tsimlyanskoye, 32 – Cherepovetskoye

salmonid (and/or coregonid) fish fauna to a cyprinid-dominated one is possible, even at altitudes of 1000 m above the sea level (Holčík and Nagy, 1985).

During the development of several Czechoslovak water supply reservoirs, there has occurred an artificially-induced salmonid phase by stocking of 300–500 individuals. ha^{-1} of brown and rainbow trout thus inhibiting any natural succession to other faunal types. Typical examples of such ephemeral stock composition are given by the Stanovice Reservoir in its first two years 1980–81 (with 51–71% of salmonids and 25–49% of perch; Křížek, 1987) or Opatovice Reservoir in its second year 1973 (with 207 inds × ha^{-1} of rainbow trout and 474 inds × ha^{-1} of brown trout, and a few minow *P. phoxinus*; Lusk, 1978). An artificial salmonid phase of development can be observed in a reservoir when salmonids are stocked to a nonsalmonid fish stock – in Záskalská Reservoir in 1976, stocked rainbow trout represented 68% of the present fish stock, but in the next year these individuals disappeared completely (Švátora, 1989). In all these examples, the salmonid phase lasted for only a few months or years. For other examples see Vostradovský et al. (1989) and Lusk (1978) for Hubenov, Landštejn and Přísečnice Reservoirs.

Only one example of a relatively stable fish stock dominated by salmonids can be found in Czechoslovakia (Lojkásek, pers.comm.). This is in Morávka Reservoir after 25 years of existence, where 51% of the catches by night shore seinings were brown trout, 3% were rainbow trout and 46% were perch. Perch-salmonid systems seem not to be stable (comp. Křížek, 1987; Vostradovský et al., 1990).

Examples of salmonid-cyprinid stocks are equally rare but may be more stable as shown by Wajdowicz (1989) on the coexistence of salmonids with the bream. An example from Czechoslovak territory is given by Václavík (1956) from Bystřička Reservoir which after 43 years contains a fish fauna with 13% brown trout, 54% chub and 28% of carp. One example of trout stock (with only some eel present) in the reservoir is given by Zintz et al. (1991).

3.2. Fish stock with extra large pike populations

A phase with extra large populations of the northern pike appears to be a regular phenomenon early in reservoir succession according to Poddubnyi (1971), Hrbáček (1981), Starmach (1986), Vostradovský et al. (1990). Unfortunately, there are only a few examples in the literature of the species composition of this type of stock (Table III). Pike may either be dominant fish species as in Nizhnekamskoye Reservoir, or the subdominant one, as in Tscheboksarskoye and Klíčava Reservoirs. The only common feature is its percentage occurrence is much higher than is usual in later years of the existence of the reservoir. The percentage of pike is also much larger than the percentage occurrence for pike in perch- and cyprinid-dominated stocks (1%, 1.5%, 0.5%, 1.5%, 3.1%, see Tables IV–VIII). The pike-phase is associated with the expansion of a spawning substratum usually formed from the flooding of terrestrial vegetation during the filling of the reservoir. Later, the spawning substratum becomes more restricted and the population is further decreased by cannibalism, predation and fishing by man. Subsequently, perch and cyprinid species increase their abundance as the pike population declines, both relatively and absolutely. The boundary between this pike-population phase and any subsequent phase is not a sharp one. For example, the 7% of pike recorded in Jesenice Reservoir (Vostradovský, 1964) or the 6% of pike recorded in Římov Reservoir (Vostradovská & Vostradovský, 1983), both during their third year of existence, represents the transition between pike-phase and a subsequent one, with about 1% of pike.

In contrast to the situation amongst the other fish fauna types, predatory control by pike of the other fish species is very important in this type of stock.

Similar phases with extra-large pike populations appears whenever terrestrial vegetation is flooded. For example, peak pike catches were recorded in Lipno Reservoir after its water level was raised after several years of lower levels due to dam repair during early 1980s (Vostradovský & Kubečka, pers.comm.).

Fish stocks with extra large pike populations cannot be considered as a mature stage.

3.3. Perch-dominated fish stock

Table IV provides the known data base on fish stock composition of eight perch-dominated reservoirs. A fish stock is defined as perch-dominated when perch represent more than 50% of the fish. The variability of perch occurrence is rather small.

A feature of perch-dominated fish stocks is the cyclic appearance of strong and weak year classes

Table III. Percentage occurrence of selected fish species in five examples of fish stocks with extra large pike population. Reservoirs: R = Římov, N = Nizhnekamskoye, G = Goczalkowice, T = Tscheboksarskoye, K = Klíčava.

	Percentage				
Species	R	N	G	T	K
Pike	68.2	58.8	40–50	22.9	17.9
Perch	15.0	2.8	3–12	0.6	57.0
Roach	–	7.8	30–40	39.7	12.6
Bream	–	12.5	1– 2	15.7	–
Pikeperch	–	2.5	0.5	6.1	–
W.bream	–	4.2	?	4.4	–
age of reservoir (years)	2	8	2– 4	5	2

Table IV. Species composition and percentage occurrence of adult fish in the eight reservoirs with perch-dominated fish stocks (type 3).

Species	Mean	± SD	Number of reservoirs with species present at % frequencies of		
	%	%	any	> 5%	> 20%
Perch	74.2	9.7	8	8	8
Roach	7.5	7.2	7	4	4
Bream	1.7	2.3	4	0	0
Chub	0.2	0.4	4	0	0
Pike	1.0	2.0	8	1	0
Pikeperch	0.6	1.1	6	0	0
Asp	0.02	0.04	3	0	0
Carp	2.1	5.4	4	1	0
Catfish	0.001	0.004	1	0	0
B.trout	3.2	7.8	4	1	1
R.trout	2.0	4.3	5	1	0
Bleak	0.06	0.15	2	0	0
Tench	0.04	0.07	2	0	0
G.carp + Bighead	0.22	0.60	2	0	0
Coregonid	0.001	0.004	1	0	0
Ruffe	2.4	3.3	4	3	0
Dace	1.4	3.0	5	1	0
Rudd	0.7	1.8	4	1	0
W.bream	2.3	5.7	3	1	0
Other spp.	0.6	0.7	7	0	0

in the perch population. This was recorded first by Alm (1946) and later by Menshutkin & Zhakov (1964), Kelso & Bagenal (1977), Kuderskii *et al.* (1983) and Craig (1987). Cyclic perch populations were also reported in all the more intensively studied reservoirs, such as Římov, Záskalská, Husinec, Stanovice and Klíčava, during their perch-dominated phase.

In this type of a fish stock, perch is the main predatory species, despite the stunting of individuals belonging to the dominant year class and despite the fact, that fish is not the main item of diet, as was demonstrated in Římov Reservoir (Kubečka, pers. comm.). Other predatory fish species are relatively insignificant in this kind of fish stock (Fig. 4, Table IV).

Various accessory species can be found in perch dominated fish stock. For example, roach was represented by 10–20% (Sulejow, Římov and Bautzen Reservoirs), white bream by 16% in Gorkovskoye Reservoir (Galkin, 1965), carp by 15% in the Záskalská Res. (Švátora, 1989). In general, the perch-dominated fish fauna is less diverse than subsequent ones.

Another feature of perch dominated stock is its capacity to develope into a cyprinid-dominated fauna in relatively short periods of time. Hrbáček (1980) thought that the abundance of larger zooplanktonic forms was the main factor enabling perch to retain their dominance. The perch-dominance is supported by earlier sexual maturity (Pivnička & Švátora, 1988), by the absence of adult perches and other predators apart from pike and by low cyprinid representation in the fish fauna of the original impounded stream or river. For example Klíčava, Stanovice, Landštejn and Lučina Reservoirs received no initial inoculum of roach and bream, according to Oliva (1949), Lusk (1978), Vostradovská & Vostradovský (1983), Křížek (1987).

However, succession of many reservoirs with perch dominated stock towards cyprinid-domination has been successfully achieved in Klíčava (Holčík, 1977; Pivnička & Švátora, 1988), Římov (Kubečka, 1990), Gorkovskoye (Galkin, 1965 v/s Isayev & Karpova, 1989), Kremenchugskoye (Volkov, 1969 v/s Volkov & Vlasenko, 1978), Šance (Lojkásek, 1986 v/s Lojkásek & Kubečka, pers. comm.). It appears that cyprinids can attain dominance in reservoirs whose fish fauna contained any cyprinid species during their perch phase. The advantage of such shift in fish stock composition is the competitive elimination of perch by the more plastic cyprinids which are able to operate more effectively under food limited conditions, as shown by Persson *et al.* (1988) in Scandinavian lakes.

There is a clear relationship between the age of reservoir and the percentage frequency of perch in its fish stock, which is illustrated in Fig. 3. The younger the reservoir, the higher the proportion of perch and in reservoirs older than 10 years, there are fewer than 10% perch. Some exceptional older reservoirs contain rather higher than expected proportions of perch for various reasons. These reservoirs often satisfy criteria for type 4 fish stock. In Záskalská Reservoir, Pivnička & Švátora (1988) suggest that roach are limited by lack of spawning substrata which may also be the cause of the extreme case of long-term perch dominance (together with ruffe) characteristic of the London water supply reservoirs (Duncan, Kubečka & Bubb, pers. comm.). In these reservoirs, cyprinid reproduction is inhibited by a littoral consisting of

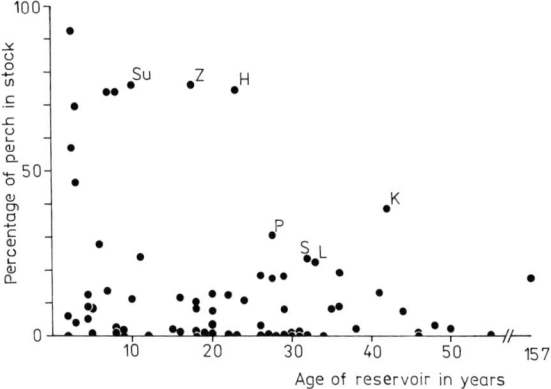

Fig. 3. The percentage occurrence of perch in relation to age of reservoir. Older localities with very high relative abundance of the perch are identified: Su - Sulejow, Z - Záskalská, H - Husinec, K - Kníničská, P - Podhora, S - Slapy, L - Lipno.

concrete slabs with a 3 : 1 slope and no emergent vegetation. Pond experimentation has shown that roach are inhibited when kept in very high perch densities (Pivnička & Švátora, 1983). Kubečka (1989) suggests that equal proportions of cyprinids and perch are a sign of reservoir biomanipulation.

Large fluctuations in water level, characteristics of hydroelectric reservoirs, are another likely cause of high proportions of perch in the fish stock; Slapy Reservoir is a good example of this (Hanel & Čihař, 1983, Hrbáček, 1984). In Slapy Reservoir at the end of the 1950's, about five years after filling, a typical cyprinid fish fauna was established with bream (37–54%) and roach (32–34%). Later, bream was almost eliminated (0.2–0.6% in 1985–87; Hanel, 1988) and perch increased from 10–15% in the late 1950's to 11–34% at present. This occurred after the construction upstream of the large Orlík Reservoir which was accompanied by large fluctuations in water levels and subsequent to the ageing of Slapy Reservoir itself. The actual proportion of roach did not change but its population density was lower (Hanel, 1988). A similar situation occurred probably in Sulejow Reservoir (Zalewski et al., 1990) and in Husinec Reservoir. A return to a fish stock with high perch proportions has occurred in Lipno Reservoir when the water level was increased after several years of low water levels (Vostradovský & Kubečka, pers. comm.).

3.4. Transient fish stocks with perch and cyprinid fish

Transient fish stocks dominated by perch plus cyprinid fish are defined by the presence of more than 20% perch (ranging 20–50%) and by a predominance of cyprinid fish, such as *Rutilus*, *Abramis* and *Blicca* (Table V).

Table V. Species composition and percentage occurrence of adult fish in the seven reservoirs with a transient fish stock (type 4).

Species	Mean	± SD	Number of reservoirs with species present at % frequencies of		
	%	%	any	> 5%	> 20%
Perch	30.4	9.7	7	7	7
Roach	36.7	8.4	7	7	7
Bream	7.3	7.4	7	4	1
Chub	0.4	0.7	3	0	0
Pike	1.5	2.1	6	1	0
Pikeperch	1.3	2.0	6	1	0
Asp	0.2	0.5	3	0	0
Carp	0.2	0.4	4	0	0
Catfish	0.001	0.004	1	0	0
B.trout	0.02	0.04	2	0	0
R.trout	0.001	0.004	1	0	0
Bleak	6.1	9.7	4	3	1
Tench	0.3	0.5	4	0	0
G.carp + Bighead	0.2	0.4	3	0	0
Coreg.	1.9	4.9	4	1	0
Ruffe	9.7	14.6	4	3	2
Dace	1.1	1.5	5	0	0
Rudd	1.3	2.2	5	1	0
W.bream	0.5	0.8	3	0	0
Other spp.	3.5	8.3	5	1	1

There are two types of transient stocks:
1. the composition of the fish is that characteristic of the shift in dominance from perch to cyprinids, such as occurred in Římov Reservoir in 1981 and in Kremenchugskoye Reservoir in 1965 and
2. with a composition of an 'enjuvenated' fish stock formed during large water level fluctuations or by biomanipulation (i.e. decreasing the cyprinids and increasing the perch).

Other characteristic features of transient fish faunas are lower percentages of bream than expected with cyprinid-dominance and higher percentages of pike, pike-perch, bleak and ruffe

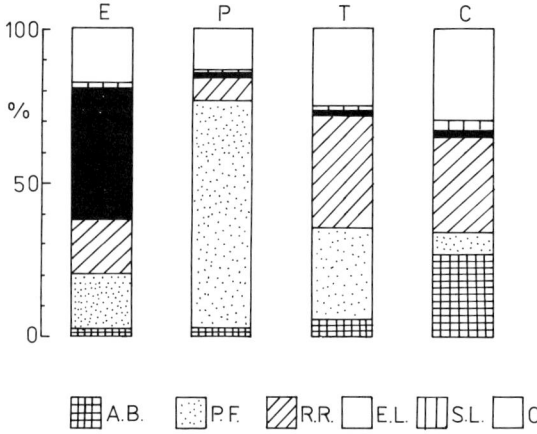

Fig. 4. Percentage frequency of species in four fish stock types given. Fish species: AB = *Abramis brama*, PF = *Perca fluviatilis*, RR = *Rutilus rutilus*, SL = *Stizostedion lucioperca*, EL = *Esox lucius*, O = other fishes. Fish stock types: E – fishstock with extra large pike population, P – perch-dominated stock, T – transient stock, C – cyprinid-dominated stock.

than expected with perch-dominance (Fig. 4, Tables IV–V).

3.5. Cyprinid-dominated fish stock

In this type of fish stock which is most frequently occurring in Central and Eastern Europe (Table XI), the three cyprinid species of *Abramis brama*, *Rutilus rutilus* and *Blicca bjoerkna* are relatively more abundant than perch, pike and salmonids, as is shown in Table VI. In most reservoirs with this type of fish stock (84%), these three species contribute more than 50% of the fish composition. Where they contribute slightly less than 50%, then the following species become important: bleak, rudd, ruffe, pikeperch and *Stizostedion volgense*. In cyprinid-dominated stocks, ruffe appears to be a very important but numerically under-estimated species. Table VI shows only the adult fish composition, because inclusion of fish fry from shore seinings would over-emphasise littoral species (rudd and bleak fry) and under-estimate the deeper and open water species (bream, pikeperch and coregonids); a comparison of Tables VI and VII show this. On the other hand, the value of studying composition of fry assemblages is that absent species can be distinguished which cannot recruit in reservoirs (carp, European catfish, trout, grass carp, bighead carp and coregonids).

Table VI. Species composition and percentage occurrence of adult fish in the fifteen reservoirs with a cyprinid-dominated fish stock (type 5).

Species	Mean	± SD	Number of reservoirs with species present at % frequencies of		
	%	%	any	> 5%	> 20%
Perch	7.9	6.0	14	9	0
Roach	26.1	23.0	15	14	9
Bream	40.6	29.2	13	13	11
Chub	0.7	1.3	8	0	0
Pike	1.5	2.0	15	2	0
Pikeperch	2.7	3.7	12	3	0
Asp	0.4	0.6	6	0	0
Carp	2.8	4.7	12	3	0
Catfish	0.6	2.1	6	1	0
B.trout	0.004	0.01	2	0	0
R.trout	0	0	0	0	0
Bleak	4.9	8.4	9	3	1
Tench	0.6	1.1	10	0	0
G.carp + Bighead	0.3	0.6	4	0	0
Coreg.	1.2	4.8	2	1	0
Ruffe	3.4	11.4	3	1	1
Dace	0.3	0.8	4	0	0
Rudd	0.7	1.3	7	0	0
W.bream	2.5	5.4	6	3	1
Other spp.	2.8	4.5	13	2	0

Table VII. Species composition and percentage occurrence of fish fry communities in the fourteen reservoirs with cyprinid dominance (type 5) from Central and Eastern Europe.

Species	Mean	± SD	Number of reservoirs with species present at % frequencies of		
	%	%	any	> 5%	> 20%
Perch	9.5	5.7	14	12	0
Roach	43.0	22.6	14	14	11
Bream	11.4	7.5	14	11	3
Chub	0.8	2.4	7	1	0
Pike	0.5	0.5	13	0	0
Pikeperch	1.5	4.2	11	1	0
Asp	0.2	0.3	9	0	0
Carp	0.02	0.08	2	0	0
Catfish	0				
B.trout	0				
R.trout	0				
Bleak	9.4	11.3	12	6	1
Tench	0.003	0.006	3	0	0
G.carp + Bighead	0				
Coreg.	0				
Ruffe	3.0	8.0	12	2	1
Dace	1.7	3.2	9	2	0
Rudd	7.3	13.9	8	5	2
W.bream	6.3	9.4	11	8	1
Other spp.	5.5	8.7	13	4	1

Table VIII. Species compositon and percentage occurrence of adult fish from commercial catches in twenty Russian reservoirs with cyprinid dominance.

Species	Mean	± SD	Number of reservoirs with species present at % frequencies of		
	%	%	any	> 5%	> 20%
Perch	1.4	1.7	18	1	0
Roach	27.7	22.0	20	19	10
Bream	36.6	24.0	19	19	15
Chub	0.006	0.02	2	0	0
Pike	3.1	3.4	19	5	0
Pikeperch	5.9	4.7	19	10	0
Asp	1.3	3.4	18	1	0
Carp	4.3	9.7	14	3	1
Catfish	0.6	0.8	14	0	0
B.trout	0				
R.trout	0				
Bleak	1.1	2.1	14	1	0
Tench	1.0	2.9	10	2	0
G.carp + Bighead	2.4	3.4	14	4	0
Ruffe	0.2	0.3	10	0	0
Dace	0.008	0.03	2	0	0
Rudd	0.02	0.03	7	0	0
W.bream	7.7	9.5	14	8	3
Other spp.	11.1	11.1	19	12	4

Table IX. The mean percentage abundance and coefficient of variation of summed values for (roach plus bream) or (roach plus bream plus white bream) from the same set of cyprinid-dominated adult and fish fry stocks and in Russian commercial catches (RCC).

	Mean and coefficient of variation			
	Adults	Fry	RCC	Total
Roach and bream				
Mean abundance (%)	69.8	64.2	56.6	–
Coefficient of variation (%)	28.2	29.0	36.5	–
Roach, bream and white bream				
Mean abundance (%)	72.4	71.7	61.5	68.6
Coefficient of variation (%)	21.4	19.2	30.0	24.1
Ratio roach : bream	7 : 9	12 : 3	8 : 12	27 : 24

The variability of species' percentage frequencies in cyprinid-dominated fish stocks is much wider than in the perch-dominated ones, even when the dominant fish species themselves are considered (Tables VI, VII & VIII). This is associated with the ability of ecologically similar cyprinid species to complement each other in species composition; the coefficient of variation can be reduced from nearly ± 100% (in Tables VI, VII & VIII) to ± 20-30% (in Table IX) by summing together the values for roach and bream and for roach, bream and white bream.

Tables VI, VII & VIII show that perch still contributes a sizeable proportion (8.0–9.5%) to cyprinid dominated fish stocks but not to the commercial catches from the USSR (1.5%; Table VIII); this is surprising, especially as some of their commercial nets were small-meshed which would over-estimate this species. Less surprising is the high proportion of predatory and large-sized fish in the commercial catches (pikeperch, pike, asp, catfish, grass carp and bighead carp) and the generally adequante representation of the main cyprinid species.

Table X gives the probabilities of occurrence of fish species in cyprinid-dominated stocks within the different levels or bands of percentage frequencies which are arranged into a geometric series. The main obligate predators, pike and pikeperch, are ubiquitous within this faunal type. Largely absent are trout, dace, grass carp, bighead carp, catfish, coregonids and rudd but, where present, one or other species may be quite numerous. For example, in Table X, coregonids are largely absent (98%) but when present in 2.4% of the reservoirs, they fall into the 5–20% frequency band, which is quite high. That is, their probability of occurrence in cyprinid-dominated stocks is low, only 2.4%.

Table X is also useful to distinguish the dominant and sub-dominant species. For example, within the frequency band of 20-100%, there are seven species plus a category called 'others' which consists of *Abramis brama, Clupeonella delicatula, Stizostedion volgense, Leucaspius delineatus, Lota lota* and which is important of reservoirs of the USSR. Table VIII demonstrates that only roach and bream are true dominants in this faunal type. Why are cyprinids dominant in so many reservoirs? It is probably because they are successful competitors for the kind of food that is abundant in reservoirs, namely crustacean zooplankton. Hrbáček *et al.* (1986) and Seđa *et al.* (1989) point out that the planktonic crustaceans of cyprinid-dominated reservoirs are under heavy

Table X. The probability of individual fish species occurring at six levels of percentage abundance in cyprinid dominated reservoir fish stocks.

Species	Probability (%) = % of Reservoirs Frequency bands (%)					
	0.0–0.05	0.05–0.25	0.25–1.0	1.0–5.0	5.0–20.0	20–100
I. Dominant species						
Roach	–	–	2.0	6.0	36	56
Bream	6.0	2.0	2.0	4.0	30	57
II. Sub-dominant ubiquitous species						
Perch	6.1	8.2	16.3	22.4	46.9	–
Pike	5.7	13.2	34.0	32.1	15.1	–
Pikeperch	19.9	8.9	13.3	27.0	31.1	–
Other spp.	18.6	2.3	9.3	30.2	25.6	14.0
III. Species sometimes abundant, often absent						
Catfish	65.2	6.5	15.2	10.9	2.2	–
B.trout	97.3	2.7	–	–	–	–
G.carp	63.4	4.8	12.2	12.2	7.3	–
+Bighead Coreg.	97.6	–	–	–	2.4	–
Dace	69.2	10.3	2.6	12.8	5.1	–
Rudd	62.6	7.5	5.0	12.5	7.5	5.9
Chub	61.7	10.2	5.1	10.2	2.6	–
Ruffe	63.4	9.7	12.2	7.3	2.4	4.9
IV. Transient species, between (II) and (III)						
W.bream	40.8	2.0	8.2	10.2	26.5	12.2
Asp	44.4	22.2	20.0	11.1	2.2	–
Carp	45.5	6.8	15.9	18.2	11.4	2.3
Bleak	37.8	5.4	8.1	18.9	24.3	5.4
Tench	59.5	13.5	16.2	8.1	2.7	–

predation pressure from most of the fish species. The juvenile perch can become severely food-limited and competitively unfit. They, and also older fish, may be forced to take benthic food or even to capture fish as food (Bergstrand, 1990; Persson et al., 1988; Kubečka & Böhm, 1991). Under these circumstances, the perch population is diluted by other fish species and the characteristic strong year-class cycling, so typical for perch-dominated type 3 faunas, becomes suppressed. This suppression returns the age distribution of the reservoir perch populations to the theoretical distribution termed 'harmonical' by Kubečka & Böhm (1991), meaning the exponential decrease of the densities of age classes.

Both perch and obligate predators, especially pikeperch, can be the main predatory species in different reservoirs with type 5 stocks but at too low a pressure to prevent overbreeding of planktivores which form the main bulk of the fish stock (Persson et al., 1988; Kubečka & Böhm, 1991).

Table IX shows that roach and bream occur in almost the same proportions and compete with each other. The bream has some competitive properties which are better than roach: more efficient predation of benthos; a better body shape for escaping from predators; intermittent and multiple spawning. High bream:roach ratios indicate successful bream competition in the following reservoirs: Seč Reservoir, 82%:2%; Buchtarminskoye Reservoir, 83%:1%; Mingechaurskoye Reservoir (Azerbadjan), 93%:2% (Isayev & Karpova, 1989). The bream is at a disadvantage in relation to roach because of its greater need for a weedy littoral for spawning and because of its inability to use tributaries as spawning sites (Kubečka, 1990). Hanel (1988)

Table XI. Percentage of the 84 studied reservoirs with fish stocks of different types.

Stock type and dominant species	% reservoir
1. River species	3.6
2. With large pike populations	5.9
3. Perch-dominated	9.3
4. Transient species	8.2
5. Cyprinid-dominated	60.7
6. With coregonids + *Clupeonella*	3.6
6. With *Carassius* spp.	2.4
6. *Ctenopharyngodon* & *Hypophthalmichthys*	2.4
6. *C. carpio*	2.4
6. *Pelecus cultratus*	1.2

reports on an extreme example of the almost total elimination of bream in Slapy Reservoir.

It appears that cyprinid-dominated fish stocks form the most stable fish community in Central and Eastern Europe because it is the most frequently occurring faunal type (Table XI; 61% + 2.4% + 2.4% + 2.4% + 1.2% = 69% are dominated by cyprinids of all kinds). Within these cyprinid fish stocks, there appears to exist some kind of a cyprinid succession with ageing and further eutrophication of the reservoir but which has not been sufficiently studied. The changes involved include an enhancement of cyprinid dominance and the decline of pikeperch (Oglesby et al., 1987). It is likely that most of the 84 reservoirs studied have not yet reach the level of hypertrophy at which can be expected a decrease in fish biomass (Holčík et al., 1989).

3.6. Fish stocks dominated by various other species

There were ten reservoirs with fish stocks which did not fit into the above classification; these are listed in Table XII together with the names of dominant species and their percentage abundance. In only three reservoirs are clupeids or coregonids dominant. The level of eutrophication in the reservoirs of Central and Eastern Europe seems to be too high for these inhabitants of oligotrophic or mesotrophic lakes. The coregonids are much more abundant in pristine Siberian reservoirs (Isayev & Karpova, 1989).

Grass and bighead carp are dominants in a few places largely because of artificial stocking and not because of natural breeding; compare this statement with Table VII. None of the fish stocks dominated by *Carassius* spp. and by tench occur in

Table XII. Percentage occurrence of fish from ten reservoirs with Type 6 fish stocks with different dominant species (marked *).

Species	Reservoirs				
	Ka	D	I	Kr	Ku
Perch	0.28	0.55	0.34	?	–
Roach	30.0	28.6	0.40	?	2.42
Bream	12.2	5.12	1.56	6.6	0.36
G. carp + Bighead	15.1	0.07	–	65.0*	94.7*
Clupeonella	31.7*	48.1*	–	–	–
Coregon.	–	–	89.7*	?	–

Species	Reservoirs				
	U	G	V	F	L
Perch	0.73	–	0.22	–	d.44
Roach	0.36	12.9	19.2	–	–
Bream	19.4	–	18.2	0.6	d.44
Carp	–	13.4	?	32*	47*
W. bream	–	–	?	65#	d.44
Carassius	56*	50.8*	?	–	–
Pelecus cultratus	–	–	56.4*	–	–

Key to reservoir names: Ka = Kakhovskoye, D = Dneprodzerdzhinskoye, I = Iriklinskoye, Kr = Krasnodarskoye, Ku = Kukhurganskoye, U = Ust-Manykhskoye, G = Goluchow, V = Votkinskoye, F = Fryšták, L = Luhačovice.
\# = more probably the mixture of common and white bream; d.44 = perch + common + white bream together represented 44%; ? = either absent or sparse.

Czechoslovakia but two have been recorded by Wajdowicz (1989) for Poland and another two, with only *Carassius* spp, from the USSR. Lack of spawning substrata and absence of extensive areas of littoral weed in Czechoslovak reservoirs probably explains the absence of these fish species.

4. Discussion

If we compare the species composition of the better documented reservoirs, some succession of fish communities can be detected (Fig. 4). During the change-over from perch-dominated to cyprinid-dominated fish communities, there occurs both:
(i) an increase in the proportions of bream, pikeperch, chub, asp, catfish, tench, rudd, pike and the fish in the 'others' category and
(ii) a decrease in the proportions of perch, salmonids and dace. Different species are highly represented in the transient type of fish stock, namely, roach, bleak, coregonids and ruffe.

Several examples of succession in fish stock types from different reservoirs are illustrated in Fig. 5. Typical for Central European reservoirs is the development with age towards a cyprinid-dominated fish fauna, as illustrated by Klíčava,

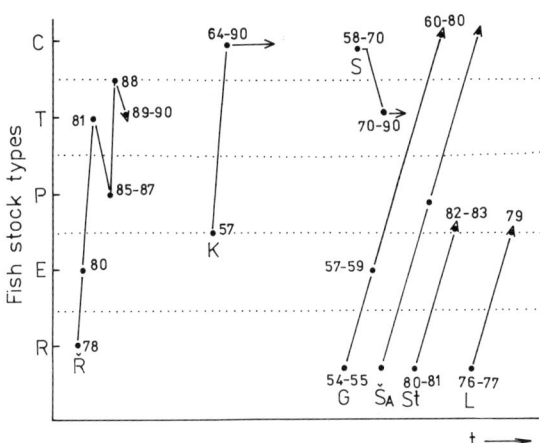

Fig. 5. The succession of fish stocks in different reservoirs. Reservoirs: Ř – Římov, K – Klíčava, S – Slapy, G – Goczalkowice, Ša – Šance, St – Stanovice, L – Lučina. The numbers represent the year(s) of study. Fish stock types: R – River species dominated stock, E – fish stock with extra large pike population, P – perch-dominated fish stock, T – transient stock, C – cyprinid dominated stock.

Goczalkowice, Stanovice, Lučina and Šance Reservoirs.

But this development is modified in Římov Reservoir by biomanipulative removal of non-predatory fish, by lowering the water level to dry out the spawned eggs and by stocking of predators (Kubečka, 1990) or in Slapy Reservoir by the worsening of cyprinid spawning conditions as described above.

But, Fig. 3 shows that such modification of succession are rare.

The phases of succession described here correspond partly with the phases of reservoir ontogeny given by Holčík *et al.* (1989). River species dominated fish stock and fish stock with extra large pike population can be usually considered as a part of initial phase. The fish composition of the depression and maturity phase is influenced by specific conditions of the reservoir. Scarcity of long termed quantitative data such as Holčík (1977), Pivnička (1982), Starmach (1986) or Kubečka (1990) causes the scarcity of good quantitative examples of the reservoir fish stock ontogeny as described by Holčík *et al.* (1989). The only problematical point is the existence of depression phase that need not to be necessarily developed (Starmach, 1986) especially if the concentration of the nutrients does not drop.

It is obvious from Table XI that native fishes of the region strongly dominate in Central and Eastern European reservoirs. Considerable effort is usually needed to ensure the dominance of introduced species such as carp, rainbow trout, grass carp, white carp or bighead.

There are probably two main strategies of the fish management of the reservoirs compared:
1. Maximization of the yield of economicaly important species such as *Abramis* and pikeperch in Russian reservoirs or carp, pike and pikeperch in Czech non-salmonid and non-water supply reservoirs. These fish are often stocked and protected by minimal size and catching season limits, by legalisation only certain fishing gears etc.
2. Minimization of the fish predatory pressure on the zooplankton in water supply reservoirs (especially Czech reservoirs). This strategy includes stocking of predators, meliorative catches of undesirable species, lowering of the water level after spawning etc.

In most of the cases discussed, the management seems not to be able to influence the succession in a fundamental way. Few apparent exceptions are described above.

5. Conclusions and recommendations

It proved possible to define four types of fish stock and one transient type for Central and East European reservoirs based largely on published studies. The definitions are not equally precise or of equal value; maybe the definitions cannot be made more precisely. It is evident, however, that the differences between stock types are sufficiently large for the proposed classification to be used as a practical framework for theoretical thinking on succession in reservoir fish communities.

The biggest source of error in this analysis lies in: diverse sampling methods adopted by different authors, the problem of representative sampling where pelagial : littoral area ratios are so different, not distinguishing numerical abundance and biomass for species compositions. This is always the fate of a first synthesis. Qualitative information about fish stock composition is less interpretable than quantitative data on population densities and their food resources. Further improvement of any theory on community succession will not be possible without standardisation of methodology and better understanding about the ecological niches occupied by the different fish species.

Acknowledgements

Author is grateful to Dr. Annie Duncan (Royal Holloway and Bedford New College) for great help in preparing the manuscript and to Mr. Z. Prachař for drawing the figures.

References

Alm, G., 1951. Year class fluctuations and span of life of perch. Rept. Inst Freshwat. Research Drottningholm 33: 17–38.
Anonymous, 1956. (Fishing out of water reservoir near Luhačovice). Čsl.rybářství, Praha, 3: 49 (In Czech).
Bergelson, B. O. & M. P. Boitsov, 1981. (Effectivity of the natural reproduction of the fish of Ivankovskoye and Uglichkoye Reservoirs in 1976-1977). Sborn. nauch. trud. Gosniiorkh, Leningrad. 165: 16–29 (In Russian).
Bergstrand, E., 1990. Changes in the fish and zooplankton communities of Ringsjon, a Swedish lake undergoing man made eutrophication. Hydrobiologia 191: 57–66.
Boytsov, M. P., 1975. (Abundance of fish fry of the Ivankovskoye Reservoir). Izvestiya Gosniiorkh, Leningrad. 93: 30–37 (In Russian).
Colby, P. J., G. R. Spangler, D. A. Hurley & A. M. McCombie, 1972. Effects of eutrophication on salmonid communities in oligotrophic lakes. J. Fish. Res. Bd. Can. 29: 975–983.
Colby, P. J., P. A. Ryan, D. H. Schupp & S. L. Serns, 1987. Interactions in north-temperate lake fish communities. Can. J. Fish. Aquat. Sci. 44: 104–128.
Craig, J., 1987. The Biology of Perch and Related Fish. Croom Helm, Timber Press, London, 333 pp.
Christie, W. J., C. R. Spangler, K. H. Loftus, W. L. Hartman, P. J. Colby, M. A. Ross & D. R. Talhelm, 1987. A perspective on great lakes fish community rehabilitation. Can. J. Fish. Aquat. Sci. 44: 486–499.
Dohelský, J., 1956. Fishery management of the Seč Reservoir. In: J. Hanzal (ed.), Fishery Management of Water Reservoirs. Brázda, Praha: 81–92 (In Czech).
Evans, D. O., B. A. Henderson, N. J. Bax, T. R. Marshall, R. T. Oglesby & W. J. Christie, 1987. Concepts and methods of community ecology applied to freshwater fisheries management. Can. J. Fish. Aquat. Sci. 44: 448–470.
Fernando, C. H. & J. Holčík, 1989. Origin, composition and yield of fish in reservoirs. Arch. Hydrobiol. Beih. Ergebn. Limnol. 33: 637–641.
Fernando, C. H. & J. Holčík, 1991. Fish in reservoirs. Int. Revue ges. Hydrobiol. 76: 149–167.
Galkin, G. G., 1965. Species composition, distribution and growth of fish fry of the Gorkovskoye Reservoir in the first years of existence. Izvestiya Gosniiorkh, Leningrad, 59: 98–122 (In Russian).
Gaumiga, R. J., 1968. The biology of fish fry in Kegumskoye Reservoir. In G. P. Andrushaytis (ed.), Ichthyology and Lake Fishery Management. Znaniye, Riga: 22–26 (In Russian).
Hanel, L., 1988. The Fish of Slapy Reservoir and its Watershed. PhD Thesis, Charles University, Prague, (In Czech).
Hanel, L. & J. Čihař, 1983. The fish of Slapy Reservoir. Sborník Vlastivědných prací z Podblanicka, Praha 24: 29–70.
Henderson, P. A., 1985. An approach to the prediction of temperate freshwater fish communities. J. Fish Biol. 27: 279–291.
Holčík, J., 1966. Development and forming of the ichthyofauna of the Orava Reservoir. Biologické práce SAV, Bratislava XII: 5–75.
Holčík, J., 1977. Changes in fish community of Klíčava Reservoir with particular reference to eurasian perch (*Perca fluviatilis*), 1957-72. J. Fish. Res. Board. Can. 34: 1734–1747.
Holčík, J. & S. Nagy, 1985. The changes of the ichthyofauna of Štrbské pleso and their relatinships to the eutrophisation of

the lake. Sborník VII konferencie Československej limnologickej spoločnosti Nitra: 268–271 (In Czech).

Holčík, J., P. Banarescu & D. Evans, 1989. A. General introduction to fishes. In: J. Holčík (ed.) The Freshwater Fishes of Europe Vol. 1/II, Aula Verlag, Wiesbaden: 18–147.

Holčík, J. & K. Pivnička, 1972. The density and production of fish populations in the Klíčava Reservoir (Czechoslovakia) and their changes during the period 1957–70. Int. Revue ges. Hydrobiol. 57: 883–894.

Hrbáček, J., 1962. Species composition and the ammount of zooplankton in relation to the fish stock. Rozpravy ČSAV, Ser. Mat. Nat. Sci. 72: 1–117.

Hrbáček, J., 1980. The structure and interrelationships of stagnant water organisms community and their importance in the evaluation of production processes and eutrophication. Dr.Sc. thesis. ÚKE ČSAV, České Budějovice, 152 pp. (In Czech).

Hrbáček, J., 1981. Production relationships and their importance for evaluation of the eutrophication of water reservoirs. Studie ČSAV 24: 58 pp. (In Czech).

Hrbáček, J., 1984. Ecosystems of the european man-made lakes. In F. B. Taub (ed.), Ecosystems of the World 23: Lakes and Reservoirs. Elsevier, Amsterdam: 267–290.

Hrbáček, J., O. Albertová, B. Desortová, V. Gotwaldová & J. Popovský, 1986. Relation of the zooplankton biomass and share of large cladocerans to the concentration of total phosphorus, chlorophyll-a and transparency in Hubenov and Vrchlice reservoirs. Limnologica 17: 301–308.

Isayev, A. I. & E. I. Karpova, 1989. Fishery management of reservoirs. Agropromizdat, Moscow, 250 pp. (In Russian).

Jelonek, M. & A. Amirowicz, 1987a. Composition, density and biomass of the ichthyofauna of the Goczalkowice Reservoir (Southern Poland). Acta Hydrobiol., Krakow 29: 253–259.

Jelonek, M. & A. Amirowicz, 1987b. Density and biomass of fish in the Roznow Reservoir (Southern Poland). Acta Hydrobiol., Krakow 29: 243–249.

Kalinowski, S., 1989. Perciform fry pressure on large filtrators, Cladocera in Sulejow Reservoir. Ms. thesis, Univ. Lodz, (In Polish).

Kelso, J. R. M. & T. B. Bagenal, 1977. Percids in unperturbed ecosystems. J. Fish. Res. Board Can. 34: 1959–1962.

Kitchell, J. F., M. G. Johnson, C. K. Minns, K. H. Loftus, L. Greig & C. H. Olver, 1977. Percid habitat: the river analogy. J. Fish. Res. Board Can. 34: 1936–1940.

Konobeyeva, V. K., A. G. Konobeyev & A. G. Poddubnyi, 1980. About the mechanisms of perch *Perca fluviatilis* L. fry accumulations in the open part of lake-type reservoir (on the example of Rybinskoye Reservoir). Vopr. ichtiol. Moscow 20: 258–271. (In Russian).

Křížek, J., 1987. Development of the ichthyofauna and the growth of perch (*Perca fluviatilis* L.) in the Stanovice Reservoir during the first five years after filling. Práce Výzkumného ústavu rybářského a hydrobiologického, Vodňany 16: 18–31.

Kubečka, J., 1984. Determination of some aspects of ecology of fish fry of the Klíčava Reservoir. MSc. thesis, Charles Univ., Prague, 207 pp. (In Czech).

Kubečka, J., 1989. Development of the ichthyofauna of the Římov Reservoir and its management. Arch. Hydrobiol. Beih. Ergebn. Limnol. 33: 611–613.

Kubečka, J. (ed.), 1990. Ichthyofauna of the Malše River and Římov Reservoir. South-Bohemian Museum, České Budějovice, 151 pp. (In Czech with English Summaries).

Kubečka, J. & M. Böhm, 1991. The fish fauna of the Jordan reservoir, one of the oldest man-made lakes in central Europe. J. Fish. Biol. 38: 935–950.

Kuderskyi, L. A., G. P. Rudenko & V. J. Nikanorov, 1983. Age of sexual maturity and culmination of ichthyomass in the perch populations in small lakes. Sbor. nauch. trud. Gosniiorkh, Leningrad 207: 139–149 (In Russian).

Leach, J. H., L. M. Dickie, B. J. Shuter, U. Borgmann, J. Hyman & W. Lysack, 1987. A review of the methods for prediction of potential fish production with application to the Great Lakes and Lake Winnipeg. Can. J. Fish. Aquat. Sci. 44: 471–485.

Lesnikova, T. V. & E. D. Kharitonova, 1979. Distribution and growth of fish fry in the Gorkovskye Reservoir. Sbor. nauch. trud. Gosniiorkh, Leningrad 142: 144–150 (In Russian).

Lojkásek, B., 1986. The growth of the perch *Perca fluviatilis* in the Šance water supply reservoir. Živočišná výroba, Praha 31: 921–926 (In Czech with English summary).

Lusk, S., 1978. Development of aimed fish populations in water supply reservoirs Opatovice and Landštejn. Vertebr. zprávy, Brno 1978: 41–46 (In Czech with English summary).

Lusk, S., 1984. Fishery management of the upper part of the Nové Mlýny reservoirs on the Dyje River. Živočišná výroba, Praha 29: 1043–1051 (In Czech with English summary).

Mastynski, J., 1984. Fish biomass of drained small reservoirs. Pol. Archiw. Hydrobiol. 31: 69–76.

Menshutkin, V. V. & L. A. Zhakov, 1964. The experiment of mathematical estimation of the abundance dynamics of the perch under given ecological conditions. In: I. I. Nikolayev & E. A. Popov (eds.), Ozera Karelskogo Peresheyka. Nauka, Moscow, Leningrad: 140–155 (In Russian).

Mirotvortsev, C. P., 1983. Estimation of the fish fry survival during the first year of life in the Volgogradskoye Reservoir. Sbor. nauch. trud. Gosniiorkh, Leningrad 199: 76–83 (In Russian).

Oglesby, R. T., J. H. Leach & J. Forney, 1987. Potential Stizostedion yield as a function of chlorophyll concentration with special reference to Lake Erie. Can. J. Fish. Aquat. Sci. 44: 166–170.

Oliva, O., 1949. Partial list of the ichthyofauna of the Klíčava Brook. Akvar. listy, Praha 21: 94–96 (In Czech).

Opuszynski, K., 1987. A feed-back dependence between the eutrophication preocess and changes in the fish community. The theory of ichthyoeutrophication. Wiad. Ekologiczne 33: 21–30 (In Polish with English summary).

Persson, L., G. Andersson, S. F. Hamrin & L. Johansson, 1988. Predator regulation and primary production along the productivity gradient of temperate lake ecosystems. In: S. L. Carpenter (ed.), Complex Interactions in Lake Communities. Springer, New York: 45–65.

Pivnička, K., 1982. Long-termed study of fish populations in the Klíčava Reservoir. Acta Sci. Nat. Brno XVI 10: 1–46.

Pivnička, K. & M. Švátora, 1977. Factors affecting the predominance from the eurasian perch (*Perca fluviatilis*) to roach (*Rutilus rutilus*) in the Klíčava Reservoir,

Czechoslovakia. J. Fish. Res. Board Can. 34: 1571–1575.

Pivnička, K. & M. Švátora, 1983. Competitive interactions of perch and roach in small drainable reservoir. Živočišná výroba, Praha, 28: 817–824 (In Czech with English summary).

Pivnička, K. & M. Švátora, 1988. Living together of roach and perch with respect to their competition in the Klíčava Reservoir between 1964 and 1986. Universitas Carolina, Environmentalica, Praha, II, 1–2: 17–85.

Poddubnyi, A. G., 1971. Ecological Topography of the Fish Populations in Water Reservoirs. Nauka, Leningrad, 307 pp. (In Russian).

Prejs, A., 1978. Eutrophisation of lakes and the ichthyofauna. Wiad. Ekologiczne XXIV, 3: 201–208 (In Polish with English summary).

Pushkina, N. P., 1969. The distribution of fish fry in the nearshore zone of the Kamskoye reservoir. Biol. vnutr. vod, Nauka, Leningrad 4: 19–24 (In Russian).

Reshetnikov, Y. S., O. A. Popova, O. P. Sterligova, V. F. Titova, L. G. Bushman, E. P. Ieshko, N. P. Makarova, R. P. Malakhova, I. V. Pomazovskaya & Yu. A. Smirnov, 1982. The Changes of the Fish Stock of the Eutrophicated Waterbody. Nauka, Moscow, 246 pp. (In Russian).

Rozmajzlová, V., J. Vostradovský, O. Albertová, J. Křížek & M. Vostradovská, 1986. Investigation of the influence of aimed fish stock on the water quality. Report VÚV (Zpráva výzkumného ústavu vodohospodářského), Praha, 50 pp. (In Czech).

Schultz, H., 1988. An Acoustic fish stock assessment in the Bautzen Reservoir. Limnologica (Berlin) 19: 61–70

Seďa, J., J. Kubečka & Z. Brandl, 1989. Zooplankton structure and fish population development in the Římov Reservoir, Czechoslovakia. Arch. Hydrobiol. Beih. Ergebn. Limnol. 33: 605–609.

Spanovskaya, V. D., V. A. Grigorash, T. V. Lebedeva & V. A. Rekurbratskyi, 1980. Fish spawning effectivity in Mozhayskoye Reservoir in different years 1969–1975. In: V. A. Bykov, I. M. Kisina & K. K. Edelshteyn (eds.), Komlexnyje Issledovanija Vodokhranilisch. MGU, Moscow 5: 198–209 (In Russian).

Starmach, J., 1986. Development and the structure of the Goczalkowice Reservoir ecosystem XV: Ichthyofauna. Ecol. Pol. 34: 515–521.

Švátora, M., 1989. The dynamics of fish populations in Zaskalska Reservoir, Czechoslovakia, with special regard to eurasian perch. Acta Univ. Carolinae – Biol., Praha 33: 141–255.

Václavík, B., 1956. Fishing out of the Bystřička Reservoir. Československé rybářství, Praha 2: 27–28 (In Czech).

Václavík, B., 1956b. Fishing out of the main Fryšták Reservoir. In: J. Hanzal (ed.), Fishery Management of Water Reservoirs. Brázda, Praha: 131–141 (In Czech).

Vasyanin, K. I., 1958. The growth of the economically important fish in the first and second years of existence of the Kuybyshevskoye Reservoir. Trudy Tatar. otd. Vniiorrkh 8: 206–217 (In Russian).

Volkov, A. N., 1969. The changes of the abundance of the fry caused by the building of the Kremenchugskoye Reservoir. Rybnoye Khoz., Kiev 20: 71–78 (In Russian).

Volkov, A. N. & V. I. Vlasenko, 1978. Species composition and standing stock of fry in Kremenchugskoye Reservoir after building of Kanevskoye Reservoir. Rybnoye Khoz., Kiev 27: 62–66 (In Russian).

Vostradovská, M. & J. Vostradovský, 1983. The prognosis and actual development of the fish stock in water reservoirs (Římov and Lučina). Živočišná výroba, Praha 28: 801–807 (In Czech with English summary).

Vostradovská, M. & J. Vostradovský, 1986. On the ichthyofauna of the Lipno Dam Lake after 25 years with special respect to whitefish and pike-perch. Bulletin Výzkumného ústavu rybářského a hydrobiologického Vodňany 4: 22–35 (In Czech with English summary).

Vostradovský, J., 1964. The common bream *Abramis brama* L. in the Lipno and Jesenice Reservoirs in the first years after filling. Živočisná výroba, Praha 37: 593–600 (In Czech).

Vostradovský, J., 1965. Some notes on the occurence of fish fry along the banks of the Lipno Valley Dam in the cours of the day and at night. Práce Výzkumného ústavu rybářského a hydrobiologického Vodňany 5: 221–230. (In Czech with English summary).

Vostradovský, J., 1968. Contribution to the knowledges about the ichthyofauna forming of the Lipno Reservoir 1958–1965. Zpravodaj Chráněné krajinné oblasti Šumava, Krajské středisko památkové péče a ochrany přírody, České Budějovice and Plzeň: 7–29 (In Czech).

Vostradovský, J., O. Albertová, J. Křížek & L. Růžička, 1988. Monitoring of water supply reservoirs and of selected lotic waters from the wiew of biomanipulation. Report No. N03 329 842 Du 03, Výzkumný ústav rybářský a hydrobiologický Vodňany, (In Czech).

Vostradovský, J., J. Křížek, O. Albertová, L. Růžička & M. Vostradovská, 1989: The changes of fish communities and biomanipulation in water supply reservoirs. Arch. Hydrob. Beih. Ergebn. Limnol. 33: 587–594.

Vostradovský, J., M. Hlaváček, J. Křížek, J. Kubečka, L. Liška & K. Stach, 1990. The composition of the fish stock samples in the Římov Reservoir during its development. In: Kubečka, J. (ed.), Ichthyofauna of the Malše River and Římov Reservoir. South-Bohemian Museum, České Budějovice: 55–60 (In Czech with English summary).

Wajdowicz, Z., 1989. V. Forming of fish stocks in reservoirs. *Gospod. rybna* 41, 6: 12–13 (In Polish).

Yerko, V. M., 1975. The influence of the condition change on the effectivity of their reproduction in the Kievskoye Reservoir. Rybnoye Khoz., Kiev 20: 55–61 (In Russian).

Zalewski, M., B. Brewinska-Zaras & P. Frankiewicz, 1990. Fry communities as a biomanipulating tool in a temperate lowland reservoir. Arch. Hydrobiol. Beih. Ergebn. Limnol. 33: 763–774.

Zavěta, J., 1990. The changes of the abundance and species composition of fish of Orlík Reservoir in 1980–1989. In: S. Nagy (ed.), Zborník referátov z konferencie ichtyologickej sekcie SZS, Bratislava: 12–14 (In Czech).

Zintz, K., U. Dost & H. Rahmnab, 1991. Management of a trout population in a new reservoir in the Black Forest, Germany. Bull. Zool. Museum, Universit. Amsterdam, Spec. Issue: 98.

Chapter XII

Framework for investigation and evaluation of reservoir water quality in Czechoslovakia

M. Straškraba,[1] P. Blažka,[2] Z. Brandl,[2] P. Hejzlar,[2] J. Komárková,[2] J. Kubečka,[2] I. Nesměrák,[3] L. Procházková,[2] V. Straškrabová,[2] & V. Vyhnálek[2]

[1] *Biomathematical Laboratory, Czechoslovak Academy of Sciences, Branišovská 31, 37005 České Budějovice, Czechoslovakia;* [2] *Hydrobiological Institute, Czechoslovak Academy of Sciences, Na sádkách 7, 37005 České Budějovice, Czechoslovakia;* [3] *Hydraulic Research Institute, Rohanský ostrov, 18000 Praha 8, Czechoslovakia*

Key words: reservoirs limnology, water quality, sampling, eutrophication, plankton, nutrients, fish

Abstract

Framework for reservoir characterization, water quality sampling, methods for analysis and data evaluation for different parameters are given. The most important drinking water reservoir water quality parameter groups are covered: those related to stratification, to eutrophication, to organic matter changes and to bacterial populations. Simple models and/or quantitative evaluation tables are developed for most parameters. Fish populations are considered important regulator of zooplankton which subsequently regulates phytoplankton. Methods of results recording, processing and evaluation are given, with emphasis on computer storage, basic statistics, trend evaluation, dependence of water quality parameters on flow and material budgets. A new simple method for characterization of reservoir stratification is outlined. Consequences of the results for water quality management are divided into three groups: consequences for the watershed, for the reservoir and for the drinking water treatment plant.

1. Introduction

This paper is an extension of Straškraba *et al.* (1986) to cover in addition to methods of reservoir water quality investigations also the approaches to evaluation of study results. Although the experiences summarized here are largely based on reservoirs from Czechoslovakia, it is expected they will be also useful for other conditions, at least in the temperate region. In world standards Czechoslovakia is a small country and so are its reservoirs. Temperate regions of Asia, America and other countries have some reservoirs much larger than that discussed here. These will need more complex evaluation schemes based on the same principles as given here. For other geographical latitudes where seasonal cycles and circulation periods the result evaluation will be different from those given. The paper is intended for limnologists and water quality engineers in river and water supply and management, but it may also be useful for public health laboratories investigating drinking water quality. The object is to unify and improve the methods for measuring and assessing the quality of water in reservoirs, and to enable limnologists to generalize their results. Although the methods have been worked out for drinking water reservoirs they can be used for reservoirs of other purposes. However, they are designed for investigations of raw water and processes taking place in the watershed and reservoir, not for the treatment of water and inspection of the quality of the resulting drinking water, as other recommendations are available for this purpose.

Procedures described here can also be used for investigations into long-term changes in the water quality, as well as for short-term prognoses of water quality in the course of the year, for management purposes.

The methods of investigation are the same regardless of whether they are used for long-term prognoses or suggestions of measures within the watershed of a reservoir. However, the predictions of water quality need additional methods and the use of mathematical models of water quality. Such methods and models have not been included in this paper.

Also included in this report are definitions of constant and variable characteristics of reservoirs and watersheds, used in water quality evaluation and guidelines for determining sampling sites, sampling frequency, and water quality parameters. The largest part of this paper deals with individual parameters, analytical methodology, and evaluation of results. The recording of results and basic methods of their processing and evaluation are discussed as well as techniques by which conclusions can be drawn from assessments made according to various parameters. This part is divided into conclusions for the watershed, reservoir, and waterworks.

1.1. Reservoir characterization and sampling, reservoir and investigation types

Figure 1 shows the characteristic types of Czechoslovak reservoirs. According to their size and importance they are divided into two categories:

(A) small – volume of water under 10^6 m^3 or offtake under 50 l s^{-1}) and
(B) large – volume of water greater than 10^6 m^3 or offtake < 50 l s^{-1}).

Based on relative morphological and hydrological criteria they are also divided into:

(a) shallow non-stratified reservoirs (except for temporary stratification lasting a few days) – because of the effects of wind and convective stirring,
(b) through-flowing non-stratified reservoirs because of high flow rates, (retention time shorter than 10 days), and
(c) deep stratified reservoirs.

Stratification of a reservoir results in the formation of water layers with vertically differentiated values of water quality parameters, (e.g. temperature and oxygen concentration). Three horizontal zones often can be distinguished as follows:

(a) the inflow mixing zone,
(b) zone of coves,
(c) the main body of the reservoir.

Inflowing water flows (plunges) into a reservoir water layer of the same temperature (density). Local temporary horizontal differences also can occur in the surface layer of deep reservoirs, due to

RESERVOIR TYPES

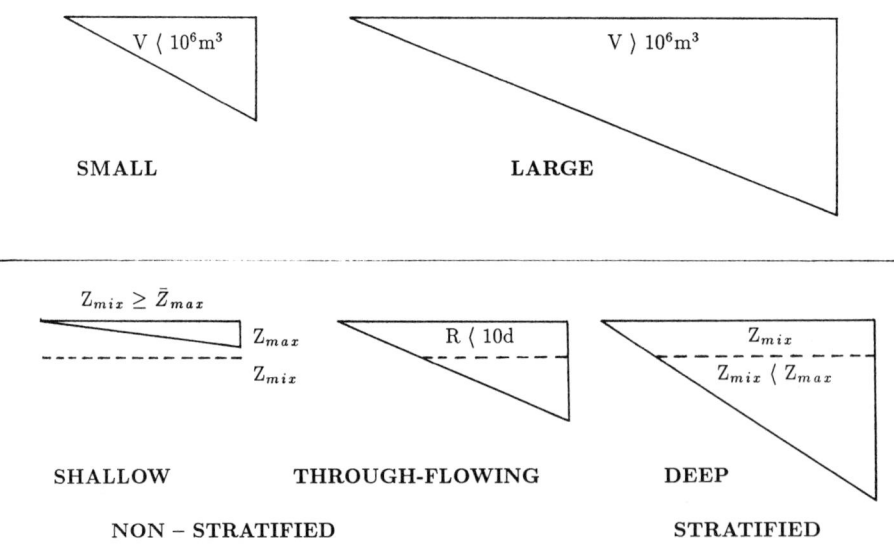

Fig. 1. Reservoir types.

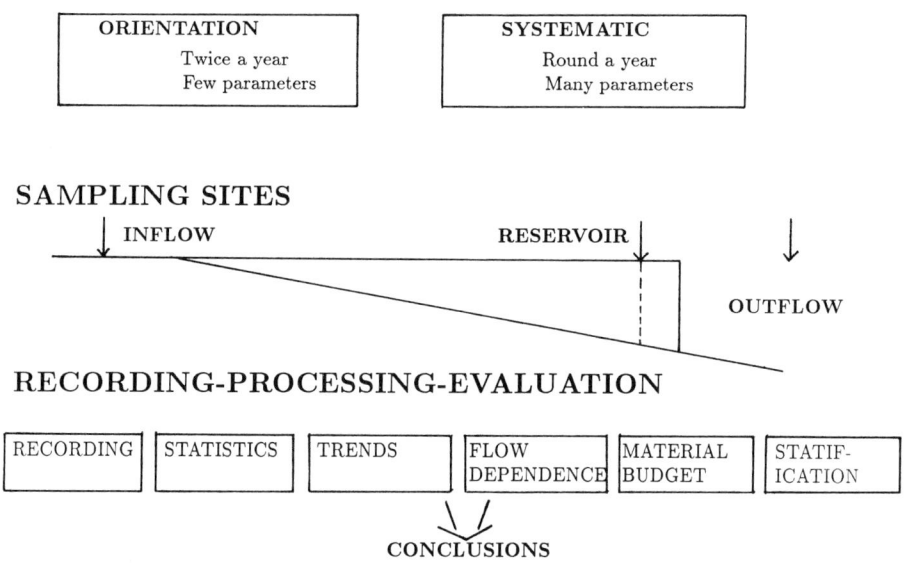

Fig. 2. Flowchart of the reservoir water quality investigation and evaluation.

accumulation of water blooms and related changes.

Two schemes of investigation are suggested (Fig. 2), depending on the extent of the water quality parameters studied and frequency of sampling:
(a) orientation – consisting of a limited extent and sampling frequency; usually are conducted in small reservoirs, and
(b) systematic – involving a higher sampling frequency at more sites, with more parameters.

All large reservoirs are investigated systematically, as well as the small ones experiencing difficulties with water treatment. It is unlikely such difficulties would arise in a small reservoir if the chlorophyll-a content does not exceed 10 μg m^{-3} in the summer and the BOD$_5$ is less than 3 mg l^{-1}.

In addition to water quality, knowledge of the constant and variable parameters indicative of the state of a reservoir and its watershed also is important. This knowledge is required in order to plan a sampling strategy, do water quality evaluations, and assess short- and long-term control measures.

1.2. Constant and variable characteristics of a reservoir and its watershed

Constant and variable characteristics of a reservoir and watershed are not of immediate value in diagnosing and evaluating reservoir water quality. However, they are necessary for assessing the causative factors of reservoir water quality and for making predictions.

Constant characteristics of a reservoir (Table I) consist of parameters, which do not change over the long term, or else they change very little (e.g. owing to mud silting).

Constant characteristics of a watershed are given in Table II. These characteristics may change over the long term, due to factors such as changes in farming techniques.

Variable characteristics of a reservoir (Table III) are determined or calculated on an annual or more frequent basis.

Variable characteristics of a watershed (Table IV) are determined or calculated annually for systematically investigated reservoirs, and at five-year intervals for smaller ones.

Table I. Constant characteristics of a reservoir.

Characteristics	Explanation	Notation	Units
Year of filling			year
Volume curve	relating volume to depth		m³,m
Area curve	relating area to depth		m²,m
Shoreline length curve	relating shoreline length to depth		m,m
Curve of backwater length	relating length to depth		m,m
Reservoir volume		V	10^6 m³
permanent storage		V_s	
reserve storage		V_z	
total controllable volume		V_{oc}	
Reservoir surface area		A	m²
at permanent storage		A_s	
at reserve storage		A_z	
Guaranteed minimum flow		Q_{min}	m³ s⁻¹
Elevation		H	m a.s.l.
Mean depth		$\bar{Z}$	m
Maximum depth		z_{max}	m
Average theoretical retention time corresponding to Q_n and V_s		R_{I-XII}	days

Table II. Constant characteristics of a watershed.

Characteristics	Notation	Unit
Area of watershed	A	km²
Long-term average precipitation		mm y⁻¹
Long-term average rate of flow at the dam	Q_n	m³ s⁻¹
Rate of flow	Q_{355}	m³ s⁻¹
Forested area		%
Farmland in the watershed		%
Arable land in the watershed		%

Table III. Variable characteristics of a reservoir.

Characteristics	Notation	Units
Daily level of water surface		m a.s.l.
Daily inflow	QI_i	m³ s⁻¹
Monthly average	QI_m	m³ s⁻¹
Annual average inflow	QI_a	m³ s⁻¹
Daily, monthly, annual average outflow	QO_i, QO_m, QO_a	m³ s⁻¹
Annual average offtake by waterworks		m³ s⁻¹
Regulation of outflow and offtake throughout the year		
Horizons of outflows and offtakes		

Table IV. Variable characteristics of a watershed.

Characteristics	Units
Number of inhabitants using public sewerage	No
BOD₅ discharge from settlements and industry	kg y⁻¹
COD discharge from settlements and industry	kg y⁻¹
Livestock heads by species	No
Livestock in large-scale breeding farms	No
Amounts of NPK fertilizers used	kg y⁻¹

1.3. Sampling sites

For the orientation scheme of investigations, water samples are collected at only one site, usually the reservoir proper. For the systematic scheme, the inflow, the main body of the reservoir and the water diverted to waterworks or flowing out of the reservoir are sampled (Fig. 2). It is necessary to examine the water quality at these three sites in the reservoir in order to make accurate assessments and for management purposes.

Inflow: The tributary sampling site must be located above the potential maximum flooding and below the last pollutant point source. The sample collection should comprise the entire inflow tributary cross-section. Water flowing directly into the reservoir from a major pollutant source must be sampled separately. If several tributaries feed the reservoir each tributary supplying more than 10% of the total inflow should be sampled for the quantity and inflow water quality. Heterogeneous watersheds should be sampled more thoroughly.

Main body of the reservoir: The reservoir must be sampled at the deepest spot, usually located near the dam. Construction dike should be avoided. The number and spacing of vertical sampling depths will depend on the reservoir depth and the degree of stratification. The 'surface' is sampled 20–30 cm under the water surface. It should not be sampled near the shoreline because of possible accumulation of wind-blown scum. The 'bottom' is sampled 1–2 m above the bottom. Individual vertical sampling depths should not be farther apart than 10 m, with the inter-sampling decreasing as one moves toward the surface (e.g., 0, 3, 5, 10, 20, 30, 40 m below surface bottom). A reliable tool for determining the spacing of individual depths is a detailed temperature profile, using an electric thermometer (e.g., thermistor or other resistance thermometer) at the time of maximal stratification. Different water outlet elevations require samples taken at their middle depths.

One depth profile usually is sufficient for a large deep reservoir. However, it should be complemented by samples taken from the inflow (mixing) zone of the reservoir. In shallow, (non-stratified) reservoirs, samples should be taken along several vertical profiles. The reason is that horizontal differences of shallow reservoirs are dominantly wind driven and the situation at the offtake point changes in dependence on short-term changes of wind direction and speed.

If more information about the movements of water layers is required it is necessary to take samples along several vertical profiles. Since water quality can vary from site to site in a reservoir, no single sampling will be fully representative of the whole lake. The knowledge of horizontal differences in the vertical stratification of water quality allows for correct interpretation of the causes of vertical changes in qualitative parameters at the dam. Water quality changes occurring throughout the year also must be known, especially during summer stagnation when conspicuous depth and horizontal changes can take place, particularly in eutrophic reservoirs. It also is important to know horizontal differences in water layers in anomalous hydrological situations, example being floods and falling the water levels due to low water inflows. Conclusions regarding the individual factors causing the quality changes of water in stratified reservoirs are generally valid only if the longitudinal water quality profile of a reservoir have been investigated for at least one full year.

Outflow: The reservoir outflow consists of:
(i) outflowing stream, and
(ii) water offtake.

Both outflows should be checked regularly for the quantity and quality of water. The quantities of water diverted or discharged water from a reservoir, as well as the layers from which the water was taken, should be determined. The quantities of water, and the layers from which raw water was taken, usually are recorded daily by waterworks operators, while the quantity and depths of discharged waters usually are recorded by reservoir operators.

1.4. Sampling frequency

Generally, the more frequently samples are taken, the more representative and reliable are the results and subsequent conclusions.

For orientation investigation samples should be taken at least twice a year, once at the time of maximal stratification (in the northern temperate region July and August, preferably form mid-July to mid-August) and once during the autumnal isothermal period (November).

Samples for systematic investigations should be taken monthly from the inflow, the outflow, any water discharges or withdrawals, and in the individual zones of a reservoir. Samples should always be taken at a fixed time of day.

Sampling and measurements should be done at regular intervals in order to obtain a true picture of water-quality changes. Regular intervals also allow an easier statistical evaluation (e.g., the calculation of 'time-representative' averages) and data summarization (e.g., trend calculation, including seasonal components). In addition, diurnal sampling can be done at critical periods.

1.5. Water quality parameters

Figure 3 characterizes the reservoir water quality parameters and their evaluation in subsequent Tables and Figs. In addition to this classical grouping, the recorded and measured parameters for basic assessment of reservoir water quality can be divided into:
(a) Flow rate at the time of sampling, and information on the regulation of outflow and offtake by waterworks,

PARAMETERS AND THEIR EVALUATION

GROUP	PARAMETERS	NOTATION	TABLES + FIGURES
PHYSICAL:	RETENTION TIME	R	Fig. 5
	TEMPERATURE	T	6, Fig. 5 to 10
	TRANSPARENCY	SD	7, 8
CHEMICAL:	NUTRIENTS	N, P	9, 10
	ORGANIC MATTER	COD, BOD	11, 12
	OXYGEN	DO	13
	OTHER	—	14
BACTERIOL.:	PSYCHROPHILIC	—	15, 16
	MESOPHILIC	—	16
	COLIFORM	—	16
	ENTEROCOCCAL	—	16
BIOLOGICAL:	PHYTOPLANTON	PB, CHA	17, 18, Fig. 11, 12
	ZOOPLANKTON	ZOO, ZB	19
	(BENTHOS)	—	—
	FISH STOCK	FB	20, Fig. 13

Fig. 3. Reservoir water quality parameters and their evaluation in Tables and Figs of the present paper.

(b) water quality parameters indicative of stratification (temperature, dissolved oxygen, pH, sulphides and hydrogen sulphide, iron and manganese),
(c) water quality parameters indicating eutrophication (primary production, transparency, chlorophyll-a (CHA) and phytoplankton composition, zooplankton, fish stock, nitrogen and its forms, phosphorus and its forms),
(d) organic matter (COD, BOD, colour),
(e) microbiological parameters, and
(f) mineral budget (conductivity, alkalinity, sulphates, chlorides).

The degree of investigation of the various parameters during orientation and systematic investigation are given in Table V.

Table V. Water quality parameters and the extent of their determination in systematic and orientation investigations. Legend: S – sampling near water surface, Z – zonal sampling, B – sampling at the bottom, + – data necessary.

Parameter	Section	Systematic			Orientation	
		IN	RES	OUT	IN	RES
Rate of flow	1.1	+	+	+	+	+
Temperature	2.1	S	Z	S	–	Z
DO	3.3	S	Z	S	–	Z
pH	1.6	S	Z	S	–	–
Total Fe, Mn	3.4	–	B	–	–	–
Sulphides and hydrogen sulphide	3.4.6	–	B	–	–	–
Transparency	2.2	–	S	–	–	S
Chlorophyll-a	5.1.2	–	S	–	–	S
Phytoplankton	5.1	–	S	–	–	–
Zooplankton	5.2	–	S	–	–	–
Fish stock	5.4	–	+	–	–	+
Ammonia	3.1.1	–	S	–	–	–
Nitrates	3.1.2	S	S	S	–	–
Nitrites	3.1.3	–	B	–	–	–
Total P	3.1.4	S	S	S	S	S
Colour	3.2	S	S	S	–	–
COD_{Mn}	3.2	S	S	S	–	–
COD_{Cr}	3.2	S	S	S	–	–
BOD_5	3.2	S	S	S	–	–
Psychrophilic bact.	4.1	S	S	S	–	–
Mesophylic bacteria	4.2	S	S	S	–	–
Coliform bacteria	4.3	S	S	–	–	–
Alkalinity	3.4.1	S	S	–	–	–
Sulphates	3.4.2	S	S	S	–	–
Chlorides	3.4.3	S	–	–	–	–
Conductivity	3.4.4	S	S	–	–	–

In orientation investigations, water temperature and dissolved oxygen are measured by depths, while pH, COD, CHA, transparency, total phosphorus and nitrates are measured at the water surface. In addition, total phosphorus and water flow rate are measured in the inflowing streams.

In systematic investigations, all the parameters listed above are recorded, to the degree indicated in Table V. Special attention is given to parameters of major significance in deep reservoirs, i.e. those of groups (b) and (c). Low values of some parameters requiring sensitive measurement methods (e.g., BOD_5, COD, phosphorus), and a close correlation of chemical changes with biological processes (particularly phytoplankton photosynthesis) are characteristic of drinking water supply reservoirs. Therefore special attention is paid to water quality parameters indicating the degree of eutrophication and, for some of the other parameters, to methods specific for reservoirs. Methods designed for other types of waters (rivers, effluents) often need some modification or alteration before they can be applied to these parameters in reservoirs. Necessary modifications are described briefly in the following sections.

1.6. Sampling methods

Zonal sampling bottles (samplers) of any available type can be used. Water must flow freely through the open descending bottle so that samples are taken from the required depth. The sampler must be equipped with a discharge tube sufficiently long to touch the bottom of the bottle used to determine the dissolved oxygen concentration and pH. Reagents for determining of dissolved oxygen are added to the bottle; the pH is directly measured.

Samples must be protected from sunlight, overheating and undercooling. Portable polystyrene boxes (or polystyren-insulated boxes) are suitable for this purpose.

Samples for determining phytoplankton (including CHA) are taken either zonally (which is time-consuming), or with a simple tube which samples the entire production layer at one time. The 3–5 cm diameter tube of hard plastic with the length of 4–6 m is equipped with a conical rubber stopper fixed on a rope passing through the tube; the handle at the upper end of the rope serves for pulling the stopper into the submerged end of the tube.

Zooplankton can be sampled with Apstein-type nets (see e.g. Hrbáček, 1966). For determining the zonation of zooplankton, a large sampler (preferably the Schindler type – see Schindler, 1969) is required.

2. Measurement and evaluation of physical parameters

The principal physical parameter is the theoretical time of water retention in the reservoir (R, shortly called retention time). It is calculated by a simple formula,

$$R \text{ [days]} = V/Q \tag{1}$$

where:
V = the volume of the reservoir [m³] and
Q = flow rate [m³ d⁻¹].

The periods for which R has been calculated must be distinguished. If only Q has been changing, its average value can be used (e.g., for February, for the summer half of year or the whole year: R_{II}, R_{IV-IX}, R_{I-XII} respectively). If V also has been changing substantially, it is necessary to calculate R for short periods and average the results:

$$R_{av} = \Sigma(V_i/Q_i)/n \tag{1a}$$

where V_i and Q_i are average volumes and flow rates for shorter time intervals and n is the number of time intervals.

There is no simple relation between the theoretical retention time (R) and the actual retention time of inflowing water in a reservoir, (R_{real}). The real retention time depends on the ratio of inflow to the volume of layers to which water is plunging in. The top layers of a strongly stratified reservoir are thin. Therefore, for the thermal stratification period, $R_{real} \ll R$, and density current can be created. It is very difficult to estimate R_{real}, because it depends on the manner in which inflowing water moves to individual layers in a reservoir, on the rate of its mixing with adjacent layers, etc. Most of these variables can only be determined indirectly, primarily on the basis of a theoretical model. In fact, $R = R_{real}$ only in non-stratified reservoirs with a freely-flowing water regime.

Temperature is another physical parameter which is of immediate importance and indicative of the stratification of the water masses. Water density S_w, decisive for stratification, depends primarily on temperature T (Fig. 4) and can be calculated from Eq. (2):

$$\varrho_w = 1 - 6.63 \times 10^{-6} (T - 4)^2 \tag{2}$$

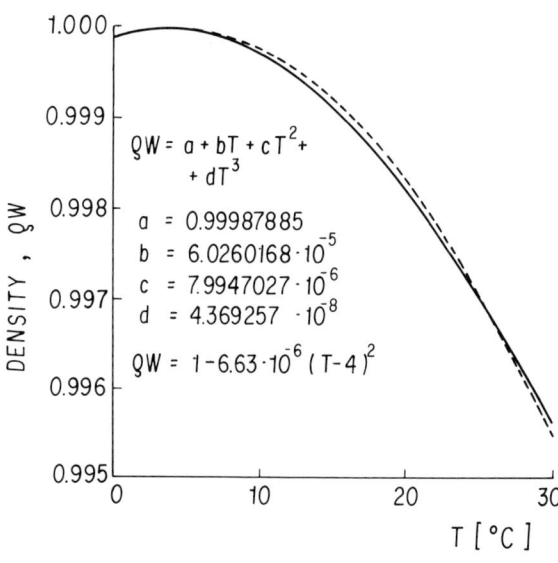

Fig. 4. Dependence of water density, S_w, on temperature, T. Values tabulated in Hutchinson are very exactly approximated by the polynomial (full line) $S_w = a + bT + cT^2 + dT^3$, where $a = 0.99987$, $b = 6.0260168 \times 10^{-5}$, $c = -7.9947027 \times 10^{-6}$, $d = 4.369257 \times 10^{-8}$. A more simple approximation (dashed line) is obtained from $\varrho_w = 1 - 6.63 \times 10^{-6}(T - 4)^2$. In the range $0 < T < 25°$ also the equation $\varrho_w = 1 - 6.8 \times 10^{-6}(T - 4)^2$ holds true. According to Straškraba & Gnauck (1985).

or read from tables in Hutchinson (1957). A high degree of mineralization or turbidity also can play a role in the determination of ϱ_w in some water bodies.

Transparency measured by Secchi disc is a simple means of categorizing water quality. Observations of long-term changes in transparency are used worldwide as an easy means of getting a relative indication of lake water quality.

2.1. Measurement and evaluation of temperature and temperature induced stratification of reservoirs

Measurement of temperature changes with depth is done easily with a thermistor, resistance or other electric thermometer. However, the values obtained with such devices should be confirmed during each sampling with an accurate mercury thermometer.

Temperature also can be taken with a mercury thermometer fixed inside a zonal sampling bottle (sampler), however, the sampler and thermometer must be tempered at each sampling for 3–5 minutes, and the temperature read immediately upon pulling the sample out of the water body.

The water temperature in reservoirs can be affected by many constant and variable characteristics of the reservoir and watershed. The main constant features are elevation, surface area, and depth. There are other characteristics as well, including the exposure of the surface to wind (especially in small reservoirs). The principal variable characteristics are hydrometeorological conditions in a given year, the outlet depth, and water manipulation in the reservoir. It is noted that the characteristic numbers for certain types of reservoirs given below apply only to reservoirs where the temperature of the inflowing water is more or less natural, and where the water is not pumped into, or flowing out, of another reservoir.

The results of temperature measurements are evaluated differently for orientation and systematic (all-year) investigations. For the former, it is necessary only to determine the type of stratification and estimate the expected changes in stratification due to different flow rates in individual years. The object of systematic investigations is to make a detailed comparisons of stratification in a given reservoir and year with other years and other reservoirs.

2.1.1. Assessment of orientation investigations

The simplest characteristic is the depth profile of temperature at a fixed sampling site at the time of maximal stratification (i.e., late July – early August). This profile has three basic features: surface temperature (T_O), bottom temperature (T_B), and the mixing layer depth (equal to the depth of the thermocline, z_{mix}). T_O and T_B can be determined directly. However, a plot of the depth profile is necessary to estimate z_{mix}. The parameter z_{mix} is the depth (in meters below surface) where the change in water temperature is abrupt. It is not always easy to determine z_{mix}, primarily because several anomalous 'bumps' may appear in the depth profile. The smallest temperature difference indicative of the thermocline location is considered 1 °C m^{-1}.

The type of stratification is determined on the basis of T_O and T_B values (the difference being indicated as T_{O-B}) and z_{mix}, in Table VI. The table

Table VI. Assessment of stratification types by temperature depth profile during the period of maximal stratification (late July–early August).

T_{O-B} (°C)	CDS	z_{mix}	Type of stratification	Corresponding reservoir type
<3	1–1.2	lacking	a) 'pond'	shallow
			b) non-stratified	deep, high flow, $R_{I-XII}<10$
<5	<1.5	difficult to distinguish	fluvial	deep, $10<R_{I-XII}<20$
>5, <14	<3.3	distinct	transitional	deep, $20<R_{I-XII}<100$
$T_B = 4$ °C	5	distinct	lake	deep, $R_{I-XII}>100$

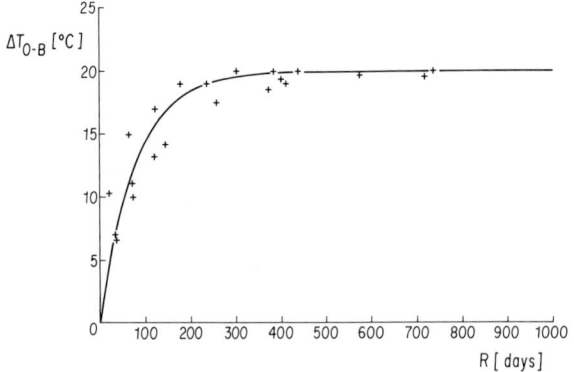

Fig. 5. Dependence of reservoir stratification on R. Stratification is expressed as the temperature difference surface – bottom (T_{0-B}) during late July – early August, for the representative values of retention time the period April to September is taken (R_{IV-IX}). Data for Czechoslovak and Sachsian reservoirs.

also contains the values of a coefficient of distinctness of stratification (CDS, see Section 6.6.1), on the assumption that $T_O = 20\,°C$.

Pond-type stratification is marked by the creation of temporary temperature gradient when the weather has been warm and windless for some time. The critical values of the annual average theoretical retention time given in Table II are only approximate value, because the stratification can change dramatically within a few days of an altered flow rate.

Differences among years are due largely to different discharges. However, other hydrometeorological variables can change as well. For example, R may be different, even at the same rate of flow if the water level rises or falls.

Changes in stratification due to R can be evaluated on the basis of the relation of T_{O-B} to R (Fig. 5):

$$T_{O-B} = 16\,(1 - \exp(-0.0126\,R_{IV-IX})) \quad (3)$$

where R_{IV-IX} covers the period of formation and duration of summer stratification (April to September). Substituting respective R_{IV-IX} values in Eq. (3) can produce a rough estimate of expected changes in stratification in different years or situations. For instance, in the Slapy Reservoir, where the long-term R_{IV-IX} differences are 18 to 90 days, the result is $6.5 < T_{O-B} < 14.8\,°C$. This means that, in a year with extremely high flows, the stratification of the reservoir is of the fluvial type. In a dry year, it is of the transitional type.

2.1.2. Evaluation of systematic investigations

Systematic evaluation can be either numerical or graphic. The annual variations of temperature in a

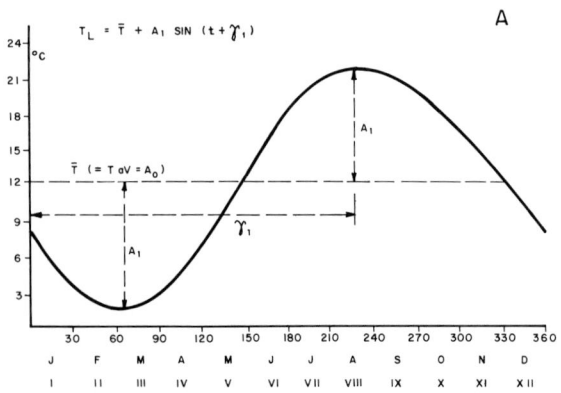
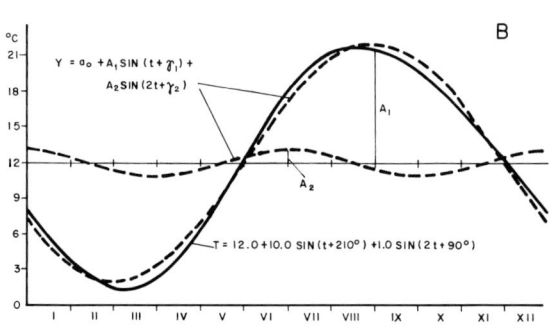

Fig. 6. Characteristic quantities of the annual temperature changes: A – T_{av} ($= A_0$) = average annual temperature, A_1 = annual semiamplitude, τ_1 = coefficient expressing in trigonometric degrees the position of temperature maxima and minima (in Northern Hemisphere usually around 230–245°, corresponding to maximum temperatures occuring between mid August and beginning of September). The numerical values of T_{av} and A_1 can be estimated directly from the graph, while the estimation of τ_1 is not as straightforward. Time t is in degrees, one year representing 360°, which approximately corresponds to 1 day = 1°. Temperature at any time t can be obtained by substituting t in the equation given. B – The figure shows how distortion of the regular sinusoidal wave (dashed line) is expressed by adding a second wave with twice the frequency of the primary wave. The resulting approximation (full line) is described by the sum of two waves. A_1 and A_2 are the semiamplitudes of the first and second wave, the additive term represent the annual average temperature (T_{av}).

certain layer of the reservoir are numerically best characterized by three quantities (Fig. 6): average annual temperature (T_{av}), annual semiamplitude of temperature (A_1) and τ_1, a coefficient expressing the time of maximal temperatures.

These values can be accurately determined if measurements have been regular throughout the year. If there has been a gap (e.g., in some winter months) a more complex calculation must be made, or else the assessment can be seasonal.

Calculation of T_{av}, A_1 and τ_1 by a computer can be obtained by using the procedure developed by Bliss (1970) — the mainframe FORTRAN programme as well as a PC version are available from the senior author. A rough estimate of these values also can be obtained as follows: T_{av} is calculated as the mean of all measurements, A_1 is estimated as $(T_{max} - T_{min})/2$. An approximate estimate of τ_1 is derived by using the ordinal number of the day in the calendar year (expressed in degrees: 1 year = 360°) on which the maximum temperature was reached.

Surface $T_{av}(0)$ values of reservoirs in Czechoslovakia depend primarily on elevation a.s.l., H, (Fig. 7), approximately by the relation

$$T_{av}(0)(H) = 34.2 - 9.29 \log H,$$
$$(200 \leq H \leq 600)$$

but also on R (increasing with growing R), and on the size of the reservoir (with smaller reservoirs being warmer). $A_{1,0}$ values for the surface waters of Czechoslovak reservoirs (provided they are not affected by upstream reservoirs, and are not pumped-storage reservoirs) are relatively constant (within the range of 10.2 ± 0.7 °C). The range of $\tau_{1,0}$ is 236 ± 8° (Straškraba & Straškrabová, 1975; Straškraba, 1976b).

Similar relations were developed for other water-body types and countries (Horváthová & Dávid, 1969 — rivers in Slovakia; Szumiec, 1969 — ponds in Poland, Collings, 1973 — rivers in USA; Straškraba, 1980 — world lakes).

T_{av} and A_1 gradually decreases below z_{mix} with increasing depth, whereas τ_1 increases. The rate of decrease in T_{av} and A_1 with growing depth (z) depends on R. The decrease is slow in reservoirs with a short R, where the difference between the temperature of the surface and bottom waters is small. The largest, and most regular, decrease occurs in natural lakes (with $R >$ few years), where the temperature at greater depths is practically constant throughout the year, so that the annual amplitude of temperature is $A_1 = 0$ °C and $T_{av} = 4$ °C. The decrease in T_{av} with depth (Fig. 8) can be expressed approximately by equation (4),

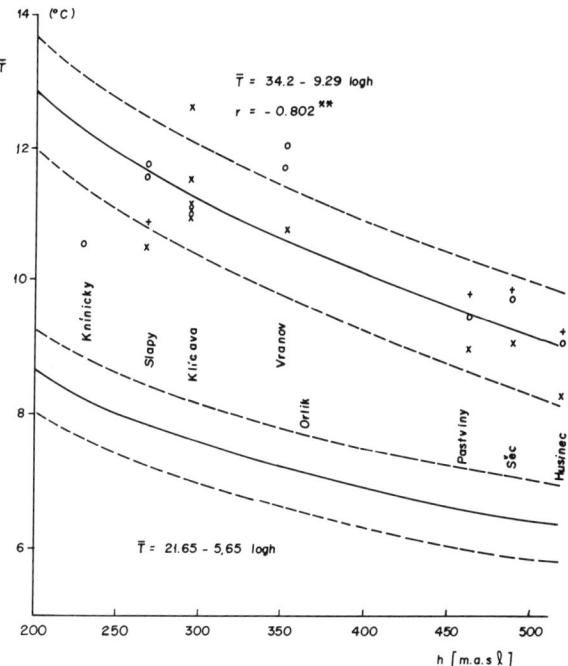

Fig. 7. Dependence of the average surface temperatures ($T_{av,0}$) of Czechoslovak reservoirs (upper curve) and streams (lower curve) on elevation (H, m a.s.l.). The dashed lines indicate the approximate intervals of ± standard deviation. From Straškraba & Gnauck (1985).

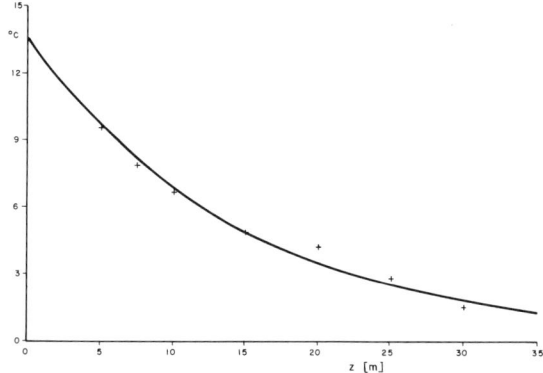

Fig. 8. The decrease of the annual average temperature in the given depth ($T_{av,z}$) with depth below the mixing depth ($z - z_{mix}$) in Saidenbach Reservoir (Germany) 1969.

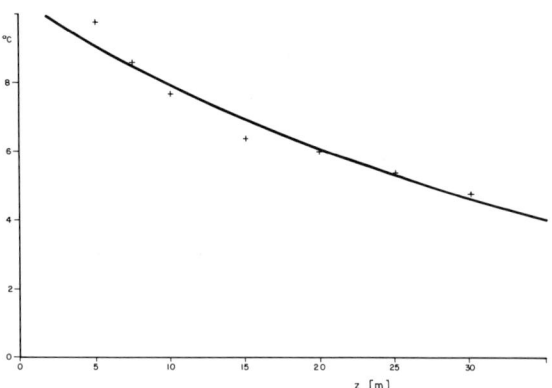

Fig. 9. The decrease of the annual semiamplitude in the given depth ($A_{1,\,z}$) with depth below the mixing depth ($z - z_{\mathrm{mix}}$) in Saidenbach Reservoir, 1969.

$$T_{\mathrm{av}}(z) = T_{\mathrm{av}}(z_{\mathrm{mix}}) \cdot \exp(-a(z - z_{\mathrm{mix}})) \qquad (4)$$

where:

$T_{\mathrm{av}}(z_{\mathrm{mix}})$ = the average temperature at depth z_{mix}, and

a = the rate of decrease in average temperature. Similarly (Fig. 9),

$$A_1(z) = A_1(z_{\mathrm{mix}}) \cdot \exp(-b(z_{\mathrm{mix}})) \qquad (5)$$

The values of a and b can be related to R, because in throughflowing reservoirs stratification is far less pronounced and therefore both a and b are lower. Figure 10 demonstrates this relation and makes it possible to estimate temperatures at different depths in a reservoir when retention time varies.

2.2. Transparency

A black-and-white chequered disc, measuring 20 × 20 cm, which is hung horizontally on a rope marked by depth intervals of 0.25–0.5 m. The depth at which the outlines of the squares of plate disappear from sight is recorded. The effects of reflections on water surface (especially when it has ripples), dazzling, and shading by the boat or structure from which the measurement is done should be avoided.

Transparency depends primarily on the combined effects of water colour (due to dissolved substances), mineral turbidity, and the presence of algae. Water usually is coloured in reservoirs in peatbog regions, or by pollution caused by paper mills. The colour usually does not change significantly in the course of a year. In contrast, a pronounced seasonal variation is characteristic of algae, whose occurrence in a waterbody is minimal in autumn to early spring and culminates in the spring to summer season. In temperate regions turbidity due to floods usually disappears within a short time.

The evaluation objective is to classify a reservoir on the basis of its transparency (Table VII) and to estimate the degree of eutrophication (Table VIII). Transparency differences between early spring and spring-summer are indicative of the degree of eutrophication.

Both tables are based on original data; estimation of the effects of eutrophication is based on the relation of transparency to the extinction of

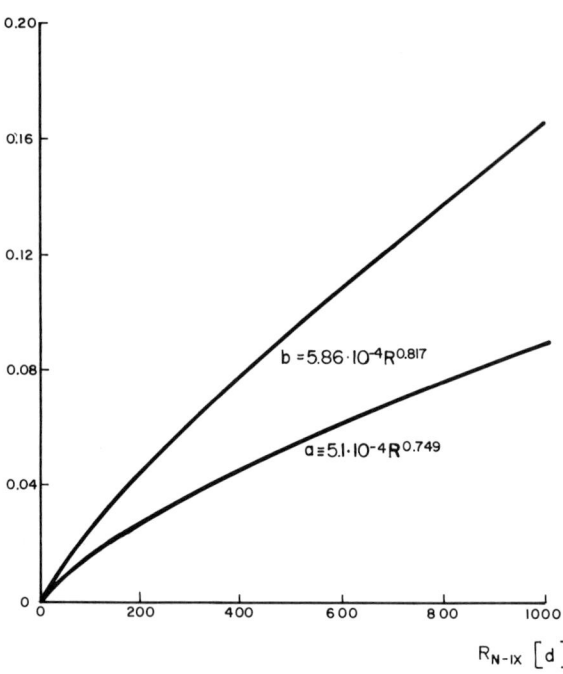

Fig. 10. Average dependence of the values a and b in Eq. (4)–(5) on the theoretical retention time, R. Data for Czechoslovak reservoirs.

Table VII. Classification of reservoirs by transparency.

Transparency in early spring or in autumn [m]	Reservoir
<0.3	not transparent
1–2	poor transparency
3–6	semi-transparent
>6	clear

Table VIII. An estimate of the effects of eutrophication based on changes in transparency in the course of the year.

Difference in transparency [m] between spring-summer and early spring or autumn	Presumed effects of eutrophication
0–0.5	none
0.5–1	slight
1–3	medium
>3	strong

light in water, and on the effects of phytoplankton (measured by CHA) on light extinction. The estimate is approximate, because the difference in transparency between autumn or early spring and the spring-summer season may have other causes, in addition to changes in phytoplankton. If transparency has been measured systematically, only measurements without significant flood-related turbidity are used in assessments.

3. Chemical parameters

3.1. Nutrients

Only nitrogen and phosphorus are evaluated, because of their role as a limiting factor for primary production of phytoplankton in reservoirs. In industrial countries the likelihood of limitation by microelements is low.

A conspicuous decrease of nitrogen and/or phosphorus concentration in the upper layers of reservoirs during the summer phytoplankton growing season indicates their intensive utilization, which may eventually limit phytoplankton biomass.

Very low concentrations, often only a few $\mu g\, l^{-1}$, result from a balanced consumption and supply, ensured by processes taking place in the reservoir, including excretion of ammonia and phosphates by zooplankton and fish, transport of nutrients from the hypolimnion, etc.

3.1.1. Ammonia, or ammonia ions

The phenolate method which is often used is not quite specific because of nitrite interference. Phenolate, being toxic and unstable has recently been replaced by salicilate (Bowen & Holm-Hansen, 1980). However, this does not make the method more specific. The bis-pyrazolon method (Procházková, 1964) is considered sufficiently sensitive and specific.

The average load and outflow of ammonia from a watershed are shown in Tables IX and X. The ammonia enters streams from sewage treatment plants and large-scale breeding farms, as well as industry: gasworks, food industry and viscose rayon manufacturers. Ammonia is effectively retained in soil and in water, it is preferred by phytoplankton to other nitrogen compounds (Procházková *et al.*, 1970). Its concentration in the

Table IX. Average specific outflow of compounds from a watershed. The table is based on data by Moldan & Dvořáková (1987) and Procházková *et al.* (1984)

Parameter	1	2	3	4	5
NH_4^+ (NH_3)	5.9–9.4 (13.8)	3.3 (6.6)	36–49	–	–
NO_3^-	18–24 (29)	29–66 (133)	220–265	–	–
N_{organ}	2.7	–	10	5	14
N_{total}	9.4–13 (17)	7.3–20 (35)	100	5	14
P_{total}	0.15–0.70	–	35	0.7	3
SO_4^{2-}	34–54 (81)	63–105 (370)	105	–	–
Cl^-	2–3 (6)	1–2 (3)	50	3.3	5
CA^{2+}	4–6 (13)	1–2 (5)	100–200	–	–
K^+	1–2 (3)	(3)	65	–	–

1 – Total vertical precipitation [kg ha^{-1} y^{-1}] (Data in parentheses apply to the most exposed parts of the Ore Mts.).
2 – Horizontal precipitation + gas deposition [kg ha^{-1} y^{-1}] (Horizontal precipitation, i.e., fog and aerosols intercepted by the uneven and varied landscape, is particularly frequent in forested areas).
3 – Fertilizers [kg ha^{-1} y^{-1}] (This applies only to farmland in the watershed, 1980's).
4 – Inhabitants [kg inhabitant^{-1} y^{-1}] (without detergents).
5 – Livestock [kg (100 kg)$^{-1}$ y^{-1}].

Table X. Average outflow of compounds from homogeneous forested and agricultural watersheds, and limits of the dependence of concentration on the rate of flow in agricultural watersheds. The table is based on data given in Černý (1987), Jehlička (1987), Kinkor (1987) and Procházková et al. (1984).

Parameter	kg ha^{-1} y^{-1} from a wooded watershed^a	kg ha^{-1} y^{-1} from agricultural watershed^1	Maximal concentration flowing out of an agricultural watershed [mg l^{-1}]^e
NH$_4^+$ (NH$_3$)	< 0.5	> 0.1	–
NO$_3^-$	2.2–3.1	17–180	90 (270)^b
N$_{org}$	0.4–2.0	0.5–2.0	–
N$_{total}$	1.5–3	4.5–43	–
P$_{total}$	< 0.15	< 0.2^c	–
SO$_4^{minus}$	5–300	30–150	60
Cl$^-$	8–20	20–80	40
Ca$^{2-\ d}$	15–90	30–100	90
K	1–5	2–7	–

^{a} without livestock kept in stables.
^{b} in winter (non-vegetative season).
^{c} except for leaching in particles of clay (erosion).
^{d} outside karst and cretaceous regions.
^{e} at a high outflow Q_{sp}, when coefficient b in the Eq. (6) is negligible compared with Q_{sp}, maximal concentration equals the sum of coefficients $a + d$.

surface layer of reservoirs is highest late of winter and early spring; it may reach up to 500 μg l^{-1} NH$_4$. An average annual concentration exceeding 150 μg l^{-1} in the surface layer indicates a high supply from the watershed or, more often, decomposition of nitrogenous organic compounds (autochthonous as well as allochthonous) and the release of ammonia in the reservoir. Ammonia concentrations higher than approximately 250 μg l^{-1} are chronically toxic to fish and invertebrates at pH ≥ 9.

3.1.2. Nitrates

Over the last approximately 15 years, nitrates have become the dominant anion in Czechoslovakia (Procházková & Blažka, 1986). Their proportion has grown to about 10%, and in areas of intensive farming, up to 20% of the sum of anions. They are most detrimental to human health. Toxic nitrates result from nitrate reduction. In the presence of nitrogenous organic compounds, the nitrites can become precursors of carcerogenic nitrosamines. Data on the load and outflow of nitrites are summarized in Tables IX and X. They are produced in soil by nitrification of ammonia and, mediately, of organic nitrogen as well. The process occurs more slowly in water. Nitrates easily leach out of soils, especially during heavy rains and thaws. The dependence of concentration on flow rate in rivers is positively hyperbolic in agricultural watersheds and can be expressed by the equation:

$$NO_3^- \ [mg\ l^{-1}] = a\ Q_{sp}/(b + Q_{sp}) + d \qquad (6)$$

where:
Q_{sp} = the specific outflow of water from the watershed [l s^{-1} ha^{-1}],
$a + d$ = the limit of the respective function (see Table X),
b and d = other parameters.

In addition to the quantities of fertilizers, all three parameters in Eq. (6) are also affected by the soil type, season and hydrological regime of the previous season (dry or wet season, increasing or decreasing flow rate, etc.). The nitrate concentration diminishes at very high flow rates, becoming lower than the limit of the hyperbola.

Nitrates from precipitation (Blažka & Procházková, 1987) are retained in forested watersheds, provided the forests are not dying because of the impact of emissions or the forest vegetation is not saturated by nitrogen. Nitrates also are released from forests for several years after timber cutting, and from soil washed away by rain in which the soil cohesion has been impaired.

In Czechoslovakia, the concentration of nitrates in natural waters increased in the 1960s and 1970s, as a result of increasing fertilizer use. Fertilizer application has stabilized in the second half of the 1980s (Nesměrák, 1986a). Therefore changes in nitrate concentrations now depend mainly on changes in the flow rate. The concentrations of nitrates in reservoirs vary slightly with depth. In most Czechoslovak reservoirs with high nitrate levels, their concentration decreases slightly in the surface layer in summer, particularly in reservoirs with long retention times. The decrease is due to the utilization of nitrates by phytoplankton if the in-lake ammonia concentration is low (i.e., concentrations < 50 μg l^{-1} NH$_4$—N). A low nitrate concentration at the reservoir bottom usually is due to denitrification.

3.1.3. Nitrites

It is recommended that the concentration of nitrites be determined within 24 hours after collection.

Nitrites are especially toxic to organisms with haemoglobin. Nitrites can arise from nitrate reduction in anoxic environments, including e.g., ground water, reservoir hypolimnion, human intestines with unstabilized bacterial flora (e.g., in intestines of babies), or as an intermediate product of nitrification.

3.1.4. Phosphorus

Phosphorus is the element most frequently found to limit primary production. In unlimited populations of phytoplankton the N:P weight ratio in biomass is approximately 16 : 1. In Czechoslovak reservoirs, however, it often is as much as 500 : 1. Total phosphorus concentrations in forests drainage and properly cultivated fields usually do not exceed an annual average of 50 μg l^{-1} P. Table IX shows that sewerage systems and large-scale breeding farms are the main phosphorus sources of P in water. In contrast, soil can retain phosphorus as long as soil particles are not washed away. The percentage of phosphorus retained in reservoirs usually is in direct proportion to its concentration in the inflow, and is related to R. Phosphorus concentrations usually are much lower in the downstream ends of reservoirs, than in the upstream ends. During the vegetative season, phosphorus can decrease to a few μg l^{-1}, especially in the surface water layer. This value represents a balance between the consumption of phosphorus, particularly by phytoplankton, and its supply from various sources. Phosphorus also can be exchanged between organisms and water up to 2–3 times a day. A relation between a spring concentration of total P and an average summer concentration of chlorophyll-a is discussed in Section 5.1.2. Determination of reactive phosphorus (so-called soluble inorganic orthophosphates) is most accurate if the sample is filtered through a fibreglass filter upon its collection, and the assessment is made within two hours after its collection. A delay in the analysis can produce biased results because of transformations of phosphorus in the water sample, due primarily to microbial activity. That is why total phosphorus often is measured and reported in the literature. As a result many useful relations for assessing water quality are based on total phosphorus.

The phosphorus load to reservoirs and, consequently, the production of organic matter (e.g., phytoplankton can be reduced by an increased retention of phosphorus in municipal wastewater treatment plants and watershed, including prevention of erosion (permanent meadows on slopes > 5°, grass borders along fields, entry to fields from the side or downhill, etc.), and prevention of the manure leaching.

3.2. Organic compounds

3.2.1. Determination

The content of organic compounds in reservoirs is evaluated by measuring the chemical consumption of oxygen, using either the permanganate (COD$_{Mn}$) or dichromate (COD$_{Cr}$) method. Their proportion in different water types is shown in Table XI.

Only a small proportion of the organic compounds determined by measuring the chemical consumption of oxygen is easily decomposed biologically in surface water. A standard method of biochemical consumption of oxygen in 5 days at 20°C (BOD$_5$) is used for estimating decomposable organic compounds. A semimicromethod (Hejzlar & Kopáček, 1990) saves chemicals and improves analytical reproducibility; a mineralization block (Schindler *et al.*, 1984) can be used.

The proportion of easily-degraded compounds (BOD$_5$) in COD$_{Cr}$ usually ranges between 0.10–

Table XI. The range of usual concentrations and proportions of COD$_{Cr}$ and COD$_{Mn}$ in different water types in Czechoslovakia according to Hejzlar & Kopáček (1990).

Type of water	COD$_{Cr}$ (mg l^{-1})	COD$_{Mn}$ (mg l^{-1})	COD$_{Mn}$/COD$_{Cr}$
ground water	0.1–10	0.2–3	0.15–0.4
precipitation	2–20	–	–
brooks and rivers	5–50	2–20	0.25–0.5
oligotrophic lakes and reservoirs without humins	3–10	–	–
eutrophic lakes and reservoirs	15–30	5–15	0.25–0.5
surface water treated by coagulation (drinking water)	3–10	1–3	0.2–0.4

0.15. Values lower than 0.10 are found in waters with high contents of slowly decomposing organic substances (e.g., the Vltava River, which is polluted by papermill waste). Values over 0.15 indicate the presence of large amounts of algae or recent pollution.

When algae are abundant, the BOD_5 value not only indicates the amounts of dissolved organic compounds, but also is affected by the respiration and decomposition of algae. If this factor might interfere with evaluation, a correction can be made on the basis of the chlorophyll-a content (Straškrabová et al., 1983); 1 μg of CHA equals 0.025 mg of BOD_5.

It is noted that water colour depends in part on the contents of organic compounds. Unpolluted surface water is coloured mainly by humins and ferrocompounds. Colour is determined in a filtered or centrifuged sample by comparison with standard coloured solutions. Finer distinctions are made by spectrophotometry (Dolejš, 1990). Absorbency at 250–370 nm can be used to a certain extent as a measure for COD concentration in water. This latter method, however, is unsuitable for industrial waste water and for natural waters coloured by non-humic substances.

3.2.2. Evaluation of organic substances in reservoirs

Both dissolved and undissolved organic substances enter the reservoirs. The amount and quality of both forms depends on the characteristics of the watershed, including the soil, vegetation, proportion of forests and farmland, industry, settlement, etc. Dissolved organic compound concentrations are usually 5–10 times higher than undissolved compounds, at standard flow rates. With increasing flow rates the proportion of undissolved organic substances increases and it may become higher than that of dissolved substances. This depends largely on erosion in the watershed and the amount of sediments in the streams.

Undissolved organic compounds enter the reservoir and do sediment dominantly at its upstream end, partially decomposing and partially becoming permanent components of sediments. In contrast, allochthonous dissolved organic substances take a long time to decompose in reservoirs. For example, the decomposition of dissolved organic compounds (determined as COD_{Cr}) in samples from various localities in the Malše River basin (outflows from forests and fields, fish pond, Římov Reservoir) does not surpass 20% during a 100-day incubation in darkness at 20 °C. Although these compounds do not significantly affect the oxygen regime, they may spoil the taste or smell of treated water, and they may be precursors of haloforms and other mutagenic organochlorides resulting from the disinfection of water by chlorination.

Other undissolved and dissolved compounds in reservoirs (autochthonous source) result from the production of phytoplankton and other organisms, e.g., bacteria (Hejzlar, 1989). The autochthonous portion of the total supply of organic compounds increases with increasing eutrophication of the reservoir and with a prolonged water retention (Straškrabová, 1975). The total load of BOD_5 to a reservoir consists of the inflow load and load by primary production of organic matter by phytoplankton in the reservoir. Primary production is converted to BOD_5: 0.5 of gross primary production equals BOD_5. The percent reduction of BOD_5 of the total reservoir load depends on the retention time and the depth of the impoundment (Table XII).

Organic compounds produced within the water body decompose easily (up to 95%) and are relatively quickly consumed. For example, in the slightly eutrophic Římov Reservoir, with a theoretical retention time of about 90 days, the majority of dissolved organic compounds in the epilimnion are allochthonous even in summer. Though the autochthonous organic compounds usually make up only a small proportion of all the

Table XII. Percentual reduction of BOD_5 of total load (average values for the warm half-year) in shallow and deep reservoirs at different retention times ($\bar{Z}$ = average depth, m; R = retention time, days).

R	$\bar{Z} < 7$	$\bar{Z} > 7$
0.5	15	20
1	40	35
5	60	45
10	80	55
50	75	70
100	75	70
200	70	75

organic matter, their presence in drinking water reservoirs is undesirable with regard to public health, e.g., malodorous compounds and toxins of some algae and blue-green algae (Carmichael, 1972).

Biochemical oxygen consumption includes the oxidation of decomposable organic compounds in a reservoir, as well as the respiration of organisms present in the sample and, possibly, oxygen consumption due to nitrification. Respiration and decomposition of algal biomass may contribute substantially to the BOD value in reservoirs and their tributaries (e.g., 15–70% of BOD_5 in mesotrophic reservoirs in summer). Nitrification usually is negligible in open waters of reservoirs.

High BOD_5 values in inflow waters can impair the quality of water in the entire reservoir if the retention time is short. They will affect only the upper part if the reservoir has a long retention time. BOD_5 value exceeding 4–5 mg l^{-1} in the warm season suggest the possibility of oxygen depletion in stagnant regions of the reservoir. It makes no difference whether the BOD in the inflow waters is due to respiration and decomposition of algae or to bacterial oxidation of dissolved organic matter. Algae in the inflow usually die in the reservoir and represent allochthonous pollution.

The proportion of decomposable compounds (BOD_5) in COD_{Cr} is usually within the range of 0.10–0.15. Values lower than 0.10 occur in waters containing large amounts of slowly decomposing organic compounds (e.g., those polluted by papermill waste). Values over 0.15 indicate the presence of large amounts of algae or recent pollution by easily decomposable compounds (e.g. sewage from settlements, food-industry wastes, etc.).

3.3. Evaluation of oxygen concentration in reservoirs

Samples for oxygen determination should be taken with a sampler through which water can flow freely when it is descending. In its lower lid there is a tap with a tube reaching to the bottom of the bottle. Another portion of the sample can be used for determining pH. Collecting sample it is advisable to flush the sample bottle at least twice its full volume, letting the discharge tube of the zonation sampler reach to the bottom of the bottle. The bottles to which reagents have been added should be transported submerged in water tank, or else reagent bottles for highly-volatile substances should be used, filling their caps with water from the sampling site.

If contact of the water sample with air is not prevented, or if reagents are not added immediately, analytic results can be wrong, especially for oxygen concentrations lower than 2 mg l^{-1}. However, if contact with air is prevented as described above, oxygen concentrations lower than 0.1 mg l^{-1} can be reliably measured. Low values usually occur at the reservoir bottom, although they also may occur at shallower depths. Another methodological problem can occur when a steep oxygen gradient exists immediately above the reservoir bottom. Values lower than 2 mg that have been measured in samples taken with a common sampling bottle 'above the bottom' (which may mean 0.5 m, but also could be a few meters above the bottom) may indicate an anaerobic condition at the bottom.

At such low concentrations some electrodes do not measure oxygen reliably. Using polarographic methods (electrodes) it is necessary to make frequent intercalibration of the polarographic and standard iodometric (Winkler method) determination.

Anoxia (absence of oxygen) near the bottom is one of the most serious phenomena affecting reservoir water quality. Under anoxic conditions, some substances are released from bottom sediments at a rapid rate, including phosphorus, iron, and manganese. Sulphides (hydrogen sulphide) also can develop. When the oxygen content in the bottom waters decreases to zero, it can be expected with certainty that iron and manganese occur at higher concentrations than under oxic conditions. Therefore oxygen can then serve as a reliable indicator of water quality degradation in reservoirs. Critical limits of O_2 concentration in the hypolimnion, indicating the suitability of a reservoir for drinking water supply, are given in Table XIII.

The simplest evaluation of oxygen can be made during the autumnal mixing. At the onset of complete mixing, the measured oxygen concentration roughly corresponds to the weighed average of the entire water column.

The oxygen concentration is of critical

Table XIII. Indication of the suitability of reservoirs for drinking water supply by a critical concentration of oxygen in the hypolimnion (after Dillon & Rigler, 1975).

Concentration of O_2 [mg l^{-1}]	Suitability of reservoir
> 5	excellent
< 5	suitable
< 2 briefly	not very suitable
< 2 for a long time	unsuitable

importance to aquatic organisms; fish and other organisms die at very low concentrations. A concentration of 2 mg l^{-1} O_2 is considered critical for fish; however, it also depends on the duration of fish exposure to a certain concentration. Fish need oxygen concentrations of 4–8 mg l^{-1} O_2 – depending on temperature – for their normal activities (swimming, feeding and growth). They can survive much lower oxygen concentrations if the exposure is brief. The specific fish species also is a factor; salmonids are the most susceptible to low oxygen concentrations. The oxygen content at the reservoir bottom also can affect the composition of benthos (see Section 5.3.).

Anoxia of bottom waters can be prevented by reducing eutrophication. Metalimnetic oxygen minima can occur when there is a sharp thermocline and high primary production. This situation can produce deterioration of water quality in the middle water layers. Such layers must be avoided in diverting water to waterworks. The occurrence of metalimnetic oxygen minima also can be prevented by reducing eutrophication in the reservoir.

The occurrence of anaerobic zones in the upstream, inflow portion of a reservoir usually indicates an oversupply of biologically decomposable, undissolved organic compounds from the watershed.

3.4. Evaluation of other chemical parameters

3.4.1. Hydrogen carbonates

These compounds are determined as total alkalinity. In the case that the pH of a sample is > 8, carbonates are included in the determination. The standard method is considered applicable for concentration > 0.05 mmol alkalinity. However, our experience indicates that the values are unreliable at an alkalinity < 0.2 mmol. A method described by Gran (Mackereth *et al.*, 1978) is recommended for such cases.

Concentrations of hydrogen carbonates depend primarily on the subsoil. Long-term trends of the concentrations on granite and other crystalline subgrades indicate the intensity of acidification processes in watersheds. The concentration of HCO_3^- can drop to zero (e.g., in some tarns in the High Tatra Mountains, lakes in the Bohemian Forest, brooks in the Iser Mts.). In contrast, in most of deep Czechoslovak reservoirs, the annual average concentrations have been approximately the same level for years (at least 30 years in Slapy – Procházková & Blažka, 1989); acid rain has so far been neutralized in watersheds and reservoirs. Nevertheless, the equivalent portion of HCO_3^- in the total sum of anions has been decreasing substantially, due to steeply increasing concentrations of all the anions of strong acids (sulphates, chlorides and nitrates).

The proportion of hydrogen carbonates used to be about 75% in the majority of Central European waterbodies in the first decades of this century. At the present time however, it does not exceed 30% of the total sum of anions in most cases (except in karstic regions). Expressed as the equivalent ratio of strong acid anions to anions of hydrogen carbonates, the value has increased from 0.35 at the beginning of this century to the present 2.5–5.0 in Czechoslovak lakes and reservoirs. A ratio exceeding 3 indicates a high degree of diffuse pollution from agricultural sources, or strong acidification. The ratio is approximately 3–5 in the Tatra Mountain tarns, which are in acute danger of acidification. In lakes that already have become acidic, and alkalinity is negligible, the ratio cannot be expressed.

3.4.2. Sulphates

The principal sources of sulphates are fertilizer leachates from agricultural fields, and precipitation (Table IX). Sulphur is applied to field mainly as a component of phosphate fertilizers and, occasionally, as ammonium sulphate. Some developed countries have stopped using ammonium sulphate because of its strong acidifying effects on soil, and we recommend this practice also for Czechoslovakia. The atmospheric supply of sulphates may be high, especially in mountains receiving industrial emissions (e.g., in northern Bohemia) (Table IX). The dependence of the sulphate concentration on water flow is usually positively hyperbolic in agricultural and forested watersheds, to the limit values given in Table X.

The positive dependence of concentration on flow indicates that SO_4^{2-} comes predominantly from surface sources. Average annual sulphate concentrations have been increasing significantly in Czechoslovak reservoirs both absolutely, and as a relative component of the sum of anions – Table XIV.

Table XIV. A comparison of equivalent proportions of individual anions in 1959 and 1986, and the increase in annual average concentrations during the same period – Slapy Reservoir.

	Equivalent proportions of			
	HCO_3^-	SO_4^{2-}	Cl^-	NO_3^-
1959	50%	31%	16%	3%
1986	26%	46%	19%	9%
	mmol l^{-1}	mg l^{-1}	mg l^{-1}	mg l^{-1}
1959	0.63	19	8	2.7
1986	0.84	65	20	16

3.4.3. Chlorides

Chlorides enter watersheds and reservoirs mainly as leachates from fields to which they are applied, usually in the form of potash fertilizers. Other sources include sewage from urban settlements and large-scale breeding farms, and salt washed away from salted road surfaces.

The dependence of chloride concentration on farm drainage is usually positive if potash fertilizers are applied annually (see Table X). As with other strong anions, maximal annual averages in mixed watersheds are recorded in years with the highest flow rates. Therefore, the effects of surface sources also prevail, although less conspicuously, in this case.

3.4.4. Conductivity

Conductivity is determined in the laboratory at 25 °C. It indicates the total salt content, and also can be used to test the accuracy of principal ions analyses.

3.4.5. Iron and manganese

In anaerobic waters, iron and manganese are released from complexes that do not dissolve easily. Their concentrations then can increase to 200 mg l^{-1} Fe and 100 mg l^{-1} Mn, the approximate limits of normal concentrations. Any further increase beyond these concentrations can cause difficulties in water treatment process. These substances must be measured if the concentration of dissolved oxygen decreases below 3 mg l^{-1} O_2 in the water layer 1–3 m under the offtake level. Water aeration will decrease the reservoir concentrations of Fe and Mn. A permanent solution can be achieved by limiting the inflow of organic compounds and their in-lake production.

3.4.6. Hydrogen sulphide

The presence of hydrogen sulphide is easily recognized by its smell. This indication is more sensitive than the analytical methods. Hydrogen sulphide is in balance with other sulphides, being regulated by pH and the concentration of individual cations. The presence of sulphides indicates an anoxic water environment (see Section 3.3), and subsequent difficulties with water treatment (Section 7.4) due to iron and manganese (Section 3.4.5).

4. Bacteriological parameters

All standard bacteriological parameters are indicative of allochthonous pollution (from tributaries and banks), its location in a reservoir, and its subsequent die off. They are of little value in assessing autochthonous bacterial flora resulting from production processes in the reservoir. Counting methods (total microscopic counts) and measurement of the activity of in-lake autochthonous bacteria are rather time-consuming and require special techniques that they are unsuitable for the operation and supervision of water-supply reservoirs at present time.

4.1. Psychrophilic bacteria

The method involves determination of the number of colonies after a short incubation time in a rich nutrient medium (such as beef pepton agar) at 20 °C. The short incubation time is critical, because other more slowly growing colonies, would appear during prolonged incubation. Psychrophilic bacteria counts are mere fractions of the percent of the total number of bacteria in water (determined microscopically). The ranges of psychrophilic bacteria counts in streams and reservoirs (excepting locations immediately polluted by municipal wastewaters) are given in Table XV, together with the corresponding values of total bacterial counts.

Table XV. Ranges of the usual counts of psychrophilic bacteria (beef-pepton agar, 2 days) and total counts of bacteria in reservoirs and streams (Straškrabová, 1973).

	Psychrophilic in 1 ml	Total in 1 ml
Reservoir	$1-10^4$	10^5-10^7
Lowland streams	$10-10^6$	10^5-10^7
Mountain streams	$1-10^4$	10^3-10^6

Higher psychrophilic bacteria counts indicate a high content of easily decomposing organic matter, which need not be of faecal origin. These compounds can make water treatment difficult, because of their odour. High psychrophilic bacteria counts are found in polluted in-flowing streams and at high flow rates, decreasing slowly in a reservoir due to sedimentation and elimination by zooplankton. In stratified reservoirs (and in large non-stratified ones with retention times greater than one week) psychrophilic bacteria counts are only hundreds per ml near the dam. If bacterial counts comparable to the inflow or higher are found in a water layer near the dam, they usually indicate a rapid progress of polluted water from the inflow. Only at a very high primary production in heavily eutrophied reservoirs psychrophilic bacteria may increase to counts of 10^3 and higher near the dam, owing to decomposition of a large phytoplankton biomass.

4.2. Mesophilic bacteria

The common method of determination is an evaluation of bacterial strains that grow rapidly in a nutrient-rich medium at 28°C. Mesophilic bacteria counts in surface waters usually are by 1–2 orders of magnitude lower than psychrophilic bacterial counts. They indicate the presence of easily decomposing organic matter, and inflow of bacteria from an environment that is warmer than the surface water. Even when not of faecal origin, the high proportion of mesophilic bacteria compared to psychrophilic ones indicates contamination from other sources (manure, silage). Mesophilic bacteria originate exclusively from allochthonous sources, and their number in a reservoir diminishes as a result of sedimentation and elimination by zooplankton.

4.3. Coliform bacteria

Coliform bacteria include the members of family Enterobacteriaceae capable of producing acid and gas from lactose. They are cultivated on Endo agar (containing beef pepton, lactose, basic fuchsin and thiosulfite) at 37°C for 48hrs and all deep-red colonies are counted. If the concentration of coliform bacteria is lower than 30 per ml, they must be concentrated on a membrane filter (0.3 μm pores) before incubation.

In addition to coliform bacteria, the commonly used method partially includes the genus *Aeromonas* which is not of faecal origin. Under exceptional conditions, *Aeromonas* also can multiply in water and sediment at higher temperatures and high concentrations of organic compounds.

Coliform bacteria are exclusively of faecal origin (from man and warm-blooded animals), entering reservoirs from allochthonous sources. Their numbers in reservoirs decrease as a result of dying, sedimentation, and elimination by zooplankton. These processes are slower in the cold season (sedimentation is the least affected by temperature); however, the processes also depend on flow, concentration of organic compounds, content of sedimenting suspended matter (which reduces the bacterial counts) and other factors. Therefore a general relationship for the decrement of faecal bacteria in a reservoir was not yet derived.

Faecal bacteria counts are always lower in the surface layer of reservoirs than in the inflow waters. In deep, stratified reservoirs with retention time of one month and more, the in-flowing concentrations decrease by two or more orders of magnitude. Their concentrations near the dam usually are so low that their determination is not accurate and therefore not recommended. On the other hand, in shallow not stratified reservoirs with a short retention time and a polluted inflow high concentrations of coliform bacteria in some layers near the dam are always probable, especially after sediment perturbation (bathing).

The difference of coliform concentration in inflow and in the reservoir is negligible in shallow reservoirs. In deep stratified reservoirs the occurrence of high concentrations of coliform bacteria near the dam is very rare (a sudden increase of inflow and consequently a rapid progress of polluted water to the dam in a particular layer or a polluted side-tributary near the dam).

4.4. Faecal streptococci

They are cultivated at 37 °C on Slanetz-Bartley medium and all deep-red colonies are counted. They are exclusively of faecal origin, do not grow in external environment and most species die more rapidly than coliform bacteria – so they indicate very fresh pollution. In the inflows they usually reach numbers by one order lower than coliform bacteria. In the reservoirs their counts decrease more rapidly towards the dam – they are by two orders lower than coliform bacteria near the dam. The other characteristics are identical with those for coliform bacteria.

4.5. Evaluation according to bacteriological parameters

The values of bacteriological parameters in the reservoir inflows, which indicate substantial pollution are given in Table XVI.

Table XVI. Values of bacterial indicators suggesting substantial pollution of the inflow streams.

Indicator	Counts per ml
Psychrophilic	10^5–10^6
Mesophilic	10^4–10^5
Coliform	10^2–10^3
Faecal Streptococci	$> 10^2$

The consequences after surpassing this values are different according to the type of the reservoir:
1. In small, shallow and unstratified reservoirs there is always a risk of occurrence of these high 'inflow' concentrations near the dam. If these concentrations are regularly being surpassed in the inflow it is necessary to decrease pollution sources in the watershed. Sediment perturbation during bathing also means an increased risk of higher faecal bacteria in water. Recreation should not be allowed in drinking water reservoirs of this type (Blažková, 1986).
2. In deep stratified reservoirs it is possible to forecast the layer into which the inflowing water plunges (according to temperature stratification and inflow water temperature) and to regulate the intake to water treatment plant correspondingly.

5. Biological indicators

The principal bioindicators of reservoir water quality are phytoplankton, zooplankton and fish. Benthos is not of equal significance. In some cases, however, benthos species composition can reveal the long-term chemical conditions existing at the bottom of a reservoir.

5.1. Phytoplankton

Phytoplankton is an autotrophic constituent of bioseston. Its determination is essential for evaluation of the trophic condition of reservoirs. It is measured either as the concentration of chlorophyll-a or as biomass determined by microscopic counting and sizing. The species composition of phytoplankton also has information value (e.g., saprobic index and evaluation of predominant components). In estimating raw water quality, important phytoplankton parameters include the depth of the reservoir, its geomorphology and the depth from which water is taken by waterworks. Phytoplankton undergo photosynthesis only in the clear euphotic layer. Dead phytoplankton enters dark deeper layers as a result of sedimentation, and during homothermy also by circulation. Phytoplankton contribute to the content of organic matter in raw water, impairing the organoleptic qualities of drinking water at high concentrations. Water bloom also may produce allergens.

5.1.1. Evaluation based on phytoplankton species composition
The phytoplankton species composition is used to determine the saprobic index for surface waters. Reservoir water quality cannot be evaluated by species composition, because the composition is

fairly similar in all Czechoslovak reservoirs (Komárková, 1989). However, a qualitative analysis will reveal which species are growing rapidly and may reach their maximum levels in a short time, so that the water treatment method can be appropriately modified. Examples of the qualitative composition of spring and summer phytoplankton populations in reservoirs are given in Table XVII.

Table XVII. Assessment of the qualitative composition of phytoplankton in drinking water reservoirs. Predominant species and maximal concentrations of CHA are listed.

	Spring aspect (sampling in April – May)
Composition suitable for treatment with standard technology	Chrysophyceae (*Chromulina, Chrysococcus, Mallomonas, Dinobryon*) Cryptophyceae (*Cryptomonas curvata, C. reflexa, C. ovata, Rhodomonas*) not more than 50 mg m^{-3} Bacillariophyceae (*Asterionella, Melosira, Stephanodiscus, Cyclotella*) not more than 40 mg m^{-3}
Unsuitable species composition, requiring special treatment	Bacillariophyceae (same species) over 40 mg m^{-3} Chrysophyceae (*Uroglena, Synura*) over 50 mg m^{-3} Chlorophyceae (Volvocales: *Chlamydomonas, Chloromonas*) over 60 mg m^{-3}
	Summer aspect (sampling in July – September)
Composition suitable for treatment with standard technology	Bacillariophyceae (*Asterionella, Fragilaria, Stephanodiscus, Cyclotella, Melosira*) Cryptophyceae (*C. reflexa, C. marssonii, C. ovata*) not more than 40 mg m^{-3} Dinophyceae (*Ceratium, Peridinium*) Chlorophyceae (Chlorococcales: *Scenedesmus, Crucigenia, Planktosphaeria, Pediastrum, Coelastrum*) not more than 50 mg m^{-3}
Unsuitable species composition requiring special treatment	Cyanophyceae (*Aphanizomenon, Microcystis Anabaena, Oscillatoria*) Dinophyceae (*Peridinium*) over 40 mg m^{-3} Chlorophyceae (Volvocales: *Chlamydomonas, Chloromonas*, Chlorococcales) over 50 mg m^{-3}

5.1.2. Evaluation based on CHA

Spectrophotometric analysis of a mixture of pigments extracted from seston in 90% acetone is the recommended method. The seston is retained on fibreglass filters (Whatman GF/C – GF/F), or in less suitable membrane filters. Similar spectrophotometric techniques are described in the UNESCO Monograph on Oceanographic methodology No 1 (1966), IBP-PF Handbook No 12 (Vollenweider, 1969), IBP-PF Chemical Handbook No 8 (Golterman & Clymo, 1969), COMECON manual of water quality methods (1975) and Straškraba *et al.* (1979).

In addition to the classic determination of CHA concentrations by extraction methods, a more effective fluorometric technique is used (Yentsch & Menzel, 1963; Lorenzen, 1966; Slovacek & Hannan, 1977; Heaney, 1978). Standard equipment is a TURNER fluorimeter. The main advantage of this technique is its speed, with complete determination requiring only one minute. However, the sample should not contain large phytoplankton particles.

Evaluation of the trophic state of the reservoir and subsequent prediction using CHA depends on its relationships to biomass of phytoplankton and to the concentration of phosphorus, which is often the limiting nutrient in European waters.

Critical limits of CHA for individual trophic states and the adequacy of water for treatment, are given in Table XVIII.

The ratio CHA : total phosphorus can be used to predict whether reservoir water quality can be improved by reducing the external phosphorus load or alternatively, what the consequences of its increaser will be. The correlation between the two parameters is usually considered to be exponential

Table XVIII. Trophic grades of reservoirs according to CHA concentrations.

CHA [mg l^{-1}]		Trophic grade	Raw water	Treatment
Summer average	Annual maximum			
0.3–5	< 10 (15)	oligotrophic	excellent	standard
5–10	10–30	mesotrophic	suitable	standard
10–25	30–60	slightly eutrophic	not very suitable	exceptions
> 25	> 60	highly eutrophic	unsuitable	special treatment

(expressed by a power function), increasing rapidly at low concentrations (Dillon & Rigler, 1974). However, when the TP reaches a certain concentration, the CHA concentration will stop increasing, and the correlation will be sigmoid in the whole range of values (Straškraba, 1976a; 1978). This correlation, derived from previous studies, has been confirmed in other studies (e.g., Jindra & Porcalová, 1984; Straškraba, 1985; Prairie et al., 1989). Recent investigations have shown that at a given concentration of TP, the CHA concentration will depend on the optical properties of water (CHA is lower in coloured or turbid waters), the mixing depth (CHA is higher in shallow waters), the water retention time (phytoplankton is washed up at $R < 10$ days), and on the fish stock (CHA is low if the stock is controlled and includes a minimum number of predatory species – see Section 5.4.2). CHA also is low in highly calcareous waters, due to coprecipitation of phosphorus with calcium (Koschel 1987).

Figure 11 shows average summer concentrations of CHA in relation to average summer concentrations of TP. Phytoplankton is not very sensitive to TP concentrations up to 20 mg m^{-3} its response is very strong between 20 mg m^{-3} and 50–60 mg m^{-3}. Other factors (light and CO_2) appear to have limiting effects at TP concentrations above 70–80 mg m^{-3}. However, not all reservoirs are phosphorus-limited. At the weight ratio N:P < 10 phytoplankton appear to be limited by nitrogen (provided the system has not been saturated both by P and N). During N-limitation, the effect of a changed nitrogen concentration on CHA can be roughly estimated with Fig. 11, plotting the value of one-tenth the total nitrogen concentration on the x-axis. Other elements also can be limiting in special cases, although never recorded in Czechoslovakia. According to other authors (e.g. Smith, 1983), specific N:P ratios can be predictive of mass growths of blue-green algae, more frequently found in N-limited lakes.

5.1.3. Evaluation on the basis of phytoplankton biomass

Most reliable method for estimation of phytoplankton is the sedimentation method by Utermöhl (1958). The sample preserved with several drops of Lugol's solution* is sedimented in plexiglass cylinders of suitable volume (Kolkwitz's chambers) with bottom made of cover glass (description of the method in Vollenweider, 1969; Anonymus, 1970; Sournia, 1978). Another recommended method is the concentration of preserved phytoplankton on membrane filters. After drying, the membrane filters (Synthesia, Czechoslovakia; Cellafilters, BRD) can be cleared with immersion oil (for details see McNabb, 1960).

The amount of phytoplankton is up to now often expressed as total counts, or as the proportion of individual species in the total count. Estimation of phytoplankton abundance in cell counts, without distinguishing between large and small cells does not give a true image of the actual composition (the difference in volume can be up to 50,000 μm^3).

Biomass (fresh weight, fresh mass – FM) must be based on average volumes of species. Total FM must be calculated multiplying abundance of phytoplankton species by their biovolumes and subsequently summing the partial results.

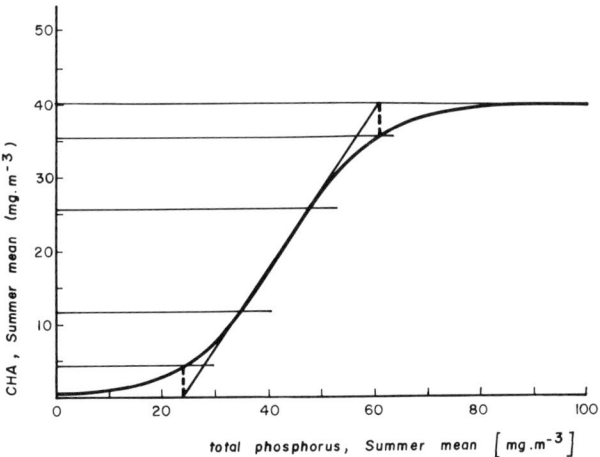

Fig. 11. The average relationship of the summer mean CHA concentrations to the summer mean total phosphorus concentration in lakes and reservoirs expressed by a sigmoid dependence. The horizontal lines correspond to CHA concentrations where starts and ceases the near-linear increase of CHA with increasing total phosphorus (~5 and 35 mg m^{-3}) and the fully linear increase (~12 and 26 mg m^{-3}) and where phosphorus saturation is reached. From Straškraba et al. (1979).

---

* 10 g of pure iodine, 20 g of KI, 20 ml of distilled water and 20 g of glacial acetic acid, mix a few days prior to using, keep in a dark glass bottle.

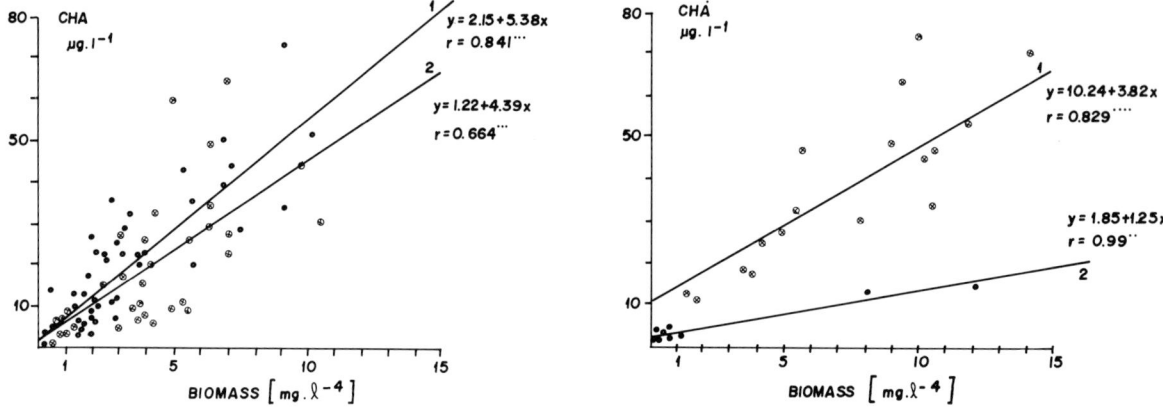

Fig. 12. Average relations between CHA and phytoplankton biomass obtained by microscopic sizing, for samples with domination of four groups of algae. Data obtained for Czechoslovak reservoirs. Upper panel: 1 – Cryptophyceae, 2 – Bacillariophyceae. Lower panel 1 – Chlorococcales, 2 – Cyanophyta. From Desortová, 1981.

Accurate volumes of spices can be received when measuring the basic dimensions of organisms (at least 30 specimens randomly chosen) with an eyepiece micrometer and calculating respective volumes according to approximate cell models (sphere, ellipsoid, cone, cylinder etc.). Fresh mass is estimated with a sufficient accuracy even if we use only approximate cell volumes which we may select of some lists (Nauwerck, 1963; Vollenweider, 1969; Anonymus, 1970). The way is not as precise as the measuring of individuals.

For calculation it is handy to use computer programms which facilitate also subsequent classification of phytoplanktonts according to different criteria.

At a range of phytoplankton biomass between 0--15 mg l^{-1} in stratified reservoirs the biomass determined by one of the above methods can roughly be converted to the concentration of CHA. A rough nomogram for this conversion is given in Fig. 12. The conversion always is a source of some error, as the content of CHA in biomass can change because of various factors (Desortová, 1981); the variance involved in determining the biomass volume is considerable.

5.2. Zooplankton

Zooplankton can directly affect the composition and quantity of phytoplankton in a reservoir. Zooplankton size and species composition are determined naturally by fish-feeding pressure, a factor indicative of the long-term character of reservoir biology. It also can be artificially regulated.

5.2.1. Sampling of zooplankton

Basic data can be obtained from net zooplankton samples collected with Apstein-type plankton nets (0.2 mm meshes) towed upward from the bottom to the surface. Samples should be collected no nearer than 50 m from the reservoir outlet and the waterworks' withdrawal site. Two parallel samples are taken at least monthly; one to determine the zooplankton biomass and one for microscopical counts analysis.

If data on the vertical distribution of zooplankton are needed, a sampler sufficiently large to sample a volume of at least 30–50 l from each layer is required e.g., the Schindler type (Schindler, 1969). The samples must cover all layers in their succession. The ability of plankton animals to escape from the net also must be considered.

5.2.2. Processing of zooplankton samples

Samples are immediately divided into size fractions by rinsing in a sieve of 0.7 mm meshes (Hrbáček et al., 1986). The fractions of a sample intended for biomass determination are subdivided after Straškraba (1964) into water fleas and other zooplankton. Air bubbles are injected into the sample with water from a rinsing bottle. Zooplankton is narcotized by chloroform. The biomass of each portion is determined by biuret reaction in a homogenate made in a glass homogenizer (Blažka, 1966). Absorbance of the stained reaction product is measured in a spectrophotometer at 530 nm (0.1 mg ml^{-1} of protein N in the homogenate corresponds to 0.115 absorbance at 1 cm optical length of the layer).

Parallel samples of both size fractions (above and below 0.7 mm) preserved in formalin are analysed qualitatively, and when all species have been identified the quantitative microscopical analysis follows. Counting is done in Sedgwick-Rafter chambers at a low magnification with the objective to examine as many samples as to find about 400 specimens of the principal species (Javornický, 1958).

5.2.3. Evaluation of zooplankton in water supply reservoirs

Crustaceans and rotatorians in water supply reservoirs can either be placed in a beta-mesosaprobic category, or else are of no value as indicators of saprobic conditions. However, the zooplankton species composition can be a reliable indicator of the state of the fish stock and of the pressure of zooplankton on phytoplankton. The presence of certain species of *Daphnia* and *Bosmina*, or *Diaphanosoma* and *Ceriodaphnia*, is a critical factor. A correct identification of species is indispensable at least within these genera.

(A) Reservoirs with low fish-feeding pressure on plankton are characterized by the predominance of large species of *Daphnia*, especially *D. pulicaria*, *D. longispina* and, if present, large-sized, round-headed populations of *D. galeata*. *Bosmina* spp. are rare or absent.

(B) Reservoirs with high fish-feeding pressure on plankton are characterized by the prevalence of small, helmetshaped populations of *Daphnia galeata*, *D. cuculata* and a high proportion of *Bosmina* spp., in particular *B. longirostris*. Species of *Ceriodaphnia* and *Diaphanosoma* also are abundant. Evaluation of summer samples is indispensable for a correct assessment because, at low temperatures (especially winter), the zooplankton composition may shift to large species due to a diminished fish-feeding pressure. Transparent species of large water fleas (*Leptodora*, *Holopedium*), are of no value as a reservoir trophic indication, even in reservoirs with abundant fish stocks. This also applies to large, motile copepod species and large, transparent (*Asplanchna*) or colonial (*Conochilus*) rotatorians.

The zooplankton biomass varies substantially during the year, usually within two orders of magnitude. The most conspicuous changes occur in April and May when zooplankton biomass increases from almost zero values to the annual maximum (usually at late May). This factor must be taken into account in planning sampling schedules and in assessing information obtained by sampling at intervals longer than one month or if sampling has not covered spring months. While the representation of zooplankton species indicating the state of fish stock can be reliably assessed by orientation samples taken at the time of maximal stratification, quantitative evaluation of zooplankton requires systematic sampling.

5.2.4. The size structure of zooplankton

Fish which feed on plankton can select large individual crustaceans with great precision. The discriminating ability of fish is reported within the range of a few hundredths of mm. This selective pressure makes size-structure the most important characteristic of zooplankton. Reservoirs differing in the species and sizes of the main water fleas also differ in the development of phytoplankton.

The proportion of large water fleas in the total zooplankton biomass is critical (Hrbáček et al., 1986). This information is difficult to obtain from microscopic samples, especially if the zooplankton have not been divided into size fractions. It can be obtained relatively easily during biomass determination of a sample that has been divided into size fractions, and the fractions subsequently subdivided into Cladocera and other zooplankton.

Table XIX. The size structure of zooplankton in reservoirs with overpopulated and balanced fish stock

	Annual averages for reservoirs with fish stocks	
	Balanced	Overpopulated
Proportion of large water fleas not passing < 0.7 mm in the total biomass of zooplankton	> 20 % (15–40)	< 5 % (1–10)
Proportion of large water fleas > 0.7 mm in the biomass of water fleas	> 40 % (30–50)	< 15 % (1–20)

Nets of 0.7 mm mesh are suitable for separating large and small water fleas. The proportion of the total zooplankton biomass in reservoirs with balanced and overpopulated fish stocks is given in Table XIX.

5.3. Benthos

Dredges are recommended for qualitative sampling, and Birge-Ekman's sampler is recommended for quantitative evaluations. The long-term oxygen availability at the bottom of a reservoir can be deduced from the species composition of the benthos. Samples are best taken from the deepest spot of the reservoir in the warmest season (Zelinka & Kubíček, 1985), however, benthos should not be used for assessing the saprobic situation in reservoirs (Zelinka et al., 1959; Zelinka & Kubíček, 1985).

5.4. Fish stock

5.4.1. Determining the fish stock in a reservoir

Of the many kinds of available fishing equipment (e.g., see survey of Kavalec, 1980), sweep seines are most suitable for assessing the state of fish stock at any given time. The seines with small meshes (10–15 mm) can catch all fishes aged 1+ present, regardless of their specific ethological and morphological features which can so markedly affect passive fishing methods. Another advantage of sweep seines is their availability. In impoundments deeper than 5 m, sweep seines can only be used for estimating the quality and quantity of fish stock at night from July 1 to September 15. Sampling cannot be done either during a full moon or in a changing weather. The seine is laid in water from a rowing-boat. The seine (of any length, 4 m deep) is at 45° with waters edge, and is let out parallel with the waters edge where water is about 5 m deep. It is hauled at both ends, always at a 45° angle. It is most important to pay out the ropes briefly and regularly after a few seconds of strong hauling, and to tread consistently on the lead line of the retrieved seine. Fishing with sweep seines should be done only at sites without snags (e.g., not wooded before flooding, or adapted to fishing). It is possible to ascertain the state of fish stock in one night (350–500 m of seining yielding 1000–3000 fishes), with the result used to evaluate further reservoir management. Lusk et al. (1983) present a detailed description of defined seining.

5.4.2. Evaluation of fish stock from a water quality perspective

The presence of zooplankton-feeding fish and their predators is significant in both evaluating the water quality of water supply reservoirs and as a biomanipulation tool (e.g. Pütz & Benndorf, 1981; Křížek et al., 1989; Seďa et al., 1989; Vostradovský et al., 1989 and Gulati et al., 1990). Fish species which have significant undesirable effects, via filtering zooplankton, include roach, bream, silver bream, bleak, rudd, whitefish, perch (up to body length of about 17 cm) and pope. There are two basic types of reservoir fish populations, the second type being subdivided into three subtypes (Table XX). Each type is characterized by the predominant group of predatory fish (Lusk et al., 1983):

(A) Trout-type reservoirs. These are the most desirable reservoirs, from the perspective of water supply. Predominant predators in these reservoirs are brown and rainbow trouts, American brook trout or hucho, as the case may be. The stock may include indifferent species (e.g., bullheads and minnows); however, it is often contaminated by cyprinids, perches and pikes. The surface elevation of trout-type reservoirs usually is above 500 m. However, up to 1200 m there is a danger of its changing into the following type of reservoir if any unsuitable fish groups are present.

(B) No-trout reservoirs. These reservoirs contain a broad range of fish species. At least 3 subtypes

Table XX. Four basic types of reservoirs according to their fish stocks.

Type	Dominant fish (accompanying fishes)	Fish biomass
Trout	brown & rainbow trout, American brook trout, hucho (bullheads, minnows)	
Pike	pike > 10% of total biomass	deep res. = 15 kg ha^{-1} shallow r. = 25 kg ha^{-1}
Perch	perch up over 50% of biomass	150– 500 kg ha^{-1}
Stable cyprinid	roach, silver bream, pike, perch, barbel, silver salmon, burbot, eel, salmonids	200–1000 kg ha^{-1}

of reservoirs can be defined, usually appearing in the following succession as the fish stock develops:

(B_1) Pike reservoirs without trout, with a large biomass of pike (over 15 kg ha^{-1} in riverbed reservoirs, over 25 kg ha^{-1} in shallow ones, over 10% of all fishes). This reservoir type is mostly transitional (2–5 years in newly filled reservoirs), usually being replaced by type B_2 or B_3.

(B_2) Perch reservoirs, with perch making up over 50% of the fish abundance and biomass. Total biomass ranges from 150 to 500 kg ha^{-1}. Abundant year classes of perch appear cyclically, always being followed by several sparse ones decimated by cannibalism. Most of the perches are undersized and grow slowly. This state of fish stock tends to change into one described below under type B_3. However, it may remain the same for decades, especially in elevations above 500 m. Perch is the principal predator, although other species also may occur in small numbers (see Section 7.2).

(B_3) Stable cyprinid fish stock without trout, with a prevalence of scarp cyprinids. Biomass 200–100 kg ha^{-1}, ratio of the biomass of cyprinids (roach, silver bream, bream, and others) to the biomass of perch 5–20 : 1. A diversity of predators exists, including pike, perch, barbel, perch, silver salmon, burbot, eel, and salmonids. The perch population is balanced in regard to age (with a majority of small individuals), consisting of quickly growing predators, without conspicuous variations in the year class abundance. If there is no intervention, such fish stock can remain the same indefinitely.

The biomass of plankton-feeding fish should not exceed 150 kg ha^{-1} in any of these stocks. Fig. 13 illustrates desirable and undesirable states of fish stock, which can be assessed by the fish caught at night in seines (meshes up to 15 mm). Another clue useful for developing appropriate regulation measures is the proportion of the predator biomass. It should not be less than 10% of the night-seine catches; (12.5–25% is desirable

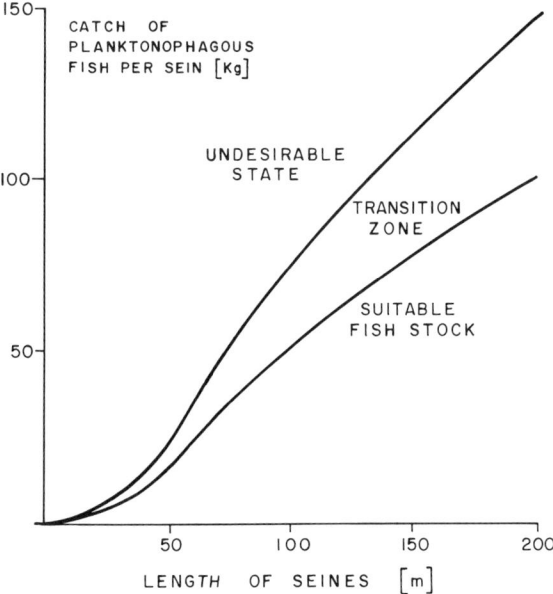

Fig. 13. Desirable and undesirable states of fish stock in reservoirs from the water quality point of view.

(Lusk *et al.*, 1983). If these two criteria are met, and a shift from a trout stock to a stable cyprinid stock is prevented, the filtering effect of zooplankton is ensured.

6. The recording, basic processing and evaluation of results

An outline of data recording, processing and evaluation is given in Fig. 2.

Data obtained by determining individual water quality parameters must be processed before they can serve as a basis for further assessments. Tabulation of data is the first step. Entering the results into a microcomputer in spreadsheat form will simplify further data analyses. One basic data processing techniques is calculation of annual or seasonal averages, as well as determination of minimal and maximal values, both observed in the data set and expected. Temporal trends also are determined, in order to determine if the water quality of a reservoir is improving or deteriorating. This requires the results of observations conducted over many years, because many qualitative parameters display conspicuous seasonal variations. Another factor complicating the evaluation of water quality changes is the dependence of parameters on the flow rate. Changes due to the effect of storage, measured as the difference between the quality of in-flowing water and outflowing or diverted waters, are affected by stratification and by processes causing horizontal water quality gradients along a reservoir. Assessment of changes is facilitated by estimating the budget of chemical substances in the reservoir (i.e., differences between their in- and outflowing quantities). This difference depends on long-term trends, as well as flow rates and stratification in the reservoir. It is clear that these individual factors are interrelated; therefore, their division here is only a compromise among several possibilities.

6.1. Recording of water quality data

Data on the quantity and quality of water in the inflow, outflow and the reservoir have to be recorded. Spreadsheet forms are best for this purpose, rows representing parameters and columns dates. Special tables must be prepared for the stratification of each parameter, with rows representing individual depths, and columns the sampling dates. All depths are recorded, even if they were measured only once in a given period. When a depth is not measured, the corresponding cell is left blank.

It also is advisable to record the data on a computer medium. The advantages of this storage method are easy calculation of basic statistics, pseudographic expression of results and a link-up with graphic representation possibilities. Spreadsheats like LOTUS 123 (for review see Neethling, 1986) allow many possibilities.

6.2. Basic statistical evaluation

The evaluations included in this manual are based on basic statistics and a simple frequency evaluation.

Basic statistics, and more detailed evaluation, are based on knowledge of the frequency distribution of measured water quality values. The distribution of qualitative parameters is not always normal. However, normal distribution of data is a prerequisite for the application of common statistical procedures. This often can be achieved by converting the data values to logarithms, with

Table XXI. Basic statistics of a water quality data set.

Statistics	Notation	Equation
Number of measurements	n	
Measured value	x_i	
Average	$\bar{x}$	$\bar{x} = \Sigma x_i / n$
Standard deviation	s	$s = \sqrt{[\Sigma(x_i-x)^2/(n-1)]}$
Standard error of the mean	$s_{\bar{x}}$	$s_{\bar{x}} = s/\sqrt{n}$
Minimum value	x_{min}	
Maximum value	x_{max}	
Median	x_m	x_m (n odd) $= x[(i = (n+1)/2)]$ (n even) $= [x(i = n/2) + x(i = (n+2)/2)]/2$
Confidence interval	$x_u(\alpha), x_l(\alpha)$	$\bar{x} \pm s_x \cdot t_{\alpha, n-1}$
Student's t-test critical value	$t_{\alpha, n-1}$	
Probability	α	

the distribution of the logarithms of values being normal (called log-normal distribution).

Basic statistics are listed in Table XXI.

The calculation of $\bar{x}$, s and $s_{\bar{x}}$ can be performed on any calculator. The standard error of the mean ($s_{\bar{x}}$) is used to evaluate the deviation of other relevant sets of data (e.g., obtained from the same locality, but on other days of measurement).

For evaluation of drinking water technology, the arithmetic mean is inadequate. Both extreme values x_{min} and x_{max}, occurring at a given locality, also must be known. Such information can be obtained directly from measurements, if the data were measured frequently and over a very long period of time. However, most generally it is theoretically and statistically incorrect to use measured values because one cannot always be certain that the actual extreme values were recorded during the measurement period.

For the practical estimation of probable extreme values the distribution curve can be used. The distribution curve describes the probability of a measured value not exceeding a given value, F(x), defined by:

$$F(x_0) = P(x \leq x_0) \qquad (7)$$

which states, that the value of the distribution function F(x) for a specific value x_0 is equal to the probability, that the value randomly selected from the evaluated data set will be lower or equal to this specific value. In hydrology the values of probability for a value to be exceeded are used. From the water quality point of view the probability of *not* being exceeded is more convenient.

For practical evaluation of reservoir water quality we suggest to use the value x_{90}, i.e. a value with 90% probability of not being exceeded. This value we consider as useful indication of the probable extreme values reached in a given data set.

The estimation of x_{90} is easily possible under the assumption that the statistical distribution of the measured values can be reasonably approximated by normal (resp. log-normal) distribution. Several computer programms are available to perform normality testing. However, often a considerable number of observations is necessary to use this tests which is not always the case in reservoir observations. Therefore, some less accurate tests whether the statistical distribution of the given data set does not contradict the normality assumption have to be used.

One possible procedure is based on the values of the third moment of the data set, m_3, obtained from:

$$m_3 = [(1/n) \Sigma(x_i^3)] - [(3/n) \Sigma(x_i) \Sigma(x_i^2)] + [(2/n^2) \Sigma(x_i^3)] \quad (8)$$

and the value of the coefficient of assymetry, r_3, derived from it:

$$r_3 = m_3 / s \quad (9)$$

Nesměrák (1978) tested for a number of water quality data sets from Czechoslovakia that their distribution can be reasonably considered normal if

$$|r_3| \leq r_{3,\text{krit}} \ (n, \alpha = 0.1) \quad (10)$$

The values of $r_{3,\text{krit}}$ are given in Table XXII. For log-normal distribution the procedure is identical except that logarithms of the observed values x_i are used for calculating the third moment according to Eq. (8).

For data sets not contradicting normal (log-normal) distribution, the value of x_{90} can be obtained from the average value, $\bar{x}$, and the standard deviation, s, of the data set:

$$x_{90} = \bar{x} + 1.28 \, s \quad (11)$$

where the coefficient 1.28 is derived from the Gaussian integral.

For Slapy Reservoir it was found that the value x_{90} for DO concentration at a depth of 30 m fluctuated in individual years from slightly negative values to 3.1 mg l^{-1}, depending on the average flow. Therefore in dry years, zero concentrations will appear for a period longer than 10% of the days over the year (i.e., more than one month). In contrast, during wet years, such concentrations likely do not occur.

Table XXII. Values of r_3 krit according to Bolshev – Smirnov. sr_3 is the standard deviation (from Nesměrák, 1978).

		α	
n	sr_3	0.05	0.01
25	0.4354	0.711	1.061
30	0.4052	0.611	0.982
35	0.3804	0.621	0.921
40	0.3596	0.587	0.869
45	0.3418	0.558	0.825
50	0.3264	0.533	0.787
60	0.3009	0.492	0.723
70	0.2806	0.459	0.673
80	0.2638	0.432	0.631
90	0.2498	0.409	0.596
100	0.2377	0.389	0.567
200	0.1706	0.280	0.403
300	0.1400	0.230	0.329
400	0.1216	0.200	0.285
500	0.1089	0.179	0.255
600	0.0995	0.163	0.233
700	0.0922	0.151	0.215
800	0.0863	0.142	0.202
900	0.0814	0.134	0.190
1000	0.0772	0.127	0.180
>1000	$2.45/\sqrt{n}$	$4.02/\sqrt{n}$	$5.70/\sqrt{n}$

6.3. *Evaluation of water quality trends*

A trend describes a systematic unidirectional change in values over a longer period of time. If the time series is short, however, one cannot be sure that a suspected trend is not just a component of a slow periodic change.

Long-term trends can be identified with some degree of certainty only when the data extend over at least a ten year period. Simple functions of time (linear or polynomial equations) are used to express the trend.

The existence of a long-term trend can be determined by calculating annual averages (in order to extract seasonality), using a parametric or nonparametric test. The nonparametric Spearman's and Kendall's test for calculation of the coefficients of rank correlation of annual means of time is recommended. A possible parametric test is Sach's test, for calculation of sum of squares of differences between successive values of annual means. See e.g. Nesměrák (1978), Conover (1980) and Nesměrák (1984) for details.

Trends can be calculated by the method of minimizing the sum of squares of deviations from annual averages. When a time series exhibits a seasonal component (fluctuation with a one-year period) the periodicity can be calculated simultaneously with trends. Some calculation possibilities are given in Kozák & Seger (1975). For this illustration, a simple additive model with constant seasonality is satisfactory.

For this model, the measured values of time series y are given two index values: the index of year ($r = 1,2,..., N$) and the index of the 'season' during the course of the year ($m = 1,2, .., M$). The condition is M identical for all years of observation. Measurements usually are done weekly ($M = 52$), monthly ($M = 12$), or quarterly ($M = 4$). However, other values and intervals also can be used. A simple example of quarterly evaluation is shown in Table XXIII. The description of the time series, using the simple additive model with constant seasonality, is given by Eq. (12):

$$y_{rm} = T_r + S_m + e_{rm} \qquad (12)$$

where

y_{rm} = measured value in r-th year and m-th 'season';

T_r = the so called annual trend (assuming a constant value for the whole year);

S_m = the seasonal component (accepting in all years the same value in the same 'season'); and

e_{rm} = residuum (including all other values)

For linear trends Eq. (12) has the form:

$$y_{rm} = a + b\,r + S_m + e_{rm} \qquad (13)$$

where a and b = regression coefficients of the linear trend.

The calculation of Eq. (13) is done in three steps:

Table XXIII. Calculation table for determination of trends with seasonality (for quarterly evaluation).

r	Y_{rm} for m				Sum	$\bar{Y}_r$	D
	1	2	3	4			
1971	7.2	8.3	3.5	8.8	27.8	6.95	1104
1972	8.6	6.1	7.1	7.9	29.7	7.42	1059
1973	12.8	7.6	4.1	5.9	30.4	7.60	1112
1974	6.5	5.9	5.7	7.7	25.8	6.45	1272
1975	9.3	9.0	8.2	9.0	35.5	8.88	1246
1976	24.0	11.6	8.5	7.3	51.4	12.85	1469
1977	23.7	15.5	7.7	8.5	55.4	13.85	1556
1978	24.0	12.8	12.3	17.7	66.8	16.70	1580
1979	22.3	10.5	10.1	10.1	53.0	13.38	1719
1980	19.0	12.3	10.9	11.3	53.5	13.38	1719
Sum	157.4	99.6	78.1	94.2	439.3	107.33	
$\bar{Y}_m$	15.74	9.96	7.81	9.42			
s_m	5.01	−0.77	−2.92	−1.31			
$1s_m$	1.42	−0.08	−1.74	0.40			
$2s_m$	8.59	−1.47	−4.11	−3.03			

(a) the measured values are tabulated (see Table XXIII and Fig. 14 as an example);
(b) the long-term trend (T_r) is calculated from the annual averages y_r (given in Table XXIII), by the method of minimization of squares; and
(c) the seasonal trend (S_m) is calculated from seasonal means y_m (row y_m in Table XXIII), according to Eq. (14).

$$S_m = y_m - \bar{Y} \qquad (14)$$

where $\bar{Y}$ = the mean of all values of y_{rm}.

Quarterly averages of the nitrate concentrations for a profile (Klabava-Chrást) for the period 1971–1980 are given in Table XXIII. The values are plotted in Fig. 14. From the values ($\bar{Y}_r$), the following linear trend was calculated:

$$\bar{Y}_r = 5.05 + 1.032 \,(\text{year} - 1980)$$

This equation approximates fairly well the measured values (see Fig. 14). However, the constant seasonal component S_m does not adequately describe the whole time period 1972–1980. It is evident from Fig. 14 that the whole period has to be subdivided into two subperiods: 1971–1975 and 1976–1980. The table also shows nitrogen fertilizers applied in the watershed, D (10^3 kg N y^{-1}). For the first period, fertilizer application values fluctuated around $1160 \cdot 10^3$ kg N y^{-1}, while for the second period 1976–1980 it has risen to $1560 \cdot 10^3$ kg N y^{-1}.

6.4. The dependence of water quality parameters on flow

The simplest equation for approximating the dependence of a water quality parameter c on flow (Q) is:

$$\log c = \log a + b \log Q \qquad (15a)$$
$$c = a' \cdot Q^b \qquad (15b)$$

The values a and b are obtained by linear regression. For calculation purposes, the following transformations are made: $y = \log c$, $x = \log Q$. The value of a' is then calculated from $a' = 10^a$. The significance of the dependence is evaluated by means of the correlation coefficient r. A necessary condition for a reasonable calculation is that observations represent a range of flows, for the given locality, with the extreme values of the range close to actual minimal and maximal flows. Using only a narrow range of Q results in inaccurate values. In this case we can maximally judge whether concentration sharply rises or drops with flow, or is insensitive to flow (Nesměrák, 1986a).

A sense of the use of the values of b and r is evident from Table XXIV. The evaluation in the table is far from being absolute; particularly relative is the validity of the given critical values of r.

In some instances, a more adequate approximation can be obtained by:

$$c = a + b/Q \qquad (16)$$

This equation expresses both a direct and indirect dependence; concentrations rise with flow when b is negative, and they decrease when b is positive. In contrast to Eq. (15), Eq. (16) expresses some type of concentration saturation during direct dependence (maximum concentration a is reached). During indirect dependence, Eq. (16) also expresses limitation: values cannot decrease below a even at the largest flows. The direct saturation dependence often holds for nitrates

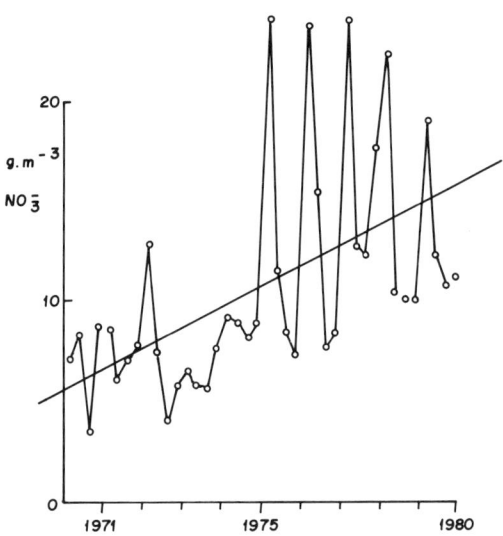

Fig. 14. An example of the water quality trend. Quarterly average nitrate concentrations, river Klabava at Chrást (Czechoslovakia), period 1971–1980. The seasonal periodicity and two different several year periods (1971–1975, 1976–1980) are seen.

Table XXIV. Evaluation of the dependence of water quality variables on flow according to Eq. (15).

b	Significance	r	Evaluation
+	During high flows, the concentrations increase. The probable source: a diffuse source	$r > r_{crit}$*	High ratio of diffuse sources in the watershed. The sources are regularly distributed over the watershed
		$r < r_{crit}$	Low ratio of diffuse sources in the watershed. The sources are irregularly distributed over the watershed
−	During high flows, dilution takes place; concentration, decreases with flow The probable parameter source: point source	$r > r_{crit}$	The source is probably close to the locality observed
		$r < r_{crit}$	The source is probably far from the locality observed. Sources with variable release

* r_{crit} is the critical value of the correlation coefficient for $\alpha = 0.05$

(e.g., according to Eq. (6)). In some instances, during extremely high flows (particularly during long-term floods), a dillution of concentrations takes place.

The relationship of concentration to flow tends to smooth out at the reservoir surface, in deeper layers, and in the reservoir outflows, due to the effect of water retention. In addition, mixing processes, currents and chemical changes are critical for concentration relationships. A linear direct dependence on longer term (several months to annual) average flows is generally seen. Whereas reservoir stratification depends on flows or retention time, and subsequent water quality changes depend on stratification, the relation for some parameters can be opposite to the relation observed in the inflow. For additional information see Weber (1972) and Suffet & MacCarthy (1989).

6.5. Material budget of reservoir inflow and outflow

The basis of the water quality budget is the hydrological budget. The amount of water entering the reservoir should be measured directly, or derived from hydrologic analogy, at all reservoir inflows; at the outflows both the quantity and quality of water released or taken out should be measured. As a minimum, the inflow water quality should be measured at all inflows which contribute more than 10% of the total inflow.

The material load L, [10^3 kg y^{-1}] can be calculated from the equation:

$$L = \sum_{i=1}^{n} 31.54 \, (Q_i \cdot c_i) \cdot n^{-1} \quad (17)$$

where
n = the number of water quality measurements per year,
c_i = i-th value of the water quality parameter [g m^{-3}],
Q_i = water flow during ith measurement of water quality [m^3 s^{-1}]

When only average values of Q and c are available, the load can be calculated from the equation:

$$L = 31.54 \, (Q \cdot c + \delta) \quad (18)$$

where δ = estimated error of L; its estimated value is equivalent to the covariance of the values of flow and concentration. It is positive for direct dependence and negative for indirect dependence, of the water quality parameter on flow (Nesměrák, 1986b).

The difference between the inflow and outflow loads indicates whether the water quality parameter of concern is retained or released in a reservoir. This can be a permanent situation, or alternatively, the budget can switch between positive and negative values due to external and internal effects (e.g., formation of anoxia at the bottom).

6.6. Evaluation of stratification and horizontal differences in water quality parameters

6.6.1. The degree of stratification

The degree of stratification is based on the difference between the maximal and minimal value of a water quality parameter in a given water column profile. As a measure we introduce the coefficient of the distinctness of stratification (CDS) to be calculated as:

$$CDS = x_{max} / x_{min}$$

This value is independent of the absolute values; however, it depends on the total range of variation for given parameter. Table XXV contains critical values for parameters with a possible range of variation below 10 (e.g., pH, alkalinity) and above 10 (e.g., most chemical and biological variables). The table illustrates approximate critical values for individual degrees of distinctness of stratification. One can differentiate between CDS for a given vertical profile, CDS_{max} during maximal stratification, average CDS_{IV-IX} for the summer half of year, and annual CDS_{aver}. Evaluation of a long period with average values can be misleading. CDS_{max} or CDS_{IV-IX} are recommended as the most relevant variable.

Table XXV. Evaluation of stratification, based on the critical value of the coefficient of distinctness of stratification (CDS_{crit}).

For the range of variation		Stratification
≥ 10	< 10	
CDS_{crit}		
≥ 1.5	≥ 1.2	stratification conspicuous
1.1–1.5	1.05–1.2	non-stratified or indistinctly stratified

Two groups of water quality parameters can be distinguished by the distinctness of their stratification in deep reservoirs:
(A) Markedly and consistently stratified. This category includes temperature, oxygen, pH, dissolved organic matter, total phosphorus and phosphates, phytoplankton counts and biomass, chlorophyll-a concentration, and zooplankton. In reservoirs where the concentrations of other nutrients (e.g., nitrogen, silicon) are lower than values considered critical for primary production, their stratification also is marked. The stratification of these parameters undergoes substantial seasonal variations, usually being deranged only at the time of autumnal isotherm.
(B) Unstratified or only temporarily stratified. Temporary stratification for a water quality parameter can occur if its inflow concentration has differed from that in the reservoir for a period of time (see Section 6.6.2). Another common cause of temporary stratification of some chemical parameters is seasonal anoxia at the bottom of the reservoir. Its consequences are either higher (e.g., Fe, Mn, NO_2) or lower (e.g., NO_3) concentrations than in oxygen-rich water layers.

6.6.2. The shapes of depth profiles

Stratification occurs because of many processes. Individual types of depth profiles are distinguished by the following characteristics (Fig. 15):
(A) A marked change in the upper productive layers – this type of curve usually denotes the effects of biological production processes in the surface water layers. The change is due to production of the organic matter by phytoplankton and incorporation of nutrients, carbon dioxide consumption (changes in pH and alkalinity) and oxygen production. A change can be positive if it is due to production, or negative if due to nutrient incorporation. The degree of change depends on the intensity of production processes and density differences in the thermocline layer. Therefore, the depth distribution of all parameters should be compared with the temperature profile. The distribution of concentrations often is relatively even in the mixed water layer, due to intensive stirring by convection and wind. The unevenness of production processes which depend on light intensity work against physical homogeneity of the water. Light intensity diminishes rapidly with depth; however, production does not reach its maximum values at very intensive light levels. The maximal production of oxygen and maximal consumption of CO_2 and other

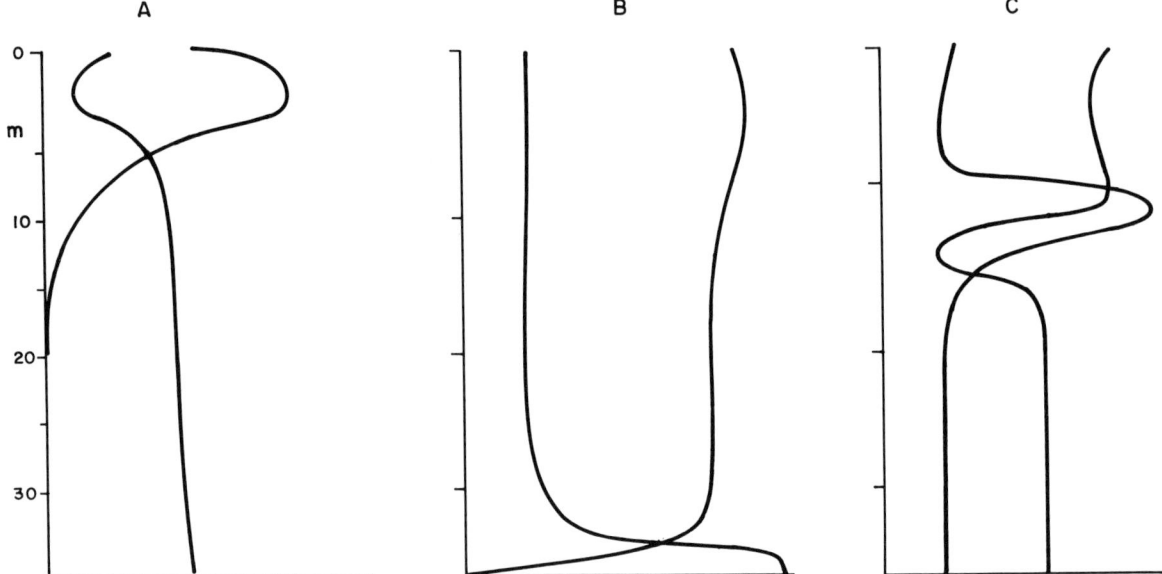

Fig. 15. Reservoir stratification. Types of depth profiles according to dominant processes. A – Profiles in case of domination of productivity processes at the surface. B – Profiles in case of domination of decomposition processes at the bottom. C – Profiles in situation with predominating processes of production and decomposition at the thermocline (in the metalimnion).

nutrients in Czechoslovak reservoirs usually are observed in the 0.5–2 m water layer.

(B) A marked change at the bottom, indicating the effects of decomposition. The cause of diminished oxygen concentration or enhanced carbon dioxide concentration can be either direct or indirect. The indirect cause is an anaerobic environment at the bottom of the reservoir, resulting in the release of various substances at a higher rate (e.g., Fe, Mn, sulphides, ammonia, phosphorus). The most detrimental parameters in regard to water treatment, are high concentrations of Fe and Mn. Production processes are affected by a higher rate of P release during anoxic conditions. Phosphorus can accumulate in bottom sediments over a long period of time, being released suddenly when anoxia occurs. The concentration of nitrates also can decrease markedly in anoxic environments.

(C) A pronounced change in middle depths, indicating either the effects of decomposition in the metalimnion (metalimnetic minima) or the inflow of water layers with different concentrations. Physical processes (retention of decomposing phytoplankton or downward migration of in-flowing waters to layers of equal density) play a major role in this case.

These main types of curves should be regarded only as being the most frequently observed. There are many modifications and combinations. Because the three principal processes, production, decomposition and oxygen deficit, can affect most parameters simultaneously, there are many modifications and combinations of these curves.

6.6.3. Evaluation by depth-time isoline charts

In addition to depth profiles, isoline charts also can be used to evaluate stratification. Several PC programming packages are useful for computer-generation of isoline charts (e.g. GRAPHER, Golden Software Inc.).

Isoline charts (Fig. 16) clearly show the depths, and periods of high and low values. The depths at which concentration changes are most pronounced are indicated by the highest density of isolines. The maxima and minima resulting in vertical profiles of type (C) of Section 6.6.2 will appear as closed areas. The course of stratification in a given reservoir and year is then evaluated by:

(a) The time of autumnal circulation. Circulation is considered early if it occurs at the end of

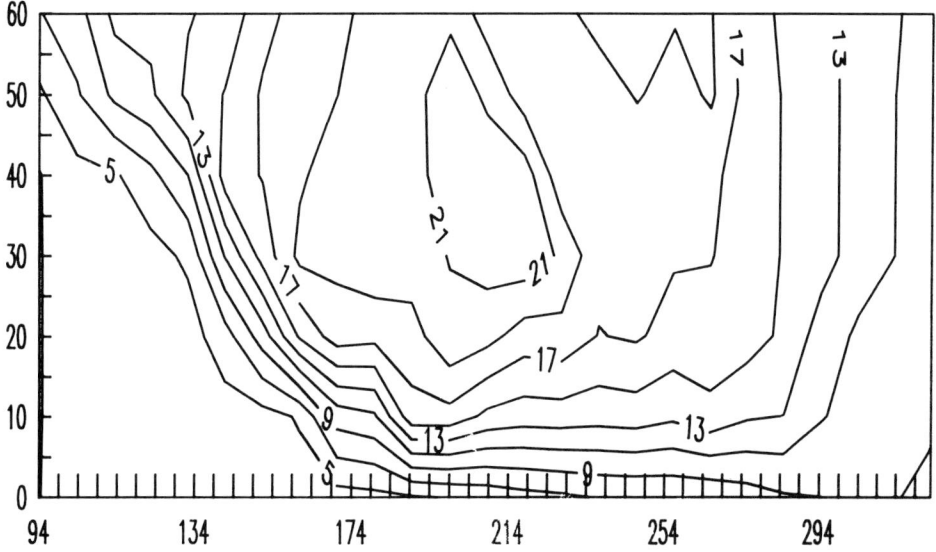

Fig. 16. Isoline charts for the annual depth distribution of temperature created by the program Grapher. Data for Slapy Reservoir. On the x-axis days of the year, on the y-axis the elevation above the bottom.

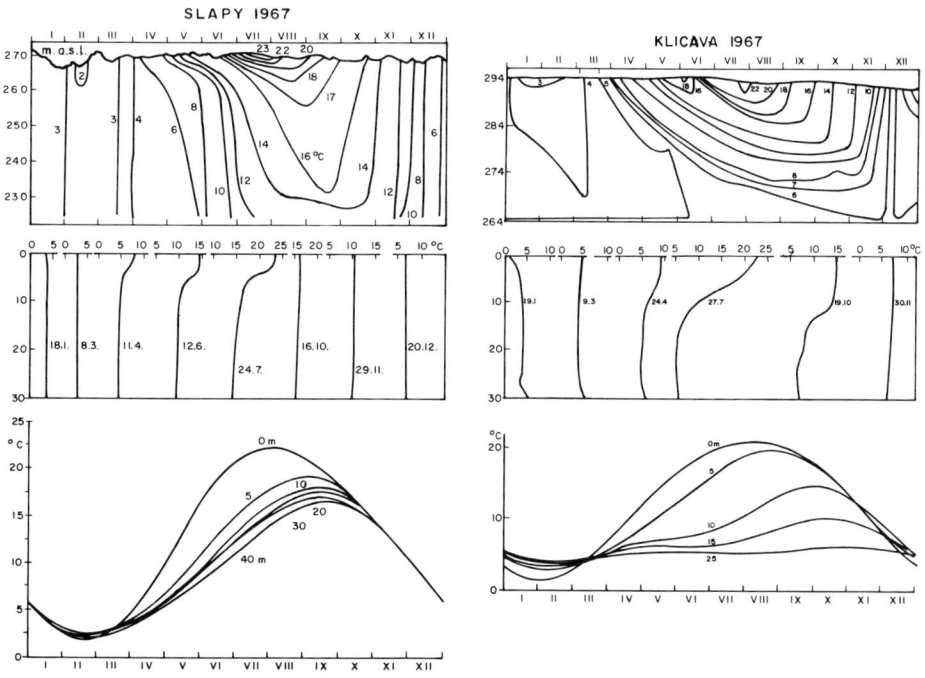

Fig. 17. Annual course of temperature stratification in Slapy and Klíčava reservoirs (Czechoslovakia) expressed in three different ways: by means of the isoline chart, depth profiles for different sampling dates and sinusoidal curves for different depths.

October, and late if it occurs in late November or December. Spring circulation is either very short (1–3 weeks), or does not occur at all;
(b) The density of isolines in a certain layer; and
(c) A decrease or increase in values on the surface or at the bottom – the time is evaluated for which a certain isoline intersects the bottom or water surface, and the depth it reaches.

Figure 17 confronts the representation of annual temperature stratification by the depth-time isoline chart with two other formats of the same data: standard profiles and temperature curves for different depths. The isoline charts give the best possibility of quantitative evaluation of differences between years and reservoirs.

If stratification is evaluated solely on the basis of temperature, a reservoir isotherm may not always mean equability of chemical parameters; however, this often is the case.

6.6.4. *Horizontal concentration differences*

Pronounced, irregular horizontal differences occur in shallow reservoirs, as a result of differences among individual tributaries, and differences in the intensity of processes in various parts of a reservoir. The irregularity of horizontal distribution in shallow reservoirs is due to the domination of the effect of wind. Wind drifting of water masses, internal seiches, blowing of surface waters with their constituent plankton, accumulation of scums and their associated decomposition produce local changes, which can rapidly move when wind speed and/or direction change.

In deep, stratified reservoirs, horizontal distinctions are regular, particularly in the direction from the main in-flowing stream to the dam. The inflow zone shows the greatest difference (Fig. 18). In riverine reservoirs with short retention

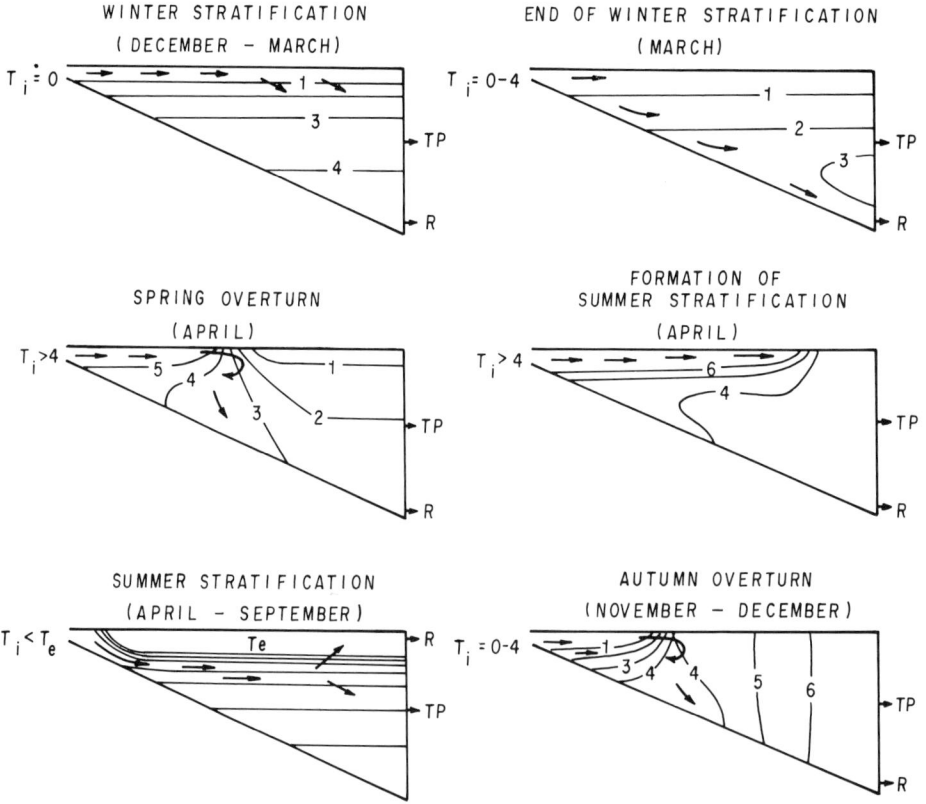

Fig. 18. Annual course of horizontal distribution of the inflowing water within the reservoir Římov (Czechoslovakia). Temperature isolines and flows are shown. TP indicates the location of the intakes to waterworks, R the location of outflow to the river. Bottom release is operating except summer months when the needs of recreation in the river below the reservoir are respected. From Hejzlar & Straškraba (1989).

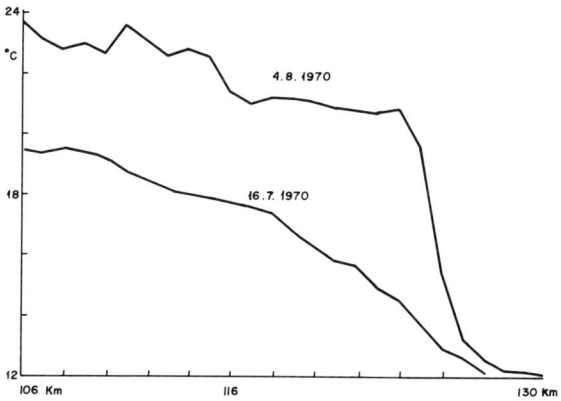

Fig. 19. Horizontal differences in a reservoir fed by cold water from the hypolimnion of the upperlying reservoir. Measurements by thermistor thermometer in Slapy Reservoir, Czechoslovakia. On the x-axis kilometers from the river mouth, i.e. the inflow is located on the right side of the figure. On August 4, 1970 the plunging of surface water in the inflow zone (km 126–128) is seen, due to high flow rates. During low flow (July 16, 1970) a gradual mixing of the inflow with the reservoir water is seen.

times, the colder, in-flowing waters plunge deep at a certain spot. On the surface a sharp boundary between the in-flowing and reservoir water usually is apparent, especially during floods (Fig. 19). The boundary is often underlined by differences in the degree of turbidity of the two water masses, or by floating objects (e.g., branches, leaves, etc.) brought in by the in-flowing stream. This may accelerate biological production processes, increase the concentration of phytoplankton and reduce the concentration of phosphorus at the time of summer stratification (Chalupa *et al.*, 1985). At the end of the inflow zone where the water flow rate decrease substantially, sedimentation is highest, affecting decomposition processes at the bottom and related chemical changes.

Regularly occurring differences can be evaluated on the basis of annual or seasonal average values. A t-test is used for determining the difference between a locality and standard sampling site; horizontal differences are evaluated by the analysis of variance. More detailed assessments are beyond the framework of this manual.

7. How to use water quality data to make conclusions

7.1. Types of conclusions

The objective of this section, is to:
(i) outline approaches by which conclusions can be drawn from data determined by methods described in the previous two chapters, and
(ii) make decisions about necessary measures for improving water quality.

The types of conclusions can be divided into three groups (Fig. 20):
1. Conclusions concerning the watershed;
2. Conclusions concerning management of the reservoir; and
3. Conclusions concerning water-treatment plants.

7.2. Assessing the effects of the watershed on reservoir water quality

The entire watershed determines the water quality and chemical composition of a reservoir. In order to estimate the probable composition of water that will flow into a planned reservoir (if reliable data are not available), or the causes of changes noted in an existing reservoir, it is advisable to begin with the least variable parameters. These parameters include data on loads in the watershed, are particularly valuable for compounds that are briefly or not at all retained in the watershed (e.g., nitrates, sulphates and chlorides), the dependence of watershed loads on the flow rate, the proportion of forests and farmland in the watershed, the size of point sources, and the geology of the watershed. Information on weathering (especially for mixed subsoils), and on the effects of individual types of soil, is less definite.

To estimate the total concentration of salts (conductivity) and of hydrogen carbonates (alkalinity), it is useful to estimate the proportion of areas with sedimentary and crystalline subsoils. As an approximation, the alkalinity of water from a crystalline subsoil will be < 0.2 mmol l^{-1} and from a sedimentary subsoil > 1.5 mmol l^{-1}. Conductivity values < 100 μS cm^{-1} can be expected in water from crystalline subsoils and values > 300 μS cm^{-1} in water from sedimentary subsoils. An estimate of the resultant concentrations (not including anthropogenic input) then can be based

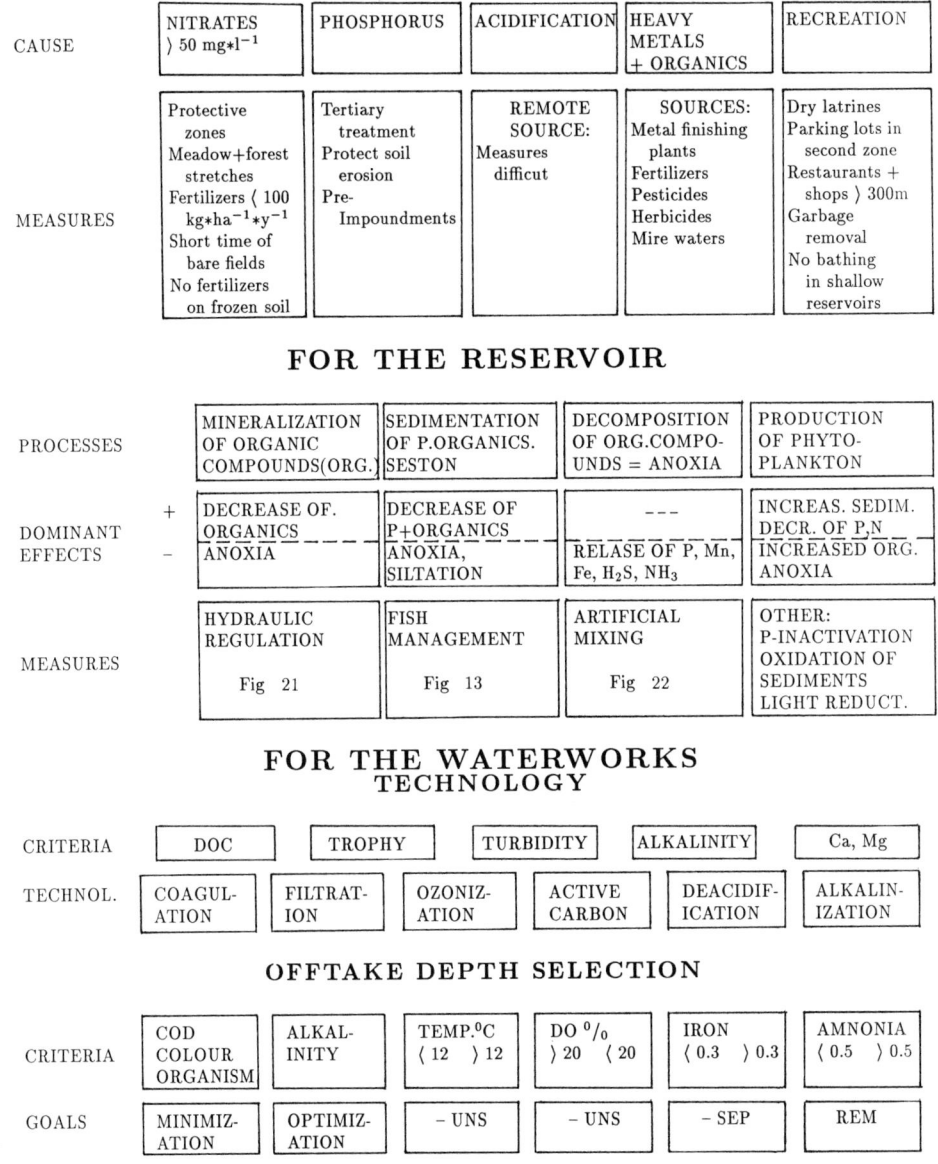

Fig. 20. Schematic representation of drawing conclusions from water quality investigations of reservoirs for the watershed, the reservoir and for the waterworks.

on the proportion of the two main types of subsoil in the watershed area.

An estimate of anthropogenic effects on the concentrations of water quality parameters can be made as follows:
1. the watershed is divided into forest- and farmland,
2. the material loads are summed up, according to data in Tables IX and X, and
3. average concentrations then are calculated from specific loads values.

The estimate is complemented by the size of point sources e.g., sewerage, number of inhabitants, heads of cattle in stables, industrial sources with their respective technologies; in all cases, the effectiveness of sewage treatment plants must be taken into account in regard to a given parameter. Temporal changes based on the dependence of concentrations on flow then can be envisaged.

If the highest nitrate concentration is close to 50 mg l^{-1} NO$_3$ (especially in high-flow periods), the following measures must be taken in the watershed, in cooperation with appropriate agricultural enterprises:
(a) To strictly conserve the second zone of public-health protection;
(b) To use nitrogen fertilizers in quantities not exceeding 100 kg (ha of farmland)$^{-1}$ y^{-1};
(c) To keep the time during which fields are left without vegetation after harvest as short as possible (at most 1–2 weeks), and to use catch crops; and
(d) To not apply nitrogen fertilizers to frozen soil or to fields where nothing has been sown or planted.

Protective stretches formed by grassland and forests prevent surface soil particles being washed down into stream and reservoirs. However, they may not reduce the leaching of dissolved pollutants (e.g., nitrates) from distant, elevated places (Zajíček, 1985).

If concentrations of SO$_4^{2-}$ are higher than estimated by the above procedure, and no simultaneous increase in NO$_3^-$ and Cl$^-$ is seen and subsoil sources can be excluded, an atmospheric source of sulphate is indicated. The impact of atmospheric acidification is particularly strong in forests, especially coniferous, and reaches its maximum at altitudes about 1000 m a.s.l. Section 3.4.1 (evaluation of the concentrations of hydrogen carbonates) discusses advanced acidification effects.

Approximately simultaneous changes in the concentrations of NO$_3^-$, SO$_4^{2-}$, Cl$^-$, and usually Ca^{2+}, which are indirectly proportional to the flow rate, indicate a substantial load from farmland. In large, mixed watersheds, the dependence of concentration on flow is less regular (e.g. shifts in time and weight, a slight asynchrony with rainfall, water retention in impoundments), however, annual average concentrations are proportional to annual average flow in such watersheds. Point sources, and substances present in the watershed only in small quantities, are indicated by an indirect dependence of concentration on flow.

If the contents of heavy metals (especially Cd, Hg and Be) exceed the WHO norm for drinking water they most often originate from three sources:
(a) Waste from metal-finishing plants. In these cases, discipline must be enforced and waste discharge into sewerage or streams must be prevented;
(b) Some fertilizers, especially imported phosphates. In these cases, different types of fertilizers must be used; and
(c) Mine water or pit heaps, even those which have not been worked since the Middle Ages or even earlier times.

The function of the *protective zones* is to prevent direct pollution of water at the offtake site; however, because of their small area in a reservoir, they cannot protect the reservoir from the load of compounds from the watershed. Chemical water quality parameters are affected by the entire watershed.

In order to be useful, *pre-impoundments*, must be deep enough for temperature stratification (i.e., average depth $z > 8$ m, length $L > 1$ km, average water retention time $R > 2$–3 weeks). In contrast, shallow reservoirs with short retention times seem to serve primarily as precultivation tanks, enhancing the development of phytoplankton. The short retention time prevents the growth and reproduction of herbivores which normally could control the phytoplankton. In addition, dredging, prior to 1/4 of the reservoir being filled with mud, is necessary to prevent accumulation of sediments. If these prerequisites are met, pre-impoundments will retain P and sediment organic pollutants, except for nitrates.

Recreational use of drinking water-supply reservoirs can be a serious public health problem, as a result of direct contamination of the reservoir by human pathogens and pollution by trodden mud. The public-health aspect must receive the top priority. Nevertheless, bathing is permitted in a few reservoirs in Czechoslovakia, although raw drinking water is taken from the outflow or the reservoir proper (e.g., Brno Dam, Husinec and Domaša). In such cases, certain minimum measures must be taken. One measure is the installation of dry latrines for bathers as far from the reservoir as possible, using concrete cofferdams insulated by 10m layers of sand mixed with iron swarf (1–2% of the weight of sand). Furthermore, parking lots and stalls must not be closer to the reservoir than the second zone. Restaurants and shops must not be closer than 300 m from the waters edge, and their sewage must be diverted from the reservoir watershed. A regular removal of all kinds of garbage must be strictly supervised. The hazard of bacterial contamination is particularly serious in shallow reservoirs, whose sediments are intensively disturbed by bathers. Faecal bacteria including pathogenic species, can survive in sediments for several months, even in winter. Recreation use should be prohibited at small, shallow, water-supply reservoirs of the fish-pond type.

7.3. Conclusions concerning the reservoir

Reservoir water quality depends primarily on the quality of the in-flowing waters. Therefore, the watershed of the reservoir is of prime importance and must be treated accordingly (see Section 7.2). The quality of in-flowing water changes in the reservoir, proportional to the retention time (R). Measures for improving reservoir water quality must incorporate the nature of these changes, which involve both the in-flowing water quality and the processes taking place in the reservoir.

7.3.1. Water quality changes within a reservoir
The most important changes include the following:
(A) Mineralization of organic compounds, indicated by a decrease in BOD, COD and water colour. The effects of mineralization are positive only up to a certain limit. If oxygen is used up in some layers (usually at the bottom) paragraph (C) takes place. The intensity of mineralization depends on the load, concentration and composition of organic compounds in the in-flowing stream, and the retention time, as well as circumstances that either enhance or suppress the development of organisms supporting mineralization.
(B) Improvement of water quality (a decrease in P, organic substances, non-dissolved (particulate) matter), due to sedimentation of particulate matter, as well as dissolved substances that become particulate only in the reservoir. As with mineralization, the effects are desirable only up to a certain level, primarily when the reservoir becomes silted or when anoxia develops – see paragraph (C). The impact of these factors depends on the quantity and nature of in-flowing substances and on flow, convection and mixing in the reservoir.
(C) Deterioration of water quality, due to intensive decomposition of organic compounds, especially if anoxia develops, as well as the related processes of either releasing or binding certain substances (e.g., release of Fe, Mn, NH_3, hydrogen sulphide, and nitrites; binding of nitrates). In addition to factors given sub (A) and others, the concentration of released and bound substances can also play a decisive role in defining the water quality in a stream or reservoir.
(D) Deterioration of water quality, due to an excessive production of organic matter in the form of phytoplankton. The load of phosphorus (and other critical nutrients), the optical qualities of water (colour, turbidity), pH and concentration of CO_2^+ in the inflowing stream are of prime importance. The critical nutrient concentration in the reservoir depends on its internal cycling (e.g., binding or release in sediments, incorporation by organisms), the development of phytoplankton, temperature, light and mixing, and biotic relations (i.e., consumption of phytoplankton by higher trophic level organisms in the food chain).

Other processes which must be considered when attempting to improve reservoir water quality include extinction or proliferation of pathogenic microorganisms, toxic effects on organisms, internal circulation of substances, and acidification.

7.3.2. Measures to improve reservoir water quality
Primary reservoir management options for water quality improvement, and principles that have to be observed in their use are discussed in this section. Different in-lake techniques that can be applied for water quality improvement in a reservoir are treated in Chapter XIII (this volume).

Table XXVI. Tentative criteria for the design of the technology treating reservoir water based on the quality of the inflow.

Criterion	Approx. range	Technology
Dissolved organic carbon (DOC) (mg l^{-1})	< 4	–
	4–10	coagulation + filtration
	10–20	coagulation + ozonation + activated carbon adsorption + filtration
	> 20	unsuitable drinking water production
Trophic status	oligotrophic (load of P or N of less than 1 and 7 g m^{-2} y^{-1})	filtration
	mesotrophic (load of P and/or N 1–2 and 7–15 g m^{-2} y^{-1})	coagulation + filtration
	eutrophic (load of P and/or N more than 2 and 15 g m^{-2} y^{-1})	coagulation + ozonation + activated carbon adsorption + filtration
non-settling colloids (as turbidity)	< 5	–
	> 5	coagulation + filtration
Alkalinity (mmol l^{-1})	< 0.2	deacidification, or coagulation with a prepolymerized coagulant
	> 0.2	–
Contents of Ca^{2+} Mg^{2+} ions (mmol l^{-1})	< 0.4	alkalinization, augmentation of the contents Ca^{2+} and Mg^{2+} ions
	0.4–5	–
	> 5	unsuitable for drinking water production

7.4. Assessment of water treatment plants

Water treatment technology must be planned in such a way that raw water of any quality occurring in a reservoir in the course of a year can be treated and desirable quality of the produced drinking water can be guaranteed (WHO, 1984). The reservoir water quality depends primarily on inflowing water quality and on processes taking place in the reservoir in different seasons.

Five criteria, based on analyses of inflow waters can aid in determination of treatment technology (Table XXVI).

The principal criteria for planning treatment technology are:
(i) the content and composition of organic substances inflowing water, and
(ii) the potential development of phytoplankton, a producer of autochthonous organic compounds (depending on the quantity of total phosphorus in the inflowing waters.)

Dissolved organic compounds are a particular problem in treating water, primarily because suspended compounds are easily removed in most cases (Žáček et al., 1984); in addition, their amounts can diminish as a result of sedimentation.

A specific technological procedure must be chosen with special regard to the composition of dissolved organic compounds in the raw water (Žáček et al., 1984). Natural coloured substances originating in the watershed are the easiest compounds to remove, especially the macromolecular compounds.

The theoretical retention time is of primary importance in regard to the ability of the reservoir to equalize water quality variations of the in-flowing waters, and to stratification (see Section 2.1). If the long-term annual average water retention time (R) is less than 20 days, it is difficult to assess the ability of a reservoir to equalize the quality of inflowing water at high flow rates. If R is longer than 100 days, the effect of water quality variations of the in-flowing water is not significant (excepting floods).

Maximum depth and retention time, as well as the morphology of the reservoir and its surroundings (which affects wind mixing) determine reservoir stratification (see Table VI) and, therefore, the possibility of optimizing the offtake depth.

In all water-supply reservoirs with maximum depths more than 10 m and retention times greater than 20 days it should be possible to choose the offtake depth. The criteria for optimizing the offtake of raw water are given in Table XXVII.

In deep, stratified reservoirs, water of optimal

Table XXVII. Raw water criteria for selection of offtake depth in water-supply reservoirs.

Criterion	Requirement
$COD_{Cr}(COD_{Mn})$ [mg l^{-1}]	Minimization, or minimization of the dose of coagulum in coagulation test
Alkalinity [mmol l^{-1}]	Optimization with regard to COD concentration
Colour, absorbency at 250–370 nm	Minimization
Temperature [°C]	< 12 suitable
	> 12 unsuitable
Dissolved oxygen [% of saturation]	> 20 suitable; must be increased at least to 50 during treatment
	< 20 unsuitable; danger of an increase in Fe, Mn, NH_3 and H_2S concentration
Iron [mg l^{-1}]	< 0.3 = suitable
	> 0.3 = separation during treatment necessary
Ammonia and ammonia ions [mg l^{-1}]	< 0.5 = not noxious
	> 0.5 = must be removed during treatment
Organisms (counts of bacteria, phytoplankton and zooplankton)	Minimization

quality for treatment usually is found throughout the year at a depth corresponding to the depth of the upper part of the hypolimnion during summer stratification.

Acknowledgements

Many people have been helpful during various stages of this work, too numerous to be mentioned. During preparation of the English manuscript we are indebted to Dr. Walter Rast (Austin) for kindly improving the English and to Dr. Albrecht Gnauck (Berlin) for improvements of the text.

References

Anonymus, 1970. Ausgewählte Methoden der Wasseruntersuchung, Band II. – Plankton. Biologische, mikrobiologische und ökologische Methoden. VEB G. Fischer, Jena.
Blažka, P., 1966. Bestimmung der Proteine im Material aus Binnengewässern. Limnologica (Berlin) 4: 387–396.
Blažka, P. & L. Procházková, 1987. Relationship of sulphate and nitrate concentrations and pH in precipitations. In: B. Moldan & T. Pačes (eds.), Proc. Internat. Workshop on Geochemistry and Monitoring in Representative Basins, GEOMON, Czechoslovakia: 26–28.
Blažková, S., 1986. Implications in recreation on reservoirs. Limnologica (Berlin) 17: 223–231.
Bliss, C. J., 1970. Statistics in Biology. McGraw-Hill, New York, Vol. II: 219–287.
Bowen, C. E. & T. Holm-Hansen, 1980. A salicylate-hypochlorite method for determining ammonia in sea water. Can. J.Fish. Aquat. Sci. 37: 794–798.
Carmichael, W. W. (ed.), 1972. The Water Environment: Algal Toxins and Health. Plenum Press, New York and London.
Chalupa, J., L. Fiala, J. Popovský, V. Sládeček & P. Vašata, 1985. Water quality control in impoundments. Acta hydrochim. et hydrobiol. 13: 3–16.
Collings, M. R., 1973. Generalization of stream temperature data in Washington. Geol. Surv. Wat. Supply, Paper 2029B.
COMECON, 1973. (Unified Methods of Water Quality Investigations III. Methods of Biological Water Analyses – In Russian). Comecon, Moscow.
Conover, J., 1980. Practical Nonparametric Statistics. Wiley, New York, 493 pp.
Černý, J., 1987. Importance of runoff episode for sulphur budget in forested catchment. In: B. Moldan & T. Pačes (eds.), Proceedings of the Internat. Workshop on Geochemistry and Monitoring in Representative Basins – GEOMON, Czechoslovakia: 22–25.
Desortová, B., 1981. Relationship between chlorophyll-a concentration and phytoplankton biomass in several reservoirs in Czechoslovakia. Int. Rev. ges. Hydrobiol. 66: 153–169.
Dillon, P. J. & F. M. Rigler, 1975. A simple method for prediction the capacity of a lake for development based on lake trophic status. J. Fish. Res. Bd. Canada 32:1519–1531.
Dillon, P. J. & F. M. Rigler, 1974. The chlorophyll-phosphorus relationship in lakes. Limnol. Oceanogr. 19: 767–773.
Doleјš, P., 1990. A simple test for determination of optimal doses in the treatment of humic waters. In: H. H. Hahn & R. Klute (eds.), Chemical Water and Wastewater Treatment. Springer Verlag, Berlin: 377–390.
Goltermann, H. L.& R. S. Clymo (ed.), 1969. Methods For Chemical Analysis of Fresh Water. IBP Handbook No 8. Blackwell Sci. Publ., Oxford and Edinburgh, 172 pp.
Gulati, R. D., E. H. R. R. Lammers, M.-L. Meijer & E. Van Donk, 1990. Biomanipulation. Tool for Water Management. Kluwer Acad. Publishers, Dordrecht, 628 pp.
Heaney, S. I., 1978. Some observations of the use of the *in vivo* fluroescence technique to determine chlorophyll-a in natural populations and cultures of freshwater phytoplankton. Freshwater Biology 8: 115–126.
Hejzlar, J., 1989. Dissolved amino sugars in the Římov Reservoir (Czechoslovakia). Arch. Hydrobiol. Beih. Ergebn. Limnol. 33: 291–302.
Hejzlar, J. & J. Kopáček, 1990. Determination of low chemical oxygen demand values in water by dichromate semi-mikro method. Analyst (London) 115: 1463–1467.
Hejzlar, J. & M. Straškraba, 1989. On the horizontal distribution of limnological variables in Římov and other stratified Czechoslovak reservoirs. Arch. Hydrobiol. Beih.

Ergebn. Limnol. 33:41-55.

Horváthová, B. & A. Dávid, 1969. Ročný rytmus zmien teploty riečnej vody (Annual changes of river temperature). Vodohosp. časopis SAV 17: 109-130. (In Czech)

Hrbáček, J. (ed.), 1966. Hydrobiological Studies I. Academia Praha, 408 pp.

Hrbáček, J., O. Albertová, B. Desortová, V. Gottwaldová & J. Popovský, 1986. Relation of the zooplankton biomass and share of large Cladocerans to the concentration of total phosphorus, chlorophyll-a and transparency in Hubenov and Vrchlice Reservoirs. Limnologica (Berlin) 17: 301-308.

Hutchinson, G. E., 1957. A Treatise on Limnology. Vol. I., Wiley, New York, 1115 pp.

Javornický, P., 1958. Revize některých metod pro zjišťování kvantity fytoplanktonu (Revision of some methods for determination of phytoplankton quantity). Sb. VŠHT, odd. fak. technol. paliv a vody 2: 283-367. (In Czech)

Jehlička, J., 1987. Characteristics of two small basins in the Sumava Mts. (South Bohemia). In: B. Moldan & T. Pačes (eds.), Proceedings of the Internat. Workshop on Geochemistry and Monitoring in Representative Basins – GEOMON, Czechoslovakia: 207-209.

Jindra, J. & P. Porcalová, 1984. Přísun veškerého fosforu a dusičnanů do nádrže Římov, Husinec a Jordan. (Total phosphorus and nitrate load of reservoirs Římov, Husinec and Jordán). In: M. Straškraba, Z. Brandl & P. Porcalová (eds.), Hydrobiologie a kvalita vody údolních nádrží. Vts Jivak a VTS JBC ČSAV, Č. Budějovice. (In Czech).

Kavalec, J., 1980. Rybářské nářadí a sítě. (Fishery Gear and Nets). Severografia, Chomutov, 28 pp. (In Czech)

Kinkor, V., 1987. Acidification of two small basins in Krušné Hory Mts. In: B. Moldan & T. Pačes (eds.), Proceedings of the Internat. Workshop on Geochemistry and Monitoring in Representative Basins – GEOMON, Czechoslovakia: 204-206.

Komárková, J., 1989. Changes of phytoplankton assemblage during the spring period in the moderately eutrophic Římov reservoir (Czechoslovakia). Arch. Hydrobiol. Beih. Ergebn. Limnol. 33: 419-433.

Koschel, R., 1987. Pelagic calcite precipitation and trophic state of hardwater lakes. Arch. Hydrobiol. Beih. Ergebn. Limnol. 33: 78.

Kozák, J. & J. Seger, 1975. Jednotné statistické metody v prognostice. (Unified Statistical Methods in Prognostic. – In Czech). Státní nakladatelství technické literatury, Praha.

Křízek, J., O. Albertová, L. Růžičková & M. Vostradovský, 1989. The changes in fish communities and biomanipulation in water supply reservoirs. Arch. Hydrobiol. Beih. Ergebn. Limnol. 33: 587-594.

Lorenzen, C. J., 1966. A method for the continuous measurement of in vivo chlorophyll concentration. Deep-Sea Res. 13: 223-227.

Lusk, S., J. Heteša, L. Hochman & K. Král, 1983. Účelové rybí osádky v údolních nádržích. (Fish stock for reservoir biomanipulation). Technickoprovozní rozvoj vodního hospodářství, Hydroprojekt, Brno, 110 pp. (In Czech)

Mackereth, F. J. H., J. Herron & J. F. Talling, 1978. Water Analysis: Some Revised Methods for Limnologists. FBA Scientific publication No 36, 120 pp.

McNabb, C. D., 1960. Enumeration of freshwater phytoplankton concentrated on the membrane filter. Limnol. Oceanogr. 5: 57-61.

Moldan, B. & M. Dvořáková, 1987. Atmospheric deposition into small drainage basins studied by geological survey. In: B. Moldan & T. Pačes (eds.), Proceedings of the Internat. Workshop on Geochemistry and Monitoring in Representative Basins – GEOMON, Czechoslovakia: 127-129.

Nauwerck, A., 1963. Die Beziehungen zwischen Zooplankton und Phytoplankton im See Erken. Symb. Bot. Upps. 17: 1-163.

Neethling, J. B., 1986. Review of generic software for environmental applications. In: P. Zannetti (ed.), ENVIROSOFT 86. Proceedings of the International Conference on Development and Application of Computer Techniques to Environmental Studies, Los Angeles, U.S.A., November, 1986, Computational Mechanics Publications, Southampton: 3-17.

Nesměrák, I., 1978. Hodnocení a modelování jakosti vody v tocích (Evaluation and Modelling of River Water Quality). Státní zemědělské nakladatelství, Praha, 364 pp. (In Czech).

Nesměrák, I., 1984. Analýza časových Řad jakosti vody v tocích. (Time Series Analysis of River Water Quality Data). Práce a studie VÚV, Praha, 160, 225 pp. (In Czech)

Nesměrák, I., 1986a. Nitrates and mineral fertilization. Limnologica (Berlin) 17: 273-281.

Nesměrák, I., 1986b. Errors of determining material load from a watershed. Limnologica (Berlin) 17: 251-254.

Prairie, Y. T., C. M. Duarte & J. Kalff, 1989. Unifying nutrient-chlorophyll relationships in lakes. Can. J. Fish. Aquat. Sci. 46: 1176-1182.

Procházková, L., 1964. Spectrophotometric determination of ammonia as rubazoic acid with bis-pyrazolone reagent. Anal. Chem. 35: 865-871.

Procházková, L. & P. Blažka, 1986. Long-term trends in water chemistry of the Vltava River (Czechoslovakia). Limnologica (Berlin) 17: 263-271.

Procházková, L., P. Blažka & Z. Brandl, 1984. The output of NO_3—N and other elements from small homogeneous watersheds. In: G. Jolankai & G. Roberts (eds.), Land Use Impacts on Aquatic Systems. Proceedings MAB 5 Workshop, 1983, Budapest: 291-306

Procházková, L., P. Blažka & M. Králová, 1970. Chemical changes involving nitrogen metabolism in water and particulate matter during primary production experiments. Limnol. Oceanogr. 15: 797-807.

Procházková, L. & P. Blažka, 1989. Ionic composition of reservoir water in Bohemia: long-term trends and relationships. Arch. Hydrobiol. Beih. Ergebn. Limnol. 33: 323-330.

Pütz, K. & J. Benndorf, 1981. Die zielgerichtete Wassergütebewirtschaftung von Talsperren und Speichern. Information zum Fachbereichstandart TGL 27885/03. Acta hydrochim. et hydrobiol. 9: 35-36.

Schindler, D. W., 1969. Two useful devices for vertical plankton and water sampling. J. Fish. Res. Board Canada. 26:, 1948-1955.

Schindler, V., J. Popovský, J. Chudoba & J. Hejzlar, 1984.

Mineralizační technika pro vodohospodářské laboratoře. Bull. metod stred. vodohosp. laboratoří MLVH SR, DT SVTS Pardubice,, 1984. (In Czech)

Seďa, J., J. Kubečka & Z. Brandl, 1989. Zooplankton structure and fish population development in the Římov Reservoir, Czechoslovakia. Arch. Hydrobiol. Beih. Ergebn. Limnol. 33: 605–609.

Slovacek, R. E. & P. J. Hannan, 1977. *In vivo* fluorescence determinations of phytoplankton chlorophyll-a. Limnol. Oceanogr. 22: 919–925.

Smith, V. H., 1983. Low nitrogen to phosphorus ratio favors dominance by blue-green algae in lake phytoplankton. Science (Washington, D.C.) 221: 669–677.

Sournia, A. (ed.), 1978. Phytoplankton Manual. UNESCO Monographs on oceanographic methodology 6, UK, 335 pp.

Straškraba, M., 1964. Preliminary results of a new method for the quantitative sorting of freshwater net plankton into main groups. Limnol. Oceanogr. 9: 347–366.

Straškraba, M., 1976a. Empirical and analytical models of eutrophication. Proc. Eutrosym., Karl-Marx-Stadt, Vol.3: 352–371.

Straškraba, M., 1976b. Limnological models of reservoir ecosystems. Limnologica (Berlin) 10: 513–516.

Straškraba, M., 1978. Theoretical considerations on eutrophication. Verh. Intern. Verein. Limnol. 20: 2714–2720.

Straškraba, M., 1980. The effect of physical variables on freshwater production: Analyses based on models. In: E. D. Lecren & R. H. Lowe-McConnell (eds.), The Functioning of Freshwater Ecosystem. Cambridge Univ. Press, Cambridge: 13–84.

Straškraba, M., 1985. Managing eutrophication by means of ecotechnology and mathematical modelling. In Lakes Pollution and Recovery, International Congress, Rome 15th–18th April, 1985: 17–28.

Straškraba, M., B. Desortová & J. Fott, 1979. Zur Methodik der Bestimmung und Bewertung des Chlorophyll-a in Oberflächenwässern. Acta hydrochim. hydrobiol. 7: 569–590

Straškraba, M. & A. Gnauck, 1985. Freshwater Ecosystems. Modelling and Simulation. Elsevier, Amsterdam, 309 pp..

Straškraba, M., I. Nesměrák, N. Štybnarová, P. Blažka, Z. Brandl, B. Desortová, P. Dolejš, L. Procházková & V. Straškrabová, 1986. Recommendations for water quality investigation of drinking water supply reservoirs in Czechoslovakia. Limnologica (Berlin) 17: 201–212.

Straškraba, M. & V. Straškrabová, 1975. Management problems of Slapy Reservoir, Bohemia, Czechoslovakia. Proc. of the Symp. on Effects of Storage on Water Quality. Reading 1975:449–484.

Straškrabová, V., 1973. Methods for counting water bacteria – comparison and significance. Acta hydrochim. et hydrobiol. 1: 133–154.

Straškrabová, V., 1975. Self-purification capacity of impoundments. Water Research 9: 1171–1177.

Straškrabová, V., B. Desortová, K. Šimek, V. Vyhnálek & B. Bojanovski, 1983. Ovlivění biochemické spotřeby kyslíku v povrchových vodách přítomnosti řas. (The influence of algae on biochemical oxygen demand in surface waters). Vodní hospodářství B 33: 165–168. (In Czech).

Suffet, I. H. & P. MacCarthy (eds.), 1989. Aquatic Humic Substances. Influence on Fate and Treatment of Pollutants. Advances in Chemistry Series 219, Am. Chem. Society, Washington D.C.

Szumiec, M., 1969. Heat balance of carp rearing ponds at the Golysz Experimental Farm. Acta hydrobiol. 11: 137–182 (In Polish).

UNESCO, 1966. Monograph on Oceanographic Methodology, Vol. 1.

Utermöhl, H., 1958. Zur Vervollkommung der quantitativen Phytoplankton-Methodik. Mitt. Int. Ver. Limnol. 9: 1–38.

Vollenweider, R. A. (ed.), 1969. A Manual on Methods for Measuring Primary Production in Aquatic Environments. IBP Handbook No 12, Blackwell Sci. Publ., Oxford and Edinburgh, 213 pp.

Vostradovský, J., J. Křížek, O. Albertová, L. Říha & M. Vostradovská, 1989. The changes in fish communities and biomanipulation in water supply reservoirs. Arch. Hydrobiol. Beih. Ergebn. Limnol. 33: 587–594.

Weber, W. J. Jr., 1972. Physicochemical Processes for Water Quality Control. Willey Interscience, New York.

WHO, 1984. Guidelines for Drinking-Water Quality. Volume 1,Recommendations. World Health Organization, Geneva.

Yentsch, C. S. & D. W. Menzel, 1963. A method for the determination of phytoplankton chlorophyll and phaeophytic by fluorescence. Deep-Sea Res. 10: 221–231.

Zajíček, V., 1985. (Zonation of plots in drainage catchment areas and its significance for water quality – In Czech). Sborník konference 'Optimalizace zemědělského využívání pásem hygienické ochrany vodních zdrojů'. ČSVTS VKPS Pečky 1985. (Proceedings of the Conference 'Optimization of Agricultural Use of Zones of Hygienic Protection of Water Resources', Pečky).

Zelinka, M. & F. Kubíček, 1985. Základy aplikované hydrobiologie.(Basic Applied Hydrobiology). SPN Praha. (In Czech).

Zelinka, M., P. Marvan & F. Kubíček, 1959. Hodnocení čistoty povrchových vod.(Surface Water Quality Evaluation). ČSAV, Slezský ústav, Opava. (In Czech).

Žáček, L., 1981. Chemické a technologické procesy úpravy vody. (Chemical and technological processes of water treatment – In Czech). Státní technické nakladatelství, Praha.

Žáček, L., J. Šorm, & L. Jusko, 1984. (Influence of the character of humic substances on the selection of optimal water processing technology – In Czech). Sborník XXI. semináře 'Nové analitické metody v chémii vody – Hydrochémia '84'. ČSVTS Bratislava, 167–182. (Proceedings of XXI. Seminar 'New analytical methods in water chemistry – Hydrochémia '84' ČSVTS Bratislava.)

Chapter XIII

State-of-the-art of reservoir limnology and water quality management

M. Straškraba,[1] J. G. Tundisi[2] & A. Duncan[3]
[1] *Biomathematical Laboratory, Czechoslovak Academy of Sciences, Branišovská 31, 370 05 České Budějovice, Czechoslovakia;* [2] *University of Sao Paulo, School of Engineering at Sao Carlos, Centre for Water Resources and Applied Ecology, Sao Carlos, 13560, Brazil;* [3] *Royal Holloway and Bedford New College, University of London, Egham Hill, Egham, Surrey, TW20 0EX, England*

Key words: reservoirs, limnology, water quality management, ecosystems, biomanipulation, ecotechnology, mathematical models, eutrophication

Abstract

The chapter presents state-of-the-art of reservoir limnology and water quality modelling and their use in reservoir water quality management. Reservoirs are classified into dam reservoirs and impoundments. The features of dam reservoirs and lakes are confronted, both qualitative (absolute) and quantitative (relative) differences being specified. Different consequences of reservoir construction for the river are outlined. Methodological problems concerning sampling and mathematical modelling are dealt with. Theoretical aspects of reservoir limnology cover the position of reservoirs in the river continuum concept, retention time as a key factor of reservoir limnology and reservoir aging (short term trophic upsurge) and long term reservoir ecosystem evolution. Pulse effects are treated as a theoretical problem with considerable consequences for reservoir management. Specific attention is devoted to multiple reservoir systems. Reservoir water quality management approaches and options as well as their use for specific reservoir uses are analyzed. The perspective for future reservoir investigations is treated.

1. Introduction

By reservoir limnology is meant limnological aspects specific to reservoir rather than lake ecosystems and taking into consideration qualitative as well as quantitative aspects. 'Comparative' reservoir limnology deals with those features common to reservoirs, such as geographical location, construction, usage and management as well as those features that differ from lakes.

1.1. Limnological definition of reservoirs

A reservoir is defined for the present purposes as an artificial basin with a riverine source capable of storing more than 1×10^6 m³ water. This volume has been decided somewhat arbitrarily. In the USA, Martin & Hanson (1966) propose a volume of 6×10^6 m³ as a lower limit in order to exclude small fish ponds, local water stores or cattle watering areas, even if they are formed from damming a stream, but the 1×10^6 m³ limit would permit the inclusion of many European supply reservoirs which are smaller than 6×10^6 m³. We do not include here the lake reservoirs – natural lakes transformed into a reservoir by rising levels and using them for power generation (see e.g. Lindström, 1973).

Reservoirs are constructed either by damming a river or stream or by building up walls on an impermeable site with a channelled riverine water supply. To distinguish them, the former are called 'dam reservoirs' and the latter 'impoundments'. Historically, impoundments were invented first in the 6th century BC or earlier (Fernando, 1984) whereas dam reservoirs appeared about 50 years ago and have been much more intensively studied. Table I lists some of the basic differences between these two types of reservoirs, and Fig. 1 illustrates these differences schematically. Although most of this book is devoted to studies on dam reservoirs, a few examples of impoundments are also considered later.

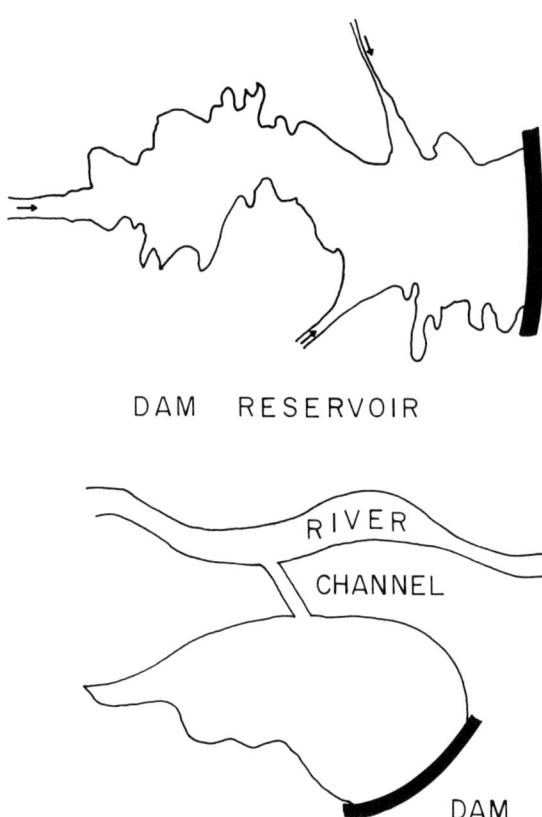

Fig. 1. Basic differences between dam reservoirs and impoundments. Orig.

Table I. Differences between dam reservoirs and impoundments.

Characteristics	Dam Reservoir	Impoundment
Location	on the river	besides the river
Dam Construction	dammining a valley	surrounding walls
Depth	deep to shallow	shallow
Form	dendritic	more regular
Inflow	river(s)	channel
Outflow	river	channel

1.1.1. Dam reservoirs: function affecting limnology

Dam reservoirs are normally constructed for one primary function which has a fundamental influence upon their morphology and limnology. Chronologically, reservoirs with single functions were built for flood protection, crop irrigation, drinking water supply, fisheries, industrial water supply and, more recently, power generation and recreation. With time, most reservoirs have had secondary functions imposed upon them, which are largely unplanned and which they perform more or less successfully.

There have been three historical phases of building dam reservoirs, usually for different purposes. The earliest reservoirs were constructed in the 19th century for storing large volumes needed to control flow, as in flood protection, flow augmentation or power generation. Later, water storage reservoirs were used for irrigation or industrial processing, for which good quality water was needed, and for navigation, for which quality was not important. By the middle of the 20th century, water storage reservoirs were forced to satisfy multiple functions as well as a growing concern about water quality.

The need for large quantities of water was easily satisfied in temperate regions like Europe and Canada which have a balanced hydrology. In semi-arid and dry regions like southern Australia, north-eastern Brazil or the Dry Zone of Sri Lanka, the need for water led to the creation of extensive reservoir systems capable of transferring water over large distances by existing river courses, by specially constructed channels or through mountain tunnels (Fig. 2). Later, increasing industrialisation generated an energy demand which could only be satisfied by large-scale networks of multi-functional reservoirs and even in parts of the world with plenty of water.

The quality as well as quantity of water became more important during the second phase of dam reservoir construction and not only for drinking water supply reservoirs. For example, Strycker (1988) reports on damage to concrete and steel structures which led to serious leakages in the Willow Creek Reservoir (USA) caused by high concentrations of free CO_2 and H_2S in 'aggressive water'. Similarly, J.G. Tundisi records early replacement of turbines in the Curua Una Reservoir (Brazil, Amazon region) after only four years and, in the El Cajon Reservoir (Honduras), interference with electrical transmission due to gas ebolitions from the sediments.

Quality of water became an even greater concern during the third phase of multi-functional reservoir building. Although human demand for a healthy environmental ensured that the quality of the reservoir water itself was good, some damage

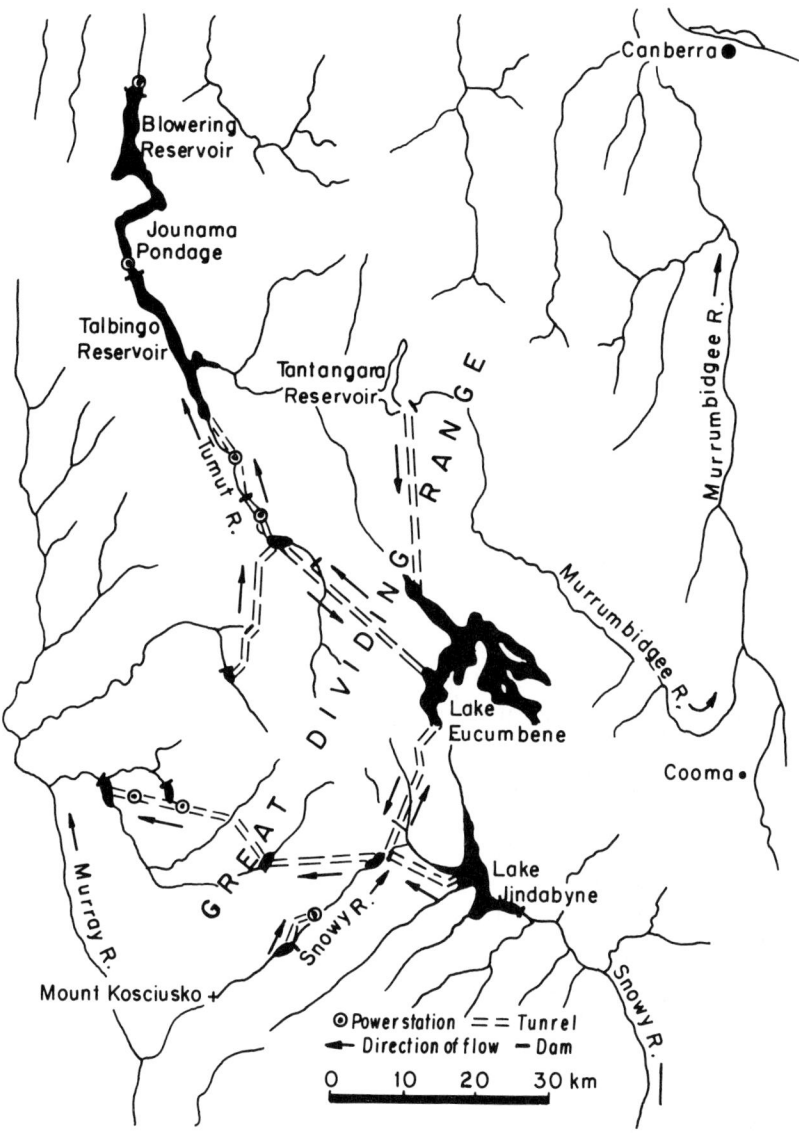

Fig. 2. Example of an extensive reservoir system: The Snowy Mountain Hydroelectric Scheme in Australia. Melt from snowfall collects in Lake Eucumbene and other reservoirs. One tunnel takes water into the upper Murrumbidgee River, another to the Murray River. From Pigram (1986).

could be caused to different water usages downstream: domestic or industrial water supply, fisheries or recreation. Multiple functions in tropical reservoirs may create conditions that facilitate the dispersal of water-borne diseases, thus exacerbating health problems and extending their geographical range (Stanley & Alpers, 1975; Panday, 1977; Moss, 1980; Symoens *et al.*, 1981). Increased public awareness and concern about the deterioration of famous lakes such as Lake Constance or Lake Washington (Edmondson, 1991) has arisen from people's aesthetic appreciation of pristine waters.

Specific features related to size, water depth, depth of outflow and retention time are built into single function reservoirs and affect their limnological characteristics (Table II). These are modified by the stream order of their riverine source and by the operational strategy of their management. Key factors influencing the

Table II. General features of reservoirs according to different major uses.

Primary use	Size	Depth	Retention time	Outflow depth
Flood protection	small to medium	shallow	regionally dependent	surface
Water storage	small to medium	–	extremely variable	below surface
Hydro-electricity	medium to large	deep	variable	near-bottom
Drinking water supply	small	preferably deep	high	intermediate to deep
Fisheries	small	shallow	low	surface
Pump storage	small to medium	deep	extreme variability	near-bottom

small – about $V < 100$, medium – about $100 < V < 1000$, large – $V > 1000$. V in 10^6 m^3.

limnology of the reservoirs are retention time and where the reservoir is sited on a river in relation to the river continuum concept; both of these are discussed more fully later.

Construction features strongly affect reservoir limnology; for example, the reservoir water volume in relation to river flow or in relation to the location of intakes and outflows. These features are not necessarily tied to particular primary functions as reservoirs of different function often have features in common. However, the primary function will affect the reservoir size, as given by site selection for dam construction, its height as determined by the valley morphometry, the volume stored and the capacity relative to flow which determines the reservoir's retention time.

1.1.2. Impoundments – exploitation of big lowland rivers

In contrast to dam reservoirs whose river waters are impounded *in situ* by the construction of a dammed up lake along the course of a river, it is possible to distinguish impoundments which are reservoirs some distance from a river whose waters are being abstracted (Ridley & Steel, 1975). The lowland water supply reservoirs of south-east England are a series of such impoundments of the River Thames or its tributaries. Other example in The Netherlands is the Biesbosch Reservoir (Oskam, 1982; 1987). In both types of reservoirs (dam and impoundments), the ecology of the impounded water is heavily dependent upon the quality of the upstream river water but, unlike dam reservoirs, the water from impoundments passes into supply and so has no effect downstream of the reservoir site either by affecting flows or by modifying the physics, chemistry or biology.

The use of rivers as a water supply source became inevitable in densely populated and industrial lowland areas like south-east England once available underground sources were outstripped and prohibitive costs ruled out water transfers from distant lakes. The main problem was one of water quality: the greatest need for river water was where its quality was both poor and there was a health hazard due to large-scale loading of sewage and industrial effluents in the London area. The findings of Frankland (1984) that storage of polluted river water in open reservoirs for a few weeks markedly improved its pathogenic and organic loading as well as increased its concentrations of dissolved oxygen led to further research on water-borne pathogens, to the development of methods for monitoring and controlling water quality and to the acceptance of storage reservoirs as a means for improving water quality. They also provided a stored reserve for periods of drought when the river was too low for abstraction.

Function and morphology of such lowland river impoundments went hand in hand because their design was an engineering problem (Ridley & Steel, 1975). What volume of water needs to be conserved of the surplus water available in winter when river flows are maximal depends upon the duration of the summer drought period for which water reserves are needed plus a constant daily supply need, which is between 1–2% of storage reservoirs' maximum capacity in the London area. Once the capacity has been decided, the impoundment then needs to have means for directly supplying the water treatment works or for discharge back into the river for augmentation of natural flows during drought. Final details of design were determined by construction economics and availability of land.

In a lowland area, reservoirs can be built close to the river with a height of the enclosing embankment that is determined by the local availability of suitable building materials such as clay and gravel, thus fixing the impoundment morphometry somewhat arbitrarily and taking no account of the water mass as an ecosystem whose flora and fauna are influenced by the chemical and biological characteristics of the river water. Until recently, water supply impoundments were designed in ignorance of the ecological importance of the depth-volume ratio of a water mass. The same volume of water can be stored either in a shallow impoundment (10 m) over a wide area, with a wide littoral zone of macrophytes and a well mixed water column, or in a much deeper reservoir (15-20 m) requiring much less land but probably thermally stratifying during the summer. The depth of water relative to surface area in these alternative designs imposes environmental features which profoundly affect the production levels of the planktonic, littoral and benthic zones and, in consequence, the overall biomass of the impoundment ecosystem.

From a biological point of view, there is a difference between the environmental situation in dam reservoirs with periodic water level fluctuations that expose any marginal vegetation and water supply impoundments where river water constantly flows through at a rate that maintains the water depth within a narrow range for long periods. In the latter, the ecological stability enhances the biological colonisation of littoral and benthic zones whilst planktonic organisms drift passively through from inlets to outlets in a period of time depending upon the size of the impoundment. If this takes weeks or months, some riverine forms may die out and lacustrine forms may develop their own populations with large enough biomasses for their removal to be significant during any subsequent water treatment. This problem has been substantially reduced by recent development of low-cost engineering systems for improving circulation in very large impoundments (see Section 4.5.2.).

1.2. Features of reservoirs

The alternative name for reservoirs coined by Lowe-McConnell (1966) is man-made lakes which implies a limnological similarity between natural and artificial water bodies. Although this is true to some extent, there are reservoir features not possessed by lakes and there are differences of degree. Table III attempts to compare the characteristics of these two types of water bodies, excluding any features associated with geography which is dealt with later. Identification of similarities and differences is an essential prerequisite for understanding the structure and functioning of reservoir ecosystems, for making predictions about their characteristics and for managing their water quality.

Table III. Comparison of the characteristics of dam reservoirs and lakes

Characteristic	Lakes	Dam reservoirs
QUALITATIVE (ABSOLUTE) DIFFERENCES		
Nature	natural	man-made
Geological Age	old ($\geq$ pleistocene)	young (<50 years)
Aging	slow	rapid (first few years)
Formed by filling	depressions	river valleys
Shape	regular	dendritic
Shore development ratio	low	high
Maximum depth	near-central	extreme (at the dam)
Bottom sediments	autochthonous	allochthonous
Longitudinal gradients	wind-driven	flow-driven
Outlet depth	surface	deep
QUANTITATIVE (RELATIVE) DIFFERENCES		
Watershed/lake area	lower	higher
Retention time, R	longer	shorter
Coupling with watershed	lesser	greater
Morphometry	V-shaped	U-shaped
Level fluctuations	smaller	larger
Hydrodynamics	more regular	highly variable
Causes of pulses	natural	man-made operation
Water resource systems	rare	common

1.2.1. Qualitative differences between reservoirs and lakes

Table III lists some of the features whereby reservoirs are absolutely different from lakes, apart from their artificial construction.

1. Reservoirs are geologically young compared with lakes. Hutchinson (1957) cites the youngest lake as 10^4 years old and Baikal as the oldest with a continuous lacustrine history from the end of the Mesozoic or beginning of Tertiary. The Brazilian Lake Jacare is less than 5000 years old (Saijo & Tundisi, 1985; De Meis &

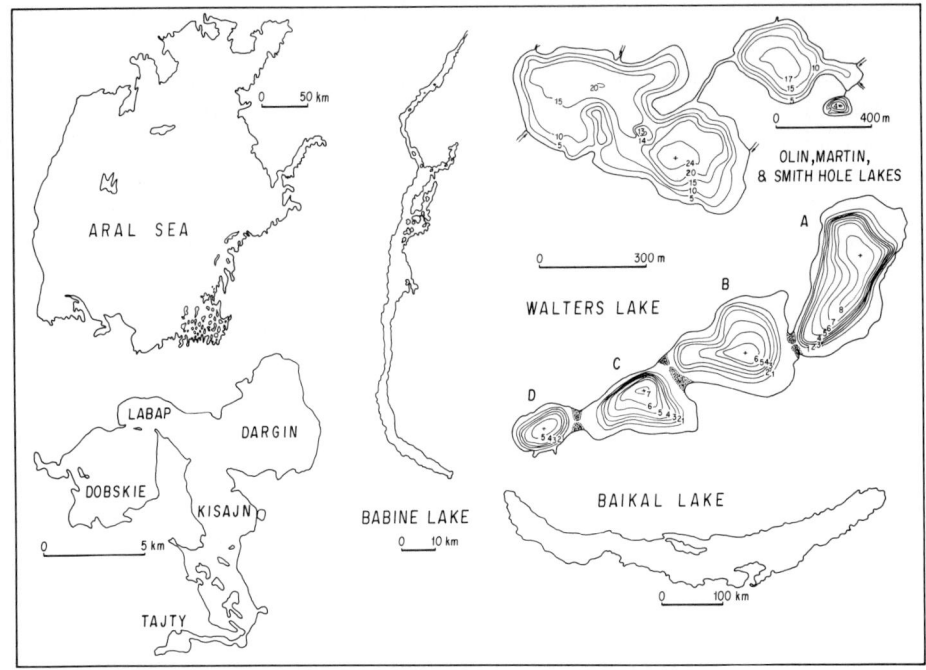

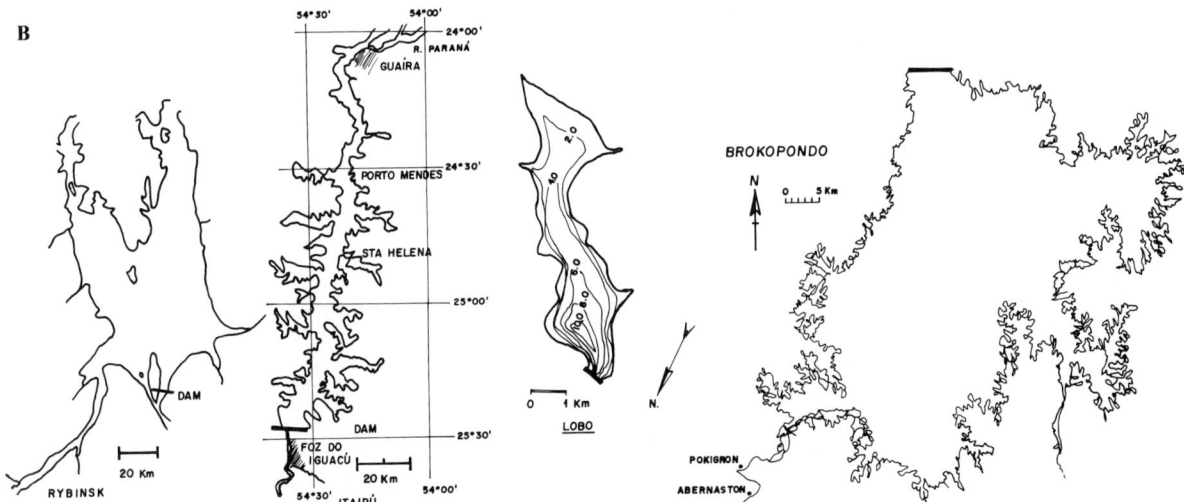

Fig. 3. Outlines of a few lakes (A) and reservoirs (B) demonstrate the higher shore development and ramification of reservoirs, although that during selection also riverine lakes were included. Compiled from different sources.

Tundisi, 1986) which is about the same age as the oldest reservoir cited by Ryder (1978). Lake Jordán (Czechoslovakia) at 500 years which is the oldest reservoir in Central Europe was built for fish culture. In general, most dam reservoirs were constructed during the last 50–100 years and can be dated exactly from historical records whereas our knowledge of the age of lakes is dependent upon indirect means such as pollen analysis and carbon dating of sediments.

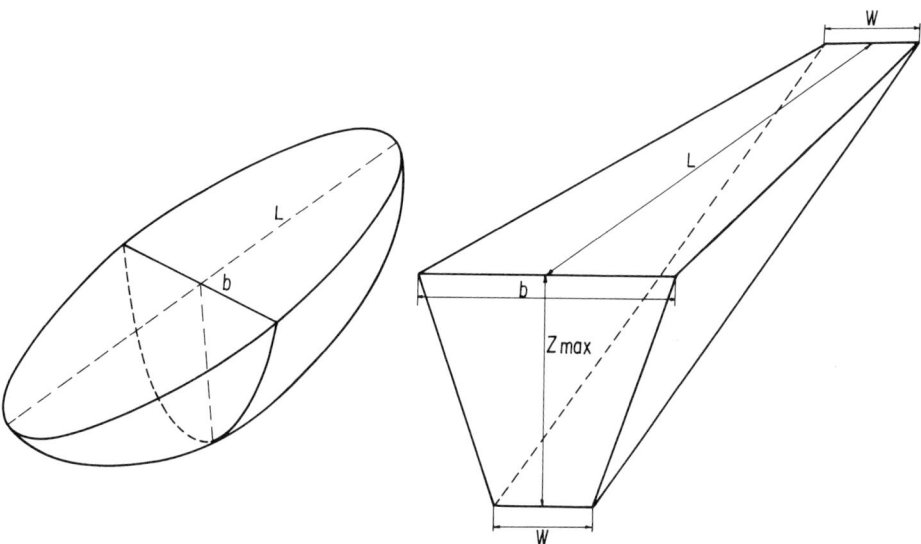

Fig. 4. Simplified representation of the morphometry for a reservoir (A) versus that for a lake (B). Orig.

2. Dam reservoirs have been formed by filling river valleys rather than by filling depressed lake basins and this gives dam reservoirs very large shore development ratios due to the highly dendritic shape of river basins. This is strikingly illustrated in the reservoir and lake outlines given in Fig. 3. The reservoir morphometry is entirely governed by the geomorphometric nature of the river basin at the construction site and may be very different. The distribution of the water volume differs from a deepest point being at one extreme end near the dam in dam reservoirs to being in or near a central point or offshore in more rounded lake basins. Some glacial lakes may have shapes similar to dam reservoirs because of their creation by damming up of a river but, usually, the point of maximal depth is not so extremely placed. Baxter (1977) talks of 'half-lakes' in this connection. Figure 4 illustrates simplified reservoir and lake morphometry. Reservoirs possess the younger more rectangular shapes and lakes older more curved ones.

3. Because of their formation in river valleys, dam reservoirs demonstrate more pronounced longitudinal gradients between inlets and outlets than do lakes.

4. Most reservoirs undergo rapid changes during their first few years of existence, often including a 'trophic upsurge'; this we term reservoir aging (Section 3.5.).

5. The bottom sediments of reservoirs are distinguishable according to their origin; allochthonous, brought in by the river, and autochthonous, coming from its own biological production (Chapter VII, this volume). The depth of sediments is greater in lakes than reservoirs because of the age difference. In general, the autochthonous material settles in the lacustrine parts of the reservoir whereas the allochthonous material accumulates near the inflow. Here, the sediment may be deep due to greater loadings from a dry and deforested catchment and, of course, land erosion accelerates infilling of the reservoir by sediments.

6. Only surface water flows out of lakes whereas reservoirs may have major outlets below the surface by design and any surface overspill is only a small fraction of the total outflow. Overspills often represent an excess flow during flood periods and are usually of short duration. Lake seepage does occur in a few glacial lakes, usually through the front moraine, and may represent the dominant outflow. It differs from the situation in reservoirs by being a diffuse outflow very different from the localised outflow of reservoir outlets. These differences will have limnological consequences which we do not fully understand as yet.

1.2.2. Quantitative characteristics of dam reservoirs

Some features distinguishing reservoirs from lakes are more subtle and can only be shown quantitatively. This is particularly true when lakes and reservoirs from the same region are compared.

1. Reservoirs in general possess larger catchment area/lake area ratios than lakes which implies shorter theoretical retention times (Fig. 5); the main exceptions are the natural riverine lakes which resemble reservoirs in this feature. Consequently, the riverine lakes differ in:
   (i) having more stable regime in water levels (that is, shorter-term water level fluctuations); and in
   (ii) having outflows from the surface layers or the whole depth profile but not through a deeply located hole.
2. In consequence of the above (1) is a tighter coupling of reservoir processes with those of the catchment area, which will include man-related ones. Reservoirs act as concentrators and digestors of chemical pollutants from the watershed as well as an information archive of economic developments and ecological changes in the catchment area (Tundisi, 1989).
3. Deep mountainous reservoirs are characterised by a V-shaped basin which is pre-disposed, according to Tyler (1980) to thermal stratification, anoxic conditions and the development of a monimolimnion of dense, chemically enriched water due to downslope migration of density currents and even more importantly of downslope migration of mineral particles (Fig. 6). There is not hard evidence for this as yet. The onset of thermal stratification is more influenced by the area:depth ratio and the retention time than the shape of the basin cross-section (Chapter XII, this volume).
4. The water level fluctuations in dam reservoirs are much greater and more frequent than in lakes.
5. Under identical climatic conditions, reservoirs are more variable hydrodynamically and more influential in modifying chemical and biological processes. Inter-annual differences in flow rates affects the mixing intensity between river and reservoir waters and their associated chemistry and biology. Changing rates of outflow or depth of outflow are operations that produce only short-term effects.
6. When dam reservoirs are managed for power generation, pulse effects are created which interrupt any natural pulses in a random or stochastic manner (Section 3.4.). Seiches as natural physical pulses have a limited duration in most reservoirs due to operational management.
7. Reservoirs are often interlinked into a water resource system of several kinds (Section 3.6.). One characteristic system is that of a cascade of reservoirs such as exists in Czechoslovakia and Brazil (Chapter II, this volume).

1.2.3. Consequences to rivers of building reservoirs

The effects of impounding river waters cause direct and indirect changes downstream of the reservoir site. The topic has been the subject of several recent books on regulated streams (Oglesby et al., 1972; Ward & Stanford, 1979; Lillehammer & Saltveit, 1984; Petts, 1984; Craig & Kemper, 1987; Petts et

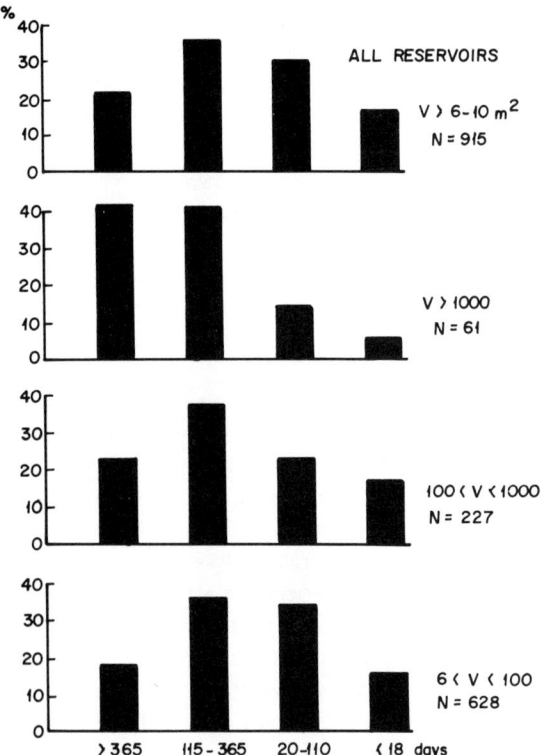

Fig. 5. Distribution of theoretical retention times of the reservoirs of different size classes in US according to data by Martin & Hanson (1966). Orig.

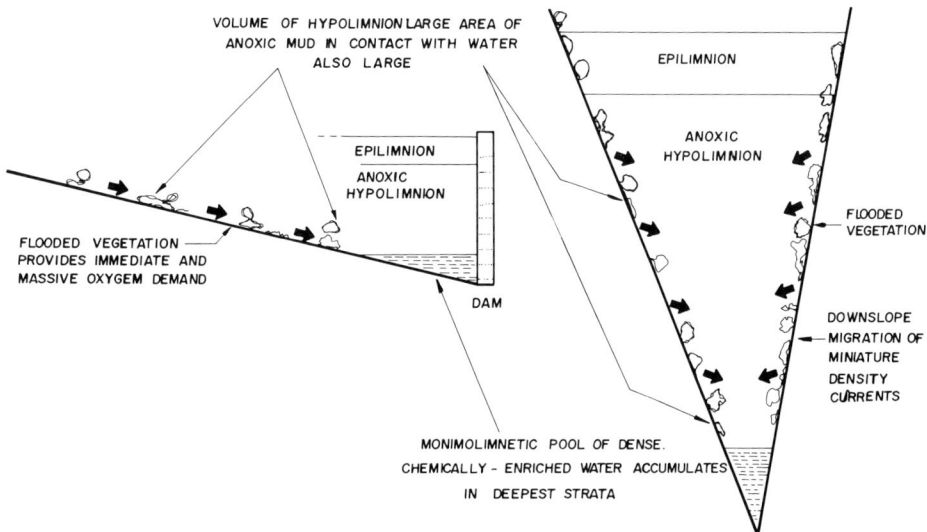

Fig. 6. Downslope migration by density currents may be of great importance also for the downward migration of mineral particles. According to Tyler (1980).

al., 1989), several summaries (Ridley & Steel, 1975; Baxter, 1977; Brooker, 1981) and of specialised journals (Regulated Streams and Regulated Rivers: Research and Management). In spite of this, Puig *et al.* (1987) write that the effect of Spanish reservoirs upon rivers is practically unknown.

What is presented here is a summary based upon what is known about reservoirs and their processes, without which it is impossible to predict the qualitative and/or quantitative effects of a particular reservoir upon its surroundings and upon its downstream river. It is also of importance in reservoir cascades, where the outflow of one reservoir becomes the inflow of the next one down. There are also significant effect of large reservoirs, particularly those with long retention time, on local climate around the lake; for example, in temperate region (Nakamura, 1967), increased winter temperatures, decreased summer ones and fog conditions may be created.

Broadly speaking, water supply reservoirs differ in their downstream effects compared with flow regulatory ones. In the former kind, the abstracted water is entirely lost to the river reach from which it has been taken, with severe consequences to the river biota if a high proportion of the river flow is removed. Whereas, flow regulation reservoirs ameliorate the extreme flow regimes to the benefit of the river biota. Different consequences result where several rivers are inter-connected for a wider distribution of the mixed water since biological dispersal is facilitated, including the dispersal of (say) the parasites of natural fish populations. The degree of disturbance downstream will differ according to whether the reservoir is:

Table IV. Major limnological characteristics of reservoirs which are decisive for changing downstream river quality.

Limnological Characteristic	Effects upon downstream river
Hydrology	Greatest effect in semi-arid regions Considerable effect in temperate regions Least effect in tropics
Stream order	Inverse with stream order – greatest effect with lowest stream order and vice versa
Reservoir Depth	No or very low depth effects in shallow reservoirs but increasing effects with increase in depth
Reservoir outlet depth	No outlet depth effects in shallow (unstratified, wind-mixed) reservoirs but increasing effects with increase in outlet depth in stratified reservoirs
Retention time	Reservoirs with short retention times do not have much effect on downstream river but effect increases with prolongation of retention time
Trophic degree	Effect on downstream river increases with greater reservoir biological productivity

(i) 'on-stream', that is in the river system with the total river discharge passing through the reservoir or
(ii) 'off-stream', where the reservoir is filled when the river has enough water or emptied when the river flow needs augmentation.

Limnological characteristics of the reservoirs themselves are very influential upon the intensity of the effects of reservoirs upon the downstream river. Some of these characteristics are outlined in Table IV. The multivariate character of reservoir ecosystems should be borne in mind so that the listed effect of changing one variable is only valid for comparable conditions: for example, a deeper reservoir will have a higher effect than a shallow one only if it retains the same retention time.

Table V. Downstream water quality effects.

Variable	Changes in downstream water quality
	Physical variables
Hydrology	Decrease in flow rate when usage of reservoir water is high, as in irrigation reservoirs or when evaporation rate is high. Periodic increases, increased variability and disruption of natural cycles.
Thermics	Decrease in temperature and the degree of temperature decrease varies with retention time and with depth. A geographical dimension: in temperate region, higher winter and lower summer temperatures.
Silt Content	Decreased silt loading.
Detritus	Composition of particles changed from abiotic to biotic, and the particle size decreases.
Light	Light penetration is increased.
	Chemical variables
Oxygen	Where reservoir is eutrophic and outlet depth is below thermocline, DO concentrations in outflow water may decline to near-zero values.
H_2S and CO_2	Increased values, especially in eutrophic, stratified reservoirs with long retention times.
pH	Decreases.
Nitrogen	Gaseous nitrogen content increases in aerated reservoirs.
Organic matter	Organic matter content decreases downstream when there are no sources of in-lake organic matter. Any phytoplankton production changes this result up to the overwhelming in-lake organic matter production by phytoplankton in highly eutrophic reservoirs with low organic matter input.
Phosphorus	Phosphorus concentration decreases, the degree of decrease increases with retention time and with trophy. An exception is when bottom waters are released from eutrophic reservoirs with anoxic hypolimnia.
Nitrates	Nitrates concentrations are usually nearly unchanged. When in the reservoir strongly reducing conditions exist the concentration downstream is decreased.
Nitrites	The concentration of nitrites is usually increased, particularly during deep-water releases from reservoirs of higher trophic degree.
Total Solids	Total solids concentration remains mostly unchanged.
	Biological variables
Plankton	In general the abundance of plankton increases downstream.
Phytoplankton Composition	The phytoplankton composition is changed downstream. In small rivers the change is from riverine (periphytic) species to pelagic ones; in large rivers more lacustrine species occur below a reservoir.
Phytoplankton Production	Specific phytoplankton production (per unit mass of phytoplankton) may be extremely high when hypolimnetic phytoplankton enriched in chlorophyll is released to the river and reaches high light conditions.
Zooplankton	A small river is below a reservoir highly enriched with zooplankton. Below a reservoir on a large river potamoplanktonic composition changes into pelagic one.
Zoomass	Zooplankton biomass usually increases downstream.
Benthos	Increased below slightly eutrophic reservoirs, usually decreased below eutrophic ones in connection with anoxia. Composition mostly highly changed. Detrimental effects of short-term water level fluctuations on benthos are high.
Fish	The reservoir represents a barrier for fish migration. Fish occurrence below reservoirs varies according to specific conditions.

Different physical, chemical and biological variables in the downstream river are affected by the upstream reservoir in various ways (Table V). What is listed in Table V represents a gross generalization which does not deal adequately with multi-variate effects. It also ignores the fact that the effect at any one time depends on the timing, intensity and location of outflows in a particular reservoir. Then, the effects differ immediately below the reservoir compared with further downstream, due to rapid changes in the river. For example, below Balbina Reservoir, deoxygenated water persisted for as much as 20 km (Tundisi, pers. observation). Moreover, the age of the reservoir has consequences since outflows from mature and young reservoirs will differ.

The biological consequences to the nature of the river downstream of large-scale 'on-stream' reservoirs are very great, as has been shown by early studies in the North America (Neel, 1953; Anderson & Pritchard, 1954; Wright, 1968; Morris et al., 1968; Spence & Hynes, 1971; Ackerman et al., 1973; Campbell et al., 1982; Anonymus, 1984) and Europe (Hrbáček, 1966; Hrbáček & Straškraba, 1973; Anonymus, 1975; Straškrabová et al., 1989). If the reservoir reduces maximal flows and scouring effects, there may be increased growths of macrophytes downstream (Fraser, 1972; Hall & Pople, 1968). Below a reservoir, there may often be observed a complete disruption of the river ecosystem which results in fish kills and an impoverishment of fish population thus causing a serious loss of protein food for man in tropical countries. There may be also a deterioration in the supply of drinking water and loss of sites of good recreational value. Sometimes, there may be positive effects as when the outflow from deep oligotrophic reservoirs improves water quality in a polluted river to such an extent that a trout-type fauna is able to survive. Straškraba & Javornický (1973) report up an incident of greatly enhanced specific phytoplankton production below Slapy Reservoir, which occurred when the phytoplankton released from deeper strata with accumulation of chlorophyll per unit weight was released to lit surface. Evaporative loss can have a significant effect where hydrological budgets are negative and evaporation exceeds precipitation (Section 3.3.) as occurs in reservoirs at 20° latitudes. The river Nile provides a good example of reservoir effects: organic silt brought down by the river has provided an annual fertilization of delta lands since the start of agriculture. After the closure of the Aswan High Dam, this ceased, the Delta agriculture became infertile and the application of manufactured fertilizers is now a necessity. A probably related effect is the decline of the sardine inshore fishery in the Mediterranean Sea off the Nile Delta reported by (Moss, 1980).

A number of quantitative evaluations exists, enabling prediction of some of the above changes. Those are listed in Table VI.

Table VI. Mathematical models for quantitative prediction of reservoir effects upon outflow water quality

Model	Reference	Features of model
RESTEMP	Chapter XII (Section 3.2.)	Empirical model to predict outflow temperature from temperate reservoirs from elevation, retention and outlet depth.
DYRESM	Chapter V (Section 2.2.)	Model to predict outflow temperature by means of a hydrodynamic reservoir model with retention greater than 10 days.
Reservoir Sedimentation	Chapter IV	Empirical model to predict sedimentation rate in temperate reservoirs.
Phosphorus Retention in Reservoirs		A model by Newbold (1987) on changes in phosphorus concentrations due to reservoir effects, with consequences for primary production. A model for pre-impoundments, with respect to phosphorus capture, by Benndorf et al. (1975).
Reservoir Outflow Organic Matter	Chapter XII	A model by Straškrabová (1976) on the effect of a temperate reservoir on the BOD of the outflowing river with different conditions of retention times, depth and phytoplankton production level.

1.3. Holistic approach to reservoir science

The dominating approach in contemporary science is reductionistic, that of understanding the behaviour of the whole from the sum of the behaviour of its individual parts. The holistic viewpoint is that the only way to understanding the system is to study the system as a whole, components and all. In nature, this means the ecosystem with all its living components, interacting with each other and with its surroundings or, in the terms of this book, the reservoir ecosystem with its dynamic biological organisation and its environment of inflows, outflows and catchment area inputs. Changing the environment (e.g. by land erosion in the watershed) can lead to fundamental changes (mineral turbidity and attenuated optical regime) to which the living components respond (e.g. reduced algae photosynthesis, food limitation to filter-feeders due to uptake of mineral particles, damage to fish gills) and which may result in a quantum lowering of the productivity level at which the reservoir ecosystem operates. Therefore, the whole is not a sum of its parts but much more.

1.3.1. Vision of holistic limnology
At the SIL Congress at Munich Straškraba (1989) presented a vision of holistic lake ecology in which he suggests that time is now ripe for substituting or fundamentally modifying current limnological paradigms. This vision has recently being placed in a more general ecological context in Straškraba (1992) and Jørgensen *et al*. (1992). Figures 7 and 8 demonstrate the differences between the old and new image of lake limnology which are as follows:
1. The notion of lakes as operating on a single time scale, namely the year, and a single space scale, namely the whole lake or zones within the lake, is extended to encompass a hierarchy of scales. A hierarchy of time scales could be: geological or paleolimnological; annual or seasonal; hydrological time scale of internal motions, currents and turbulent mixing. An example of space scales could be: region; watershed; whole lake; within lake (bays, depth zones with different mixing and flow patterns); particle parcels; turbulence cells; Langmuir cells. At one and the same time, it is necessary to recognise the influence of both global phenomena and events at the microscopic level. Moreover, there is a further complexity caused by interactions between the different hierarchies of time and space scales. Another scale is that of resolution.
2. The prevailing simplistic view that observed macrogeographical latitudinal effects are due solely to the global distribution of solar radiation and temperature is no longer tenable because of the manner in which lake ecology is so tightly coupled with hydrology. The section on comparative reservoir limnology in this book considers how regional water budgets cause geographical differentiation of limnogeographical regions. In Section 3.2. and Chapters I, II and III (this volume) the hydrological consequences for various reservoir properties are demonstrated. Weather behaves like a summarizing variable, encompassing environmental variables such as solar radiation and temperature, and influences the variability of water flow on macrogeographical scale as well as in-lake phenomena such as stratification and mixing.
3. Direct hydrological effects on lake limnology are reinforced by related indirect effects associated with the character of suspended particles, their transport and influence on aquatic chemistry. This brings in how terrestrial ecosystems influence aquatic ecosystems via the effect of vegetation cover upon physical and chemical processes in soil and surface waters. In consequence, the new paradigm in Fig. 7A illustrates latitudinal distribution of many variables as jumpy rather than smooth phenomenon.
4. The notion or paradigm of single dominant effects in lake ecology must be changed to considering lakes multivariable spaces with axes such as morphometry, optical qualities and productive level which are constrained by species composition, trophic structure, and duration of biological communities. The interaction between a particular lake's physicochemical and the biological characteristics operates simultaneously (Fig. 7B and 9) and results in its observed limnological properties (processes and variables). Neglect of the paradigm of lake as multivariable systems can occur during the pooling of data in regressions and may lead to bias. For example,

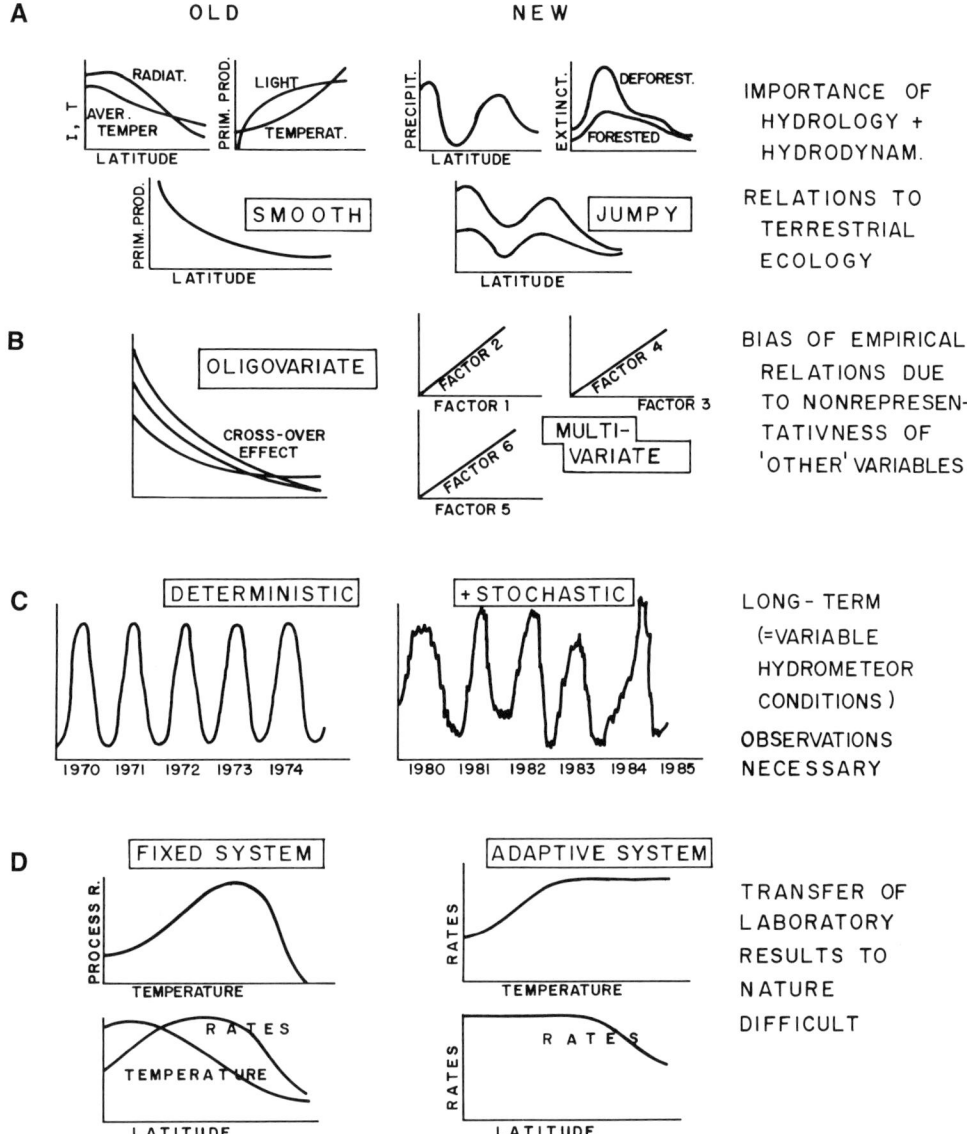

Fig. 7. Old (left) and new (right) paradigm of limnology. For explanation see text. Orig.

optimal temperatures of phytoplankton species have been determined from their seasonal time course; with Cyanobacteria, this is considered to be a high temperature because of the appearance of blue-greens in high summer but this is not supported experimentally. The same applies to their supposed high light requirements.

5. The notion of lake ecosystems as deterministic systems needs to be changed towards inclusion of stochastic components (Fig. 7C). Otherwise, experimental differences noted between a lake during the 'control' years and during the 'experimental treatment' years may be claimed for the treatment to be due to hydrological variability on an annual cycle. Longterm limnological observations must always be tested against long-term changes in hydrology.

6. The paradigm of linear or at least monotonous (continuously increasing or decreasing) effects is now outmoded by the general recognition of the ubiquity of nonlinear, particularly saturation type effects (Fig. 7D), which are often asso-

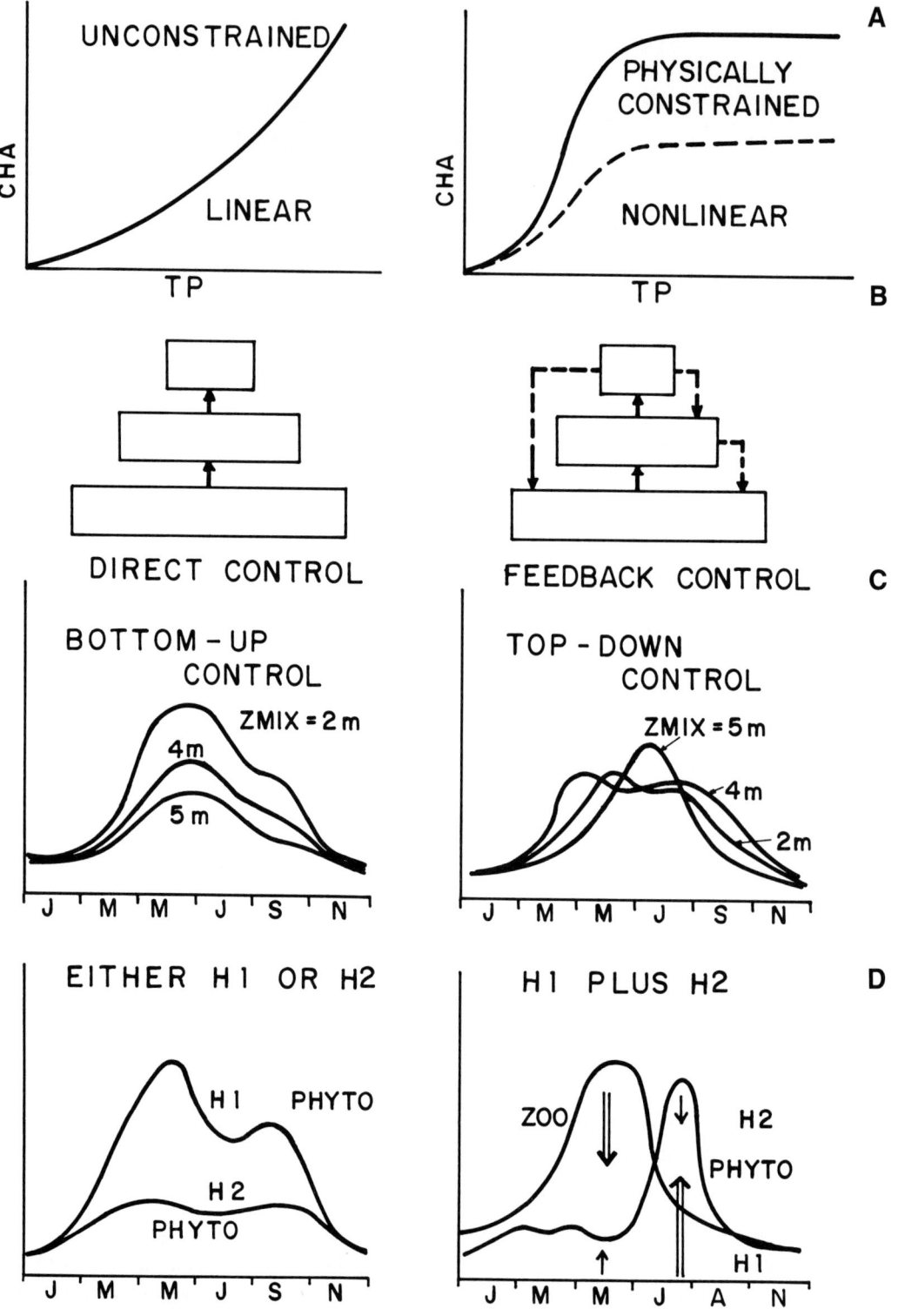

Fig. 8. Old (left) and new (right) paradigm of limnology – continuation of Fig. 7.

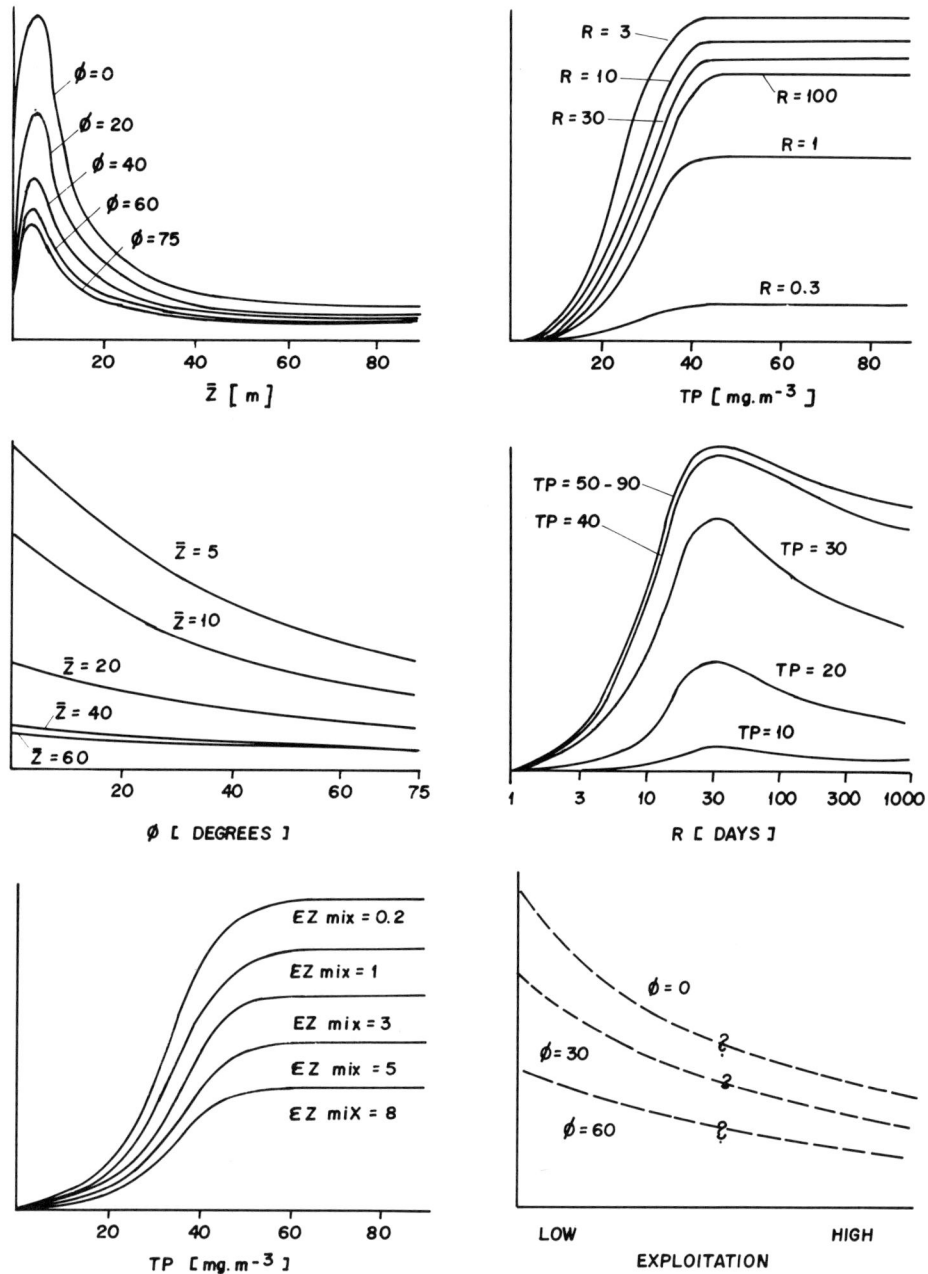

Fig. 9. Representation of the multivariable nature of lake and reservoir ecosystems. On the x axis freshwater productivity expressed as chlorophyll-a in phytoplankton. Left, top: At higher latidudes (Θ) higher values are observed, depending at otherwise similar conditions on the mean depth of the water body (z). Center: the same expressed the other way round. Bottom: Dependence on the total phosphorus concentration (TP) combined with light limitation, due to light availability expressed by the 'optical depth' ϵz_{mix}. Right, top and center: In addition to the saturation character of the dependce on phosphorus, retention time (R) plays a role as an increase of R means increased nutrient loading. Bottom: Exploitation of fish from the waterbody decreases phytoplankton concentrations via changes in zooplankton compositions due to fish feeding pressure. Orig.

ciated with thermodynamic constraints upon ecosystem function. Linear relations may adequately describe the narrow span of individual variability such as the linear dependence of algal growth on critical nutrient concentration in the range of some micrograms but this is not true when the full range of concentrations, from limiting to excess, is covered. Non-monotonous responses, that is an increase followed by a decrease, are more usual as can be seen in physiological processes in relation to environmental variables such as temperature (Fig. 8A) pH and oxygen concentration.

7. Density dependence effects seem to play a much more important role in the metabolism of both aquatic and terrestrial ecosystems than has hitherto being generally recognised (Straškrabová & Šimek, 1984; Straškraba & Gnauck, 1985; Šimek, 1986; Šantrůčková & Straškraba, 1991).

8. There has occurred an evolution of the hitherto dominant paradigm of aquatic ecosystems seen as a static system with only nutrient-based bottom-up effects into a dynamic feed-back system with only predation based top-down effects. What the model in Fig. 8B tries to demonstrate is that bottom-up and top-down influences are not 'either/or', not mutually exclusive, but both coexistent. Nor is one theoretically more important than the other, although this may be so in practice in particular water bodies at particular times as can be seen during the seasonal cycle of plankton (Fig. 8C).

9. There are physical constraints upon aquatic ecosystems not only based upon the fundamental laws of hydrophysics and hydrodynamics but also due to the size constraints imposed upon the physiological and ecological properties of organisms due to the universal biological regularities between body size and metabolism (Peters, 1983). In addition, living organisms are by no means systems wholly controlled by the laws of physics and have found means to circumvent the negative effects of physical constraints by adopting various life cycle strategies, such as diapause, migration, r and K growth and reproductive strategies (Begon et al., 1986).

10. The responses of aquatic organisms to environmental variables is not invariate, as in physical systems. Both species and communities of species possess the ability to adapt their responses to the environment appropriately. One of the earliest and best studied examples of this is the temperature adaptations of fish (Fry, 1957; McLaren et al., 1969), and phytoplankton (Straškraba, 1976). There are also examples of fast morphological adaptations to biotic interactions such as the development of crests in Daphnia and spines in Brachionus. Rapid genetic adaptations within a few generations are also possible, it seems (Lande, 1982). Our understanding of what happens in nature must take account not only of the mechanisms of biological responses, often experimentally derived, but also morphological, physiological and/or genetic adaptations, detected in longer-terms field investigations. An example of this is given in Fig. 8D.

11. During major environmental changes within a water body, there may be biological responses at a higher scale than that at the level of individual organisms, namely changes in the species composition of the biological communities within the ecosystem. This is the kind of community response often seen in massively polluted environments.

12. The notion of 'optimality' in acute responses as well as longer-term adaptive responses of organisms has been very fruitful for incorporating into ecological cybernetics. The best documented ecophysiological examples are related to ectothermic temperature responses and nutrient limitation in plants. Another more recent ecological example is that of 'optimal foraging' (McFarland, 1985; Krebs & Davies, 1991) which has generated a large number of studies as well as strong criticism of a philosophical nature (Ollason, 1980).

13. An emerging idea that is exciting and needs consideration is that structural changes in ecosystems can be treated in a manner similar to adaptations of organisms. This involves the notion that ecosystems are capable of selforganisation and evolution towards optimal performances in ways which can be predicted by analogy to optimal responses of organisms to environmental variables. Other ways at description of structural changes are followed by Lhotka & Straškraba (1987) and Nielsen (1990).

1.3.2. Holistic reservoir limnology

For reservoir limnology, several consequences arise from adopting a holistic approach:

1. The study of reservoirs within the framework of their watershed. The basic study unit must be the flowthrough system of inflow-reservoir-outflow (Chapter XII, this volume). Events in the catchment, natural or otherwise, determine environmental conditions of the inflowing river which, in turn, influence what biological growths can develop in the lacustrine environment of the reservoir and which may have consequences to the performance of the reservoir function. The resultant of reservoir biomass growths and human operational management then jointly imposes damaging or beneficial effects upon the downstream river. To understand all this complexity needs not only a simultaneous study of all the significant parts but also a manager who understands this need.

2. Although artificial man-made water bodies, like lake impoundments are tightly coupled to human society and its environment. This is because reservoirs are built by man for man and because economic consequences follow on from the construction. At first, these consequences were small scale and local but now are on a world scale with human organisations like the World Bank taking political decisions that have very long-term effects, both good and bad (see e.g. Dixon *et al.*, 1989). Switching from environmentally damaging energy production to hydro-electricity would lead to greater local or regional sustainability.

Activities in the catchment area before and after building of the reservoir affect its subsequent performance of its primary function as well affecting the downstream river. The problems that can arise are: eutrophication due to runoff from excessive application of chemical fertilisers; siltation from catchment land erosion; excess loadings of organic particles from domestic or industrial sewage effluents. Once a reservoir exists, it attracts human settlements within its catchment to exploit the new opportunities offered by the new water body, such as new fisheries, enhanced agriculture and afforestation. However, the migration of large numbers of people increases the level of human activities in the watershed and intensifies the human pressures upon the reservoir: new roads, increased sewage effluents, new types of industry. The consequence of the tight coupling between the reservoir and its watershed are seen much more easily in the new hydroelectric reservoirs built in tropical countries during the last ten years. Readers are recommended to read Tundisi (1988).

2. Methodological problems

Any investigation, whether on reservoirs or other water bodies, has to start with defining the problem to be studied and the optimal methodology for studying it. Past experience has shown that in water quality monitoring it is easy to accumulate raw data without analysis, without interpretation of results and without application of results. To coin a phrase (Ward *et al.*, 1986), monitoring tends to be 'data rich but information poor'. One way out is to treat each investigation as a system with subsystems representing the stages in analysis of the problem. The locality to be investigated, the methods to be used and how to evaluate the data have all to be considered in relation to the defined goals of the investigation. Figure 10 illustrates this system approach and brings out the following points:

1. For research on reservoir limnology, as for any investigation, the data collected cannot be independent of the objectives of the investigation.
2. Definition of the objectives depends upon the quality of the investigator (his level of education and his background knowledge of the problem) and the level of prior knowledge of the subject and the locality.
3. The following stages should be recognized in any investigation:
    (a) a clear definition of goals, preferably as a question to be asked and the results expected;
    (b) plans for data collection such as a sampling schedule (time, place and frequency of sampling), sampling handling (transportation, preservation, variables to be measured and the method to be used, including the levels of sensitivity) and the level of accuracy and precision desirable for each determinant;

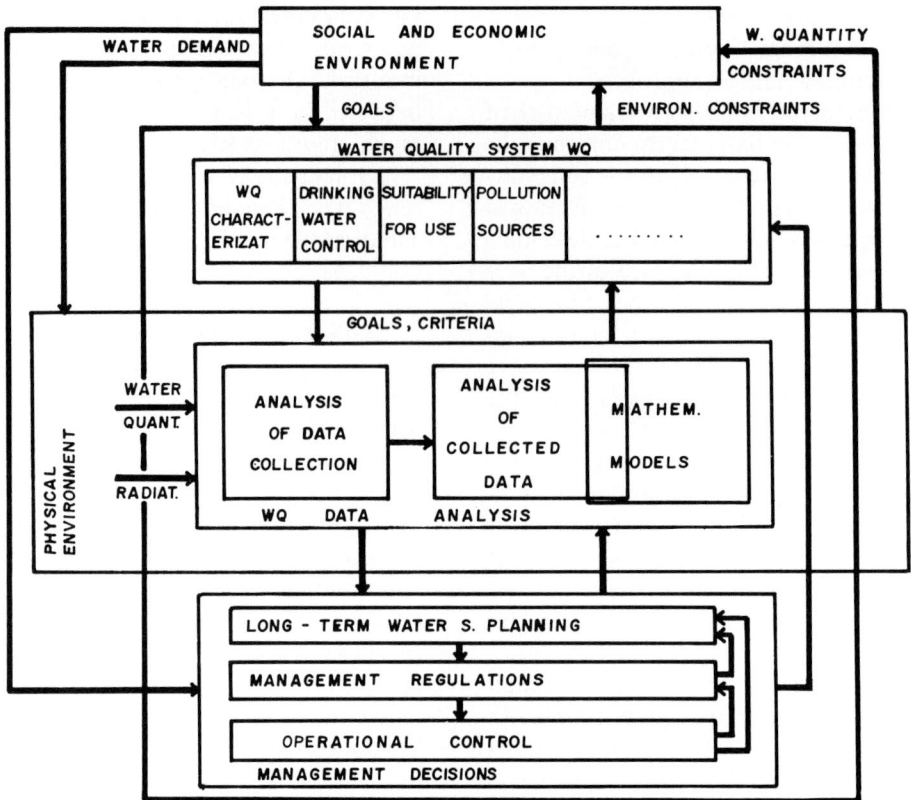

Fig. 10. Systems approach to water quality management. Primary for water quality management is the determination of goals and criteria. Two mutually related major steps are distinguished: Analysis of data collection and analysis of collected data. The goal determines not only the variables to be measured, but also the accuracy and sensitivity needed. The accuracy is the cumulative result of sampling accuracy, and of sample handling. A very exact chemical determination is useless if major changes happen during data handling. Determination of a variable to three decimal places has no value if a few metres of the locality sampled the values might be completely different. The analysis of data collected is up to the purpose, however, relations of different variables to flow rates and their seasonal trends have to be distinguish to enable data interpretation. Orig.

(c) the analysis procedures (the nature of data distributions, relationships between variables) and the nature of the final outcomes (e.g. mathematical models);

(d) interpretation of results and presentation of conclusions.

4. Although the above schedule of the investigation has been outlined in stages, it forms part of a whole and not separate entities.

5. Special attention should be paid to feedback effects from higher to lower steps. For example, any sampling schedule depends on type of data distribution and on relationships between variables or how data is analyzed depends on the expected evaluation which comes from the goals.

It is a common error in 'data rich information poor' investigations to find uneven levels of precision amongst the different determinands. It would be a better use of time, money and manpower to establish a common level of accuracy and precision.

2.1. Sampling

The complexity of sampling representatively in reservoirs depends upon the degree of spatial and temporal heterogeneity that exists there. Spill water and position of outlets cause spatial complexity during the reservoir's operation and the existence of cyclical events in the watershed adds temporal complexity on top of the seasonal cycle.

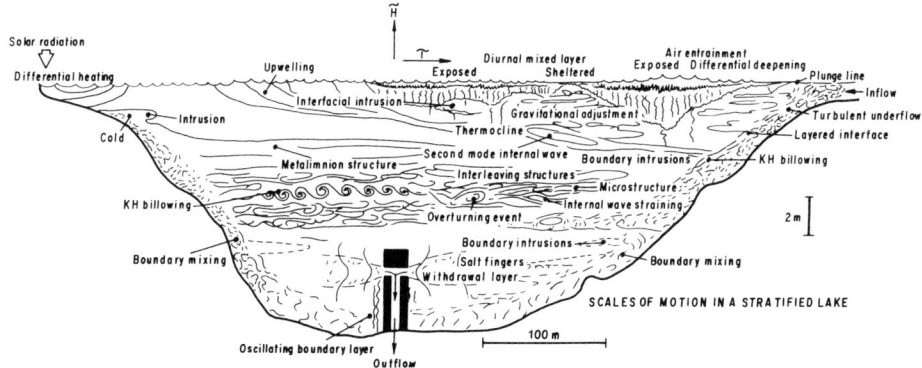

Fig. 11. Mixing modes and scales in a lake or reservoir. From Imberger (1985).

It is sensible, if possible, to sample the whole system (watershed, reservoir and downstream), especially downstream as it responds sensitively to reservoir water quality.

Sampling in large reservoirs can take advantage of new types of large scale technology like remote sensing capable of giving a synoptic view of the whole system. Chapter X (this volume) gives some information on how remote sensing can quantify events in the watershed from the level of suspended particles entering the reservoir. The number of sampling stations on horizontal and vertical scales will be a compromise between costs in time, money and manpower and the needs of the objectives, bearing in mind reservoir size, seasonality and thermal structure. Sampling representativeness is a matter of coupling spatial and temporal scales (Chapter IX, this volume). However, the information gain does not increase above a ceratin number of samples.

The notion of the existence of a hierarchy of scale has been introduced into ecology by Allen & Starr (1982). This puts forward the idea that our view of any organism or particles depends upon the scale at which we view or measure it. For example, mixing in reservoirs can be seen differently if observed from an airplane or measured by a flowmeter or detected by the depth positions of microscopic algal species suspended in open water (Fig. 11).

2.1.1. Spatial scales in reservoirs

The degree of horizontal and vertical heterogeneity existing is influences decisively by reservoir morphometry. In dam reservoirs, the following zones can be distinguished:
– riverine (or back-water)
– transitional (between riverine and lacustrine)
– lacustrine
– bays, and
– shallows with wetlands, submerged trees etc.

In extremely large reservoirs, horizontal variability is on a macro-geographical scale. Major tributaries into the reservoir may come from different geological regions and, concomitantly, will have different chemistries. For example, Kuybyshev Reservoir in European Russia receives water from a highly saline and turbid tributary whose flow varies seasonally as well as the Volga which has a more balanced hydrology and more stable chemistry. It is hydrodynamic conditions which cause horizontal variability in most reservoirs. As an example, the littoral wetlands and weed beds are a major source of horizontal heterogeneity in shallow reservoirs; the plants inhibit free water exchange, their architecture creates differences in the underwater light regime and temperature conditions and their presence influences the local chemistry. In deep reservoirs, the variability is caused by the flow conditions near their main and tributary inflows. Wind-created seiches will influence planktonic distributions in all reservoirs that possesses them. Surface scums of cyanobacterian blooms get piled up downwind in the reservoir. In river-shaped reservoirs, at any bend may be found upwelling and downwelling regions in the direction of the prevailing winds. Most of these consist of transient phenomena which may be missed by most sampling schedules.

Vertical scales also differ in their level of

precision. Only two major layers may need to be studied in stratified reservoirs, the mixing epilimnion and the hypolimnion, but in some reservoirs, the position of the outlet may impose complexities in the shape and intensity of the outflowing water that needs finer detail. This is even more likely in reservoirs with multiple outlets. In some situations, it may be necessary to study the metalimnion, with its sharp vertical gradients, in even finer detail (at a cm level).

2.1.2. Time scales in reservoirs

The time scales of interest in reservoir limnology range from reservoir succession on a scale of years, to annual seasonal cycles or to circadian scales through to seiche periods (days and hours) or stochastic events (minutes and seconds) (Reynolds, 1987; 1989). This point is illustrated in Fig. 12 with an analyzed example of a time series taken automatically at half hourly intervals through July in Slapy Reservoir. As mentioned above, temporal and spatial scales are interrelated.

Seasonal cycles are not restricted to the cold temperate regions with its pronounced variation in radiation and temperature. The dry and humid tropics also have seasons with reduced radiation during the rainy season. These natural seasonal cycles are sometime interfered with by water level fluctuations and flow modifications imposed by reservoir operation and which may be irregular (flow regulation) or regular on an annual scale (irrigation) or on a daily scale (power generation).

2.2. Mathematical modelling

Mathematical modelling is a methodological approach used as a means for studying complex systems and for transferring the scientific knowledge to engineers, managers and politicians. It represents an extension of studies done by more classical empirical approaches based on extensive field measurements of individual variables and/or experimental in-situ and laboratory research studies on isolated components. It helps to organise research efforts and the data gained by diverse specialists and using various methods into a coherent framework. Although called mathematical, it is not based upon mathematics but on detailed knowledge of the subject; mathematics is only the language used for communication and methods for solution of complex problems. Mathematical modelling is as diversified as limnology itself, various branches of mathematics being involved. The individual methods lead to different models, the same topic being also the subject matter of different models depending on the goal for its construction. However complex, the mathematical model is a gross over-simplification of reality. Nothing like a universal model exists: every model covers only selected aspects of reality depending on its purpose. When we say 'this is a model of reservoir hydrodynamics', it is only for the sake of simplicity; in fact, our model is a model of certain aspects of hydrodynamics of a particular kind of reservoir and is based upon a number of simplifying assumptions.

An extensive recent literature exists about mathematical modelling of natural systems (e.g. Hall & Day, 1977; Jeffers, 1978; Starfield & Bleloch, 1986; Jørgensen, 1986; Swartzman & Kaluzny, 1987), particularly aquatic ones (e.g. Jørgensen & Gromiec, 1989). Only a few books cover reservoir limnology specifically (Henderson-Sellers, 1984; Straškraba & Gnauck, 1985; Henderson-Sellers, 1992).

2.2.1. Goals of modelling

There are three major goals for mathematical modelling in reservoir limnology:
1. The theoretical advancement and understanding of inter-related processes in complex and dynamic reservoir ecosystems.
2. Prediction of future states of the system, its states when inputs are changed or when environmental conditions are altered. A frequent request is for prediction of conditions in reservoirs formed by newly built dams.
3. Management decisions about water quality and quantity.

2.2.2. Modelling philosophies and approaches

Modelling methodology is far from uniform. There are a number of opposing approaches which may be adopted:
- black box versus simulation modelling,
- stochastic versus deterministic models,
- continuous versus discrete,
- with quantitative as against qualitative features and
- mechanistic versus evolutionary models.

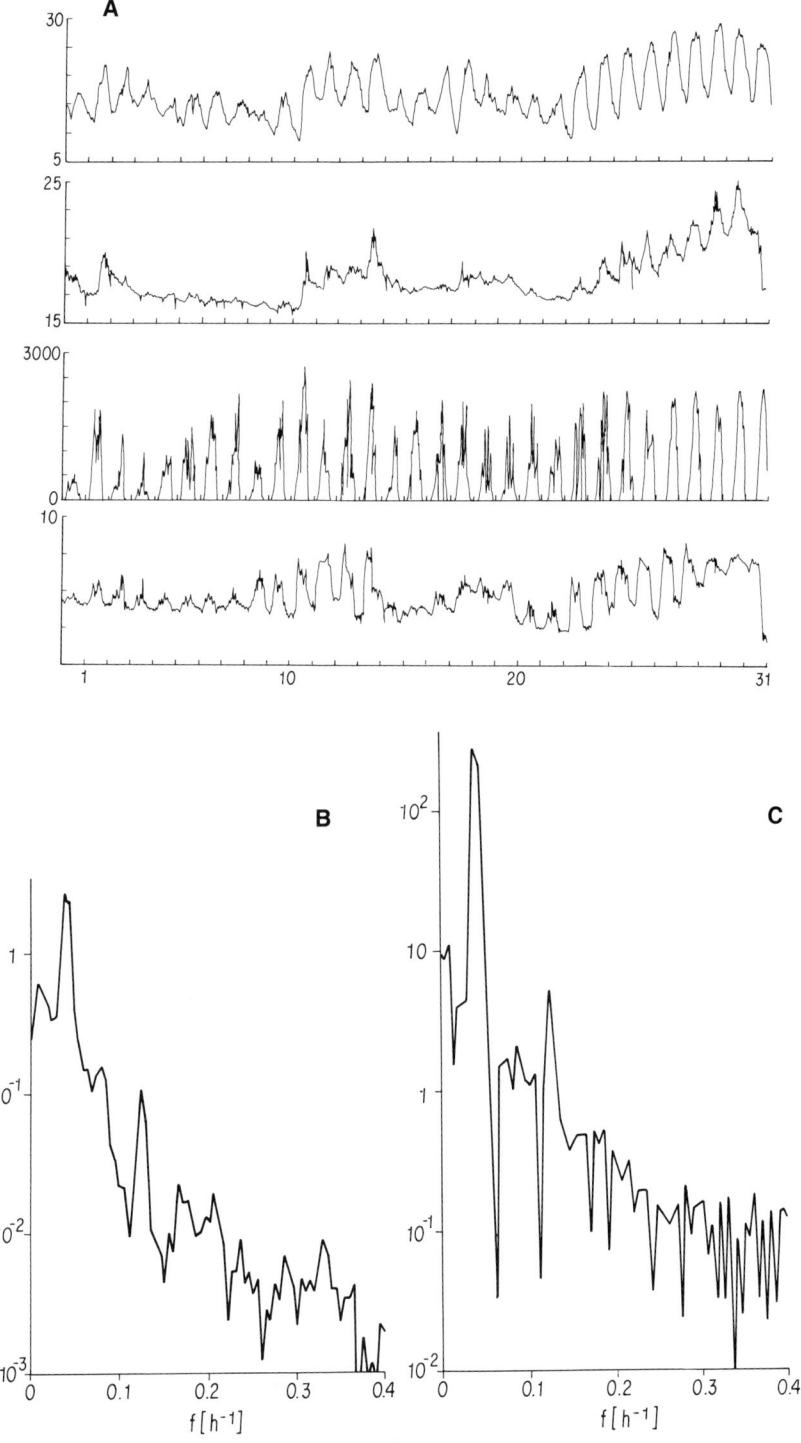

Fig. 12. Short term variability of meteorological and limnological variables. Results of automatic monitoring at Slapy Reservoir (Czechoslovakia) at half hour intervals for a selected period during July 1978. A – Variation of air temperature (T_A), surface water temperature (T_W), solar radiation (S) and pH at the reservoir surface. B – Power spectra (time series analysis of the same data) for pH. C – Same as B for T_W. The peak of daily frequency is much more pronounced for T_W, a peak at 8 hr frequencies characterizes both signals and frequencies up to 2 hrs are observed. From Nesměrák & Straškraba (1985), modified.

Which of the various methodologies we adopt depends both upon the nature of the problem and our habits and/or our level of understanding. There are no simple rules for finding and selecting the optimal approach.

(A) In the *black box approach*, the subject of the study is seen as a box with a known input and a measured output, without reference to particular processes which lead to the transformation of the input signal into an output signal. The methods commonly used for characterising the transformation of inputs into outputs are based on statistics and particularly the procedures involving simple and multivariate regression analyses and time series analyses. Examples relevant for reservoir limnology are the eutrophication models of Vollenweider (1968) and the regression models of oxygen conditions in reservoirs (Gnauck, 1975; Gnauck *et al.*, 1976; and summary in Straškraba & Gnauck, 1985). *Simulation modelling* differs in incorporating the internal processes mediating the system reaction to external inputs. Thus, the black box approach deals with extensive data sets whereas simulation modelling is more theoretically oriented where data is only needed to characterise specific situations or particularities of the processes deciding the system reaction. One common synonym of the simulation model is the mechanistic model, which refers to the internal mechanisms or processes. Simulation models of aquatic systems are usually expressed as a series of ordinary or partial differential equations and are usually solved by numerical techniques. Such models are by far the most dominant type of ecological model, as can be seen in recent numbers of the journal Ecological Modelling. From 1985–1988, about 60% of the published contributions used simulation modelling. Some examples of reservoir simulation models are given in Table VII.

(B) *Nature is stochastic*. This stochasticity lies in both the unknown and unrecognised sources of variability of ecosystem responses to specific inputs and the true stochasticity of the signals (sometimes called white noise). Random variability may be due to some uncertainty surrounding the system parameters and some associated with the dominant processes. Stochastic behaviour leads to a deterministic response in the case of the validity of the law of great numbers. There is some dispute as to whether this is true for natural ecosystems. However, for many theoretical and practical matters, the deterministic aspect of signal variability is of utmost importance. A model is *stochastic* if it incorporates this variability but is *deterministic* when variability is ignored and nature's response is represented by some non-varying average value. A deterministic model can become a stochastic one by allowing its parameters to vary within some pre-determined limits and by repeating the solution many times. This is a simple methodology much used by engineers to estimate the confidence limits of nature's response to inputs. An example of a deterministic model of a reservoir system is Parker (1968) and its stochastic version was studied by Parker (1974). Both black box and simulation models can be treated deterministically and stochastically, although the statistical procedures employed in black box modelling determine both the deterministic and stochastic components.

(C) A *continuous* process, or rather the continuous representation of a process, is one that uses infinitesimally small increments of time as a basic unit. The mathematical representation is by differential equations. This is an adequate representation for certain physical and chemical processes and for the activities of micro-organisms. In fact, most processes are *discrete* and discontinuous in that events occur only in discrete time steps. For example, an organism reproduces during its life cycle only when it is mature. One possible mathematical description of this is by difference equations. Recently, event-oriented methods originally developed for factory production are now being used in ecosystem modelling (e.g. Volokhonsky *et al.*, 1980; Hogeweg, 1988).

(D) *Quantitative* features of natural systems are often represented in models; this is true for the models discussed in points (A), (B) and (C) above. However, it is often the case that the quantitative response has not been deter-

Table VII. Some prescriptive models, management models and decision support systems applied or applicable for reservoir management.

Year	Author	Goal
		WATER RESOURCE SYSTEMS
1975	Haimes *et al.*	multiobjective optimization in water resources systems
1977	Findeisen	control methodology for water resource systems
1977	Haimes	hierarchical analysis of water resource systems
1981	Loucks *et al.*	monograph on water resource systems
1991	Bensari *et al.*	selected computational methods for water resource systems
1991	Brebbia *et al.*	models for water resource planning and management
		WATERSHEDS
1978	Knisel	system of models for evaluating non-point source pollution
1980	Krenkel & Novotny	water quality management models
1980	Procházková	relation of nitrate concentration increase to mineral fertilization
1981	Duncan & Rzoska	land use impact on lake and reservoir ecosystems including use of models (symposium)
1981	Novotny & Chesters	handbook of nonpoint pollution
1982	Zwirnmann	nonpoint sources of nitrogen from municipal water supply sources
1982	Haith	agricultural nonpoint source pollution
1982	Golubyev & Shvytov	agricultural effects on water bodies
1986	Giorgini & Gonzales	models of agricultural nonpoint source pollution
1987	Braat & Lierop	review of economic-ecological models
1987	Heatwole *et al.*	basin-scale water quality model
		RESERVOIR HYDRODYNAMICS
1969	Slotta *et al.*	modelling the effect of reservoir inflows on stratification
1970	Orlob & Selna	onedimensional reservoir stratification model based on slices
1973	Markofsky & Harleman	the most widely used semi-empirical model of stratification in a specified reservoir
1978	Henderson-Sellers	simple modell of vertical temperature variations in lakes and reservoirs
1978	Imberger *et al.*	Modell DYRESM for the specified reservoir with much improved theory over Markofsky & Harleman
1980	Ford & Stefan	thermal prediction based on an energy budget
1981	Imberger & Patterson	DYRESM5, one-dimensional model with improved inclusion of mixing processes
1982	Harleman	review of hydrothermal conditions in reservoirs
1982	Imberger	Model DYRESM6, improved version of DYRESM5
1983	Edinger & Buchak	LARM2 – a longitudinal-vertical, time-varying hydrodynamic reservoir model
1983	Orlob	review of numerical simulation models of reservoir thermics
1985	Imberger	review of processes of reservoir thermics
1986	Gray	review of physics based reservoir modelling
1986	Henderson-Sellers	review of surface energy budget for reservoir modelling
1986	Spigel *et al.*	modelling daily changes of the mixed layer
1987	Henderson-Sellers	review of one-dimensional stratification models
1987	Patterson	model for convective motions in reservoir sidearm
1988	Henderson-Sellers	sensitivity of thermal stratification models
1988	Patterson & Hamblin	modelling thermal stratification during winter ice cover
1989	Henderson-Sellers	review of thermal stratification modelling
this volume	Henderson-Sellers	hierarchy of reservoir models
		RESERVOIR WATER QUALITY
1973	Lorenzen & Mitchell	model for artificial destratification and its effect on algal production in reservoirs
1975	Kozerski	temperature and hypolimnetic oxygen in reservoirs
1975	Benndorf *et al.*	optimal reservoir size for P elimination
1975	Patten *et al.*	detailed species model of a reservoir embayment community
1976	Straškrabová	organic matter budget in relation to load, retention and mean depth

Table VII. Continued

Year	Author	Goal
1976	Gnauck et al.	recursive regression model of DO conditions in Saidenbach Reservoir
1976	Straškraba	AQUAMOD1, one-layer dynamic reservoir eutrophication model
1979	Park et al.	MS CLEANER dynamic simulation model of the reservoir ecosystem
1979	Straškraba	AQUAMOD2, two-layer reservoir eutrophication model
1980	Dvořáková & Kozerski	three-layer model, including sediment (AQUAMOD3)
1980	Jørgensen	review of lake and reservoir management and the use of models
1980	Stefan & Hanson	model for minimization of internal nutrient recycling
1981	Andrews et al.	impact of acidification on lakes and reservoirs
1981	Biswas	summary of modelling for water quality management
1981	Fontane et al.	optimal control of reservoir discharge quality through selective withdrawal
1981	Matsumura & Yoshiyuki	optimization of P-reduction of Lake Biwa
1981	Thérien & Spiller	decomposition of flooded vegetation
1981	Swartzman et al.	power plant impact assessment
1981	Canfield & Bachman	prediction of TP concentrations, CHA and Secchi depth in reservoirs
1981	Park et al.	predicting fate of coal-derived pollutants
1982	Kalčeva et al.	multiparameter optimization of measures against reservoir eutrophication
1982	Schindler & Straškraba	GIRL OLGA – dynamic optimization model for eutrophication abatement
1982	Duthie & Ostrofsky	use of phosphorus budget models in reservoir management
1982	Kerekes	application of phosphorus load – trophic response models to reservoirs
1982	Stefan et al.	RESQUAL II – shallow reservoir water quality model
1982	Anonymus	CE-QUAL-R1 – numerical one-dimensional model of reservoir water quality
1983	Beck & Van Straten	Uncertainty and forecasting of water quality
1983	Bedford et al.	CE-QUAL-RIV1 – combined hydrodynamic water quality model for reservoirs and their outflows
1983	Hoyer & Jones	empirical model of chlorophyll-a
1983	Orlob	review of water quality models for lakes and reservoirs
1983	Reckhow & Chapra	review of empirical lake and reservoir eutrophication models management models
1984	Groeger & Kimmel	organic matter retention in reservoirs
1984	Ortiz Casas & Martinez	applicability of the OECD eutrophication models to Spanish reservoirs
1985	Riley & Stefan (see also Hanson et al. 1986))	MINLAKE, an extension of the lake reservoir eutrophicatioon model RESQUAL II
1985	Martin et al.	MORDOR, a model of reservoir dissolved oxygen resources
1985	Grobler	models of phosphorus in South African Reservoirs
1985	Straškraba & Gnauck	systematic treatment of aquatic ecosystem and water quality management models
1985	Scott et al.	dissolved oxygen model
1985	Walker	review of empirical methods for predicting reservoir eutrophication
1986	Anonymus	CE-QUAL-R1 (see Anonymus 1982) – users manual
1986	Salas & Limon	empirical models for the estimates of P-load for tropical conditions
1986	Van Straten	identification, uncertainty assessment and prediction in lake eutrophication
1987	Loehr	model of hydrodynamics and primary production of an impounded river
1987	Lung	model of reservoir / lake acidification
1988	Reckhow	empirical models for trophic state in USA reservoirs
1989	Leonov	simulation model of water quality in the Ivankovo Reservoir
1990	Filho et al.	water quality management through hydraulic structures
1990b	Tundisi	perspectives of ecological modelling of tropical and subtropical reservoirs
1991	Jørgensen	modelling environmental chemistry
1991	Jørgensen et al.	handbook of ecological parameters
	FISHERIES	
1954	Ricker	classical models for fish population management
1957	Beverton & Holt	classical models for fishery management
1968	Jenkins	empirical models for fish yield in US reservoirs
1971	Jenkins & Morais	extension of the model by Jenkins, 1968
1976	Clark	population management models

Table VII. Continued

Year	Author	Goal
1976	Jørgensen	model of fish growth
1979	Skaletskaya *et al.*	optimization of fishing industry
1980	Goh	mathematical formulation of optimization models of fish catches
1981	Vincent & Skorkowski	symposium on models for fishery and other renewable resource management
1982	Jones & Hoyer	sportfish harvest predicted by summer chlorophyll-a
1984	Marshall	summary of predictive models of fish yield in African reservoirs from preimpoundment and physico-chemical data
1987	Reckhow *et al.*	empirical models of fish response to lake acidification
1987	Clark	economic fisheries models
1988	Saila *et al.*	microcomputer programs for empirical model of fish yields, catches and population dynamics
		RECREATION
1975	Dillon & Rigler	model for cottage capacity of a lake

mined, due to the complexity of the system, and the only statement we are able to make is qualitative. For example, under certain conditions, some species will not reproduce at all or reproduce weakly or strongly. On solution for coping with such variable responses is to adopt a fuzzy modelling technique (Bosserman & Ragade, 1982). This approach was adopted by Belayev (1980) for aquatic systems.

(E) *Mechanistic or simulation* models emphasis mechanics in the physical sense, with prescribed fixed reactions instead of having variable parameters representing the variable structure of natural ecosystems with a capability of adaptation and self-organisation. Models where the system responses are allowed to develop side by side with changing conditions of existence are generally called *evolutionary* or *cybernetic* models. These incorporate the possibility of decision-making within the model, that is, to change its own parameters or structure according to some rules. Optimality is the principle most often used for deriving appropriate decisions.

2.2.3. Model use

Table VII lists some usages of models in relation to reservoirs. The three goals for modelling outlined above are represented in Table VII in various models to a differing degree. There is some overlap between goals and individual models.

An example of the simulation model of reservoir hydrodynamics is the model DYRESM by Imberger & Patterson (1981) used in Chapter IV (this volume). To demonstrate the capability of the model we present in Fig. 13 the confrontation of the calculated and predicted annual pattern of the depth distribution of temperature in Slapy Reservoir, Czechoslovakia (Straškraba & Hocking, in prep.). This model uses actual reservoir morphometry and observed hydrometeorological data as input. In Section 3.2. the model is used as a tool to investigate some theoretical questions of physical reservoir limnology.

Whole lake/reservoir ecosystem models are treated in Chapter VI and VIII (this volume). Theoretical questions such as the competitive ability of blue-greens or the influence of changed phosphorus load on the biological and chemical variables are investigated. A number of recent models concerned with water quality deal with eutrophication.

3. Comparative reservoir ecology

3.1. Reservoirs in the River Continuum

According to the *River Continuum Concept* proposed by Vannote *et al.* (1980), there exists a continuous gradient of physical conditions from head to mouth of undisturbed river systems which evoke what they call 'a continuum of biological

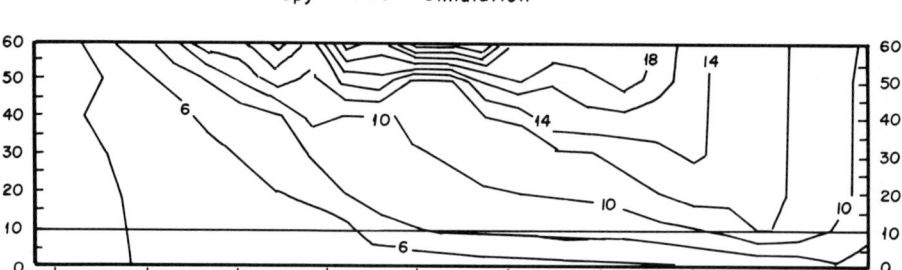

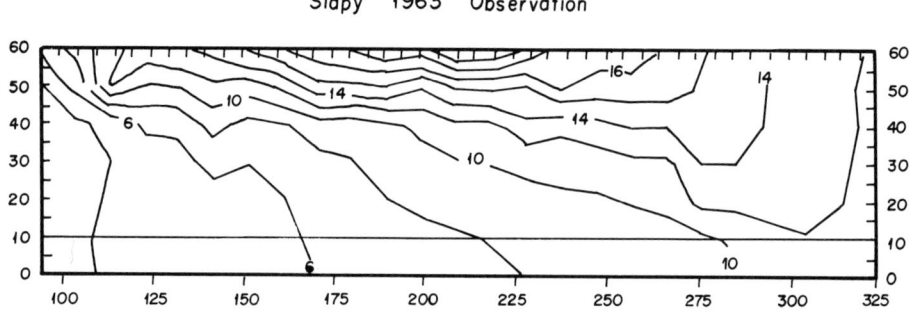

Fig. 13. Confrontation of results of a hydrodynamic reservoir simulation with reality. Model DYRESM applied to Slapy Reservoir (Czechoslovakia). On the x axis days of the year, on the y axis depth in m. The contours represent isolines of temperature. Orig.

adjustments' along the length of the river. In other words, stream communities become adapted to local conditions and establish some kind of balanced stability with the dynamic physical circumstances. This involves a continuous replacement of species both downstream and in time (in relation to a piece of moving water) thus establishing communities with differing patterns of handling energy and organic matter inputs. In general, downstream communities tend to make more efficient use of inputs than upstream ones by better resource partitioning and maybe, they have evolved to capitalise upon upstream inefficiencies. Another tendency is an evolution towards a more uniform rate of energy processing in the river as a whole throughout the year. The River Continuum Concept provides a theoretical framework for studying and understanding the functioning of undisturbed lotic systems.

This theory is relevant to reservoir limnology because the position of a reservoir on a river affects several of its basic characteristics. One practical consequence of the theory is the classification of twelve stream orders within the sequence of the environmental gradient. Of course, when a reservoir is built anywhere along the river length, the physically graded and biologically integrated River Continuum structure and functioning is disrupted to a greater or lesser degree. According to Ward's & Stanford's (1983) *Serial Discontinuity Concept*, the downstream effects of a reservoir is determined by its position along the river, that is, by the stream order. Some distance below the dam, pre-impoundment riverine conditions may be restored at a 're-set distance' which estimates the recovery distance for a particular set of variables and which expresses the degree of perturbation involved (Ward *et al.*, 1984).

From the viewpoint of the reservoir, the location of the dam in relation to its position in the stream order sequence and its elevation determines several reservoir features of importance, namely, rates of flow, landscape patterns in the river valley, temperatures of inflowing streams, level of insolation, degree of turbidity and so the underwater light regime, nutrient chemistry and so the reservoir biota. The following description shows major differences between reservoirs located on streams of different orders:

(A) A reservoir located in the upper reaches of a river and on a low order stream is usually fed by a small creek with certain predictable characteristics: low flow, low temperature, low organic matter loading, low nutrient salts, no plankton and with a characteristic assemblage of fish species. The reservoir will be located in a deeply incised valley with extreme declivity of its bottom slope. Its mountainous position implies a range of lower temperatures but higher humidities and precipitation and higher levels of insolation. This is a generalised picture relative to the middle or lower reaches of this particular region since streams of the same order but different geography will have different characteristics. Man-made effects from pollution or excess nutrient loadings will also modify the nature of the stream. Such a reservoir can only be of one type, namely, a deep, stratified flowthrough oligotrophic one. Horizontal gradients will be weak or non-existent. Any differentiation between such reservoirs in the same geographical region will be due to geology (calcareous or non-calcareous underlying rock) or degree of exposure to sun and winds (affecting temperature and mixing).

(B) Building a reservoir in the middle reaches of a river means that the river has an average flow, moderate declivity, average temperatures, increased loadings of natural organic matter and nutrient salts, often with some turbidity, a developed phytoplankton and an assemblage of fish that can survive in standing water. The limnology of the reservoir will depend mainly upon the morphology of the valley, namely whether it is shallow and unstratified or deep and stratified; also on the theoretical retention time which is influenced mainly by the size of the water body. Retention time can vary within wide limits. In small reservoirs with low retention times, horizontal gradients will be low, stratification not very pronounced and planktonic biomasses not well developed. Larger reservoirs with high retention will exhibit well-developed horizontal and vertical gradients of physical and chemical variables, reasonable growths of plankton and a lake-type fish assemblage.

(C) Reservoirs constructed on large lowland rivers with very low declivity will consist of one particular type which is characterised by inundation of large areas, extreme horizontal variability with well-developed wetlands and extensive shallows with riparian vegetation. Such reservoirs will be eutrophic, with high natural organic loadings likely to contribute to the formation of an anoxic hypolimnion. However, any shallow basins will be well mixed by wind and stratified conditions will only develop where depth exceeds the depth affected by wind-mixing.

3.2. Retention time as a key factor in reservoir limnology

The theoretical retention time R is defined as

$$R = V / Q \text{ (in days)} \qquad (1)$$

where V is the volume in m³ of water in the reservoir and Q is the average annual flow in m³ d⁻¹. Retention time is also known as water residence time, hydraulic detention time, retention rate or flushing rate.

If the retention time is calculated for periods shorter than a year, it is usual to indicate this: R_{IV-VII} is for April-July. However, this is an oversimplification since both V and Q will vary throughout the year, both seasonally and due to reservoir function. There may be particular occasions when it is better to calculate an average R from the mean of daily R's over a period of time. R is a purely theoretical concept which, for example, tells us nothing about the average retention times of water particles within the reservoir. There may be occasions when water parcels traverse the path from inflow to outflow in times much shorter than the theoretically calculated R. Nevertheless, R is a measure useful limnologically and permits comparison between reservoirs. For example, reservoirs are some kind of and intermediate state between a unidirectionally flowing river and 'still water', exemplified by a lake with diffuse inputs and a small stream output; R expresses nicely the degree difference in flowthroughput although not the difference caused by the position of the outflow (surface in lakes and deeper in reservoirs).

Given the same geography, R gives an important expression of limnological differences between reservoirs (Straškraba, 1973). This is more true of deep stratifying reservoirs than shallow ones as the riverine throughflow has a more drastic mixing effect in the former than latter. In lakes, wind energy and thermal-density gradients are the dominant physical driving forces influencing thermal stratification whereas in reservoirs the physical force of flow is added to these and may be strong enough to suppress their effects. The importance of wind and density gradients probably increases with the value of R, as in natural lakes these two forces are overwhelming. Moreower, retention time is recognized as a useful ecological indicator also for rivers and lakes (Søballe & Kimmel, 1987), particularly in respect to phytoplankton.

Dam reservoirs exhibit typical macroscale features imposed upon them by the unidirectional water flow and seen in a serial zonation: riverine zone-transitional zone-lacustrine zone (Chapter III, Fig. 1). The size of these zones varies in particular reservoirs and depends upon factors such as morphometry, retention time, thermal stratification, season and geographical location. In deep temperate reservoirs during the summer, the extent of these zones depends very greatly upon retention time. Thus, when R is less than 10 days, the whole reservoir may become a river zone whereas with an R of more than 200 days, the riverine zone and most of the reservoir lies in the lacustrine zone. Figure 14 illustrates for a reservoir with and $R < 10$ days the homogeneity of flow rates and temperature distribution both vertically and horizontally. However, the critical value of $R < 10$ is only approximate, the more precise criterion for development of stratification being:

$$320 \, Q \, L/(V \, \bar{z}) < 1/\pi \qquad (2)$$

where Q and V are as before, L is the reservoir length and $\bar{z}$ the mean depth.

In a reservoir with approximately $10 < R < 100$, thermal stratification will develop, with vertical separation of the wind-mixed surface layers but

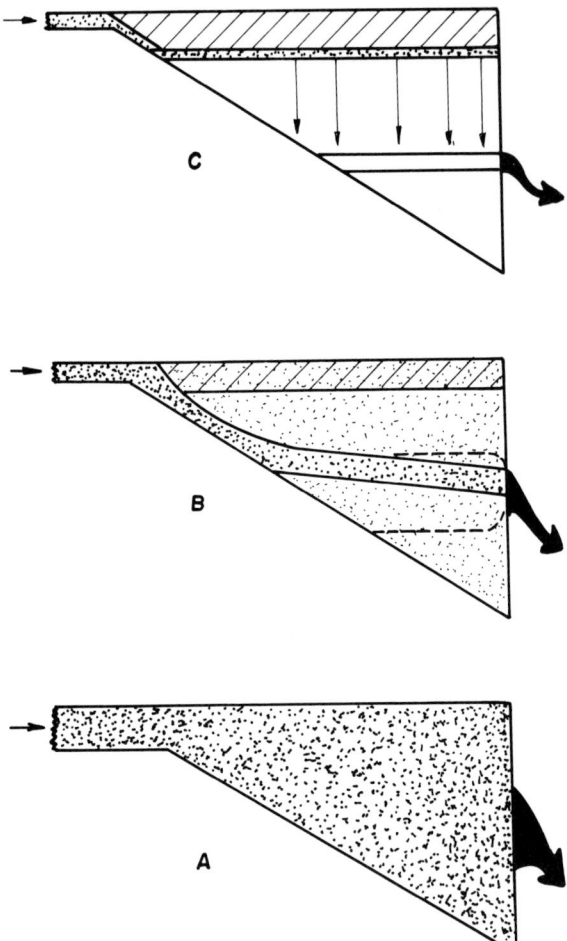

Fig. 14. For a deep temperate reservoir, in summer the dependence of horizontal and vertical (depth) differentiation on retention times is very pronounced. A – short retention time, rapidly flowingthrough reservoir ($R < 10d$) with no differentiation. B – A riverine throughflowing reservoir ($10 < R < 100$) with a long inflow zone, short transition one and dominating lacustrine zone, and with a density underflow. C – Lake-type conditions with full stratification in a long retention time reservoir ($R > 100$). Orig.

also with continuing intensive mixing of the deeper strata as well as the separation out of the three zones just described. In some cases with increase in inflow rates the riverine water forms a density current or underflow and passes down straight to the depth of equal density and rapidly out to the outlet. This is illustrated in Fig. 14; the tilting of the underflow current that is indicated was never measured but assumed to occur as it mixes with the deeper colder water layers rather than the upper warmer ones. In contemporaneous models no tilting is assumed. In a fully stratified reservoir, with $R \gg 100$ days, a typical lake stratification is developed but with inflowing water parcels being transported deeper to replace the volumes of deep water passing out of the deeply sited outlet. Sometimes the boundary between the transitional and lacustrine zones is sharpened by 'down-plunging' of the inflowing riverine water (Chapter XII, Fig. 18). The location of this boundary will not be fixed but will depend upon water levels and water flows, that is upon R. The difference between reservoirs with different R's is reflected in their chemistry and biology as well as water movements, stratification, horizontal zonation, vertical mixing and internal flows. The resulting hydro-dynamical characteristics will be decisive for the behaviour of conservative substances not readily changing. This is not so for non-conservative substances and living organisms of the plankton which will not only interact biotically but also react to the mixing regime and water flow or to the light conditions. Larger organisms like fish also react to light, temperature, dissolved oxygen conditions and food availability and so, indirectly, are also affected by R.

3.2.1. Effects of R on physical reservoir limnology

The above scheme of the relations between R and reservoir hydro-dynamics has been quantified empirically and theoretically analyzed by mathematical modelling. Whilst empirical observations are precise, they are valid for only a limited set of conditions whereas the outcomes from mathematical models are less precise because detail is omitted and because stochastic variability is not easily covered but the conditions for which the model is applicable may be specified. This is not the only advantage of mathematical models; they can also be used to simulate the effects of different variables, one by one, or to explore the course of processes not easily measured, or to speed up data generation. In the senior author's experience, one 2 man – 3 month reservoir modelling job generated more data on its hydrodynamics than did ten years of measurement!

3.2.1.1. Empirical observations of R in slapy Reservoir.
There are two approaches to studying how retention time can affect stratification: one involves long-term observations of flows and water

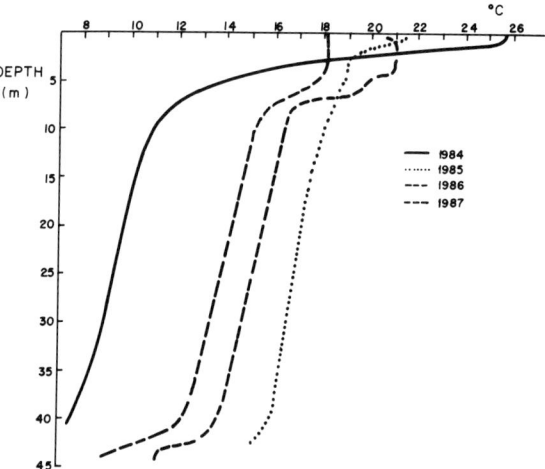

Fig. 15. Depth profiles of full summer temperatures in different years at Slapy Reservoir (Czechoslovakia). The most dry year has lowest near-bottom temperatures, the most wet year the highest near-bottom temperatures. Orig.

levels in a series of reservoirs with different conditions.

Slapy Reservoir is used here as an example of the first of the above approaches because it has been so closely observed. It was also suitable because of its variable retention times, $15 < R < 90$ days. Figure 15 illustrates how different were the temperature depth profiles over the years. In the dry year (1964), a strongly developed but shallow thermocline developed due to high surface temperatures and low deep ones. In the following wet year (1965), temperature was distributed uniformly with depth and the 1965 bottom temperature (16 °C) was almost double that of 1964 (8 °C) and the surface temperature managed to reach 21 °C when the sun shone. In other years (1966 and 1967), intermediate temperature distributions and deeper thermoclines were recorded. A cold monimolimnetic layer behind a submerged wall was also found which may mix in with the layers above in some years.

The length of observations and a good coverage of more extreme flow rates permitted regression analysis of the relationship between temperature and flow rate (which is indirectly proportional to R). Some relationships are shown schematically in Fig. 16. Although surprising, a (significant negative) relationship can be calculated between air temperature and flow rate which, on closer examination, was found to be due to less cloud and

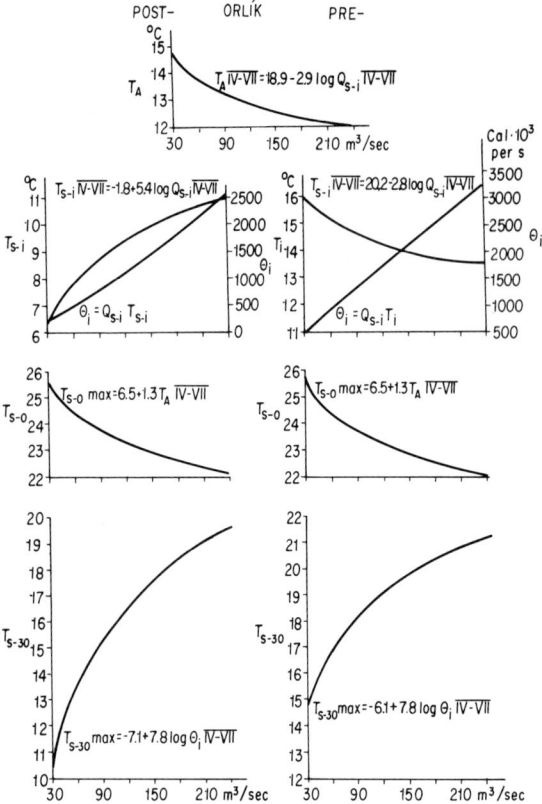

between wet and dry years are more pronounced. Bottom temperatures show an opposite trend: these increase rather than decrease with increased flow. From the given equation in Fig. 16, this is related to higher inflows of a higher temperature.

Figure 17 illustrates schematically three characteristic situations in Slapy Reservoir, namely, years when R was 15, 30 or 90 days during the period April-July (crucial period for the beginning of stratification). These three R's correspond to water flows of $\approx$ 60, 120 and 240 m^3 s^{-1}, respectively. Values for different parameters during the full summer are also given. One consequence of an increase in flow rate is that the unstratified riverine zone is prolonged. Any difference in flow and in the temperatures of the inflow and reservoir causes the 'down-plunging' jet

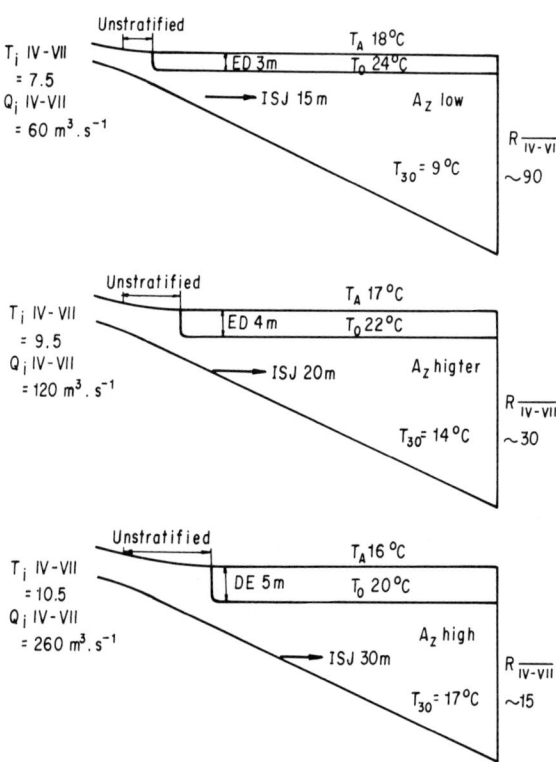

Fig. 16. Quantification of the dependence of characteristic parameters of temperature in Slapy Reservoir on flow (retention time). The left part shows situation after the upper reservoir was built, whereas the right part characterizes the solitary reservoir situation. Values are for the spring-summer period (April to July) or for the period of maxima. As shown, not only the maximum summer surface water temperatures ($T_{S-0\ max}$) but also the average spring-summer air temperatures (T_A) are related to flow. For inflow temperature (T_{S-i}) the negative dependence on flow is converted to a positive one after water flows from hypolimnion of the upperlying reservoir, as seen to exist for Slapy ($T_{S-0\ max}$). Θ_i is the thermal contribution of the inflow. Orig.

more sunshine appearing in the dry years. Differences in the heat budget between years with different flow rates seem to come back to a variation from 15 °C to 12 °C in the span of the average temperature for period April-July. Such observation seems specific for one reservoir but it may have some general validity for temperate regions. There already exists a known relationship between surface and air temperatures for ponds and lakes (Szumiec, 1975) which seem to be valid for maximal surface temperatures, higher than average air temperatures but also the differences

Fig. 17. Schematic representation of several variables during summer conditions during different years at Slapy Reservoir (Czechoslovakia). The years differ, first of all, in their average flow rates, only variables causally related to R being indicated. T_A, T_i, T_0, and T_{30} are the air, inflow, surface and deep water temperatures, respectively. Q_i – inflow rate, ISJ – inflow stream jet, ED – mixing depth and A_z – effective turbulent mixing. Orig.

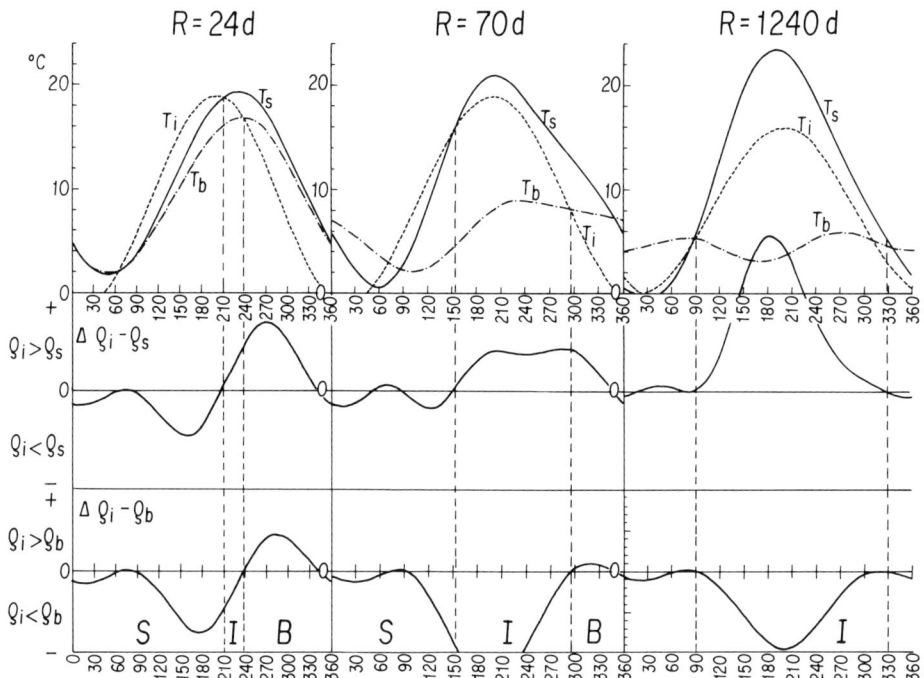

Fig. 18. Observed temperatures of the inflow, surface and near-bottom (T_i, T_s and T_b, respectively) and derived density differences and corresponding flow pattern in reservoirs with different R. For explanation see text. From Hejzlar & Straškraba (1989).

stream to go deeper, so that the mixing depth (ISJ) becomes deeper and the mixing intensity (A_z) increases.

Further discussion on R in Slapy Reservoir is given in Section 3.6. Chapter XII summarizes some results of the comparative studies on stratification in Czech and Bavarian reservoirs with different retention times. It seems that at least reservoirs in certain geographical regions the degrees of thermal stratification is closely related to R. The rather more sophisticated RESTEMP model in Chapter XII is capable of predicting the temperatures of a temperate reservoir for any depth on any day of the year, given information about surface elevation and R.

3.2.1.2. Theoretical model analysis. From sinusoidal curve fits of empirical temperature observations, it is possible to analyze the annual changes of reservoirs with different R characteristics. Figure 18 illustrates the observed temperatures for three real situations (Slapy Reservoir in 1964 and 1965 and Klíčava Reservoir in 1964) and the corresponding density differences predicted from Eq. (2) in Chapter XII. From this is derived the existence and duration of surface (S), bottom (B) and intermediate (I) flows. As R increases, the duration of the surface and bottom flows reduces and that of the intermediate flow becomes dominant. Figure 19 gives a schematic representation of the horizontal and vertical distribution of temperature, current speed and current location near the dam wall of temperate reservoirs during characteristic periods in the year and for typical levels of R. These are typical analytical results relating R to reservoir hydrodynamics based on empirically derived, simple models.

More sophisticated hydrodynamic models can predict the annual time-course of temperature from reservoir shape and outlet depth, specified components like solar radiation, temperature (air and inflow), optical quality of water and some other meteorological variables. Model DYRESM provides solutions similar to empirical observations for the same reservoir and meteorological conditions (Straškraba & Hocking, in prep.). Figure 20 illustrates that the maximal temperature difference between surface and bottom, calculated from a number of simulations (where one point

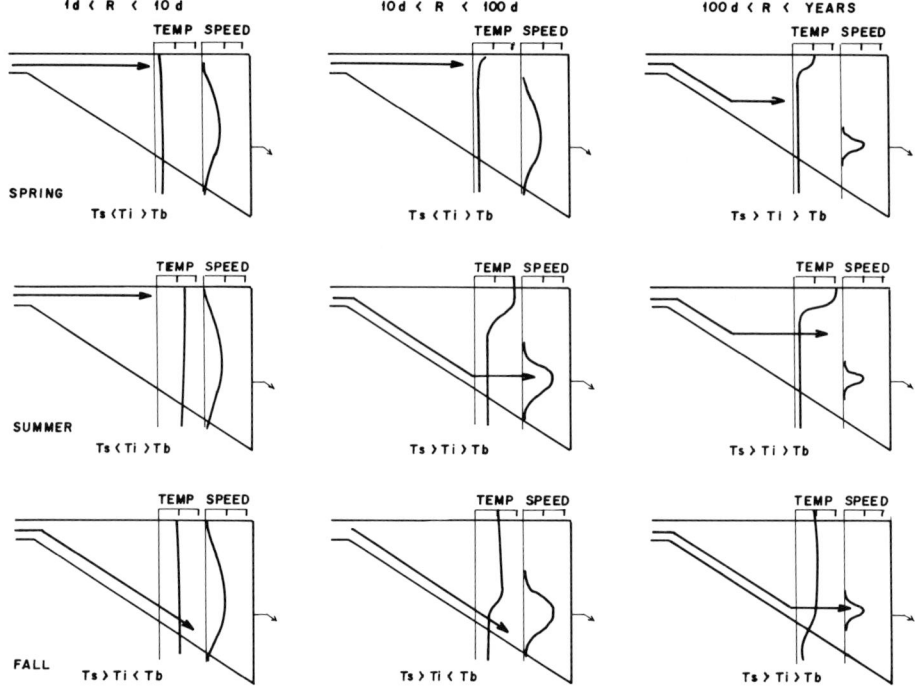

Fig. 19. Schematic representation of the annual pattern of stratification and flow (direction and speed) in reservoirs of different *R*. From Hejzlar & Straškraba (1989).

represents a simulation for the whole year), is identical to measured differences. The model differs from previous models in that it can be used to analyze the processes leading up to the final results themselves as well as to explore how different are the results when variables are changed. For example, two situations for Slapy Reservoir are illustrated in the figure: one when it has a deep outlet (as is so in reality) and another when surface water flows out, all other conditions being similar. The figure demonstrates that the position of the outlet significantly affects the results obtained, particularly for low and average values of *R*. Martin & Arneson (1978) report similar observation when comparing two reservoirs of the Madison River, Montana, one having a deep discharge and the other a surface discharge. Tundisi (1984) coined hydraulic stratification when in a reservoir stratification is caused by hydraulic structures.

Model simulations of the effect of *R* upon thermal stratification in a subtropical reservoir, Canning Reservoir Australia, shows similarities with the temperate situation (Fig. 20). For the surface water outflow condition, the sub-tropical dependence of the degree of stratification on *R* is steeper than in the temperate reservoir whereas, with a deep outlet, the effects are less pronounced and there is a wider range or *R*-values to which thermal stratification is sensitive. The surface-bottom temperature difference changes considerably between *R*-values of 400 and 800 days. In consequence, locating the outlet at the surface affects a subtropical reservoir more profoundly.

3.2.2. Effect of R on chemical variables

Retention time affects chemical composition and chemically-driven processes in reservoirs as a consequence of the fact that chemical loading must decrease with *R*. When concentration of a chemical substance remains constant in the inflow, i.e. its concentration is independent of flow rate than the decline of loading with increase in *R* is exponential:

$$LOAD = LOAD_0 \exp(-bR) \qquad (3)$$

where $LOAD_0$ is the theoretical load when $R = 0$ and *b* is slope of the relationship between Ln(*LOAD*) and *R*.

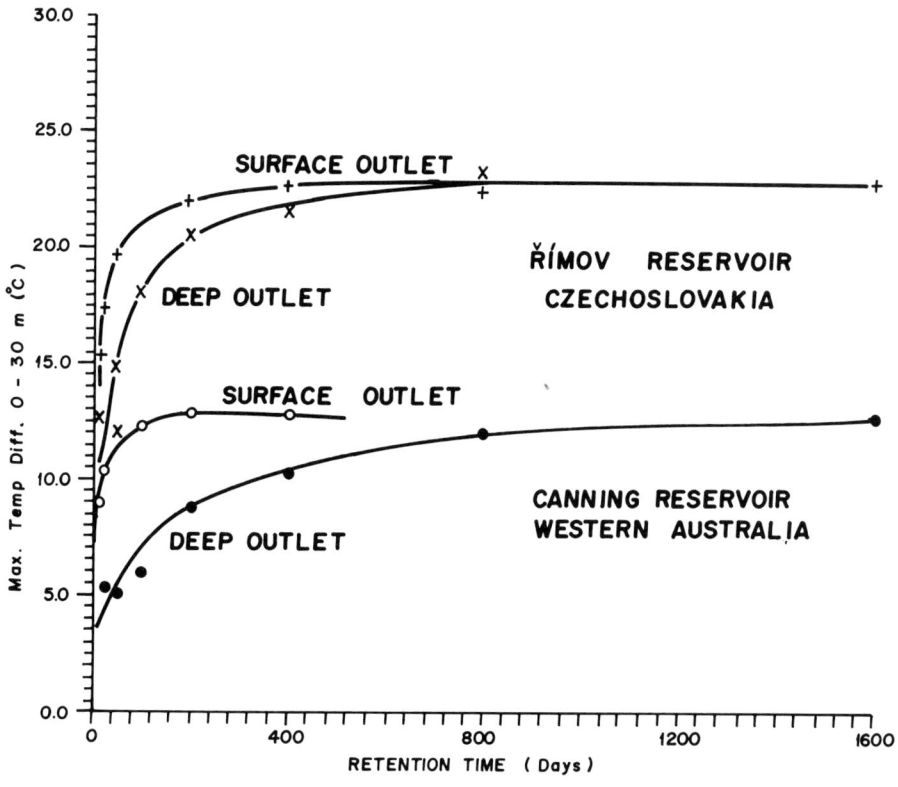

Fig. 20. Quantification of the effect of R on the maximum temperature difference surface-bottom for two locations of the outlet: surface one (a lake situation) and a deep one (a reservoir situation). Calculated with the model DYRESM for same meteorological situation, same volume but different assumed flow rates. Římov Reservoir is a temperate one while Canning a subtropical one. Curves fitted by eye. Orig.

In most instances, the inflow chemical concentration does depend upon R. If this dependence is positive, then loading will decline hyperbolically with R, as can be seen in Fig. 21 for BOD_5 in which the steepest change occurred when $R > 30$. This will apply to other chemical substances like phosphate and nitrate which have a positive relationship with flow.

The fate of a particular chemical substance in reservoirs is determined not only by its nature and quantity but also the water depths to which it has been dispersed, which is related to the R-value, as shown earlier. Thus, the fate of phosphorus as a critical nutrient depends largely on whether it is dispersed throughout the surface layers where phytoplankton can take it up or whether it ends up in the hypolimnion where active nutrient uptake does not occur. Particle sedimentation is another important process for the distribution of nutrient chemical which are influenced by retention times.

At a low retention time, little sedimentation can occur and most particles will be flushed out but particle sedimentation is important for reservoir chemistry when retention time is high.

Like lakes, reservoirs can trap both total and orthophosphate phosphorus efficiently through algal uptake of the dissolved nutrient followed by algal sedimentation. The efficiency of the phosphorus retention process in lakes has been modelled several times (Chapra & Canale, 1991). As expected, in reservoirs, it is strongly influenced by retention time (Schreiber & Rausch, 1979; Fiala & Vašata, 1982). Fig. 22 illustrates the situation in Western part of Germany (Based on data by Wilhelmus et al., 1978): when $R > 30$ days, 70–90% of total phosphorus was retained in the reservoir whereas, when $R < 30$ days, total phosphorus retention dropped to less than 10%. Turner et al. (1983) observed for the shallow reservoir Lake Talquin a decreased retention of

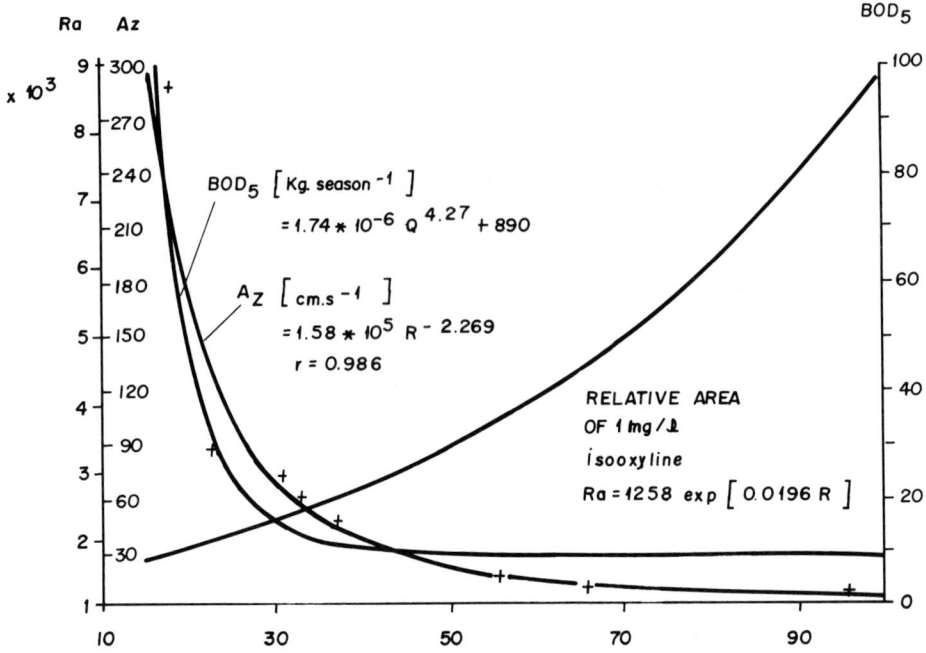

Fig. 21. Calculated empirical dependences of a few variables in Slapy Reservoir (Czechoslovakia) on *R*, based on many year investigations. Observation points are given only for the values of the effective turbulent mixing, calculated from summer temperature distribution of each year. The BOD$_5$ load as a measure of the external organic loading drops strongly with rising *R*. However, due to much reduced mixing and increasing internal organic matter production the oxygen conditions at the reservoir bottom (expressed as the relative area of the 1 mg l^{-1} isooxyline) increasis. Orig.

both total phosphorus and silica during years with low *R*. A decrease of *R* from 44 to 22 days produced a drop of total phosphorus retention from 65 to only 25%.

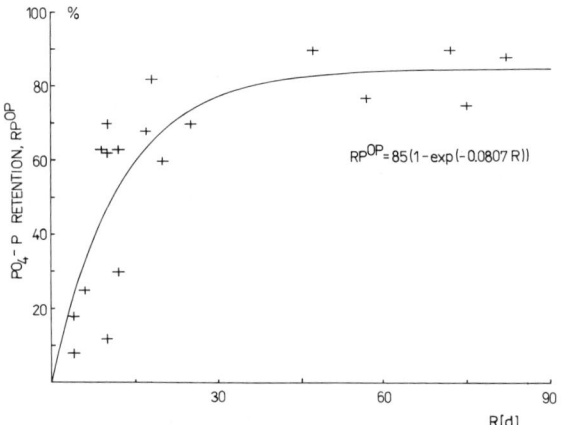

Fig. 22. Phosphorus retention of pre-impoundmens in Germany in dependence on *R*. Data from Wilhelmus *et al.* (1978). Orig.

3.2.3. Effect of R on aquatic biology

Uhlmann (1971, 1978) proposes that phytoplankton can be modelled as if a continuous culture: in continuous cultures, a constant input of nutrients supports algal growth at a rate that increases with the retention time due to the higher nutrient loading and despite the increasing mortality by washout. This continues up to a certain value of *R*, at which the gain by growth can no longer compensate the loss by washout and the whole culture is lost. In their studies on rapidly flushed lakes, Dickman (1969) and Dillon (1975) made corresponding observations in nature whereas Williams *et al.* (1977) reported on algal washouts occurring before maximal densities were reached when the reservoir had *R* < 14 days. This hypothesis was used by Benndorf *et al.* (1975) to develop a predictive model on the use of pre-impoundments for phosphorus retention and in which critical values of *R* for algal washouts were calculated for a series of conditions. Vollenweider (1975) included a term '*t*^{1}' (which is equivalent to *R*) in his well-known empirical eutrophication

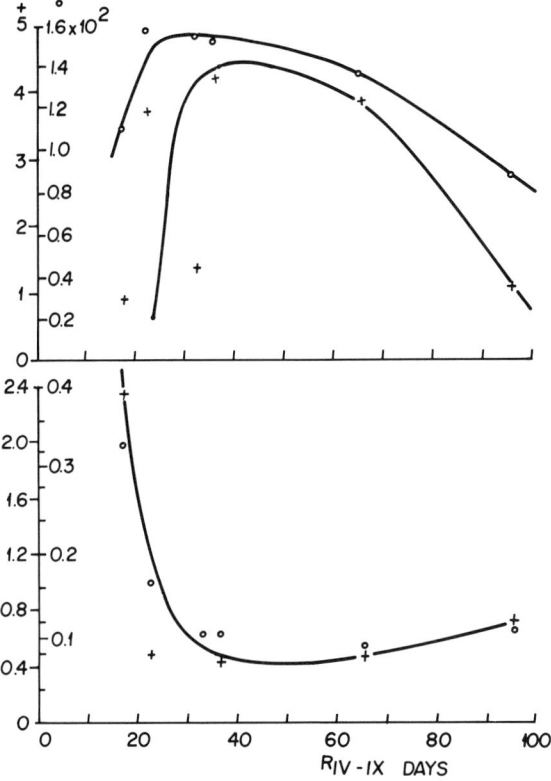

Fig. 23. Seasonal mean values of the parameters of algal productivity in Slapy Reservoir (Czechoslovakia) plotted against theoretical retention time for the April to September period. Upper panel: circles – gross algal production, crosses – algal fresh weight. Curves fitted by eye. Lower panel: crosses – algal growth rates observed, circles – algal growth rates calculated. The curve is fitted by eye to the observed values. From Straškraba & Straškrabová (1975).

models. Recent observations on negative correlations between phytoplankton biomass and R in reservoirs were made e.g. by Laberge & Mann (1976) and Turner *et al.* (1983).

Figure 23 illustrates that in Slapy Reservoir phytoplankton biomass decreased when $R < 30$ days, despite there being high algal growth rates. This has been interpreted as being due to high losses by flushing and to sedimentation. The low primary production of the phytoplankton was due to lowered biomasses which could not be fully compensated for by the high specific growth rates of the algae. In fact, the specific growth rates of the algae increased as algal biomass decreased, being density dependent (Javornický, 1980). Similar observation was made on Lake Francis by Søballe & Threlkeld (1985).

There is much less evidence that reservoir zooplankton is affected by retention time in any systematic manner although it is to be expected. One of the IBP hypotheses tested was that the efficiency of energy transfer from phytoplankton to zooplankton declined as retention time increased. Although not satisfactorily explained, observations from Czechoslovak reservoirs do show such a trend but with poor statistical significance, even with the use of double-log transformed values (Fig. 24). The study by Threlkeld (1982) seemed to suggest for 6 reservoirs an indication of decreasing zooplankton reproduction rate with increasing R, which might indicate a strategic repply of zooplankton to dillution. However, he pointed out complicating factors and moreower, only two reservoirs did have $R < 100$ days (minimum 28 days). Not much effect is to be expected in this set.

For benthos, the decisive factors is food availability as supplied by sedimentation of organic particles. This will be influenced by the shifts in the position and size of the riverine and transitional zones which will be the place for sedimenting of particles brought in by the river.

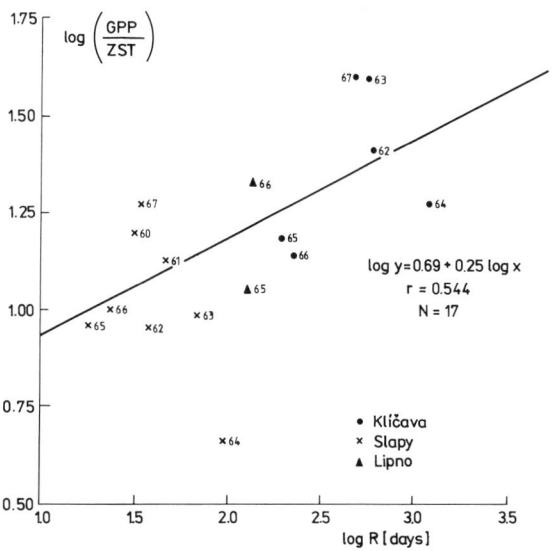

Fig. 24. A measure for the transfer efficiency between primary and secondary production (GPP – Gross primary production, ZST – zooplankton standing crop) plotted against theoretical retention time, summer averages for several year observations in three reservoirs in Czechopslovakia. Numbers relate to the years of observation. Based on summer averages. Orig.

The other source is sedimentation of epilimnetic algal production, also influenced by R.

The depth distribution of fish in a lake or reservoir has been shown by Straškraba (1974) to depend on stratification and thus also upon R (see earlier). Jenkins (1968) and Jenkins & Morais (1971) developed multiple regression equations to relate environmental factors to fish stocks and angler catches in American reservoirs. They found that retention time was a potent predictive variable along with total dissolved solids, reservoir depth and reservoir age.

3.3. Major geographical differences in reservoir limnology

The present generalisations about the latitudinal distribution of variables which affect lake or reservoir temperatures and thermal stratification are mostly based upon average values of the relevant parameters (Straškraba, 1980; Lewis, 1987). The values obtained for solar radiation, air temperatures, water budgets and wind speed are usually smoothed representation of annual variations over the whole globe and account is taken of neither the distribution of land or oceans nor of any local orographic particularities. When latitude is corrected for the geographical equator being positioned at 3.4° south, then the two hemispheres are found to be identical. Altitudinal variation was considered only for solar radiation and air temperature and only in relation to thinning of air masses due to increased radiation and adiabatic cooling with increased elevation; these causes are reflected in parallel decreases in the average surface temperatures of lakes with increased elevation. Surface temperature drop below 4 °C at 40° of corrected latitudes which makes this latitude important as a boundary for mixing pattern in lakes and reservoirs.

More insight is gained if not only average and seasonal trends are considered but also stochastic variability, given by inter-annual changes and short term changes. The present analysis, which is based upon a lecture by Straškraba (1989), takes a step in this direction but needs a more systematic analysis of gross geographical trends.

3.3.1. Major Limno-geographical Regions

Figure 25 summarises the picture obtained for solar radiation, air temperature, precipitation as well as water temperature, mixing depths, turbidity and the related light extinction for lakes and reservoirs with long retention periods of $R > 200$; also included is the underwater light regime associated with both level of turbidity and mixing. These estimates are based either on data analyzed in Straškraba (1980) without stochastic variability

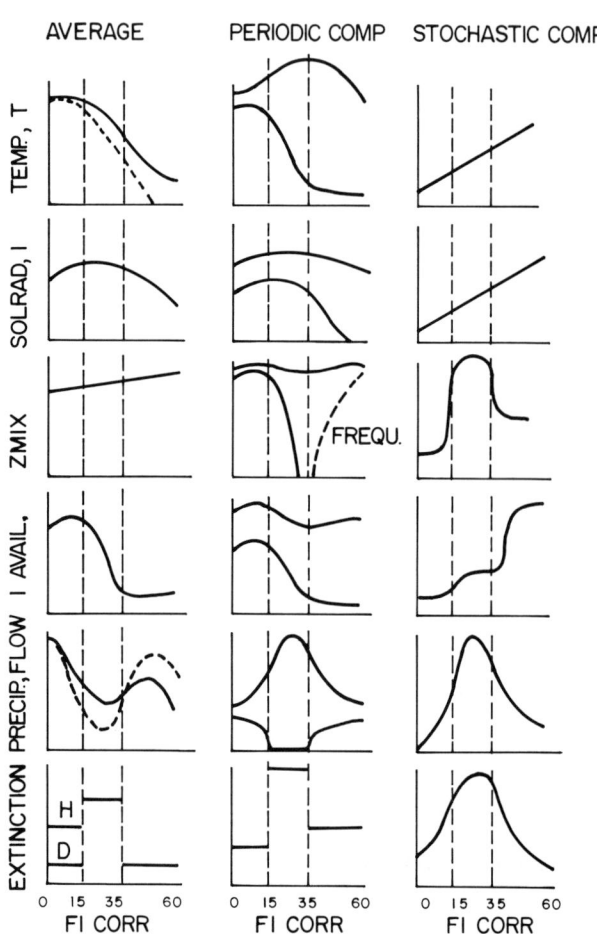

Fig. 25. An approximate estimate is given for the geographic distribution of the average values, of the periodic annual component and of the summarized stochastic component (degree of variability) for several variables most important from the point of view of reservoir and lake limnology: water and air temperature, solar radiation, mixing depth, light availability for the phytoplankton population, precipitation and outflows and light extinction of water due mainly to turbidity. H – humid tropics, D – dry tropics. The estimates are based on data in Straškraba (1980). Orig.

being considered or by Lewis (1987) plus some additional information culled from standard atlases or geographical summaries. For some variables, these estimates are only gross guesses largely gained indirectly. Altitudinal variation was neglected so that the figure is valid for low altitudes only. However, the results are strengthened by analyzed data from Chapters I and II of the present volume.

For *solar radiation*, the average values show an increasing trend up to 15° latitude, are almost constant up to 35° latitude and drop off at higher latitudes. Minima and maxima of the periodic component are rather stable (with identical annual variability) over a range of latitudes from 0 to 35° whereas, at higher latitudes, the range of regular annual variation increases rapidly, largely due to minima dropping to near-zero. From IBP data, it seems that stochastic variability shows a steady increase from the meteorological equator upwards. A similar picture is obtained for air and water temperatures.

For *average temperatures*, water and air surface values are almost identical at lower latitudes but diverge as latitude increases. Only water temperature has been analyzed for the periodic component but air temperature is probably similar as the only apparent difference is that minimal air temperatures continue to decrease whilst surface water temperature does not. The periodic component is expressed as the monthly averages of maximal and minimal surface temperatures. Minimal temperatures lie close to maximal ones between latitudes of 0° and 15° but the max-min difference increases with latitudes up to 35° and, at about 40°, the minimum temperature of the surface water remains at 4 °C. This is, of course, a crucial point for lakes despite Berg's (1963) demonstration that lakes like Cayuga in New York with surface temperatures of 4 °C need not necessarily freeze. Moreover, saline lakes will differ from freshwater ones in this respect. The stochastic component was calculated as the standard deviation of the regression estimate, the regression being a periodic one (Straškraba, 1980). Results were expressed as the ratio between the standard deviation and the mean for the locality and this ratio shows a linear increase with latitude.

For *precipitation and water flow*, representing the third kind important hydro-meteorological variables, there is an apparent extreme spread of both the periodic and stochastic components between latitudes 15° and 35°, with seasonal minima reaching zero. This is a feature that uniquely distinguishes this region from all others.

Light extinction, measured as the vertical light attenuation coefficient of visible wave lengths between 400–700 nm, can be considered as a composite variable since it is affected by mineral and organic matter suspended or dissolved in water. Although geographical differences in dissolved matter are known to occur (as in the black waters of the Amazon), there is no reason to attribute these to latitude. On the other hand the level of mineral turbidity is geographically dependent being closely associated with erosion due to rainfall, particularly with strong incidents of flash rainfall. Here, the stochastic element is rather high.

The variables *mixing depth* (z_{mix}) and *light available for the phytoplankton population* (I_{avail}) are inter-related, the later being the integral of visible light over the mixing depth. Mixing depth only relates to deep stratifying lakes and the deeper mixing in tropical lakes is a generally accepted phenomenon. The periodic component for z_{mix} indicates by means of two lines the annual minimum and anual maximum of mixing. The figure considers only two variables affecting I_{avail}, namely incident solar radiation and mixing depth, but mineral turbidity can be an important modifying factor which was discussed in Chapter I and which contributes greatly to the average periodic component and to stochastic variability. As a summary of the external driving variables affecting lake and reservoir limnology (in their average, seasonal or stochastic variable forms) and of the physical and chemical properties of lakes and reservoirs, Table VIII distinguishes three major limno-geographical zones.

Maybe a fourth zone exists in the arctic at latitudes above 60°. This should be a zone of continuously frozen lakes. Likens (1975) showed mixing below the ice.

To what extent are these lake-based geographical zones valid for reservoirs? The hydro-meteorological driving variables are probably the same for lakes and reservoirs but their consequences will be similar only in reservoirs with long retention times ($R > 200$ days). In reservoirs

Table VIII. Three major limno-geographical macro-zones.

Tropical region (0–15°)	
Radiation	Highest annual average radiation with very low seasonal and stochastic component
Photoperiodicity	Constant
Mixing	Deepest

Dry region (15–35°)	
Hydrological budget	Negative, leading to extreme stochastic variability of precipitation and flow rates
Mineral turbidity	Extremely variable in connection with the above
Chemical composition	Extremely variable

Temperate region (35–60°)	
Radiation	Largest seasonal component of the variability of incident solar radiation and of radiation available to lake phytoplankton
Air and surface temperature	Largest seasonal and stochastic components of variability
Minimum surface temperature	Reaching freezing point

with shorter R's than normally met with in lakes, thermal stratification is less pronounced and more subject to stochastic variability in the driving forces, particularly in the driving force of flow. This is the justification for emphasising the usefulness of comparing water bodies and reservoirs by their retention time (or, less precisely by their watershed/water surface area ratio). Semi-arid reservoirs may differ from temperate ones in this respect because they are designed to store water for longer periods.

Regarding the typology of lake stratification, this is based on Löffler & Hutchinson as revised by Walker & Likens (1975) and distinguishes four major mixing types:
(i) meromictic lakes with an annual vertical mixing that is only partial and with the water mass divided into three vertical zones;
(ii) monomictic lakes which are mixed for one period of the year;
(iii) dimictic lakes which mix twice a year in spring and autumn; and
(iv) polymictic lakes which are shallow water bodies fully mixed to the bottom.

This classification is applicable to reservoirs. Meromixis in reservoirs may be managed for functional purposes (e.g. the London reservoirs) as well as natural like in lakes.

3.3.2. Meso-geographical zonation in reservoirs and lakes

Comparative limnology of reservoirs must take account of differences due to nature and due to man's activities. Both are significantly influential in determining reservoir characteristics. In addition to the geographical differentiation discussed above, there are finer differences within this larger framework due to local orography, hydrology, geology and vegetation. Man's influences are related to the reservoir's operational function itself as well as the local economy. The latter is most influential when economic developments in the watershed modifies not only water quality but also the hydrology of the region. Water quality is affected by watershed land usages and the quantities of pollutants allowed to enter the reservoir. As a consequence, reservoirs develop specific characteristics resulting for this mixture of global, local, natural and man-made influences.

More detailed information is available in Chapter I (Fig. 11) and Chapter II on reservoirs in hot and dry tropics or in maritime or non-maritime regions. One useful biogeographical orientation that may be applicable to reservoir limnology are the climatic-phytogeographic regions of Walter & Breckle (1991). These regions are based upon the feedback effects between climate and vegetation which also reflect soil characteristics. These combined effects are also important for water bodies in the regions, since hydrology and aquatic chemistry is affected by vegetation and soil properties (Bormann & Likens, 1979). It may be that the link is better for the rivers of the region than for the lakes.

We try to illustrate meso-geographical characteristics of reservoirs on the example of Brazilian ones.

The Brazilian reservoirs of the Amazon Basin:
(A) The global natural factors are: humid tropics, precipitation exceeding evaporation, extremely high seasonal rainfall and a flourishing vegetational cover in the form of a tropical forest (Tundisi *et al.*, 1991).
(B) The local natural factors are: orographically

determined as one of the largest rivers of the world in a lowland area which has a valley with a low slope and sediment transport is significant.

(C) Economically, the status of the area is that of a very scattered human population, with massive clearance of the forest locally and with an increasing agricultural development.

(D) The primary reason for building of reservoirs in this region is power generation by harnessing the enormous natural forces of the River Amazon. This goal has a dominating influence upon the size of the reservoirs and the elevation of their dams. As a consequence:

1. The reservoirs are very large and highly compartmentalised, formed by inundating vast areas of the tropical rain forest.
2. The inundated forest is a source of great spatial heterogeneity, as surfaces for macrophytic growth and periphyton development as well as sites for fish reproduction.
3. Large areas of the river's flood plain are also inundated which causes the disruption of the evolved food web structures and loss of biological diversity in fishes and forest (Junk & De Mello, 1987).
4. After the construction of the reservoir, there is a characteristic development of new large areas of wetlands (Tundisi, 1990a).
5. As a consequence of the inundation of the tropical rain forest is the appearance of the chemically aggressive hypolimnetic water, described in detail in Chapter II (this volume).
6. The quality of downstream water deteriorates, with low dissolved oxygen concentrations, high organic matter content and unacceptably high bacterial contamination (Matsumura-Tundisi *et al.*, 1991).
7. The main factors causing eutrophication of the reservoirs are agricultural activities and input of untreated sewage (Bonetto *et al.*, 1987; Pedroso *et al.*, 1988; and Tundisi & Matsumura-Tundisi, 1986).
8. Although most reservoirs have been shown to be phosphorus-limited, this element attains very high concentrations in the sediments largely due to inundation and the precipitation of ferric phosphate (Pedroso *et al.*, 1988).
9. In contrast with the prevalence of polymictic reservoirs in other tropical regions, the Amazonian reservoirs are all warm meromictic in nature with anoxia in their large hypolimnions.
10. During the phase of filling, several pulses occur due to deoxygenation and decomposition of organic matter. The start of discharge increases the river contribution during rainfall.

3.4. *Pulse effects in reservoirs*

Pulses are defined as any type of sudden change of natural or man-induced origin which affects any physical, chemical or biological variable of reservoirs. Pulses may result from an input into the reservoir, such as rainfall, or from an output, such as a sudden operational withdrawal of water.

Pulses of *natural origin* come from climatic changes such as wind or rainfall and may result in direct or indirect effects. Such pulses tend to be seasonal and may be frequent and repeated or infrequent as with strong winds or off-season precipitation.

Pulses of *artificial origin* may be caused by man's manipulation of water levels during regulation of water flows for power generation. These too may be frequent and repeated in some established operations or infrequent as in short-term releases of water.

The consequences of pulsed fluctuations are various for both the reservoir and its downstream ecosystem, which maybe a river or a cascade of reservoirs. Calijuri (1988) demonstrates a clear relationship between pulses of precipitation in the Barra Bonita Reservoir and periodic discharges of inorganic seston, largely clay. The high concentrations of inorganic particles affected the underwater light regime drastically, such that the Secchi Disc Depth dropped from 2.0m to 0.3m in a few hours. Zaret & Tundisi (pers. comm.) concluded that riverine seston brought into Lake Jaceretinga during the flooding of the River Solimoes could be responsible for impaired visual predation of fish. Inorganic suspended particles can also interfere with water chemistry as they are sites of sorption and desorption for various chemical substances.

Seasonally frequent pulses induced by climatic conditions may affect water circulation and water chemistry by introducing advective currents in the tributaries and causing enrichment at certain depths.

Sudden and infrequent changes of temperature may occur downstream and initiate changes in the thermal structure of the reservoir when water is selectively withdrawn from particular depths for operational purposes. Short-term releases of low quality water can cause the deterioration in the water quality of the downstream cascade of reservoirs or river.

Infrequent pulses may come from the breakdown of vertical stratification by wind action which causes changes in the vertical distributions of oxygen, nutrients and plankton. Tundisi & Matsumura-Tundisi (1990) provide us with an example of how a change in wind speed was followed by a change in the vertical distribution of phytoplankton accompanied by a change in the dominant algal species, from the buoyant *Microcystis* spp. to the heavier *Melosira italica* which needs some turbulence for suspension. Aranha (1990) also records the breakdown of colonies of *Microcystis* spp. by infrequent and rapid manipulation of selective withdrawals (both spill water and turbine water). Supersaturation of oxygen downstream can result from fast and sudden pulses of surface water releases and can cause fish mortalities, as recorded for Itaipu Reservoir, Southern Brazil.

The magnitude of these pulses differs geographically in the case of natural ones and according to the nature of water usage in artificial ones. It varies also in relation to whether the pulse was generated upstream or downstream of a reservoir.

In relation to management of reservoir operations, it is important to conceptualise the generation and the effects of pulses. More needs to be known about the risks of pulse generation in different kinds of reservoir sites for future constructions. A more detailed understanding of how the seasonal patterns of local hydrology is translated into changing water levels in the river and, in turn, into predicting the optimal periods of dam closure in order to protect downstream fish populations. Indirect effects are also important; for example, cold water with low oxygen content discharges from an upstream reservoir can cause the release of phosphorus from the sediments of a downstream reservoir in a cascade.

3.5. Reservoir aging and evolution

Reservoir 'aging' refers to the large-scale limnological changes that take place during the years after first filling and which is also referred to as 'trophic upsurge'. Reservoir evolution refers to a much longer term limnological phenomenon which lasts for decades or centuries.

3.5.1. Reservoir aging or trophic upsurge

The marked changes in chemical and physical characteristics of large reservoirs that take place during water storage was termed 'the aging' of reservoirs by Purcell (1939). He classified a reservoir "as not having come of age until the impounded water therein has ceased to improve" which he considered to be a phenomenon of vital importance to the sanitary engineer. In later studies, the notion of aging was extended to cover biological characteristics. Sylvester & Seabloom (1965) write about aging in relation to the biochemical changes that occur at the soil-water interface or in relation to phytoplankton reactions. A list of studies covering the maturation process of reservoirs can be found in Ostrofsky & Duthie (1978), and Tyler (1980) summarises observations on aging in Tasmanian reservoirs.

3.5.1.1. Field observations. During the aging of Wanaque Reservoir, Purcell (1939) detected anoxic conditions near the bottom accompanied by the presence of hydrogen sulphide and increased concentrations of carbon dioxide, iron and manganese. He comments that such changes, which were normal events during the limnological development of reservoirs, interfere with treatment of the water for drinking purposes. The water of Wanaque Reservoir also developed a high colour which lasted for seven years. This author recognised that each period in the reservoirs' limnological succession developed stable characteristic properties which indicate conditions of water quality.

Not only water quality but also the fish fauna

and the fisheries change with reservoir aging. This can be illustrated in the Russian reservoirs where fish population changed and fisheries declined only a few years after filling; this history is summarised in English by Rzoska (1966), Frey (1969), Avakian *et al.* (1979) and Benson (1982). The sequence of events during the first years after filling are:
(i) a phase of increase leading up to
(ii) an upsurge when populations are at maximal densities; this is followed by
(iii) a period of depression of the populations and stabilization.

The fish populations exhibit maximal biomasses and yields during the maturation of the first generation to be born in the reservoir. These tend to be both riverine and lacustrine species. Reservoirs are colonised by fish species from the rivers of the reservoir's watershed but few species are adapted to reservoir conditions so that fish species diversity declines markedly immediately after closure of the dam. In addition, migratory species are isolated from their reproductive grounds. In many tropical reservoirs, it is characteristic that the biomass of certain fish species increases once the water level has stabilised (Balon & Coche, 1974). The two Amazonian reservoirs of Tucurui and Samuel provide good examples as the major catch after filling was *Cichla ocellaris* and *Cichla sp.* (common name Tucunaré) both of which were tolerant of the prevailing severe ecological conditions. *Cichla ocellaris* is a fast growing species which can reproduce at a small size and is capable of producing a large fish stock in the first five years of existence (Chapter II).

Russian reservoirs provide a good material for studying aging as, being large and shallow, it took several years to fill them. During the continuous filling process, new areas of terrestrial vegetation were being covered and later replaced by extensive aquatic weed beds. These provided spawning sites and nursery areas for many fish species and the reservoir developed high biomasses of zooplankton and macrobenthos accompanied by abundant detritus. The macrobenthos was the main food source for the fish and developed to peak values ('an upsurge') during the initial several years but stabilised to more constant levels after three to five years. In his study on the Czechoslovak Lipno Reservoir, Brandl (1963) recorded very high quantities of zooplankton in the littoral zone following the continuous flooding of terrestrial vegetation. During his studies on new African reservoirs, MacLachlan (1974) followed the effects on the land-water ecotone with very active interactions during successive flooding and drying of the littoral. Rapid build-up of floating aquatic plants such as *Eichhornia crassipes* and the fern *Salvinia molesta* is commonly observed in African waters. Both plant species can reach densities that can clog bays and inflow areas of reservoirs to such a degree that they provide a substratum for rooted emergent plants such as *Scirpus cubensis* (in Lake Kariba) to develop. Flooded trees also provide a substratum for the growth of periphyton which supports high invertebrate and young fish production.

Substantial populations of bacteria appear in the pelagic zone of some Russian reservoirs a year or so after first filling. The development of the phytoplankton appears to follow three characteristic phases:
(i) destruction of the riverine phytoplankton;
(ii) dominance (as maximal biomass) of a few species successful in the new environmental conditions; and
(iii) decline in biomass and the development of a stable lacustrine phytoplankton assemblage, usually within one to five years after filling.

The main formative factor for phase (iii) is the declining flow-through which follows the near-completion of reservoir filling. How long the phase (ii) lasts depends upon the rate of organic matter decomposition and the continuity of leaching of nutrient salts from vegetation and submerged soil. In terms of zooplankton, the characteristic phenomenon is the increasing number of species that appears during the first years of the reservoir's existence. Simultaneously with this, the ecological equilibrium between zooplankton groups may be disrupted and there occurs an increase both in facultative planktonic species and in the total zooplankton. This is followed, within five and seven years, by declining abundance due to decreased contributions from the cyclopoid copepods, as recorded in the Czech Lipno Reservoir (Brandl, 1973) and in the two American lakes, Francis Case and Lewis and Clark Lake (Benson, 1982).

Figure 26 summarises the results of studies on aging of Klíčava Reservoir in Czechoslovakia by

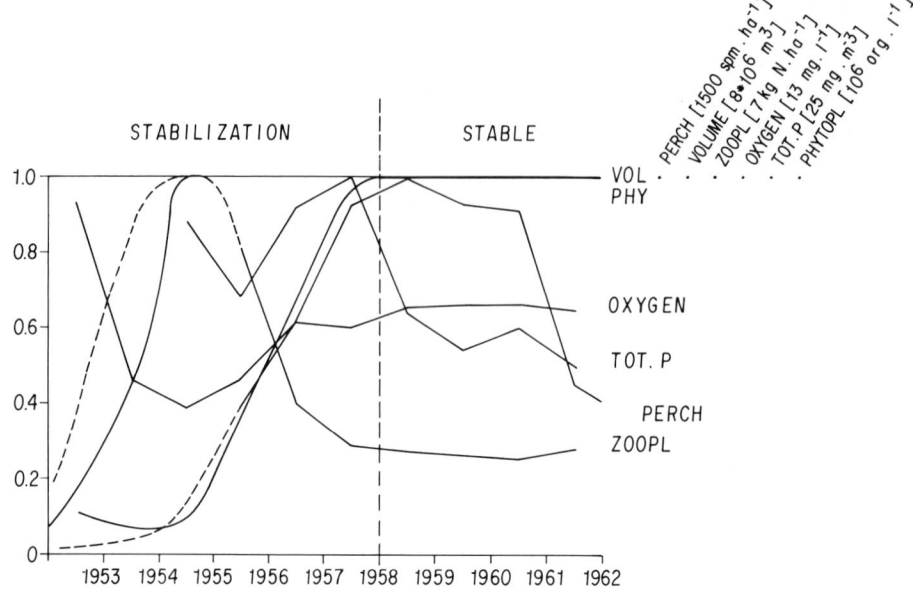

Fig. 26. The aging process in Klíčava Reservoir (Czechoslovakia).

Holčík (1977), Pivnička (1982) and Straškraba *et al.* (1990). In contrast to other studies, all factors relevant to aging were studied simultaneously: nutrient concentration, organic matter content (deduced from near-bottom dissolved oxygen concentrations) and both abundance and species composition of the phytoplankton, zooplankton and fish fauna. Ecological interactions were also investigated, from both bottom-up and top-down viewpoints. The filling of Klíčava Reservoir lasted from 1952 to early 1955, during which time the phytoplankton biomass was low and zooplankton biomass was high. The dashed line in Fig. 26 is derived from rough estimates of abundance during the early period in the life of the reservoir when direct estimates of zooplankton biomass were not done. Fish abundance was low with the dominating species being perch (*Perca fluviatilis*). The phytoplankton was not phosphorus limited, judging from near-saturation values of 20–25 μg l^{-1} of total phosphorus and similarly high levels of dissolved reactive phosphorus. A more likely interpretation for the low phytoplankton biomass during the early period of filling is high grazing pressure from the abundant zooplankton which is not subjected to much fish predation. This interpretation is supported by the subsequent decline in zooplankton abundance coinciding with increases in fish abundance. Decay of organic matter in the hypolimnion is probably responsible for the levels of about 5 mg l^{-1} dissolved oxygen in the hypolimnion. The reservoir bottom was cleared of all its trees and shrubs before filling and the high organic matter content was probably derived from soil erosion and decomposition of terrestrial plants. Levels of 5 mg l^{-1} dissolved oxygen are far from anoxia but represent the unsaturated values that persist.

3.5.1.2. Laboratory studies on reservoir aging. Table IX summarises the more important processes that occur during the aging of reservoirs and which differentiate reservoirs according to the intensity and duration of the aging process.

The effects of some of these processes listed in Table IX have been tested by laboratory studies.

Effect of soil: The leaching of organic matter, nutrients, alkalinity, conductivity and iron salts from cores of bottom sediment were studied by Sylvester & Seabloom (1965) in large laboratory columns. The exchange of solutes and leaching from organic soil was effectively stopped by a 5 cm cover-layer of mineral material. The authors conclude that 'soil effects' due to both leaching and being covered decline with aging of the reservoir. Similar studies were conducted by

Table IX. Processes during reservoir aging.

Process	Importance, changes
Leaching of nutrients from soil	Rich soils release more nutrients than poor ones. Leaching intensity decreases with time. Soil surface destruction increases leaching intensity.
Leaching of organic matter from soil	Easily decomposable organic matter is leaching more rapidly than resistant one. Rich soils produce more organics.
Decomposition of coloured organics	Colour reflects the concentration of resistant organic matter. Changes are very slow, increased values of colour are the long persisting signs of aging.
Decomposition of vegetation	Drowned grass is rapidly decomposing, providing substrate for bacteria, algae and these are a source of food for evertebrates. Detritus is formed rapidly and used as food by evertebrates, both pelagic and benthic. Shrubs and trees decompose in phases – dead leaves represent a pulse of organics shortly after drowning, wooden parts persist for years to decades.
Oxygen production/consumption	Oxygen is produced by phytoplankton and macrophyte photosynthesis and consumed by dissolved and particulate organic matter, during decay of vegetation and consumed by soils and sediments.
Oxygen consumption by sediments	Both soils and sediments consume oxygen for respiration of associated microorganisms.
Macrophyte growth	Floating vegetation boom is a burden in new tropical reservoirs. Submerged flora colonizes rapidly reservoir shallows, representing substrate for periphytic organisms, cover for fish, source of organic matter.
Retention time changes	During rising water level the retention time increases, flow rate slows down, conditions for lacustrine organisms develop. Sedimentation rates rapidly increase. The length of the aging period is directly related to retention time.
Light penetration changes	High initial turbidity and colour decrease light penetration and counteract high initial nutrients.
Structural changes of communities	The original riverine fauna and flora has to be substituted by lacustrine elements. Inocula, growth rates and productivity are decisive factors of the rate of conversion of river communities into lacustrine ones.
Differential rate of community development	Organism growth rates and life cycle durations decisive for community development are differentiated according to organism size. Bacteria react on changed conditions most rapidly, followed by phytoplankton, zooplankton, macrobenthos and fish.
Fish population growth	Fish is the most slowly developing biotic component in a reservoir. The formation of stabilized fish populations takes a few years.
Size composition of fish	Low year classes of fish exert highest grazing pressure on zooplankton. They predominate in the filling phase, their abundance being more important only in the second to fourth year of reservoir existence.
Predatory/Nonpredatory fish	During aging the participation of predatory fish in total population increases and finally a balance of predatory/nonpredatory fish is obtained.
Trophic interactions	Bottom-up effects during reservoir aging start with nutrient leaching, increasing phytoplankton production and up the trophic pyramid. Zooplankton control by fish and phytoplankton control by zooplankton become very evident in the aging period with differential development of trophic levels.
Horizontal heterogeneity	During aging an initially horizontaly more uniform river stretch starts to convert into a heterogenic water body. When water levels are rising a marked zonation of near-shore areas becomes apparent, with intensive decomposition and production processes in these zones. The inflow region starts to differentiate from the lacustrine one.
Vertical differentiation	The initially shallow, intensively mixed and vertically homogeneous river changes into a deep water body with decreasing mixing intensity. Stratification starts to develop, changing productivity conditions.
Substrate development	Drowning of terrestrial vegetation enormously increases substrate for colonization. This is followed soon by formation of aquatic vegetation. Also, reservoir sediments represent a substrate very different from the river one.
Initial set of conditions	Geographical location, geological substrate, soil types, vegetation cover, watershed development and soil uses decide on trophic conditions. Negative water quality changes are more profound in eutrophic conditions. Inoculum of fishes and other organisms decides on rates of community development.
Geographical differences	More rapid aging due to higher temperatures in tropics is often mentioned in literature but is not verified by comparative studies. Hypolimnetic anoxia seems to be more profound in tropical reservoirs.
Timing and pulses	Timing of rising water level in relation to vegetation periods, progress and retreat of water level fluctuations and pulses of flushing are decisive for the aging progress.

Davis *et al.* (1973), Kozerski (1975) and Gunnison *et al.* (1980; 1983; 1986). Hecky *et al.* (1984a) studied the effects of internal loading of sediments that arise from shore erosion.

Mathematical expressions for nutrient leaching from soils were developed by Ostrofsky & Duthie (1978), which led to the development of a hypothesis and a model of trophic upsurge in reservoirs (related to phytoplankton development) based on phosphorus leached from soils as the dominant variable (Ostrofsky, 1978; Grimard & Jones, 1982). Another quantitative model in respect to near-bottom oxygen depletion (Kozerski, 1975) takes into account soil qualities and reservoir stratification.

Effect of flooding grasslands: There are many laboratory studies which demonstrate the importance of plant nature (ferns, perennials, shrubs, trees) upon reservoir concentrations of organic matter, dissolved oxygen and nutrient salts due to differences in decomposition processes: Kaushik & Hynes (1971), Petersen & Cummins (1974), Campbell *et al.* (1975), Triska *et al.* (1975), Howarth & Fischer (1976), Reed (1979), Thérien *et al.* (1982), Mouchet (1984), Crawford & Rozenberg (1984), Gunnison *et al.* (1986) and James *et al.* (1988). In studies of the food composition of insects from newly flooded tropical and temperate reservoirs Cherry & Guthrie (1975) and McLachlan (1977) conclude that the decline in benthic invertebrate biomass that takes place after filling is associated with a switch in their diet from terrestrial detritus to algal biomass which occurs when filling is completed.

Effect of flooded trees and shrubs: It is generally believed that the fish fauna is richer and more diverse amongst submerged trees, especially in the tropics (Moss, 1980). This is rather difficult to prove since net-fishing is rather impossible. In Chapter XII, it is proposed that some areas are cleared in reservoirs predestined for a net fishery. Matsumura-Tundisi *et al.* (1991) followed a pulse of increased productivity in Samuel Reservoir that arose from the submergence of trees with their leaves still attached. Ploskey (1985) summarises the advantage and disadvantages of clearing vegetation from reservoir beds before filling in order to avoid anoxia in the hypolimnion arising from the decomposition of vegetative organic matter. In the tropics the rate of plant re-growth is so rapid that the clearing of very large areas is almost impossible.

3.5.1.3. Hypotheses. The 'trophic upsurge' with its associated changes in populations of phytoplankton, invertebrates and fish is interpreted by both Russian and American scientists as a classical bottom-up food chain effect generated by the increased flux of organic matter coming from flooded vegetation and leaching of soils. The subsequent decline in productivity in US reservoirs following such a 'trophic upsurge' was explained by Baxter (1977) and Kimmel & Groeger (1986) as follows:

(i) the large input of organic matter and nutrients from the inundated land was followed by a period of reduced detritus formation and internal nutrient loading;

(ii) this generates a decline in the quality of habitat and food resource for benthic organisms; and

(iii) rapidly expanding lacustrine environment. Five years or so after filling, the detrital input into the reservoir declines and this is followed by reduced production of phytoplankton, zooplankton and fish (Benson, 1982).

A new hypothesis was formulated to explain results from Czechoslovak reservoirs (Straškraba *et al.*, 1990; namely: that changes in the plankton assemblages of reservoirs are the consequence of time-scale differences between the functional components of the ecosystem food web. For example, the shorter-term annual zooplankton cycles can develop more strongly in the earlier phases of reservoir aging before the impact of the longer-term fish cycles. This constitutes a changeover from early bottom-up food chain influences to later strong top-down effects.

3.5.2. Limnological evolution of reservoirs

Most reservoirs stabilize after existing for a few years. Subsequently, changes do occur but largely due to the impact of man's activities which affect the earth's environment. The subsequent evolution of new reservoirs in Brazil is inevitable due to intensification of land usages and of industrial activities to which they are subject. Not only are there changes in what is cultivated in the watershed but there are also structural alterations (clearing the vegetation cover; changing the course of rivers, canalization and regulation of the banks; irrigation of crops). One consequence are much higher organic matter loadings to the reservoirs, leading to rapid eutrophication, extensive siltation, massive macrophytic growths and greater dispersal of water diseases.

Reservoirs, like lakes, undergo a natural evolutionary sequence of developmental or successional phases which lead to cumulation of bottom sediments, a diminishing of water volume and a decreasing depth, all of which augment biological productivity. Associated with this picture of succession are accompanying changes in spatial structure, in substrate quality, in water movements (speed and direction of currents). In lakes, this natural eutrophication takes millennia but, in reservoirs, it is a much shorter process, largely due to higher area ratios of watershed to reservoir. The average age for Chinese reservoirs is only 80 years whereas it is 50–350 years for some of the large reservoirs in Brazil.

3.6. Reservoir systems

Reservoir systems refer to multiple reservoirs which are connected hydrologically and whose operation is inter-related for fulfilling some common goal, such as water supply or generation of electricity. There are three basic types of reservoir systems existing (Fig. 27):

Reservoir cascades: a chain of reservoirs located on one river.

Reservoir multi-systems: a group of reservoirs located on different branches of one river system or on several river systems and whose releases of water is shared between them according to a joint goal. Where water is transferred between watersheds, hydrochemical differences in the watersheds may create specific limnological and water quality problems. The most extensive system known is in Australia. Figure 2 illustrates the Snowy Mountain reservoir system in which reservoirs are also interconnected by tunnels. According to McComb & Williams (1987), the Murray River now behaves ecologically more like an interconnected series of reservoirs than as a great river. This may also be true to the Tucurui River system (Chapter II, Fig. 4). It is of major concern how to operate hydrologically such large multi-systems and is the subject matter of many books (Haimes et al., 1975; Haimes, 1977).

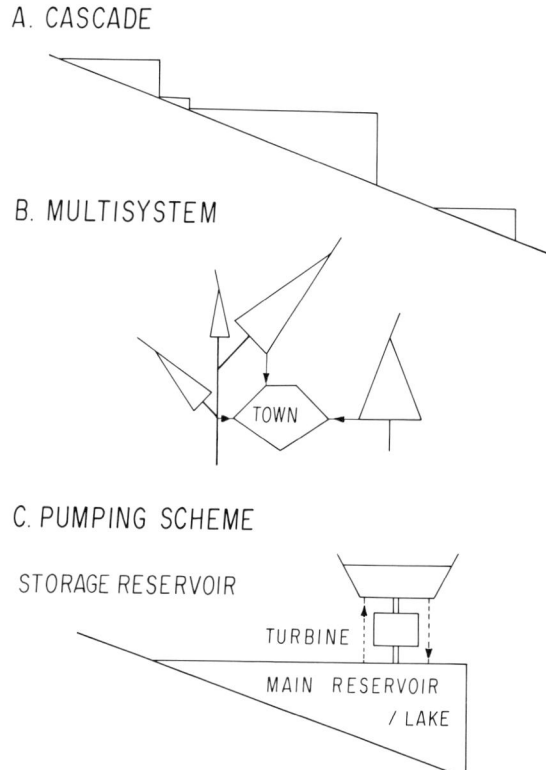

Fig. 27. Three basic types of reservoir systems. A – Reservoir cascade, B – Reservoir multisystem, C – Pumping scheme. Orig.

Reservoir pumping schemes: these are characterized by the pumped circulation of water between reservoirs. Most such schemes are concerned with the generation of power during peak demand: Wägitalersee, Switzerland (Chapter IV, this volume), Dalešice Reservoir, Czechoslovakia (Fig. 28). For situation in U.S.A see Karadi et al. (1971) and Schoumacher (1976). Others are operated for the provision of water for cooling plants.

Reservoir cascades have been more intensively studied than the other types of reservoir systems but the knowledge gained is valid for all types. The basic difference between the types is the more irregular operational regime and more dynamic and variable limnological responses of the reservoir multi-systems and pumping schemes.

3.6.1. Reservoir cascades
What makes reservoir cascades limnologically specific is the effect that the upper reservoir has on

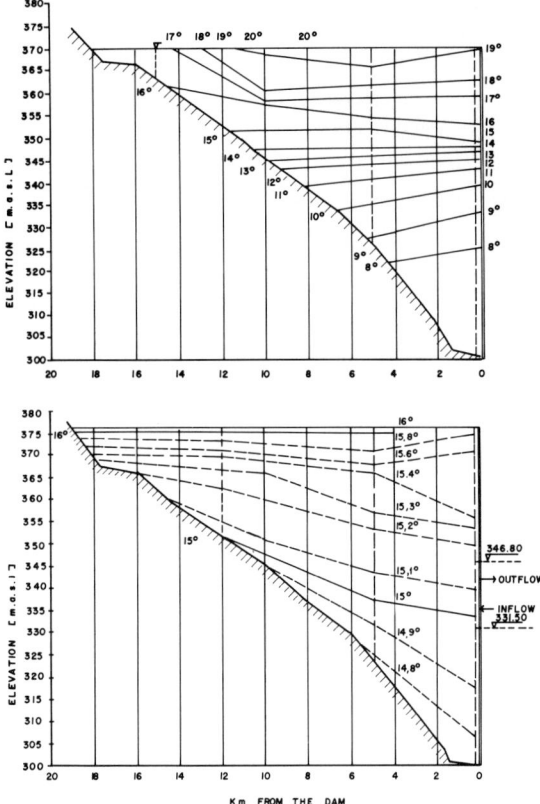

Fig. 28. Temperature conditions in Dalešice Reservoir (Czechoslovakia) with cyclic power generation. Above – before the start of pumping: isotherms nearly horizontal, bottom temperature below 8 °C. Below – after pumping started: isotherms tilted, bottom water above 14 °C. From Kratochvíl (1982).

increases, and as retention time prolongs, so the reservoir's effect upon the river becomes greater. This is because greater reservoir productivity is correlated with higher stream orders of the original river and this is reflected in the quality of the those below it, irrespective of whether they are operationally inter-connected. In a reservoir cascade, the top reservoir is usually not different from a solitary reservoir although there may be some modification if the whole cascade is operated for one function; the next and lower reservoirs, usually termed 'cascade reservoirs' are all modified to some degree. The extent of the modification caused by the upper reservoir upon the cascade one depends upon whether it is a deep stratified reservoir (which has profound effects) or a shallow one (less effects). The intensity of influences depends upon stream order, reservoir trophy and the distance between reservoirs. With stream order, the concept of the River Continuum becomes relevant: as the stream order of the river on which the reservoir has been constructed

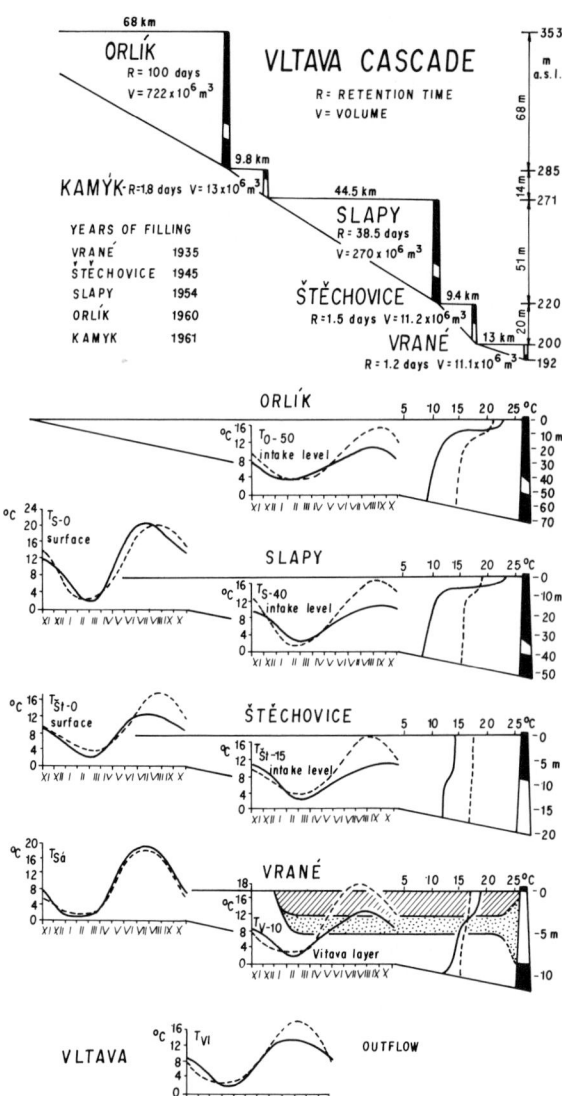

Fig. 29. Schematic representation of stratification conditions in Vltava River Reservoir Cascade in Czechoslovakia. A – Cross section of the cascade with basic infromations about the reservoirs. B – Schematic representation of temperature differences in the two principal reservoirs. The annual course of temperature at the surface and turbine intake depth (sinusoidal approximation) as well as summer temperature profiles near the dam are given for two situations: a dry year (full lines) and a wet year (dashed lines). After Straškraba & Straškrabová (1975), modified.

reservoir outflow and its effect upon the downstream river. The quality of reservoir outflow water is influenced by increased retention times (enhancing the difference between river and reservoir water); increased retention time, in turn, sharpens the degree of stratification and the chemical quality of hypolimnetic water that may be the source of outflow water. The Serial Discontinuity Concept (Section 3.1.) is also relevant here since the upper reservoir does not exceed the 'reset' distance but its effect will be maximal where reservoirs are closely adjoined.

Reservoir cascades constructed for power generation are usually operated as a single unit but, limnologically, each reservoir behaves like a single input-output system. This is because all the cascade reservoirs are influenced by the processes within the reservoir(s) above them. Due to this, we cannot understand the limnology of a particular cascade reservoir without knowledge of processes occurring in the chain of reservoirs above it. It is not easy to distinguish causal effects due to morphometry, retention time, nutrient loading, outflow structures or operational schedules or due to position in the cascade chain.

A good opportunity for determining features due to cascade position occurred in Czechoslovakia when Orlík Reservoir was built upstream of Slapy Reservoir which had been adequately studied when it was solitary (Hrbáček, 1966, 1984; Hrbáček & Straškraba, 1973). Both water bodies are deep temperate reservoirs built on medium order streams at low elevations. Although the inflowing river was rich in nutrients, productivity was depressed by low underwater light intensities caused by humic water colour coming from industrial effluents from upstream paper mills. The following account of this well-documented example of a reservoir cascade system deals first with the specific conditions existing and then tries to generalise the situation, by relating all comparisons to similar retention times.

3.6.1.1. Stratification and mixing in reservoir cascades. Summer temperatures and dissolved oxygen conditions in the reservoirs of the Vltava

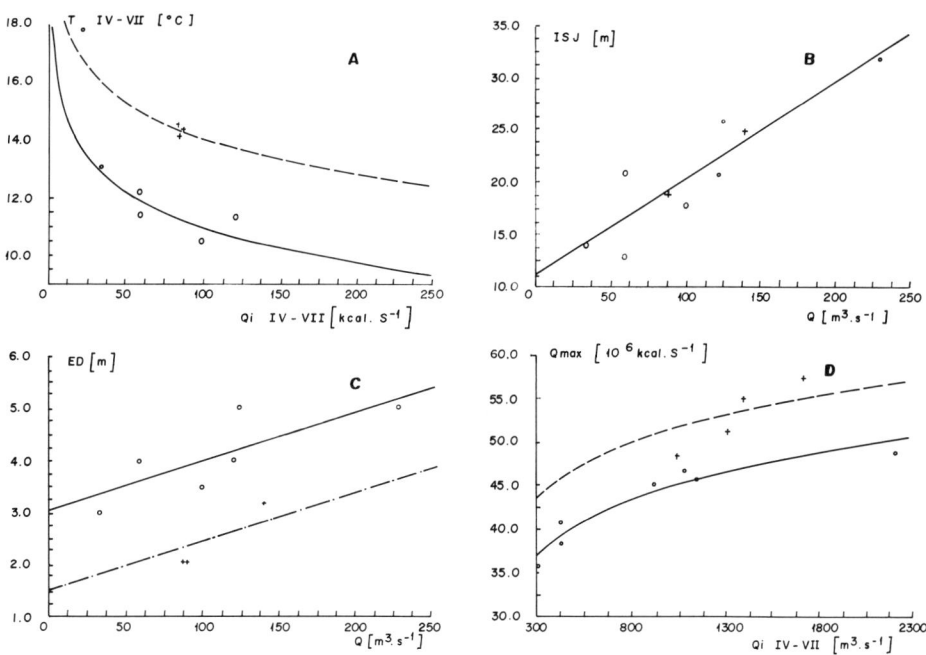

Fig. 30. Dependence of several variables on flow in a solitary reservoir (Slapy Reservoir, situation before the construction of the upperlying Orlík Reservoir) and the same reservoir when belonging to a cascade (years after Orlík construction). For each variable, the two regressions were derived under the assumption of identical slope. Full lines and o cover data for the reservoir when in cascade, dashed lines and + for the solitary reservoir. A – Average surface temperature for the period April to July. B – Inflow stream jet during summer. C – The mixing depth ($ED_{Z_{mix}}$) during summer. D – Birgean heat budget for the period of maximum surface temperature. Orig.

cascade are illustrated in Fig. 29. It would be misleading to attribute the different profiles of Orlík and Slapy reservoirs to the fact that Slapy is a cascade reservoir and Orlík is not, because the two reservoir have very different retention times due to their different volumes. Any realistic comparison must take account to retention time and this is only possible by long-term comparative studies in which variables have been measured over the full range of retention time change (Fig. 30).

The following changes were observed in Slapy Reservoir ($20 < R < 90$; average $R = 43d$) when Orlík Reservoir ($45 < R < 180$; average $R = 100d$) was constructed upstream of it (Straškraba, 1990):

(i) the average $T_{i(IV-VII)}$ decreased by 3 °C (Fig. 30A);
(ii) the $T_{S(IV-VII)}$ decreased by more than 3 °C whereas the maximal T_S remained identical;
(iii) T_B increased in spring and full summer;
(iv) this affected the depth distribution of temperature;
(v) the depth of the inflow stream jet (ISJ) was raised in spring but remained unchanged in summer due to the decrease in T_i (Fig. 30B);
(vi) the mixed depth (z_{mix}) is raised by 1.5 m due to the combined effect of a lower T_S and a higher T_B (Fig. 30C); and
(vii) the maximum Birgean heat budget (Θ_{max}) is increased by 6.5×10^6 kcal cm^{-2} (Fig. 30D) as a consequence of points (iii) and (vi) above.

The two situations in Slapy Reservoir before and after the construction of the upstream Orlík Reservoir are illustrated in Fig. 31 by sinusoidal approximations of the annual time courses of the temperature of the inflow, surface and bottom waters. An analysis of the density differences between inflow-surface and inflow-bottom waters provides some measure of the duration of periods

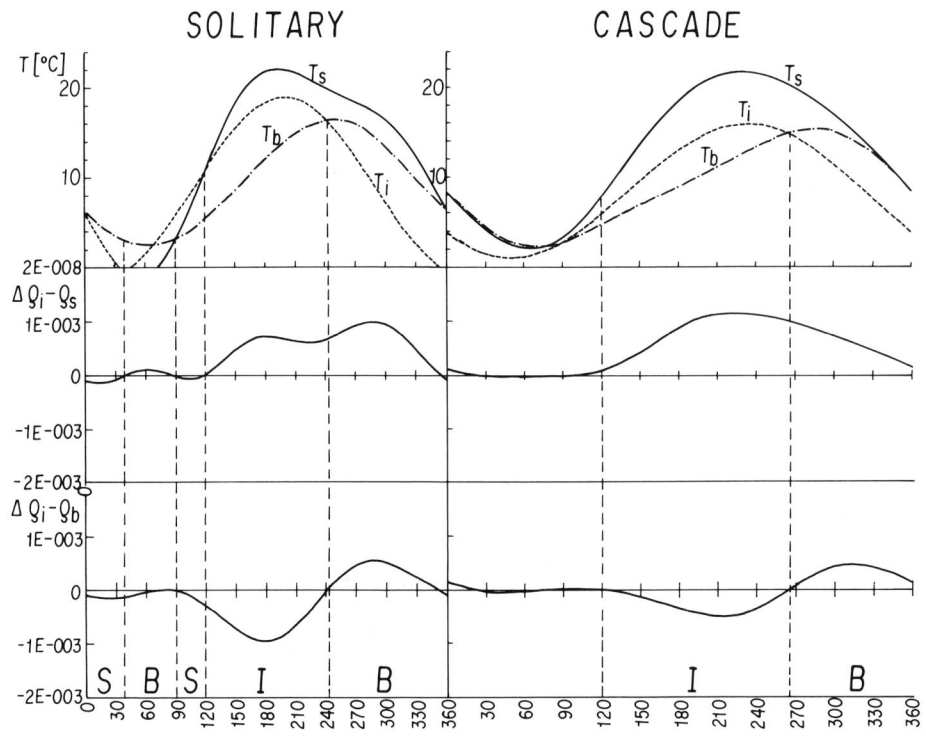

Fig. 31. Two situations of temperatures and flows in Slapy Reservoir in years with similar water flow rates, when solitary and when in a cascade. Top row – sinusoidal temperature approximation of the annual course for the observed data in 1960 and 1967 for the inflow (T_i), surface (T_S) and bottom (T_B). Center and lower rows give the density differences between the inflow and surface water and between the inflow and bottom water, respectively. Calculated from the above temperatures and the dependence of density on temperature. Dashed vertical lines separate periods of surface flow (S), bottom flow (B) as well as the intermediate flow (I). On the x-axis days of the year. Orig.

of surface flow (S), bottom flow (B) and interflows (I).

On a global scale, the geographical location of the reservoir has to be taken into account. The differences between solitary and cascade reservoirs will be less pronounced in tropical cascades than in the temperate example illustrated because of smaller difference between surface and bottom temperature. From what is known, we can predict a similar relationship of stratification to retention time in tropical cascades. In Table X, a summary is presented of the physical differences between cascade reservoirs compared with solitary reservoirs.

3.6.1.2. Eutrophication, organic matter and plankton cycles in cascade reservoirs. Slapy Reservoir no longer exhibited dense blooms of the blue-green *Aphanizomenon flos-aquae* and *Anabaena circinalis* (Hrbáček, 1966) after the upper Orlík Reservoir was filled (Hrbáček & Straškraba, 1973). The cause was the retention of

Table X. A schematic comparison of the physical limnology of a cascade and a solitary reservoir

Variable	Conditions in the cascade reservoir
Inflow temperature (T_o)	A lower temperature when temperature is increasing and a higher temperature when temperature is decreasing
Surface temperature (T_S)	A lower temperature when temperature is increasing but then identical
Bottom temperature (T_B)	A higher temperature throughout
Inflow stream jet (ISJ)	When temperature is increasing, the inflow stream plunges down to deeper water layers than in the solitary reservoir but there are not differences at other times
Mixed depth (ED)	This is higher in cascade reservoirs
Maximum Birgean heat budget (Θ_{max})	This is larger in cascade reservoirs
Intensity of mixing	This is greater in cascade reservoirs

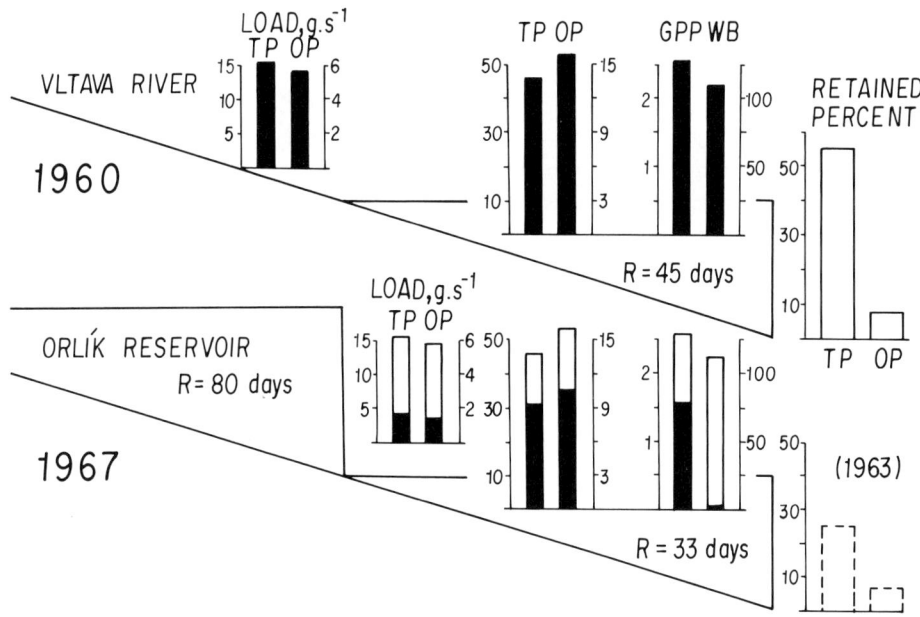

Fig. 32. Phosphorus retention and phytoplankton development in Slapy Reservoir when solitary (1960) and after Orlík Reservoir was build immediately upstream (year 1967 with flow rates similar to 1960). TP – total phosphorus, OP – dissolved reactive phosphorus, loads for both in g s^{-1}, concentrations in mg m^{-3}, GPP – gross phytoplankton production in g O$_2$ m^{-2} d^{-1}, WB – blue green biomass in g N m^{-2}. For 1967 the black histograms are valid, the white histograms representing those for 1960. For comparison of the retention which was not measured in 1967 a histogram for another year is given (a dry year, the years 1960 and 1967 being close to average). The arrows show how the situation shifted. From Straškraba & Straškrabová (1975).

phosphorus during the passage of water through Orlík Reservoir and the reduced loading of this critical nutrient in Slapy Reservoir (Fig. 32). A similar situation occurred in a tropical reservoir cascade where the time course of a blue-green was followed (Tundisi, 1990c); here the bloom was delayed due to both its transport down the cascade and its slower growth in the lower reservoirs. However, quite an opposite experience was recorded by Kimmel *et al.* (1988) for the Savannah River Cascade where phytoplankton biomass increased down the cascade from reservoir to reservoir. As this event took place during the first year of existence of the reservoir cascade, it may be due to processes associated with reservoir aging rather than anything else. The claim by Elser & Kimmel (1985) that the nutrient availability increased down the cascade being enhanced by hypolimnetic releases from the upstream reservoirs is in agreement with this assumption.

In the earlier section (Section 3.2.) on retention time as a key factor, some evidence was given for the effectiveness of reservoirs as traps for phosphorus as the nutrient most limiting freshwater production. The processes involved in this are sedimentation of and uptake by phytoplankton. Here it was shown that the concentration of both total and soluble reactive phosphorus may decline by as much as 80–90% during the passage of water through a reservoir and that the decrease was dependent upon retention time. A second reservoir lower in the cascade series will receive a substantially reduced phosphorus loading than a solitary reservoir in the same site. If phosphorus is the limiting nutrient, then this cascade reservoir's productivity will be lower than the solitary one above it. There will also be considerable differences in its chemistry, bacteriology and biology (Fig.33).

The reduction of both total and soluble reactive phosphorus is largely caused by the sedimentation of suspended particles in the reservoir. In that part of the inflow stretch of the upper reservoir where the sediments are trapped, the following marked changes occur: sediment accumulation, gas ebolition, anoxic conditions and the development of a tubificid population. Subsequent to the region of active sedimentation, conditions are again different, with high oxygenation and good quality water so that indicators of dissolved organic matter

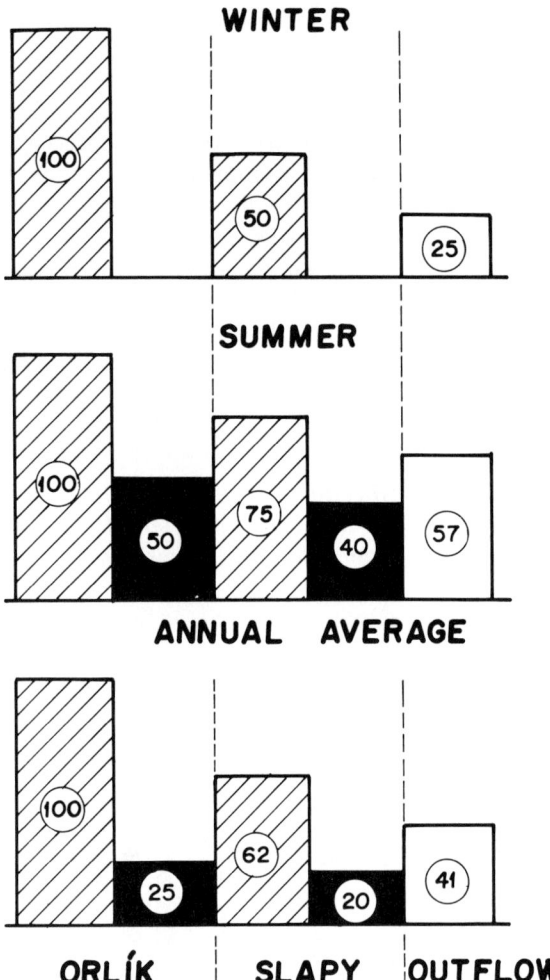

Fig. 33. The effect of two reservoirs in series on the amount of easily decomposable organic substances, expressed as BOD$_5$. The scheme is derived for a deep reservoir with R around 30d. The values indicate percentage in the inflow to the first impoundment Inflows are dashed, reservoirs black and outflows empty. Modified after Straškraba & Straškrabová (1975).

content, such as BOD, COD and water colour, all indicate improved conditions downstream. That is, water quality improves during its passage through a reservoir cascade.

The decline in biological productivity down a reservoir cascade associated with phosphorus trapping, as just described, may be counteracted by enhanced sedimentation of suspended matter. Often in the upper reservoirs but rarely in the lower ones, the presence of abundant suspended matter can cause an increased light attenuation and a

Table XI. A comparison of eutrophication changes in a cascade and a solitary reservoir.

Variable	Conditions in the cascade reservoir
1. Mineral turbidity	Lower concentrations
2. Allochthonous organic matter	Lower concentrations
3. Colour	Less colour
4. Light	Less attenuated due to items 1. and 3.
5. Phosphorus	Lower concentrations
6. Primary production	Lower rate due to item 5. but compensated by item 4.
7. Phytoplankton composition	Shifted to more oligotrophic species assemblages
8. Oxygen in inflow	Lower concentrations
9. Oxygen in the hypolimnion	Lower concentrations
10. Vertical distribution of conservative substances	More uniform

consequential light limitation in the phytoplankton. If the upper reservoir has a low retention fast flow-through and is built on a low order stream of low productivity, then mineral particles predominate over biotic ones and sediment readily, thus improving the underwater light regime downstream in the cascade. This is unlikely to happen in lower reservoirs in the cascade which tend to grow more phytoplankton, despite the reduced phosphorus loadings associated with upstream trapping. Table XI summarises some of the effects of reservoir cascades upon eutrophication.

4. Management of reservoir water quality

The management of reservoir water quality is a complex issue because of the dynamic nature of reservoirs, the interference of man with their natural phenomena and the variability of operation procedures upon the reservoirs' physical, chemical and biological processes. There is the question of how to separate management for water quality from the problems arising of water quantity, from their operation and from their multi-functional nature; these problems are diverse and may be ecological, economic and/or social.

We need to develop management approaches to problems which adhere to the central notion of the sustainable development of the planet earth. Present management solutions are largely environmentally insensitive, relying as they do on brute technological force, energetically wasteful usages and with uncaring chemical profligacy. An environmentally sensitive approach encourages the development of sustainable methodology, which saves energy and avoids environmental deterioration both locally and in far places in the world, where these chemicals are mined, processed and transported. It is not possible to develop environmentally sensitive approaches in conjunction with the environmentally insensitive engineering methods that dominate reservoir management. It is time to turn to management procedures which draw upon limnological understanding of reservoir behaviour and which seek to invent methods of control based upon sustainable biology rather than just chemicals, so often toxic.

4.1. Reservoirs in the development of regions

In South America and in Africa, reservoirs were built in order to strenghten regional development by increasing power and energy. The construction of reservoirs in remote areas not only altered the local biogeophysics but also improved the economy of the local society. These reservoirs also proved to be disruptive, ecologically, economically and socially. Such disruptions are avoidable, given some prior study of local ecology and economy. Before the construction of the reservoir, there needs to be an analysis of how the local community uses the natural resources and how the local social economy relates to the ecology of the region. For example, how man relates to the local riverine system (Garzon, 1984; Dixon *et al.*, 1989; Tundisi, 1990a).

Problems of health hazards arising out of reservoir construction have been first detected in tropical countries, for example, in the Volta Reservoir (Ghana), Tucurui Reservoir (Amazon) and Yacireta (Argentina). Such health hazards can be predicted before construction by means of an environmental impact assessment study (EIA) which brings together what information is available and what is known about local ecology. To be effective, an EIA should be started several years prior to the construction of the reservoir and

Table XII. Criteria for environmental impact assessments (EIA) associated with reservoir constructions. Problems culled from case studies on the Volta (Ghana); Kariba (Zimbabwe); Tucurui, Samuel and Balbina (Amazon, Brazil); Itaipu and Yacireta (River Parana, Brazil).

A re-location programme for the local displaced population
Health problems arising out of the reservoir construction, such as the dispersal of water-borne diseases and excessively dense population on the construction site itself.
Loss of edible native species of fish, some of which may be indigenous and unique.
Loss of native terrestrial communities of fauna and flora.
Loss of valuable timber resources.
Loss of mature agricultural land, well cared for generations, such as rice paddy.
Loss of natural wetlands, wetland species and unique land/water ecotones.
Loss of aquatic and terrestrial biodiversity from the locality.
Excessive human immigration to the reservoir region, with many attendant social, economic and health problems.
Provision of an adequate compensation for loss of agricultural lands and fisheries grounds as well as compensation for loss of fishing, recreational and subsistence agricultural activities.
The effect of the presence of the reservoir itself upon the local ecology, economy and society.
Degradation of water quality locally.

should deal with cause-and-effect relationships rather than just the structural effects. The disruption of processes integrated across land-water interfaces needs evaluation both qualitatively and quantitatively. Inadequate EIAs (Table XII) result in poor economic assessments of impacts and erroneous estimates of compensation for losses.

4.2. *Implications of ecological theory for reservoir management*

Ecological theory offers a number of hints for the proper management of reservoirs and their watersheds (Table XIII). In particular, ecological theory helps with solving water quality problems by biological rather than chemical control.

In a comparison of pre-impoundment predictions with post impoundment observations in Southern Indian Lake in Manitoba, Canada, Hecky *et al.* (1984b) were able to demonstrate which phenomena were in agreement with the predictions and which were not. Moreover, only rather a small number of good quantitative relationships could be established between pre-impoundment condition of water quantity and nature of biological response.

Mathematical modelling of the biological interactions within reservoir ecosystems offers one approach towards the management of reservoirs with limnological understanding, always bearing in mind that, in practice, models have to sacrifice precision for generality. One classical and successful approach has been to balance carbon gains and losses due to phytoplanktonic photosynthesis and respiration in the ocean or lake ecosystems (other fluxes being considered of lesser

Table XIII. Principles of theoretical ecology applied to management of reservoirs and their watersheds.

Principle	Examples of usage
Bottom-up effects	– Chemical determination of biological production
	– Fish yield determination by natural food production
Limiting factors concept	– Eutrophication abatement
	– Upper limits to production
Top-down effects	– Biomanipulation
Sub-system interactions	– Interactions (terrestrial/aquatic)
	– Watershed/reservoir interactions
Positive feedback	– Exponential (sigmoidal) reactions
Negative feedback	– Nutrient concentration depends on utilization by phytoplankton
Indirect effects	– Water temperature affected by phytoplankton development
Connectivity	– Upstream – downstream effects
Ecosystem adaptability	– Chemical pest management inactive after organisms adaptation
Ecosystem selforganization	– Unforseen reactions of reservoir ecosystems
Ecosystem spatial heterogeneity	– Conservation and management of riparian forests
	– Protection of headwaters
	– Protection of shoreline
Ecological succession	– Reservoir aging and evolution
Biological diversity	– Reforestation by native species as means to retain diversity
	– Wetland maintenance (areas of high diversity)
	– Preservation of ecotones
Competition	– Stop introducing exotic species
Pulse effects theory	– Maintenance of forests and wetlands in the watershed depress the negative pulse effects
	– Regulation of water blooms
Theory of colonization	– Few pelagic fish species in reservoirs

magnitude) which comes from Sverdrup (1953), Talling (1957a,b) and Vollenweider (1965) and was enthusiastically adopted during the years of the International Biological Programm (IBP) by Steel (1972, 1980) and Uhlmann *et al.* (1981). The procedure involved finding analytical solutions which permitted sufficient generalisation of algal population dynamics and production so that predictions could be made about the time course of future algal biomass. When applied to operational reservoirs, other losses had to be incorporated such as grazing, sedimentation and limitation of primary production by reduced light intensity or low nutrients as well as the effects of carbon inputs and outputs in flow-through reservoirs. An early example of such a model used for operational management is that of Steel (1972; 1976; 1978a,b) applied to the Thames Valley reservoirs (see also Duncan, 1990).

The newest development is called eco-technology and represents a marriage of ecological theory and mathematical modelling techniques for ecosystem management (Straškraba, 1986; Mitsch & Jørgensen, 1989). Figure 34 illustrates the general principles of ecosystems and how these are reflected in eco-technological theory:

– As energy inputs and reserves of storage matter are strictly limited, it is necessary both to minimise the wastage of energy and to close circulations of matter in order to conserve all this valuable biological store capable of providing renewable resources.
– Information stored in structures and coded in gene pools is the memory of ecosystems making them resistant to perturbation. Therefore all structures and heterogeneities have to be retained. It is also imperative to retain genetic information stored in the biological diversity as this is the potential to antientrophic growth and adaptation to varying conditions.
– As ecosystems are open dissipative systems, they are vulnerable to external inputs and imbalances; maintaining balance may need active management as does prevention of pollution.
– As multi-factorial feedback systems, ecosystems may behave in ways not yet encompassed by present ecological theory. Thus, effects in one part can have unexpected consequences in some remote part and may be the cause of dangerous large-scale damage arising out of a small, local event.
– Ecosystems are capable of homeostasis, that is maintaining a healthy steady state over a wide range of conditions. However, there are limits beyond which they cannot assimilate change. Once forced beyond these limits, homeostasis breaks down and major damage may occur.
– The capacity of ecosystems to adapt and 'self-organise' is known and should be incorporated into management strategies. Our management practices change into mismanagement procedures when the adaptability of ecosystems is ignored and when it is treated as a rigid rather than variable system.

4.3. Geographical differences and management

Principles of reservoir management as well as the basic processes operating in reservoirs are the same the world over. However, there are geographical differences amongst reservoirs, considered earlier (Section 3.3.) at several levels called macro-, meso- and micro-geographical, which lead to differences in methods for optimal management. In the same geographical context, we may have to speak about 'geographical differences' in reservoir management. There are also considerations of local experiences with certain management approaches. Despite the speed with reservoir 'know-how' spreads around the earth, there are still some obstacles to the dispersal of good management.

Without implying any sharp boundaries, there are two groups of geographical effects on reservoir management which can be distinguished:
1. Direct geographical effects. These include those specific to a particular locality and not found in other places. These are due to:
   (i) reservoir mixing patterns;
   (ii) soil type with its associated vegetation; and
   (iii) species composition of the aquatic flora and fauna.
2. Indirect geographical effects due to particular hydrology and orography, which determine water storage and retention time as key factors in the limnology of reservoirs and which can severely limit management options.

Biomanipulation is a typical example of management method with geographical dependence, namely the local species composition can

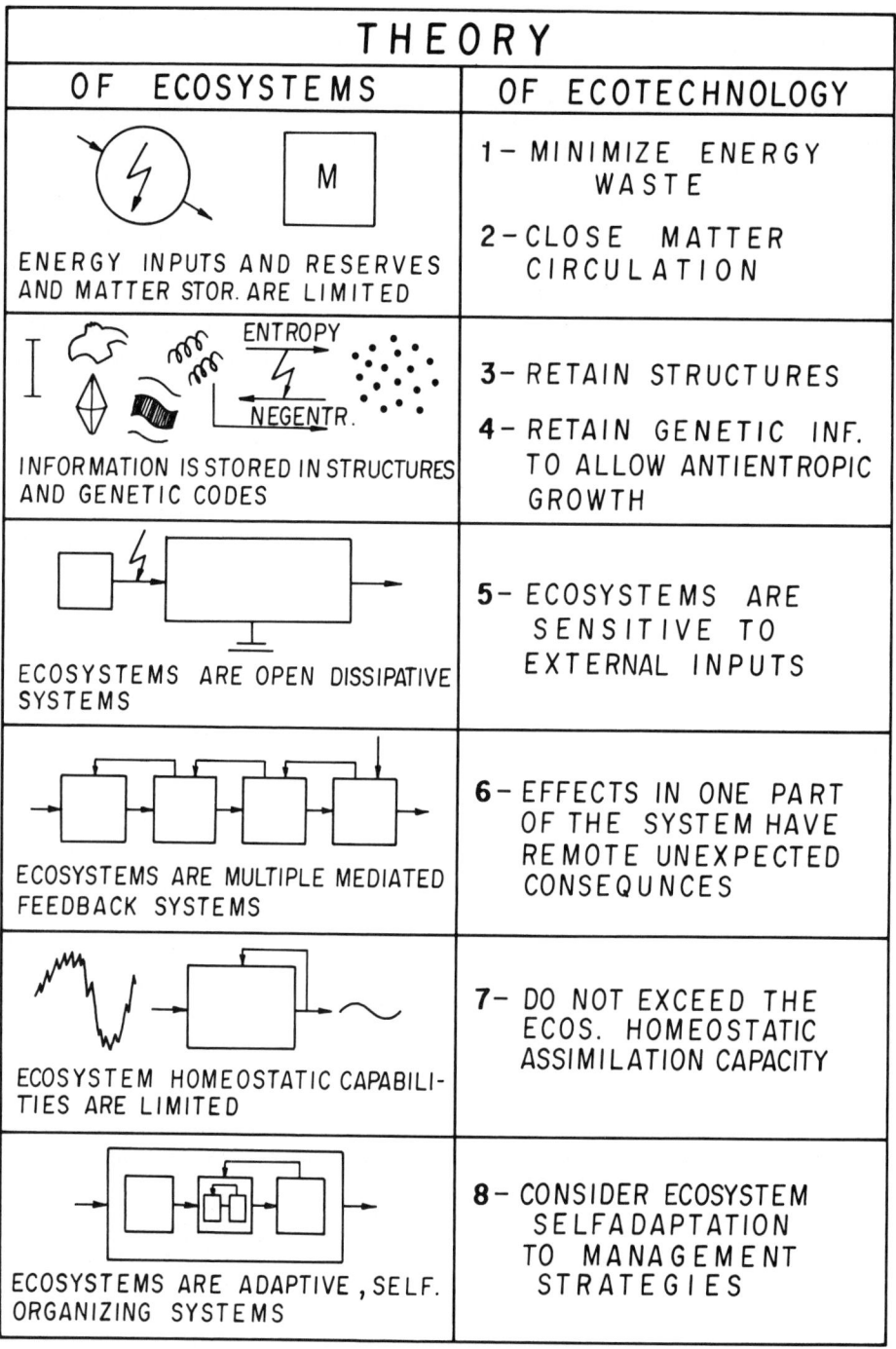

Fig. 34. Derivation of the theory of ecotechnology from ecosystem principles. On the left representation of ecosystem principles as derived by Straškraba (1992), on the right the corresponding rules for ecotechnology. Orig.

play a major role in management. What is involved is the selection of the fish species with potential for maximal effect upon the zooplankton and, consequently, upon the composition and abundance of the phytoplankton; this is clearly related to the local regional fauna. The employment of introduced (exotic) fish species is always a limited possibility because the introduction may be too successful so that it develops into a pest species in its new habitat, damaging the local fish fauna or it may bring foreign parasites or diseases which disperse to the local domestic species. Other direct geographical effects are introduction of atypical water budgets. The construction of huge reservoirs in regions of hydrological deficit with local maximal evaporation rates (e.g. Aswan Dam) has rather negative consequences as discussed in Section 1.2.3.

In Chapter I (this volume), there was a discussion on systematic indirect geographical effects in the levels of limiting nutrients, such as phosphorus which is critical in many waters for causing different productivity levels. The power of phosphorus levels to do this should not be neglected in management models and practices. In semi-arid zone reservoirs, on the other hand, nitrogen limitation of primary production and chlorophyll levels is more prevalent (Chapter II, this volume). Recently, studies on tropical reservoirs (Henry, 1990) demonstrate that phosphorus dominantly limits freshwater production. However, the initial loading of phosphorus into reservoirs may be high due to rapid re-mineralisation within the water column and precipitation in the sediments of $P-PO_4$ as ferric phosphate. Thus, a simple transfer of a successful empirical management procedure from one to another situation may not work and needs prior verification of its validity. The same applies to the use of simulation models; local estimates of relevant parameters are essential.

On a meso-geographical scale, nitrogen limitation seems to prevail in coastal areas as it does in marine waters. Probably more important than any geographical differences between management practices for a given reservoir are its own particularities. Management procedures developed for deep stratified water bodies are unlikely to be successful for shallow unstratified reservoirs, however similar the geography. Retention time of reservoirs is always decisive in selection of the most appropriate management procedures whereas, time and time again, it has been demonstrated that the introduction of chemicals into a rapidly flushing reservoir is worse than useless, is expensive and may be toxic to man and beast.

4.4. Problems of water quality and how to manage them

Because of the tight coupling of water quality in reservoirs with man's activities in their catchments, managing for water quality can be broken down to dealing with:
(i) the watershed and the reservoir inflows;
(ii) the processes occurring within the reservoir; and
(iii) the outflowing water, whether passing to the downstream river or into a drinking water treatment plant.

The nature of the water quality problems differ widely and are linked with different particular causes and with the reservoir function itself. Some common water quality problems and their probable causes are listed in Table XIV.

Various functions of reservoirs were listed in Table II. The most severe water quality criteria are imposed upon those supplying drinking water. Recently, these have included criteria for xenobiotics, substances toxic to humans and other organisms, and, with these substances, not only must present-day supply be protected but also present waters which are likely to be sources for future water supply. With such substances as xenobiotics, valuable environments may be damaged carelessly and without our knowing about it until too late.

Table XIV. Common water quality problems in reservoirs.

Classic organic matter pollution of domestic origin.
Bacterial and viral contamination.
Water-borne diseases.
Eutrophication: excessive organic matter production within the reservoir due to high nutrient inputs.
Acidification: decrease of pH and associated leaching of metals, acid rain and mass transfer of atmospheric gases.
Turbidity problems caused by siltation.
Salination due to excessive fertilizer application on land or due to soil salination in arid and semi-arid regions.
Heavy metal pollution.
Agro-chemicals and other toxic chemicals. Accumulation in sediments and bio-accumulation in organisms.
Hypolimnetic anoxia. Agressivity to structures. Increased manganese and iron concentrations. Nutrient releases from sediments.

A combination of approaches to management for water quality is likely to be the most valuable in practice: that is, a combination of field observations, experimental tests, application of engineering limnology and use of simulation modelling to predict the consequences of the operational rules. There are some success stories from the USA about such 'combination' approaches within the Clean Lakes Program, started in 1972 and now coming into fruition. South Fork Rivanna Reservoir in Virginia solved water quality problems:

(i) by implementing a strict run-off control ordonance;
(ii) by implementing agricultural, roadway and other 'best' management practices;
(iii) by the construction of a detention basins; and
(iv) by employment of a watershed manager (EPA, 1987).

4.5. Management options

Management options available can be considered according to the location where the options are to be operated which are the watershed, in the body of the reservoir itself (in-reservoir) and at the outflow(s).

4.5.1. Management at the watershed level

It is clear that any management of reservoir water quality must take into account human activities within the watershed and regulate them. In general, the environmental health of any standing water body and, in particular, of drinking water storage reservoirs will be compromised by the consequences of the following human activities in the watershed:

(i) the disposal of domestic waste water;
(ii) the disposal or runoff of agricultural waste water, particularly if it includes effluents from intensive animal husbandry;
(iii) runoff from farm lands or from land subject to erosion;
(iv) runoff from land in a region subject to atmospheric pollution in the form of acid rain;
(v) seepage from dumps of ores containing enhanced concentrations;
(vi) organic toxic compounds resulting from application of pesticides in agriculture and forestry; and
(vii) runoff contaminated by xenobiotics, by persistent organic compounds used as industrial catalysts or by minute traces of pharmaceutical compounds of unknown activity from hospital waste (Bernhardt, 1990, 1991).

How to manage these deleterious effluents from the watershed involves the development of ecotechnological 'know-how' on soil types and processes, on the time courses of agricultural and industrial activity and on the pattern of the water drainage network. It is also important to distinguish between point and non-point sources of nutrients causing an enhanced rate of eutrophication.

Table XV lists some of the eco-technological methods available for solving certain problems that

Table XV. Ecotechnological methods applied to reservoir watershed management and recovery.

Problem to be solved	Methods
Pollution by effluents	Recirculation within plants
	Efficient technology
	Diversion of effluents
	Purification plants
Excess nutrients	Diversion of wastes
	Tertiary treatment plants
	Reduced fertilization
	Meadow zones on the banks
	Wood zones on the banks
	Changed agricultural practices
	Wetlands maintenance and creation
	Pre-impoundments on inflows
	Wahnbach P-reduction plant on the inflow stream
Sediment transport to reservoirs	Erosion control
	Rehabilitation of river banks
	Reforestation
	Ground water recharge
	Pre-impoundment of inflows
	Series of small ponds with controlled throughput
Eutrophication and oxygen depletion of rivers	River restoration
	Re-oxygenation
Heavy metal contamination	Reduction of pollution at plants
	Wetlands
Decreased biodiversity due to reservoir construction	Reintroduction of native species
	Maintenance of wetlands as nursery grounds
	Maintenance of preserved areas for native species

arise in watersheds. These involve the application of certain basic principles to watershed management:

1. Taking steps to protect and improve spatial heterogeneity by ensuring the existence of riparian forests and a natural vegetation mosaic. An important aspect of ensuring forest recovery, qualitatively and quantitatively, is the planting of native species of trees rather than quick growing introduced species. The riparian forest is capable of functioning as a 'biological filter' which can remove phosphorus and nitrogen from the inflows as well as retain sediment and suspended material thus decreasing inputs to the reservoir.
2. Taking steps to maintain, protect and encourage the recovery of areas of natural wetland so that biological diversity is enhanced and denitrification is encouraged. Wetlands areas near the reservoir and within the watershed form significant buffer zones in the ecotone of the land/water interface which behave as a reserve of native aquatic species which become seed species for colonizing the reservoir. The wetlands become epicentres for the re-colonization of the reservoir biota.
3. All point sources of nutrient pollution must be treated, even to the extent as using pre-impoundments in the rivers.
4. Taking steps to avoid the input from non-point sources of sediment, dissolved nutrient salts and any form of toxic chemicals. This may involve
   (i) changing agricultural practices within the watershed (what crops are grown; the nature and manner of fertilizer application, that is, when applied, and how applied). Examples can be read in Chapter XII (this volume); or
   (ii) it may involve protection of the reservoir shore-line with marginal vegetation such as tree species capable of withstanding flooding or such as floating, submergent of emergent macrophytes.
5. Taking steps to avoid inputs of sediments to the reservoir by means of pre-impoundments along the main river courses. A specific reservoir protection procedure is the Wahnbach type phosphorus elimination plant at the reservoir inflow (Bernhardt & Schell, 1979; 1982).

4.5.2. Management within the reservoir
Different techniques are available for managing the improvement of water quality within the body of the reservoir, which are listed in Table XVI. Some of these are discusses in more detail.

Selective withdrawal: 'Dilution as a solution to polution' is a useful technique in lakes and reservoirs (Welch & Patmont, 1980). However, it is rarely to be realized for the reservoir as a whole and can best be performed for selected layers. The cheapest option for improving water quality involves hydraulic regulation which is choosing the best time and depth of the outflow. This is a good method provided the limnological situation of the reservoir is known beforehand. A simple but rarely adopted precaution is to ensure, at the planning stage, that the reservoir possesses a multiple outlet structure with the outlet pipes spaced vertically at about 5 m intervals. A minimum provision would be a surface and a bottom possibility of water

Table XVI. Management options within the reservoir water body.

Measure	Means	References
Hydraulic regulation	Selective offtake and withdrawal	Straškraba, 1986
Fish management (biomanipulation)	Zooplankton control —	
	Phytoplankton reduction	Gulam et al., 1990
Artificial mixing	1. Destratification	Symons et al., 1967
	2. Hypolimnetic aeration	Bernhardt, 1967
	3. Epilimnetic mixing	Straškraba, 1986
Phosphorus inactivation	1. Allum precipitation	Cooke & Kennedy, 1988
	2. Sediment covering	Petersen, 1980
Sediment aeration	Sediment injection	Ripl, 1976; 1980
	Sediment removal	Hanson & Stefan, 1985
Light reduction	Shading, covering, suspensions, colours	Jørgensen, 1980

discharge. A much better control is afforded by reservoirs with multiple offtake horizons so that good quality water can always be abstracted by selecting the appropriate outlet depth in relation to thermal stratification (Fig. 35). The technique of zonal chemical determinations (Chapter XII, this volume) will provide warning of the quality of water from inflowing streams or at particular depths. Automatization of monitoring water quality parameters (depth profiles of temperature, dissolved oxygen and water transparency) is provided by the 'clear layer' equipment of Pařízek (1984); this assumes that the most transparent water gives optimal quality with lowest concentrations of organic substances.

Artificial mixing: The aim of artificial mixing is

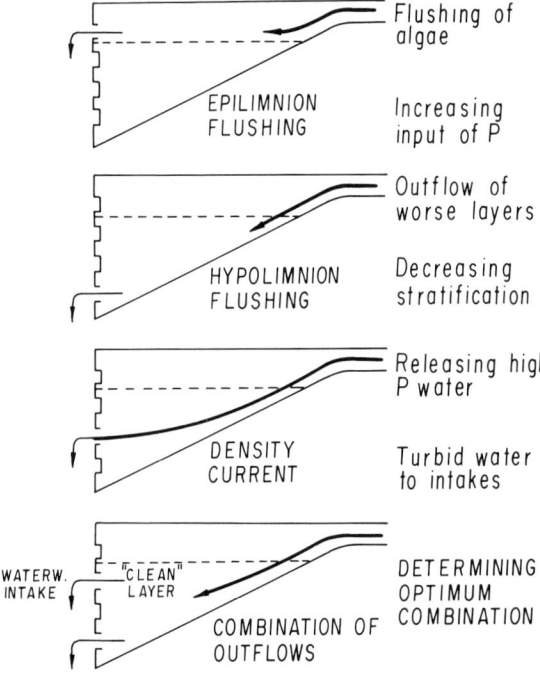

Fig. 35. The use of multiple outlet structures for reservoir water quality management. However, each positive effect is accompanied by some negative side effect and considerations have to be given which one predominates. From above: Withdrawal through the surface outlet is used for flushing of excessive algae. Simultaneously, the phosphorus input to the surface layers increasis. Hypolimnion flushing is used to release oxygenless and iron, manganese and phosphorus rich hypolimnetic waters. Simultaneously, stratification is decreased. By creating density currents a peak of phosphorus rich water (e.g. during spring floods) can rapidly pass the reservoir. Simultaneously, turbid layers can reach the raw drinking-water intakes. From Straškraba (1986).

either oxidation of a deoxygenated hypolimnion or the inhibition of phytoplanktonic growths. Fig. 36 illustrates the three mixing types widely used: – destratification by total mixing of the water column, re-aeration of the hypolimnion or epilimnetic mixing. A fourth type, the metalimnetic aerator, was described by Stefan *et al.* (1987).

Aerators capable of breakdown of a thermocline (destratification) can often also be used for oxidation of the hypolimnion. There may be a price to pay as the process of increasing the near-bottom dissolved oxygen concentrations may also result in higher phosphorus levels and increased algal growth in the warmer epilimnion. Lorenzen & Mitchel (1973) were the first authors to model the circumstances under which desirable effects of destratification were to be expected. More theoretically oriented is the treatment by Henderson-Sellers (1981). Earlier examples of practical use are summarized by Pastorok *et al.* (1980) who claim that from 40 attempts (not following the principles by Lorenzen & Mitchel) resulting in relatively full destratification in 65% there was a significant change in biomass, in 30% an increase and in 70% a decrease, accompanied by changes in species composition. Raman (1988) provides some practical examples for twelve lakes in Illinois. Burns & Powling (1981) and Burns (1990) reports some examples from Australia with negative effects as do Fast & Hulquist (1982) where compressed air used to destratify a lake caused supersaturation of dissolved nitrogen and downstream fish kills.

There are now aerators which can oxidize the hypolimnion without destroying the thermocline (Bernhardt, 1967; 1987; Anonymus, 1985; McQueen & Lean, 1986; Prien & Bernhardt, 1989; Anonymus, 1989).

Epilimnetic mixing can be recommended for the reduction of phytoplanktonic biomass growth. The surface water layers are mixed to a reservoir depth decisive for minimal net phytoplankton production, namely that z_{mix} is deeper than the depth at which photosynthesis is reduced to zero, so that the reservoir is 'optically deep' (Talling, 1957a). Such mixing systems have been developed for stratifying impoundments by designing an engineering system that controls or prevents thermal stratification. Some systems use compressed units which discharge air bubbles

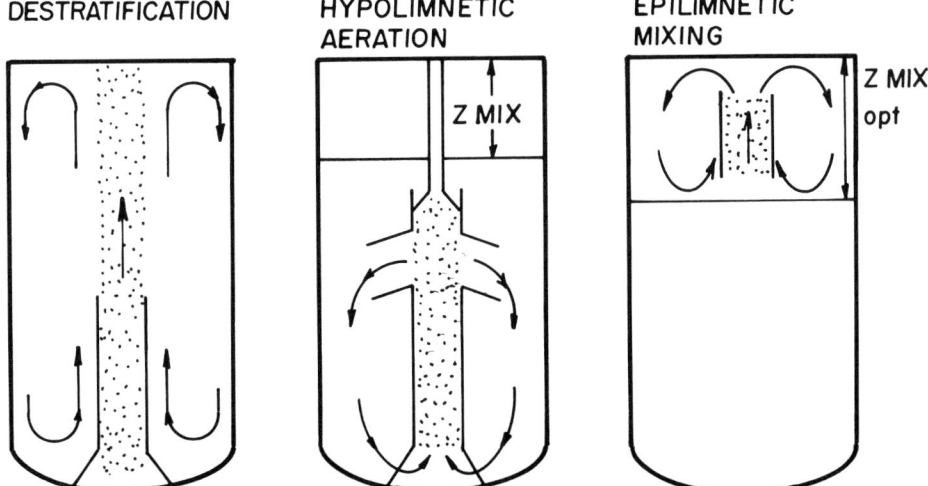

Fig. 36. Three basic modes of artificial reservoir mixing: Destratification, when the whole water layer is mixed, usually accompanied with earation. Hypolimnetic aeration prevents total destratification, only the hypolimnion being directly enriched with oxygen. Epilimnetic mixing – a preventive approach based on decreasing the light availability for algae by mixing to a calculated depth optimum from the point of view of decreasing algal production. From Straškraba (1986).

through bottom diffusers (Symons *et al.*, 1967) and others physically transport water by pumps (Ridley *et al.*, 1966). Cooley & Harris (1954) designed the mixing system of three-angled inlet jets installed in the London reservoirs by the then Metropolitan Water Board. Sufficient mixing energy could be applied to prevent the onset of summer thermal stratification or to destroy an established thermocline in 17 m deep reservoirs, often with the aid of wind energy applied to the longest axis. In practice, the degree of horizontal and vertical mixing achieved resulted in isothermy throughout the summer and a distribution of chlorophyll to the non-photosynthetic depths. Ridley & Steel (1975) even proposed the possibility of changing the optical properties of the reservoir water by using the Θ_o-angled jet to re-suspend bottom sedimented material but this option could well increase water treatment costs. The advantages of artificial mixing are many:

(i) the absence of a mid-summer hypolimnion prevents deoxygenation and the release of nutrients from the sediments;
(ii) the reduction of algal crops in highly eutrophic conditions by mixing operations rather than by chemical algicides which failed in these impoundments (Ridley, 1970, 1971); and
(iii) the improvement of put-and-take trout fisheries by transforming a stratified water body with a warm epilimnion and de-oxygenated hypolimnion into a cooler, well-mixed and well-oxygenated one favourable for salmonids (Dendy, 1945; Fast, 1972).

In the case of the London impoundments, (iii) was not an advantage as the trout decimated the reservoir summer daphnid populations which were capable of clearing the water of edible algal crops. The consequence for treatment of raw eutrophic riverine water was that the overall costs of producing a satisfactory drinking water were much reduced. The mixing system originally designed to double the supply of potable water by oxidising the de-oxygenated hypolimnion (Cooley & Harris, 1954) proved to be a major cost-effective tool for limnological management of eutrophic impoundments that had previously been bedeviled by blue-green blooms not responding to algicide treatment (Windle-Taylor, 1964; 1965; 1967; 1969; Steel, 1972; 1976; 1978a,b; Duncan, 1990).

Biomanipulation: The theoretical basis for 'biomanipulation' a term coined by Shapiro *et al.* (1982), was already established during the 1950s and 1960s by Hrbáček *et al.* (1961), Hrbáček *et al.* (1966), Brooks & Dodson (1965), Hall *et al.* (1970). Hrbáček showed that fish stock levels in fish ponds influence the level of primary production and nutrient cycling via the herbivorous grazing zooplankton at an ecosystem level what has subsequently been termed the 'top-down food chain effect' as opposed to Lindeman and Hutchinson's 'bottom-up effect'. Shapiro et al. (1975) developed the idea further by manipulation of piscivors in food webs in order to produce a desirable pelagic community structure within the ecosystem for the benefit of man. The most recent review of all aspects of biomanipulation can be found in Gulati et al. (1990).

Most effective examples of biomanipulation apply to small water bodies (Leventer, 1979) because of the difficulty of manipulating fish populations in large ones. There are only a few published examples for larger water bodies like reservoirs (Benndorf et al., 1988; Hrbáček et al., 1986; Seďa et al., 1989; Zalewski et al., 1990) and the results are usually not as convincing as those from smaller water bodies. More convincing are the results from the highly eutrophic London reservoirs (Duncan, 1990; Duncan et al., 1991) where the fish biomass was so low (5–37 kg ha^{-1}) for about 20 years or more that grazer control of algal populations is dominant throughout the summer, clearing the water and reducing the costs of subsequent treatment.

Chapter XII provides examples of biomanipulation in reservoirs from Central Europe. This chapter also deals with reservoirs in which the function of fish yield for human consumption is important which is often in direct contradiction with the aims of biomanipulation (Duncan et al., 1991).

Copper poisoning: The addition of algicides like copper salts has long been used as an emergency measure for controlling excessive algal growths, usually when already well advanced. This is not advisable because of long-term accumulation of copper in sediments or the addition of a toxic chemical to drinking water. Side effects of 58 years of copper sulfate treatment of Fairmont Lake (Minnesota) are described by Hanson & Stefan (1984).

Inactivation of phosphorus: Experience in some countries has shown that the chemical coagulation of phosphorus in lakes is highly effective for up to five years (Welch et al., 1982). This management option is only suitable for use in reservoirs with retention times longer than one year. No special equipment is needed (Cooke et al., 1986).

Sediment inactivation by removal, aeration and oxidation, and cover: Peterson (1982) reviewed the methods of sediment removal and their cost-effectivness. Ripl's method of sediment aeration and oxidation is widely used in Scandinavia and Germany and serves to decrease the phosphorus release from sediments. It does require special equipment which can only be used on flat and shallow bottoms (Ripl, 1976; 1980). An alternative cheaper technique with the same aim is to cover the bottom sediments with foil, ash or crushed bricks. A review of properties, costs and effectivness of materials used is given by Cooke & Kennedy (1988).

Manipulation of the underwater light regime to reduce photosynthesis and algal biomass: The most effective technique to reduce algal biomass is to reduce their column photosynthesis, and so the capacity for algal biomass growth, by mixing the algal population uniformly throughout the reservoir depth which is possible in some reservoirs by the techniques described in 'artificial mixing' above. Photosynthesis is then reduced to zero when $z_{eu} < z_{mix}$, where z_{eu} is the depth at which PAR light is 1% of subsurface light intensity and z_{mix} is the depth of vertical mixing. An additional manipulation might be possible by increasing the background light attenuation coefficient (ϵ_q), as has been suggested by Ridley & Steel (1975). Then, the condition of $z_{eu} < z_{mix}$ is given by $z_{eu}\epsilon_q = 3.7$ (Steel, 1980).

4.5.3. Quality management of reservoir outflow water

For functional use, water stored in reservoirs may be abstracted from the downstream river after flowing out of the reservoir or it may be taken directly from the reservoir. Where the water passes to a drinking water treatment works, it is desirable to have multiple outlets at different depths. The quality of the outflowing water is directly related to the horizontal and vertical distribution of water of good quality within the reservoir and the flexibility of operation of the multiple outlets. There is a good relationship between water quality and amount of water outflowing: with large water releases, the outflowing water changes its quality according to nature of the water from other layers

Table XVII. Ecotechnological techniques for management of reservoir outflows. The references are are either to the inventor or to a good summary of the use, or both.

Technique	Reference
Selective withdrawal	Gaillard, 1984
	Pařízek, 1984
	Cassidy, 1989
	Filho *et al.*, 1990
Aeration/oxygenation at hydropower outlet works	Cassidy, 1989
Spill-water reaeration	Cassidy, 1989
Czech method of oxygenation	Haindl, 1973
Epilimnetic pumps	Quintero & Garton, 1973
	Mobley & Harsbarger, 1987

or from the inflow to replace it.

Table XVII lists some eco-technological methods of managing reservoir outflow quality.

In the case of reservoir water being released directly to the river, problems may arise where the water comes from hypolimnion of an eutrophic reservoir because it is deoxygenated, contains large concentrations of organic compounds, iron and manganese and has a high phosphorus content. The consequence may be large scale fish kills, eutrophication of the river accompanied by odours and tastes so that the water is not treatable for drinking supply. Several management options of re-oxygenation are available such as spill water oxygenation and the use of hydraulic skis.

4.5.4. Management of reservoirs with wetlands

Wetlands associated with reservoirs have several functions:
(i) they serve as buffer zones by concentrating heavy metals from the catchment area and reducing the nitrogen and phosphorus loadings;
(ii) they are areas of high biodiversity as well as sources of seed species for re-population of the reservoir;
(iii) feeding and spawning grounds of fish and aquatic birds; and
(iv) areas of active denitrification (Whitaker, 1992).

Two types of wetlands are associated with reservoirs – macrophytic wetlands with submergent, emergent and floating plants and forested wetlands colonized by tree species capable of surviving inundation for long periods. The maintenance of these wetlands, their re-population with native species and the provision of ecological conditions for their survival and expansion are important management aims (Zalewski *et al.*, 1990).

4.5.5. Management problems with reservoir systems

In many countries, reservoir cascades are built for hydroelectric power generation, which involves complex operational procedures because of the need to integrate the management of the whole system. The management for hydroelectricity has to be further integrated with hydrological operations associated with management for water quality. Inevitably, this leads to conflicts of interest. Surface water releases are better for water quality than deep water releases which are better for generating electricity. Reduction of water levels cause havoc when fish populations are spawning; the extraction of anoxic and nutrient-rich hypolimnetic water may conflict with the need to store water for use in future. Management must be capable of dealing with reservoirs as multi-functional water bodies.

It is clear that the very design of reservoir cascades represent a useful management tool for improving water quality for drinking or for raw water treatment for drinking. It is selfevident that the intake to the water treatment works should be located in the lower (lowest) reservoir. Management options should be available for selectively withdrawing water from any reservoir in the cascade, in case of trouble.

4.6. Mathematical modelling in management

Several kinds of models are available for use in management:
1. *Prescriptive models* which include the different types of models mentioned in Section 2.2.2. These do not calculate directly the management options appropriate for a given situation. By means of a scenario analysis simulating the outcome of different management options and different possible situations of the system they can be used to indicate the adequate management possibilities.
2. *Management* or optimization *models* which

incorporate selection procedures for choosing the best suitable option according to a set of criteria appropriate to the situation. Fig. 37 indicates the steps of formulation and the major components of a water quality management model. First, the management objective for which the model is to be used has to be set. This determines the type of waterbody as well as the processes leading to given water quality. Then a model relevant for this type of problem is selected and validated by confrontation with reality. The simulation model itself is sometimes able to indicate new management options by pointing out the manageable variables to which the system strongly reacts. The variables which can be manipulated have to be selected from those, which are due to natural forces and other fixed site characteristics. The most important step in a management model formulation is the determination of the goal function and constraints, to which it is subject. The most common formulation of the goal function in water quality management consists in minimization of costs for preserving a given water quality. In the model water quality is characterized by prescribed levels of certain variables like DO or algal biomass. One basic constraint is in fact due to the model used, others are due to physical limits and capabilities of different management options. Costs for applying each management option have to be specified. For those options which assume continuous or stepwise application the estimation of unit prices is necessary. An optimization algorithm is then used for selecting among the various parameter combinations leading to the solution 'optimal' in a sense given by the goal function and constraints. It is to be understood that such procedure only selects among the possibilities included in the model and is limited by the validity of the model, its assumptions and formulations as well as the limitations imposed. Therefore, the model conclusions are to be used with caution, taking into account the limitations of the model, possible inadequacies of its formulation and the possible incompleteness of the input data. The model can be run for different situations expected to happen. A deterministic way is to assume fixed parameter values. With much higher computer expences a stochastic calculation taking into account the uncertainties of the parameters is possible. The solution may be static, time independent, or dynamic taking the annual cycle of events into account. Multiobjective formulations aim at satisfying simultaneously different objectives and multiparameter solutions select among several management options.

3. *Expert systems.* These use qualitative and quantitative expressions for guiding the user toward relevant answers to complex questions. These systems consist of two parts: the software product or expert system shell which contains the computer code handling the second part – the knowledge base. While the first is general and can be used for different problems, the second one is specific. The knowledge base includes a set of rules for a specific problem which, in the judgement of the experts serve to distinguish between different alternatives. The major advantage of expert systems is their

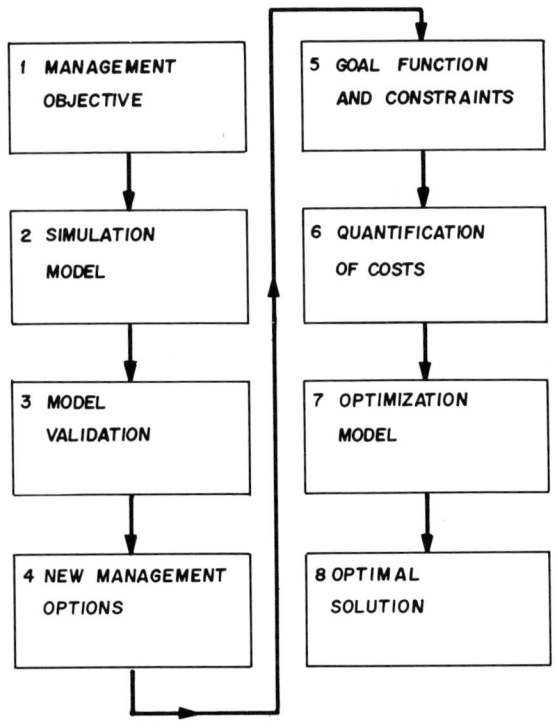

Fig. 37. Steps in a construction of a management (optimization) model. From Straškraba (1985).

ability to cover qualitative characteristics like small, medium and large and complex decision rules. For a general description of expert systems and application for environmental problems see e.g. Starfield & Bleloch (1986), for application to water quality e.g. Ford (1985).

4. *Decision support systems* represent a further extension of expert systems. They incorporate in addition to expert systems other computer software relevant for solving a decision problem. Examples are empirical and simulation models, data bases for parameter values, explanations of management options available and others. An integral part of a decision support system can be a graphical package generating explanatory drawings and texts. Another possibility is the inclusion of a geographical information system (GIS) which e.g. for a watershed generates on the basis of input data or model results maps of distribution of different variables like pollution sources. All parts of a decision support system are driven automatically on the basis of questions by the computer and answers by the user so that the decision process is maximally computer supported in a user friendly way.

For reservoir problems the two most advanced computer techniques given in (3) and (4) are summarized in Henderson-Sellers (1992).

5. Future research needs

It is our feeling that reservoir science is now coming into a period of synthesis which usually follows on from periods of data accumulation and analysis of the cumulated data which result in 'partial knowledge'. The need for comparative evaluation of the data store from previous phases is the major driving force for the preliminary syntheses attempted in the present volume. Comparative studies on reservoirs are useful for understanding the mechanisms in ecological functioning and the effects of dam operations on reservoir ecosystems. We will treat future research needs in relation to theory, practical methods and applications.

5.1. Future of theoretical reservoir limnology

General limnological topics are separated from specific reservoir ones in the discussion that follows. Wetzel (1990) believes that as theory develops, more topics will be developed at a general limnological level than at specific levels. However, we still recognise that a number of important questions are particular to reservoirs and less relevant for lakes. The following are some of the most important topics specific to reservoir limnology and needing concerted research effort:

(A) The need to study the short-term dynamics of reservoir ecosystems. Uniquely for reservoirs, we are normally able to study the limnology of reservoir from their beginnings. This means that we can investigate the whole time span of their existence, including the juvenile phase which is usually missing from lake studies. In evolutionary terms, short-term history plays a more important role in reservoir studies than is normal for lake ones. In practical terms, it is important to understand short-term non-steady state dynamics as this is what man's operation imposes upon the reservoir ecosystem with its underlying natural and regular seasonal changes. Management creates irregular pulses that are mixed with and may interfere with the regular natural pulses. This is valid for the unstable conditions created for the pelagic system when disturbed by sudden pulses of water to which the ecosystem reacts. In littoral regions, it is the difference between rapid and slow fluctuations in water level which leads to re-structuring of the complex multi-species communities living there. The profundal benthic communities are probably least disturbed by water releases but can be affected if the water is polluted.

In this context, one major question for the future is the development of methods and approaches for the study of short-term dynamics of reservoir ecosystems and the evaluation of any sudden changes. The matter is of great importance to multiple reservoir systems of the future.

(B) Transitional regimes and transitional zones, also called ecotones, are not well known, have hardly been studied but are very important for reservoirs. A typical transition is from a river to a rapidly through-flowing reservoir to a stagnant backwater and back again to a river;

not only is the water and its abiotic ingredients packaged into water parcels but so are populations and communities compartmentalised. Another important but unstudied transitional zone is the ecotone at the littoral boundary between land and water where two contrasting ecosystems meet and can influence each other. This influence will be very different in mature stable reservoirs and those subject to large water fluctuations.

(C) How the reservoir ecosystem couples with its hydrological processes is another area of research urgently needing more attention. It is possible to ignore this topic in lake studies because of their more stable hydrological regimes but it is an urgent need in reservoir studies because of their short-term dynamical nature.

(D) There is also an urgent need for much more intensive studies on reservoirs in the semi-arid regions and on those of the River Amazon. The Amazonian reservoirs are water bodies of an entirely different kind in their enormous complexity and different mechanisms of functioning. They are so unique that further studies will contribute new ideas to theoretical ecology as applied to reservoirs. It is likely that they will enforce the development of new technological approaches to management of single and multiple usage reservoirs.

(E) Moving on to more general limnological topics for future investigation, more information is needed, on the one hand, on the ecology of individual species of significance to reservoirs and, on the other hand, on the systems properties of reservoir ecosystems. Knowledge about individual species, however significant, will not of itself provide insight necessary for predicting future states of the ecosystem or how to manage it but it may be helpful in understanding topics like 'trophic upsurges' or the stability or instability of reservoir food webs.

(F) The wetlands associated with reservoirs, particularly tropical ones, are 'crying out' for a systems analysis study. They are very large, clearly of great significance but too complex for the kind of species-oriented study common in temperate wetlands (Löffler, 1979). They probably form important 'buffer zones' between land and water, represent a source of key species ('seed' species) that colonise reservoirs and are vital for the fish fauna. Wetlands are also regions of active denitrification and gas exchanges that needs quantification.

(G) The relationship between the reservoir and its watershed is known to be fundamental but has been properly investigated in very few cases; such studies will be as revolutionary as was the original study of the Hubbard Brook Experimental Forest by Likens *et al.* (1977). We need to know what are the levels of loading from point and non-point sources of nitrogen and phosphorus as well as of toxic substances in different kinds of reservoirs. This need is particularly urgent in the tropics because of the large scale of reservoir construction for power immediately followed by extensive development of catchments agriculturally, industrially and for new settlements. A common problem with new tropical reservoirs is land erosion due to poor engineering which results in greatly enhanced reservoir loadings with mineral suspended material; the consequences of this chemically and biologically needs quantification. Loadings with fine suspended matter is a characteristic of reservoirs in semi-arid regions and we know something about the consequence to nutrient availability to the phytoplankton and food availability of algae to zooplankton.

(H) The recently confirmed large scale global changes in temperature and carbon dioxide levels have consequences for reservoirs although we need to know much more about them for predictive purposes.

5.2. Methodological problems for the future

Limnology is an ecological discipline and has in common with ecology methodological problems associated with studying complex natural ecosystems. An interesting theoretical advance lies in the idea that natural systems are characterised by *deterministic chaos*. This implies that the direction of the future development of such systems is extremely sensitive to their initial conditions and any minor changes in these so that they are unpredictable to a greater or lesser extent.

This is a mathematically derived conclusion and to what extent is it valid for natural systems remains to be seen. The application of deterministic chaos to meteorology has proved very fruitful but meteorology is confined to physical systems and does not have regulating mechanisms so characteristic of biologically based ecosystems. However, the regularities so commonly observed in aquatic ecosystems gives some hope for successful mathematical descriptions.

The other fundamental methodological question concerns *dynamic versus static measurements*. Most studies deal with measurements of state rather than the more interpretable but more difficult measurements of rates of change. Rates of change are clearly the more direct measure of what is happening in dynamic systems like reservoirs. Similarly, most investigations are content with *description* of different phenomena rather than *prediction* of effects and of the direction of future changes. This is a matter of defining the question being asked in a study and developing appropriate methodology.

Mathematical modelling represents one efficient tool both helpful in the matters just discussed and an excellent language of communication between limnologists and engineers. Such modelling is open for further development which is desirable in two directions. One is the need for suitable methods to be developed for studying complex non-linear, adaptive and self-organising systems which are a feature of biology. Another area that requires some attention is how to transfer, easily and effectively, complex knowledge from one set of specialists to another set of specialists (the managers) who need it for informed decision-making; without this, reservoir limnologists are merely amusing themselves.

The matter of *scales* of study is another area that needs more attention. There is a requirement for spatial scaling to be flexibly moved up or down. Enlarging the scale of investigation is what remote sensing allows us to do. Scaling down is also essential or we are to understand the effect of micro-scale eddies upon the relative behaviours of algae and their grazers. Scaling up and down the time scales of investigations also needs flexible handling. Slapy Reservoir has been studied for 30 years or Lobo/Broa Reservoir (S. Carlos, Sao Paulo, Brazil) for 20 years which is long enough to detect the effects of global changes, the existence of long-term cycles and inter-annual differences in hydrology as well as the occurrence of extreme events as on-off phenomena. The methodology for scaling down the time scale of observations now exists, as was shown by the results of short-term automatic recording in Slapy Reservoir (Section 2.2.).

5.3. Applied problems: managing reservoir water quality

Managing water quality problems fall into two groups: those concerning general environmental water quality and their management and specific management problems related to reservoirs.

The management of natural systems is largely ineffective because of the crudity of the basic approach. Industrial production of energy and materials cause the negative effects on natural systems on the one hand and on the other hand the energy and materials used to combat these themselves compound the environmental problems further. The usual solution is to transfer the environmental problem to elsewhere: into the sea (sewage effluents or oil spills) or into the atmosphere (acid rain or nuclear waste) or from the site of extractions to the sites of manufacture.

What must be developed are more sensitive, less environmentally damaging solutions and better managerial implementation of these. Straškraba (1986) coined the term 'ecotechnology' and this notion was given a full coverage by Mitsch & Jørgensen (1989). A new journal, Environmental Engineering (a Journal for Ecotechnology) is devoted to these topics. One hopeful development is the notion of biomanipulation dealt with in detail by Gulati *et al.* (1990).

We have introduced the distinction between dam reservoirs and impoundments. Many impoundments are ancient, created for a single purpose, usually irrigation, and still exist and function properly after many centuries due to expert management by knowledgeable hydrologists (Sri Lanka). Some impoundments are modern, also with a single function of drinking water supply, and are able to operate with high nutrient sources by the application of knowledgeable limnological management (the

London reservoirs). In striking contrast are the more recent dam reservoirs which last only a few decades or occasionally for a century because of poor penny-pinching construction for quick profit rather than for the benefit of local needs (hydroelectric reservoirs in the developing countries). Management of drinking water reservoirs in large urban areas is an important and urgent problem to be followed. In many countries intensive urbanization produce quantitative impacts in the water resources and for this reason large dam reservoirs were constructed for water storage. As an example the town of Sao Paulo has 13 such reservoirs. These systems are subject to pollution and contamination. Atmospheric pollution (acid rain) is a serious problem not only in the North but starts to threaten the South, too (Tundisi, 1986).

The presence of diverse contaminant chemicals in waters everywhere represent a growing problem for limnology. How these behave in aquatic food webs is largely unknown as so, without study, forms now and will form in future a hidden hazard for the human population.

The development of effective tools for management of reservoirs must involve the coupling of diverse but important aspects, each with its complex framework of measurement technology. Modelling provides one solution for this and needs to be developed further. Automation offers some hope for obtaining records of events which operate on different time scales but are largely confined to physical and chemical variables and not to biological ones. There exists a huge store of relevant limnological knowledge compiled by scientific study of reservoirs which is not implemented in management. It is clear that watershed management and reservoir management is an area than needs to be integrated immediately by means of synoptic studies of both.

McComb & Williams (1987) and Puig *et al.* (1987) claim that the impact of reservoirs on downstream biological communities belong to future research needs. Managing impoundment releases (timing and quantity) to reduce impact on downstream aquatic communities is a particular topics of concern.

5.4. Reservoirs as teaching grounds for dynamic systems ecology

As we stated several times in this volume reservoirs are dynamic systems with high spatial and temporal variability both in the physical forcing functions and in the response of the biological community. The teaching of dynamic approaches in ecology can thus be much improved by the theoretical and practical examples provided by reservoir systems. In the development of ecology, limnology was often the discipline producing progressive ideas. Interdisciplinary fields are known to be a nursery ground for major scientific advances. The same is now true in ecology for intermediary systems like ecotones, marginal systems between land and water (Naiman & Décamps, 1990). This is because their complexity is high, dynamics rapid and communities subject to variable conditions. We are convinced that reservoirs as intermediates between rivers and lakes will play a similar role.

Also, the linkages between theoretical, applied and managerial aspects can be stressed with several examples from local/regional scales of reservoir studies. As an example, reservoir studies help to extend our solutions of importnat applied problems like Environmental Impact Assesment. This is particularly important for an improvement of methods for EIA in several regions of the world, as shown in Section 4.1. In this respect comparative studies on a regional basis are a very important background for this type of training.

Acknowledgements

J.G. Tundisi acknowledges financial support from IUNU, UNEP and ICSU for trips to Europe and Czechoslovakia respectively in May and September 1990 and December 1991.

References

Ackerman, W. C., F. G. White & E. B. Worthington (eds.), 1973. Man-Made Lakes: Their Problems and Environmental Effects. American Geophysical Union, Monograph 17, Washington, D.C., 847 pp.

Allen, T. F. M. & Starr T. B., 1982. Hierarchy: Perspectives for

Ecological Complexity. Univ. Chicago Press, Chicago, 310 pp.
Anderson, E. R. & D. W. Pritchard, 1954. Physical Limnology. In: Lake Mead Comprehensive Survey 1948-49. US Dept. Interior Geol. Survey: 186-210.
Andrews, A. K., G. T. Auble, R. A. Ellison, D. B. Hamilton, J. E. Roelle, D. R. Marmorek & O. L. Loucks, 1981. Impact of acid precipitation on watershed ecosystems: an application of the adaptive environmental assessment processes. In: W.J. Mitsch, R. W. Bossermann & J. M. Klopatek (eds.), Energy and Ecological Modelling. Proceedings of a Symposium held from 20 to 23 April 1981 at Louisville, Kentucky. Developments in Ecological Modelling 1. Elsevier, Amsterdam: 393-400.
Anonymus, 1984. Lake and Reservoir Management. Proceedings of the Third Annual Conference. North American Lake Management Society, October 18-20, 1983 Knoxville, Tennessee. US Environmental Protection Agency, Washington, D.C., 604 pp.
Anonymus, 1985. LIMNO - Cure for degraded waters. Prospectus. Atlas Copco Aqua Technique, Stockholm, 8 pp.
Anonymus, 1986. CE-QUAL-R1: A numerical one-dimensional model of reservoir water quality. User's Manual/Instruction. US Army Corps of Engineers, Environmental Laboratory, Waterways Experiment Station, Report E-82-1.
Anonymus, 1989. TIBEAN - The Revolutionary Technology of Lake Restoration. Petersen Schiffstechnik GMBH, Hamburg, 8 pp.
Aranha, F. J., 1990. Influencia de fatores hidraulicos e hidrologicos na composicao e biomassa de fitoplancton em certos períodos de tempo, na Represa de Barra Bonita, Médio Tiete, S. P. Dissertacao de Mestrado, Universidad do Sao Paulo, Escuela do Enghineria, Sao Carlos, 150 pp.
Avakian, A. B., V. A. Sharapov, V. P. Saltapkin & M. Fortunatov, 1979. Vodokhranilishcha mira (Reservoirs of the World. - In Russian). Nauka, Moskva, 284 pp.
Balon, E. K. & A. G. Coche (eds.), 1974. Lake Kariba: a Man-Made Tropical Ecosystem in Central Africa. Monogr. Biol. 24, Junk, The Hague, 767 pp.
Baxter, R. M., 1977. Environmental effects of dams and impoundments. Ann. Rev. Ecol. Syst. 8: 255-283.
Beck, M. B. & G. Van Straten (eds.), 1983. Uncertainty and Forecasting of Water Quality. Springer, Berlin, 386 pp.
Bedford, K. W., R. M. Sykes & C. Libicki, 1983. Dynamic advective water quality model for rivers. Proc. ASCE, J. Env. Engng Div. 109: 535-554.
Begon, M., J. L. Harper & C. R. Townsend, 1986. Ecology, Individuals, Populations and Communities. Blackwell Sci. Publ. Oxford, 876 pp.
Beljaev, V. I., 1980. Prognozirovanie izmenenij struktury slozhnykh ekologicheskikh sistem pod dejstviem antropogennykh faktorov. (Prognoses of structural changes of complex ecological systems. - In Russian). Avtomatika, Kiev 4: 73-79.
Benndorf, J., M. Zesch & E.-M. Wiesner, 1975. Prognose der Phytoplanktonentwicklung in geplanten Talsperren durch Kombination von wachstumskinetischen Modellvorstellungen und Analogiebetrachtungen zu bestehenden Talsperren. Int. Revue ges. Hydrobiol. 60: 737-758.
Benndorf, J., H. Schulz, A. Benndorf, R. Unger, E. Penz, H. Kneschke, K. Kossatz, R. Dumke, U. Hornig, R. Kruspe & S. Reichel, 1988. Food web manipulation by enhancement of piscivorous fish stock: long-term effects in the hypertrophic Bautzen reservoir. Limnologica (Berlin) 19: 97-100.
Bensari, D., C. A. Brebbia & D. Ouazar (eds.), 1991. Computational Water Resources. Computational Mechanics Publications, Southampton, 330 pp.
Benson, N. G., 1982. Some observations on the ecology and fish management of reservoirs in the United States. Can. Water Res. J. 7: 2-25.
Berg, C. O., 1963. Middle Atlantic states. In: D. G. Frey (ed.), Limnology in North America. Univ. of Wisconsin Press, Madison: 191-237.
Bernhardt, H., 1967. Aeration of Wahnbach Reservoir without changing the temperature profile. J. Amer. Water Works Assoc. 59: 943-964.
Bernhardt, H., 1987. Strategies of lake sanitation. Swiss J. Hydrol. 49: 202-219.
Bernhardt, H., 1990. Control of Reservoir Water Quality. In: H. H. Hahn & R. Klute (eds.), Chemical Water and Wastewater Treatment. Springer, Berlin: 285-304.
Bernhardt, H., 1991. Protection of water resources. European Water Congress: Surface Water Quality in the European Community, Antwerpen-Belgien, 19-20. 9. 1991, 21 pp.
Bernhardt, H. & H. Schell, 1979. The technical concept of phosphorus-elimination at the Wahnbach estuary using floc-filtration (The Wahnbach system). Z. f. Wasser- und Abwasser- Forschung 12: 78-88.
Bernhardt, H. & H. Schell, 1982. Energy-input-controlled direct filtration to control progressive eutrophication. J. Amer. Water Works Assoc. 74: 261-268.
Beverton, R. I. H. & S. J. Holt, 1957. On the Dynamics of Exploited Fish Populations. Fish. Invest. London, Ser. 2, 19, 533 pp.
Biswas, A. K., 1981. Models for Water Quality Management. McGraw-Hill, New York, 348 pp.
Bonetto, A. A., H. P. Castello & I. R. Wais, 1987. Stream regulation in Argentina, including the superior Paraná and Paraguay Rivers. Regulated Rivers 1: 95-100.
Borman, F. H. & G. E. Likens, 1979. Pattern and Process in a Forested Ecosystem. Springer, New York, 253 pp.
Bosserman, R. W. & R. K. Ragade, 1982. Ecosystem analysis using fuzzy set theory. Ecol. Modelling 16: 191-208.
Braat, L. C. & W. F. J. Van Lierop, 1987. Economic-Ecological Modeling. North-Holland, Amsterdam, 329 pp.
Brandl, Z., 1963. Příspěvek k biologii litorálu lipenské údolní nádrže. (Contribution to the biology of the littoral in Lipno Reservoir. In Czech). Živoč. výroba 8: 175-188.
Brandl, Z., 1973. Relation between the amount of net zooplankton and the depth of station in the shallow Lipno Reservoir. In: J. Hrbáček & M. Straškraba (eds.), Hydrobiological Studies 3, Academia, Praha: 7-51.
Brebbia, C. A., D. Ouazar & D. Ben Sari, 1991. Computer Methods in Water Resources II. Vol. 3 Computer Aided Engineering Water Resources. Computational Mechanics Publications, Southampton, 347 pp.
Brooker, M. P., 1981. The impact of impoundments on the

downstream fisheries and general ecology of rivers. Adv. appl. Ecol. 6: 91–152.

Brooks, J. L. & S. J. Dodson, 1965. Predation, body size, and composition of plankton. The effect of a marine planktivore on lake plankton illustrates theory of size, competition, and predation. Science 150, 3692: 28–35.

Burns, F. L., 1990. A decade of progress in lake mixing: Australia revisited. Lake Line 10: 6–7,13.

Burns, F. L. & I. J. Powling (eds.), 1981. Destratification of Lakes and Reservoirs to Improve Water Quality. Australian Water Resources Council, Conf. Ser. No. 2, 915 pp.

Calijuri, M. C., 1988. Respostas fisioecologicas da communidade fitoplanctonica e factores ecologicos em ecosistemas com diferentes estagios de eutrofizacao. PhD. Thesis, Univ. of Sao Paulo, School of Engineering, S. Carlos, 293 pp.

Campbell, P. G., B. Bobee, A. Caille, M. J. Demalsy, P. Demalsy, J. L. Sasseville & S. A. Vissor, 1975. Pre-impoundment site preparation: A study of the effects of top soil stripping on reservoir water quality. Verh. Int. Verein. Limnol. 19: 1768–1777.

Campbell, P. G., R. Perrier & M. Cantin, 1982. International Symposium on Reservoir Ecology and Management, June 1981, Quebec, Canada. Can. Water Res. J. 7: 1–470.

Canfield, D. E. & R. W. Bachmann, 1981. Prediction of total phosphorus concentrations, chlorophyll-a, and Secchi depths in natural and artificial lakes. Can. J. Fish. Aquat. Sci. 38: 414–423.

Cassidy, R. A., 1989. Water temperature, dissolved oxygen, and turbidity control in reservoir releases. In: J. A. Gore & G. E. Petts (eds.), Alternatives in Regulated River Management. CRC Press, Boca Raton, Florida: 27–62.

Chapra, S. C. & R. P. Canale, 1991. Long-term phenomenological model of phosphorus and oxygen for stratified lakes. Wat. Res. 25: 707–715.

Cherry, D. S. & R. K. Guthrie, 1975. Significance of detritus or detritus associated invertebrates to fish production in a new impoundment. J. Fish Res. Bd. Canada 32: 1799–1804.

Clark, C. W., 1976. Mathematical Bioeconomics: The Optimal Management of Renewable Resources. Wiley, New York, 352 pp.

Clark, C. W., 1987. Fisheries as renewable resources. In: L. C. Braat & W. F. J. Van Lierop (eds.), Economic-Ecological Modeling. North-Holland, Amsterdam: 73–86.

Cooke, G. D. & R. H. Kennedy, 1988. Water Quality Management for Reservoirs and Tailwaters. Report 1. In-Reservoir Water Quality Management Techniques. US Army Engineer Waterways Experiment Station, Vicksburg, Mississippi, Technical Report E-88-X, 181 pp.

Cooley, P. & S. L. Harris, 1954. The prevention of stratification in reservoirs. J. instn. Wat. Engrs. 8: 517–537.

Craig, J. F. & J. B. Kemper (eds.), 1987. Regulated Streams, Advances in Ecology. Plenum Press, N.Y., 431 pp.

Crawford, P. J. & D. M. Rozenberg, 1984. Breakdown of conifer needle debris in a new northern reservoir, Southern Indiana Lake, Manitoba. Can. J. Fish. Aquat. Sci. 41: 649–658.

Davis, E., R. B. Smith & G. Goos, 1973. Predicting reservoir quality changes – A laboratory investigation. In: E. R. Reinelt, A. H. Laycock & W. M. Schultz (eds.), Proceedings of the Symposium on the Lakes of Western Canada, 1973. The University of Alberta, Edmonton, Alberta: 40–55.

De Meis, M. R. & J. G. Tundisi, 1986. Geomorphological and limnological processes as a basis for lake typology. The Middle Rio Doce Lake System. An. Acad. Brasil. Cienc. 58: 103–120.

Dendy, J. S., 1945. Fish distribution in Norris Reservoir, Tennesseee, 1943. Depth distribution of fish in relation to environmental factors. J. Tennessee Acad. Sci. 20: 114–135.

Dickman, M., 1969. Some effects of lake renewal on phytoplankton productivity and species composition. Limnol. Oceanogr. 14: 660–666.

Dillon, P. J., 1975. The phosphorus budget of Cameron Lake, Ontario: The importance of flushing rate relative to the degree of eutrophy of a lake. Limnol. Oceanogr. 20: 28–39.

Dillon, P. J. & F. H. Rigler, 1975. A simple method for predicting the capacity of a lake for development based on lake trophic status. J. Fish. Res. Bd. Canada 32: 1519–1531.

Dixon, J. A., L. M. Talbot & G. J. M. Lemoigne, 1989. Dams and the Environment. Considerations in the World Bank Projects. World Bank Technical Paper No 110. The World Bank, Washington, D.C., 64 pp.

Duncan, A. & J. Rzoska, 1981. Land Use Impact on Lake and Reservoir Ecosystems. Proc. MAB Project 5, Facultas-Verlag Wien.

Duncan, A., 1990. A review: limnological management and biomanipulation in the London reservoirs. Hydrobiologia 200/201: 541–548.

Duncan, A., J. Kubečka, D. G. Hopkins & J. N. Bubb, 1991. Composition of fish stock under limiting conditions of the London water supply reservoirs. Bull. zool. Mus., Spec. Issue Aug., 1991.

Duthie, H. C. & M. L. Ostrofsky, 1982. Use of phosphorus budget models in reservoir management. Can. Water Res. J. 7: 337–347.

Dvořáková, M. & H.-P. Kozerski, 1980. Three-layer model of an aquatic ecosystem. ISEM Journal 2: 63–70.

Edinger, J. E. & E. M. Buchak, 1983. Developments in LARM2: a longitudinal-vertical, time-varying hydrodynamic reservoir model. US Army Engineer Waterways Experiment Station, Vicksburg, Mississippi, Technical Report E-83-1.

Edmondson, W. T., 1991. The Uses of Ecology. Lake Washington and Beyond. Univ. of Washington Press, Washington, 329 pp.

Elser, J. J. & B. L. Kimmel, 1985: Nutrient availability for phytoplankton production in a multiple-impoundment series. Can. J. Fish. Aquat. Sci. 42: 1359–1370.

EPA, 1987. Clean Lakes Program. Office of Water Regulations and Standards Office of Water US Environmental Protection Agency, Washington, D.C., 24 pp.

Fast, A. W., 1972. The Effects of Artificial Aeration on Lake Ecology. Environmental Protection Agency, Project No. 16010 EPA, Washington, D.C.

Fast, A. W. & R. G. Hulquist, 1982. Supersaturation of Nitrogen Gas Caused by Artificial Aeration in Reservoirs. US Army Engineer Waterways Experiment Station Hydraulics Laboratory, Vicksburg, Mississippi, Technical Report E-82-9, 77 pp.

Fernando, C. H. (ed.), 1984. Ecology and Biogeography of Sri Lanka. Junk Publishers, Den Haag.

Fiala, L. & P. Vašata, 1982. Phosphorus reduction in a manmade lake by means of a small reservoir on the inflow. Arch. Hydrobiol. 94: 24–37.

Filho, M. C. A., J. A. O. De Jesus, J. M. Branski & J. A. M. Hernandez, 1990. Mathematical modelling for reservoir water-quality management through hydraulic structure: a case study. Ecol. Modelling 52: 73–85.

Findeisen, V., 1977. Studies in Control Methodology for Water Resources Systems. Institute of Automatic Control, Technical University of Warszaw, Warszaw, 155 pp.

Fontane, D. G., J. W. Labadie & B. Loftis, 1981. Optimal control of reservoir discharge quality through selective withdrawal. Water Resour. Res. 17: 1594

Ford, D. E. & H. Stefan, 1980. Thermal prediction using integral energy model. Proc. ASCE, J. Hydraulic Engng Div. 106: 39–55.

Ford, L. N., 1985. Decision support systems and expert systems: a comparison. Information and Management 8: 21–26.

Francko, D. A. & R. T. Heath, 1981. Aluminium sulfate treatment: short term effect on complex phosphorus compounds in a eutrophic lake. Hydrobiologia 78: 125–128.

Frankland, P., 1984. Micro-Organisms in Water. Longmans, London, 527 pp.

Fraser, J. C., 1972. Regulated discharge and the stream environment. In: R. T. Oglesby, C. A. Carlson & J. A. McCanu (eds.), River Ecology and Man. Academic Press, New York and London: 263–285.

Frey, D. G., 1969. Reservoir-research objectives and practices with an example from the Soviet Union. In: Reservoir Fishery Resources Symposium. Amer. Fish. Soc., Washington, D.C.: 26–36.

Fry, F. E. J., 1957. The aquatic respiration of fish. In: M. E. Brown (ed.), The Physiology of Fishes. Academic Press, New York: 1–63.

Gaillard, J., 1984. Multilevel withdrawal and water quality. Proc. ASCE, J. Environ. Engng Div. 110: 123.

Garton, J. E., R. G. Strecker & C. R. Summerfelt, 1978. Performance of an axial flow pump for lake destratification. Proc. SE. Ass. Game Fish Commn. 30: 336–347.

Garzon, C. E., 1984. Water Quality in Hydroelectric Projects. Considerations for Planning in Tropical Forest Regions. World Bank Technical Paper 20, The World Bank, Washington, D.C., 33 pp.

Giorgini, A. & F. Zingales (eds.), 1986. Agricultural Nonpoint Source Pollution: Model Selection and Application. Contributions to a Workshop held in June 1984 in Venice, Italy. Developments in Environmental Modelling 10, Elsevier, Amsterdam, 409 pp.

Gnauck, A., 1975. Mathematische Modellierung der Veränderungen der Wasserbeschaffenheit in Trinkwassertalsperren. Acta hydochim. hydrobiol. 3: 479–487.

Gnauck, A., J. Wernstedt & W. Winkler, 1976. Zur Bildung mathematischer Modelle limnischer Ökosysteme mittels rekursiver Schätzverfahren. Int. Revue ges. Hydrobiol. 61: 609–626.

Goh, B. S., 1980. Management and Analysis of Biological Populations. Elsevier, Amsterdam, 288 pp.

Golubyev, G. & I. Shvytov, 1982. Modeling Agricultural Environmental Processes in Crop Production. International Institute of Applied Systems Analysis, Laxenburg, CP-82-S5.

Gray, W. G. (ed.), 1986. Physics-based Modeling of Lakes, Reservoirs, and Impoundments. American Society of Civil Engineers, New York, 308 pp.

Grimard, Y. & H. G. Jones, 1982. Trophic upsurge in new reservoirs: a model for total phosphorus concentrations. Can. J. Fish. Aquat. Sci. 39: 1473–1483.

Grobler, D. C., 1985. Phosphorus budget models for simulating the fate of phosphorus in South African reservoir. Water S. A. 11: 219–230.

Groeger, A. W. & B. L. Kimmel, 1984. Organic matter supply and processing in lakes and reservoirs. In: Lake and Reservoir Management. EPA 440/5/84–001. US Environ. Prot. Agency, Washington, D.C.: 282–285.

Gulati, R. D., E. H. R. R. Lammers, M.-L. Meijer & E. Van Donk, 1990. Biomanipulation: Tool for Water Management. Kluwer Acad. Publishers, Dordrecht, 628 pp.

Gunnison, D., J. M. Brannon & R. L. Chen, 1986. Reservoir Site Preparation: Summary Report. US Army Engineer Waterways Experiment Station. Vicksburg, Mississippi. Technical Report E-86-4: 22–35.

Gunnison, D., J. M. Brannon, I. Smith Jr. & G. A. Burton, 1980. Changes in respiration and anaerobic nutrient regeneration during the transition phase of reservoir development. In: J. Barica & L. Mur (eds.), Hypertrophic Ecosystems. Junk, The Hague: 151–158.

Gunnison, D., R. L. Chen & J. R. Brannon, 1983. Relationship of materials in flooded soils and sediments to the water quality of reservoirs. I. Oxygen consumption rates. Water Res. 17: 1609–1617.

Haimes, Y. Y., 1977. Hierarchical Analyses of Water Resources Systems. McGraw-Hill, New York, 478 pp.

Haimes, Y. Y., W. A. Hall & H. T. Freedman, 1975. Multiobjective Optimization in Water Resources Systems. The Surogate Worth Trade-off Method. Elsevier, Amsterdam.

Haindl, K., 1973. Suitable solution of bottom outlets of dams and oxidation outlets for the improvement of water quality in rivers. Proceedings IAHR, Istanbul.

Haith, D. A., 1982. Models for analyzing agricultural nonpoint – source pollution. International Institute of Applied Systems Analysis, Laxenburg, RR-82-17, 29 pp.

Hall, C. A. S. & J. W. Day Jr., 1977. Systems and models: terms and basic principles. In: C. A. S. Hall & J. W. Day Jr. (eds.), Ecosystem Modeling in Theory and Practice: An Introduction with Case Studies. Wiley, New York: 5–36.

Hall, D. J., W. E. Cooper & E. E. Werner, 1970. Experimental approach to the production, dynamics and structure of freshwater animal communities. Limnol. Oceanogr. 15: 839–928.

Hall, J. B. & W. Pople, 1968. Recent vegetational changes in the Lower Volta River. Ghana J. Sci. 8: 24–29.

Hanson, M. J. & H. G. Stefan, 1984. Side Effects of 58 Yars of Copper Sulfate Treatment of the Fairmont Lakes, Minnesota. Water Res. Bull. 20: 889–900.

Hanson, M. J. & H. G. Stefan, 1985. Shallow Lake Water

Quality Improvemen by Dredging. In: Lake and Reservoir Management: Practical Applications. Proceedings of the Fourth Annual Conference and International Symposium, October 16–19, 1984 McAfee, New Jersey: 168–171.

Hanson, M. J., H. G. Stefan & M. Riley, 1986. Dynamic (mathematical) modeling of lake processes for management decisions. In: Lake and Reservoir Management Volume II. Proceedings of the Fifth Annual Conference and International Symposium on Applied Lake & Watershed Management. November 13–16, 1985 Lake Geneva, Wisconsin: 225–228.

Harleman, D. R. F., 1982. Hydrothermal analysis of lakes and reservoirs. Proc. ASCE, J. Hydraulic Engng Div. 108: 303.

Heatwole, C. D., A. B. Bottcher & K. L. Campbell, 1987. Basin scale water quality model for coastal plain flatwoods. Transactions of the ASAE 30: 1023–1030.

Hecky, R. E. & G. K. McCullough, 1984a. Effect of impoundment and diversion on the sediment budget and nearshore sedimentation of Southern Indian Lake. Can. J. Fish. Aquat. Sci. 41: 567–578.

Hecky, R. E., R. W. Newbury, R. A. Bodaly, K. Patalas & D. M. Rosenberg, 1984b. Environmental impact prediction and assessment: the Southern Indian Lake experience. Can. J. Fish. Aquat. Sci. 41: 720–732.

Hejzlar, J. & M. Straškraba, 1989. On the horizontal distribution of limnological variables in Římov and other stratified Czechoslovak reservoirs. Arch. Hydrobiol. Beih. Ergebn. Limnol. 33: 41–55.

Henderson-Sellers, B., 1978. Water quality in lentic water bodies (lakes and reservoirs). Proceedings of the Baden Symposium, September 1978. IAHS-AISH 125: 192–199.

Henderson-Sellers, B., 1981. Destratification and reaeration as tools for in-lake management. Water S. A. 7: 185–189.

Henderson-Sellers, B., 1987. One-dimensional modelling of thermal stratification in oceans and lakes. Environmental Software 2: 78–84.

Henderson-Sellers, B. (ed.), 1992. Decision Support Techniques for Lakes and Reservoirs. Water Quality Modelling, Volume IV. CRC Press, Boca Raton, Florida.

Henry, R., 1990. Estrutura espacial e temporal de ambiente fisico-quimico e analise de alguns processos ecologicos na Represa de Jurumirim e sua bacia hidrografica. PhD Thesis, University of Sao Paulo, San Carlos.

Hogeweg, P., 1988. MIRROR beyond MIRROR, puddles of life. In: C. Langton (ed.), Artificial Life. Addison-Wesley, Reading, Massachussets: 297–315.

Holčík, J., 1977. Changes in fish community of Klíčava Reservoir with particular reference to Eurasian perch (*Perca fluviatilis*). 1957–72. J. Fish. Res. Bd. Canada 34: 1734–1747.

Howarth, R. A. & S. G. Fisher, 1976. Carbon, nitrogen, and phosphorus dynamics during leaf decay in nutrient-enriched stream microecosystems. Freshwater Biol. 6: 221–228.

Hoyer, M. V. & J. R. Jones, 1983. Factors affecting the relation between phosphorus and chlorophyll-a in midwestern reservoirs. Can. J. Fish. Aquat. Sci. 40: 192–199.

Hrbáček, J. (ed.), 1966. Hydrobiological Studies Vol. 1. Academia, Praha, 408 pp.

Hrbáček, J., 1984. Ecosystems of European man-made lakes. In: F. B. Taub (ed.), Lakes and Reservoirs. Ecosystems of the World. Elsevier, Amsterdam: 267–290.

Hrbáček, J., M. Dvořáková, V. Kořínek & L. Procházková, 1961. Demonstration of the effect of the fish stock on the species composition of zooplankton and the intensity of metabolism of the whole plankton association. Verh. Int. Verein. Limnol. 14: 192–195.

Hrbáček, J., O. Albertová, B. Desortová, V. Gottwaldová & J. Popovský., 1986. Relation of the zooplankton biomass and share of large Cladocerans to the concentration of total phosphorus, chlorophyll-a and transparency in Hubenov and Vrchlice Reservoirs. Limnologica (Berlin) 17: 301–308.

Hrbáček, J. & M. Straškraba (eds.), 1973. Hydrobiological Studies, Vol. 2. Academia, Praha, 348 pp.

Hrbáček, J. & M. Straškraba (eds.), 1973. Hydrobiological Studies Vol. 3. Academia, Praha, 310 pp.

Hutchinson, G. E., 1957. A Treatise on Limnology. Vol. 1. Wiley, New York, 1115 pp.

Imberger, J., 1982. Reservoir dynamics modelling. In: E. M. O'Loughlin & P. Cullen (eds.), Prediction in Water Quality. Proceedings of a Symposium on the Prediction in Water Quality, Canberra 1982. Australian Academy of Science, Canberra: 223–248.

Imberger, J., 1985. Thermal characteristics of standing waters: an illustration of dynamic processes. Hydrobiologia 125: 7–29.

Imberger, J. & J. Patterson, 1981. A dynamic reservoir simulation model – DYRESM5. In: H. B. Fischer (ed.), Transport Models for Inland and Coastal Waters. Academic Press, New York: 310–361.

Imberger, J., J. Patterson, B. Hebbert & J. Loh, 1978. Dynamics of a reservoir of medium size. Proc. ASCE, J. Hydraulic Engng Div. 104: 725–743.

James, W. F., R. H. Kennedy, W. E. Shain & R. K. Myers, 1988. Leaf litter breakdown in a recently impounded reservoir. Water Res. Bull. 24: 831–837.

Javornický, P., 1980. Density dependent effects. In: E. D. Lecren & R. H. Lowe-McConnel (eds.), The Functioning of Freshwater Ecosystems. Cambridge Univ. Press, Cambridge: 170–173.

Jeffers, J. N. R., 1978. An Introduction to Systems Analysis: With Ecological Applications. University Park Press, Baltimore, 198 pp.

Jenkins, R. M., 1968. The influence of some environmental factors on standing crop and harvest of fishes in US reservoirs. In: Reservoir Fishery Resources Symposium, Proceedings of a Conference of the American Fisheries Society, Athens, Georgia, April 1967. University of Georgia Press, Athens, Georgia: 298–321.

Jenkins, R. M. & D. I. Morais, 1971. Reservoir sport fishing effort and harvest in relation to environmental variables. In: G. E. Hall (ed.), Reservoir Fisheries and Limnology. American Fisheries Society, Special Publication 8: 371–384.

Jones, J. R. & M. V. Hoyer, 1982. Sportfish harvest predicted by summer chlorophyll-a concentration in midwestern lakes and reservoirs. Trans. Am. Fish. Soc. 111: 176–179.

Jørgensen, S. E., 1976. A model of fish growth. Ecol. Modelling 4: 303–313.

Jørgensen, S. E., 1980. Lake Management. Pergamon Press, 167 pp.

Jørgensen, S. E., 1986. Fundamentals of Ecological Modelling. Elsevier, Amsterdam, 389 pp.

Jørgensen, S. E., 1991. Modelling Environmental Chemistry. Elsevier, Amsterdam, 507 pp.

Jørgensen, S. E. & M. J. Gromiec, 1989. Mathematical Submodels in Water Quality Systems. Elsevier, Amsterdam, 408 pp.

Jørgensen, S. E., S. N. Nielsen & L. A. Jørgensen, 1991. Handbook of Ecological Parameters and Ecotoxicology. Elsevier, Amsterdam, 1300 pp.

Jørgensen, S. E., B. C. Patten & M. Straškraba, 1992. Ecosystem emerging. Towards an ecology of complex systems in a complex future. Ecol. Modelling 62: 1–27.

Junk, W. J. & J. A. S. De Mello Nunes, 1987. Ecological impacts of the brasilian hydroelectric dams on the Amazon Region. Tübingen Geographische Studien 95: 367–387.

Kalčeva, R., J. Outrata & Z. Schindler, 1982. An optimization model for the economic control of reservoir eutrophication. Ecol. Modelling 17: 121–128.

Karadi, G. M., R. J. Krizek & S. C. Czallany (eds.), 1971. Pumped Storage Development and its Environmental Effects. American Water Resources Association, Urbana, Illinois, 541 pp.

Kaushik, N. K. & H. B. N. Hynes, 1971. The fate of dead leaves that fall into streams. Arch. Hydrobiol. 68: 465–515.

Kerekes, J. J., 1982. The application of phosphorus load – trophic response relationships to reservoirs. Can. Water Res. J. 7: 349–354.

Kimmel, B. L. & A. W. Groeger, 1986. Limnological and ecological changes associated with reservoir aging. In: G. E. Hall & M. J. Van Den Avyle (eds.), Reservoir Fisheries Management: Strategies for the 80's. Southern Division American Fisheries Society, Bethesda, Maryland 1986: 103–109.

Kimmel, B. L., D. M. Søballe, S. M. Adams, A. V. Pallumbo, C. J. Ford & M. S. Belvelhimer, 1988. Inter-reservoir interactions: Effects of a new reservoir on organic matter production and processing in a multiple-impoundment series. Verh. Int. Verein. Limnol. 23: 985–994.

Knisel, W. G. Jr., 1978. A system of models for evaluating non-point source pollution – an overview. International Institute of Applied Systems Analysis, Laxenburg, CP-78-11, 17 pp.

Kozerski, H.-P., 1975. Ein mathematisches Modell des Temperaturregimes und des Sauerstoffhaushaltes im Hypolimnion von Talsperren. Acta Hydrophysica, Berlin 19: 201–250.

Kratochvíl, S., 1982. Proudění a mísení vody v nádržích při provozu přečerpávacích vodních elektráren. (On the water flow and water mixing processes in reservoirs during the operation of pumped storage hydro-plants. – In Czech with English summary). Vodohosp. časopis SAV 30: 374–391.

Krebs, J. R. & N. B. Davies, 1991. Behavioural Ecology. An Evolutionary Approach. Third edition. Blackwell Sci. Pub., Oxford, 482 pp.

Laberge, E. & K. H. Mann, 1976. The importance of water discharge in determining phytoplankton biomass in a river impoundment. Naturaliste can. 103: 191–201.

Lande, R., 1982. A quantitative genetic theory of life history evolution. Ecology 63: 607–613.

Leonov, A. V., 1989. Phosphorus transformation and water quality in the Ivankovo reservoir: study by means of a simulation model. Arch. Hydrobiol. Beih. Ergebn. Limnol. 33: 157–168.

Leventer, H., 1979. Biological Control of Reservoirs by Fish. Mekoroth Water Co. Jordan District Central Laboratory of Water Quality Nazareth Elit, Israel, 71 pp.

Lewis, W. M., 1987. Tropical Limnology. Ann. Rev. Ecol. Syst. 18: 159–184.

Lhotka, L. & M. Straškraba, 1987. Combinatorial model of ecosystem dynamics. Ecol. Modelling 39: 181–200.

Likens, G. E., F. M. Borman, R. S. Pierce, J. S. Eaton & M. M. Johnson, 1977. Biogeochemistry of a Forested Ecosystem. Springer, New York.

Lillehammer, A. & S. J. Saltveit, 1984. Regulated Rivers. Universitetsforlaget Oslo, 540 pp.

Lindström, T., 1973. Life in a lake reservoir: fewer options, decreased production. AMBIO 2: 145–153.

Loehr, J., 1987. Impact of the hydrodynamic conditions on the primary production in an impounded river. Ecol. Modelling 39: 227–245.

Löffler, H., 1979. Neusidelersee: The Limnology of a Shallow Lake in Central Europe. Monographiae Biologicae 37, Junk, The Hague, 543 pp.

Lorenzen, M. & R. Mitchell, 1973. Theoretical effects of artificial destratification on algal production in impoundments. Environmental Science & Technology 7: 939–944.

Loucks, D. P., J. R. Stedinger & D. A. Haith, 1981. Water Resource Systems. Prentice Hall, Englewood Cliffs.

Lowe-McConnell, R. H. (ed.), 1966. Man-Made Lakes. Academic Press for the Institute of Biology, Symposia of the Institute of Biology 15, London, New York, 218 pp.

Lung, Wu-Seng, 1987. Lake acidification model: Practical tool. Proc. ASCE, J. Env. Engng Div. 113: 900–915.

Markofsky, M. & Harleman D. R. F., 1973. Prediction of water quality in stratified reservoirs. Proc. ASCE, J. Hydraulic Engng Div. 99: 729–745.

Marshall, B. E., 1984. Towards Predicting Ecology and Fish Yield in African Reservoirs from Pre-Impoundment Physico-Chemical Data. CIFA Techn. Pap 12. FAO, Rome, 36 pp.

Martin, D. B. & R. D. Arneson, 1978. Comparative limnology of deep-discharge reservoir and a surface-discharge lake on the Madison River, Montana. Freshwater Biol. 8: 33–42.

Martin, S. C., S. W. Effler, J. V. DePinto, F. B. Trama, P. W. Rodgers, J. S. Dobi & M. C. Wodka, 1985. Dissolved oxygen model for a dynamic reservoir. Proc. ASCE, J. Envir. Engng Div. 111: 647–665.

Martin, R. O. R. & R. L. Hanson, 1966. Reservoirs in the United States. US Geol. Surv. Water Supply Paper 1838, 114 pp.

Matsumura, T. & S. Yoshiyuki, 1981. An optimization problem related to the regulation of influent nutrient in aquatic ecosystems. Int. J. Syst. Sci. 12: 565–585.

Matsumura-Tundisi, T., J. G. Tundisi, A. Saggio, A. L. Oliveira Neto & E. G. Espindola, 1991. Limnology of Samuel Reservoir (Brazil, Rondonia) in the filling phase. Verh. Int. Verein. Limnol. 24: 1482–1488.

McComb, A. J. & W. D. Williams, 1987. The ecology of reservoirs, lakes and wetlands. National Water Research

Seminar. Discussion papers. Canberra 17-18 September: 201-212.
McFarland, D., 1985. Animal Behaviour, Psychobiology, Ethology and Evolution. Putman, London, 576 pp.
McLachlan, A. J., 1974. Development of some lake ecosystems in tropical Africa, with special reference to the invertebrates. Biol. Rev. 49: 365-397.
McLachlan, A. J., 1977. The changing role of terrestrial and autochthonous organic matter in newly flooded lakes. Hydrobiologia 54: 215-217.
McLaren, I. A., C. J. Corkett & E. J. Zillioux, 1969. Temperature adaptations of copepod eggs from the arctic to the tropics. Biol. Bull. 137: 486-493.
McQueen, D. J. & D. R. S. Lean, 1986. Hypolimnetic aeration: An overview. Water Poll. Res. J. Canada 21: 205-217.
Mitsch, W. J. & S. E. Jørgensen (eds.), 1989. Ecological Engineering. An Introduction to Ecotechnology. Wiley, New York, 472 pp.
Mobley, M. H. & E. D. Harshbarger, 1987. Epilimnetic pumps to improve reservoir releases. In: Proceedings CE Workshop on Reservoir Releases. US Army Engineer Waterways Experiment Station. Vicksburg, Mississippi, Miscellaneous Paper E-87-3: 133-135.
Morris, L. A., R. N. Langmeier & T. R. Russell, 1968. Effects of main stem impoundments and channelization upon the limnology of the Missouri River, Nebrasca. Trans. Am. Fish. Soc. 97: 380-388.
Moss, B., 1980. Ecology of Fresh Waters. Blackwell Sci. Pub., Oxford, 332 pp.
Mouchet, P. C., 1984. Influence of recently drowned terrestrial vegetation on the quality of water stored in impounding reservoirs. Verh. Int. Verein. Limnol. 22: 1068-1619.
Naiman, R. J. & H. Décamps (eds.), 1990. The Ecology and Management of Aquatic-Terrestrial Ecotones. Man and the Biosphere Series. Parthenon, Paris, 316 pp.
Nakamura, K., 1967. City temperature of Nairobi. Japanese Progress in Climatology: 61-65.
Neel, J. K., 1953. Certain limnological features of a polluted irrigation stream. Trans. Am. Microscop. Soc. 72: 119-153.
Nesměrák, I. & M. Straškraba, 1985. Spectral analysis of the automatically recorded data from Slapy Reservoir, Czechoslovakia. Int. Revue ges. Hydrobiol. 70: 27-46.
Newbold, J. D., 1987. Phosphorus spiralling in rivers and river-reservoir systems: implications of a model. In: J. F. Craig & J. B. Kemper (eds.), Regulated Streams. Advances in Ecology. Plenum Press, London, 431 pp.
Nielsen, S. N., 1990. Recent development in structural dynamic models. In: J. Fenhann, H. Larsen, G. A. MacKenzie & B. Rasmussen (eds.), Environmental Models: Emissions and Consequences. Elsevier, Amsterdam: 177-176.
Novotny, V. & Chesters G., 1981. Handbook of Nonpoint Pollution. Sources and Management. Environmental Engineering Series, Van Nostrand Reinhold, New York, 555 pp.
Oglesby, R. T., C. A. Carlson & J. A. McCanu, 1972. River Ecology and Man. Proceedings of an International Symposium on River Ecology and the Impact of Man. Academic Press, New York, 465 pp.
Ollason, J. S., 1980. Learning to forage optimally? Theor. Popul. Biol. 18: 47-56.
Orlob, G. T., 1983. Mathematical Modeling of Water Quality: Streams, Lakes, and Reservoirs. Wiley, Chichester, 518 pp.
Orlob, G. T. & L. G. Selna, 1970. Temperature variations in deep reservoirs. Proc. ASCE, J. Hydraulic Engng Div. 96: 391-410.
Ortiz Casas, J. L. & Pena Martinez, 1984. Applicability of the OECD eutrophication models to Spanish reservoirs. Verh. Int. Verein. Limnol. 22: 1521-1535.
Oskam, G., 1982. Quality aspects of the Biesbosch Reservoirs. Aqua 6: 447-504.
Oskam, G., 1987. Light and zooplankton as algae regulating factors in eutrophic Biesbosch Reservoir. Verh. Int. Verein. Limnol. 20: 1612-1618.
Ostrofsky, M. L., 1978. Trophic changes in reservoirs; an hypothesis using phosphorus budget models. Int. Revue ges. Hydrobiol. 64: 481-499.
Ostrofsky, M. L. & H. C. Duthie, 1978. An approach to modelling productivity in reservoirs. Verh. Int. Verein. Limnol. 20: 1562-1567.
Panday, R. D. (ed.), 1977. Man-Made Lakes and Human Health. University of Suriname, Paramaribo, Suriname, 73 pp.
Pařízek, J., 1984. Využití efektu čisté vrstvy (Utilization of the "clean layer" effect. - In Czech). In: M. Straškraba, Z. Brandl & P. Porcalová (eds.), Hydrobiologie a kvalita vody údolních nádrží. ČSVTS České Budějovice: 72-83.
Park, R. A., T. W. Groden & C. J. Desormeau, 1979. Modifications to the model CLEANER requiring further research. In: D. Scavia & A. Robertson (eds.), Perspectives on Lake Ecosystem Modeling. Ann Arbor Science Publishers Inc., Ann Arbor, Michigan: 87-108.
Park, R. A., B. H. Indyke, & G. W. Heitzman, 1981. Predicting the fate of coal-derived pollutants in aquatic environments. In: W. J. Mitsch, R. W. Bosserman & J. M. Klopatek (eds.), Energy and Ecological Modelling. Proceedings of a Symposium held from 20 to 23 April 1981 at Louisville, Kentucky. Developments in Ecological Modelling 1. Elsevier, Amsterdam: 115-122.
Parker, R. A., 1968. Simulation of an aquatic ecosystem. Biometrics 24: 803-821.
Parker, R. A., 1974. Some consequences of stochasticizing an ecological system model. Lecture Notes in Biomathematics 2: 175-183.
Patten, B. C., D. A. Egloff & T. H. Richardson, 1975. Total ecosystem model for a in Lake Teome. In: B. C. Patten (ed.), Systems Analysis in Ecology. Academic Press, New York, Vol. 7: 206-423.
Patterson, J. C., 1987. A model for convective motions in reservoir sidearm. In: W. H. Graf & U. Lemmin (eds.), Topics in Lake and Reservoir Hydraulics. Proceedings AIRH-Congress-IAHR, Lausanne: 68-73.
Patterson, J. C. & P. F. Hamblin, 1988. Thermal simulation of a lake with winter ice cover. Limnol. Oceanogr. 33: 323-338.
Pedroso, F., C. A. Bonneto & Y. Zabocar, 1988. A comparative study of phosphorus and nitrogen transport in the Parana Paraquay and Bermozo Rivers. In: J. G. Tundisi (ed.), Limnologia e Manejo de Represas. (Limnology and Management of Reservoirs). Série Monografias em

Limnologia. Universidade de Sao Paulo, Escola de Engenharia de Sao Carlos, Centro de Recursos Hídricos e Ecologia Aplicada, Sao Carlos, Vol 1: 92–117.

Peters, R. H., 1983. The Ecological Implications of Body Size. Cambridge, Univ. Press, Cambridge, 329 pp.

Petersen, R. C. & K. W. Cummins, 1974. Leaf processing in a woodland stream. Freshwater Biol. 4: 343–368.

Peterson, 1982. Lake restoration by sediment removal. Water Res. Bull. 18: 423–435.

Petts, G., 1984. Impounded Rivers. Wiley, Chichester, 326 pp.

Petts, G., P. D. Armitage & A. Gustard (eds.), 1989. Fourth International Symposium on Regulated Streams. Regulated Rivers 3: 1–394.

Pigram, J. J., 1986. Issues in the Management of Australia's Water Resources. Longman Cheshire Pty Limited: 94–188.

Pivnička, K., 1982. Long-termed study of fish populations in the Klíčava Reservoir. Přírodovědecké práce, Brno 16: 1–46.

Ploskey, G. A., 1985. Impact of Terrestrial Vegetation and Pre-impoundment Clearing on Reservoir Ecology and Fisheries in the United States and Canada. FAO Fish. Techn. Pap. 258, 35 pp.

Prien, K-J. & H. Bernhardt, 1989. Belüftung der Aabach-Talsperre. Wasser-Abwasser 130: 206–213.

Procházková, L., 1980. Agricultural impact on the nitrogen and phosphorus concentration in water. In: A. Duncan & J. Rzóska (eds.), Land Use Impact on Lake and Reservoir Ecosystems. Proc. MAB Project 5, Facultas-Verlag, Wien: 78–100.

Puig, M. A., J. Armengol, G. Gonzales, J. Penuelas, S. Sabater & F. Sabater, 1987. Chemical and biological changes in the Ter River induced by a series of reservoirs. In: J. F. Craig & J. B. Kemper (eds.), Regulated Streams, Advances in Ecology. Plenum Press, London, 431 pp.

Purcell, L. T., 1939. The ageing of reservoir waters. J. Amer. Water Works Assoc. 31: 1775–1806.

Quintero, J. E. & J. E. Garton, 1973. A low energy lake destratifier. Transactions of the American Society of Agricultural Engineers. 16: 973–978.

Raman, R. K., 1988. Aeration/destratification succeeds in Illinois Lakes. Lake Line 8: 6–7.

Reckhow, K. H. & S. C. Chapra, 1983. Engineering Approaches for Lake Management, Volume 1: Data Analysis and Empirical Modeling. Butterworths Pub., Boston, Massachusetts, 340 pp.

Reckhow, K. H., R. W. Black, T. B. Stockton Jr., J. D. Wogt & J. G. Wood, 1987. Empirical models of fish response to lake acidification. Can. J. Fish. Aquat. Sci. 44: 1432–1442.

Reckhow, K. H., 1988. Empirical models for trophic state in Southeastern US Lakes and reservoirs. Water Res. Bull. 24: 723–734.

Reed, F. C., 1979. Decomposition of Acer rubrum leaves at three depths in a eutrophic Ohio Lake. Hydrobiologia 64: 195–197.

Reynolds, C. S., 1987. Community organization in the freshwater plankton. In: J. H. R. Gee & P. S. Giller (eds.), Organization of Communities, Past and Present. Blackwell Sci. Pub., Oxford: 397–425.

Reynolds, C. S., 1989. Physical determinants of phytoplankton succession. In: U. Sommer (ed.), Plankton Ecology. Springer Series in Contemporary Bioscience, Springer, Berlin, 368 pp.

Ricker, W. E., 1954. Stock and recruitment. J. Fish. Res. Bd. Canada. 11: 559–623.

Ridley, J. E., 1970. The biology and management of eutrophic reservoirs. J. Soc. Wat. Treat. Exam. 19: 374–399.

Ridley, J. E., 1971. Water Supply Lakes and Raw Water Storage Reservoirs. Report on the Phytoplankton of Selected Lakes and Storage Reservoirs in the USA. Pan. Am. Health Org., Washington, D.C., Document No. 71W4/USA/3100.

Ridley, J. E., P. Cooley & J. A. Steel, 1966. Control of thermal stratification in Thames Valley reservoirs. Proc. Soc. Wat. Treat. Exam. 15: 225–244.

Ridley, J. E. & A. Steel, 1975. Ecological aspects of river impoundments. In: B. Whitton (ed.), River Ecology. Blackwell Sci. Pub., Oxford: 565–587.

Ripl, W., 1976. Biochemical oxidation of polluted lake sediments with nitrate – a new lake restoration method. AMBIO 5: 132–135.

Ripl, W., 1980. Natural and induced sediment rehabilitations in hypertrophic lakes. In: J. Barica & L. Muhr (eds.), Hypertrophic Ecosystems. Junk, The Hague.

Ryder, R. A., 1978. Ecological heterogeneity between north-temperate reservoirs and glacial lake systems due to differing succession rates and cultural uses. Verh. Int. Verein. Limnol. 20: 1568–1574.

Rzoska, J., 1966. The biology of reservoirs in the USSR. In: R. H. Lowe-McConnell (ed.), Man-Made Lakes. Academic Press, New York: 149–154.

Saijo, Y. & J. G. Tundisi, 1985. Limnological studies in Central Brazil. Rio Doce Valley Lakes and Pantanal Wetland. 1st Report, Water Research Institute, Nagoya University, 201 pp.

Saila, S. B., C. W. Recksiek & M. H. Prager (eds.), 1988. Basic Fishery Science Programs. Elsevier, Amsterdam, 230 pp.

Salas, H. J. & G. Limon, 1986. Memoria del tercer encuentro del proyecto regional desarrello de metodologias simplificadas para la evaluacion de eutroficacion en lagos calidos tropicales. Proc. Workshop, Centro Panamericano de Ingenieria Sanitaria y Ciencias del Ambiente, Organizacion Mundial de la Salud, Lima, Peru.

Šantrůčková, H. & M. Straškraba, 1991. On the relationship between specific respiration activity and microbial biomass in soils. Soil Biol. Biochem. 23: 525–532.

Schindler, Z. & M. Straškraba, 1982. Optimální řízení eutrofizace údolních nádrží (Optimal control of reservoir eutrophication. – In Czech). Vodohosp. časopis SAV 30: 536–548.

Schoumacher, R., 1976. Biological consideration of pumped storage development. Special session. Trans. Am. Fish. Soc.: 155–180.

Schreiber, J. D. & D. L. Rausch, 1979. Suspended sediment-phosphorus relationships for the inflow and outflow of a flood detention reservoir. J. Environ. Qual. 8: 510–514.

Scott, C. M., S. W. Effler, J. V. Depinto, F. B. Trama, P. W. Rodgers, J. S. Dobi & M. C. Wodka, 1985. Dissolved oxygen model for a dynamic reservoir. Proc. ASCE, J. Envir. Engng Div. 111: 647–665.

Seďa, J., J. Kubečka & Z. Brandl, 1989. Zooplankton structure and fish population development in the Římov Reservoir, Czechoslovakia. Arch. Hydrobiol. Beih. Ergebn. Limnol. 33: 605–609.

Shapiro, J., V. Lamarra & M. Lynch, 1975. Biomanipulation. An Ecosystem Approach to Lake Restoration. Contribution 143, Limnological Research Centre, Univ. of Minnesota, Minneapolis, Minnesota, 32 pp.

Shapiro, J., B. Forsberg, V. Lamarra, G. Lindmark, M. Lynch, E. Smeltzer & G. Zotto, 1982. Experiments and Experiences in Biomanipulation. Studies of Biological Ways to Reduce Algal Abundance and Eliminate Blue-greens. Interim Report No 19 of the Limnological Research Centre, Univ. of Minnesota, Minneapolis, Minnesota, 251 pp.

Šimek, K., 1986. Bacterial activity in a reservoir determined by autoradiography and its relationships to phyto- and zooplankton. Int. Revue ges. Hydrobiol. 71: 593–612.

Skaletskaya, E. I. et al., 1979. Diskretnye modeli dinamiki chislennosti populatsii i optimizatsia promysla (Discrete Models of Population Dynamics and Optimization of Industry. - In Russian). Nauka, Moskva.

Slotta, L. S., E. M. Elwin, M. T. Mercier & M. D. Terry, 1969. Stratified reservoir currents. Bull. eng. Exp. Stat. 49: 1–61.

Søballe, D. M. & B. L. Kimmel, 1987. A large-scale comparison of factors influencing phytoplankton abundance in rivers, lakes, and impoundments. Ecology 68: 1943–1954.

Søballe, D. M. & S. T. Threlkeld, 1985. Advection, phytoplankton biomass, and nutrient transformations in a rapidly flushed impoundment., Arch. Hydrobiol. 105: 187–203.

Spence, J. A. & H. B. N. Hynes, 1971. Differences in fish populations upstream and downstream of a mainstream impoundment. J. Fish. Res. Bd. Canada 28: 45–46.

Spigel, R. H., J. Imberger & K. N. Rayner, 1986. Modeling the diurnal mixed layer. Limnol. Oceanogr. 31: 533–556.

Stanley, N. F. & M. P. Alpers (eds.), 1975. Man-Made Lakes and Human Health. Academic Press, London, 495 pp.

Starfield, A. M. & A. L. Bleloch, 1986. Building Models for Conservation and Wildlife Management. Macmillan, New York, 253 pp.

Steel, J. A., 1972. The application of fundamental limnological research in water supply system design and management. Symp. zool. Soc. Lond. 29: 41–67.

Steel, J. A., 1976. Eutrophication and the operational management of reservoirs of the Thames Water Authority. Instn. Publ. Hlth. Engrs. symp. Eutrophication of Lakes and Reservoirs. Chameleon Press, UK: J1–J9.

Steel, J. A., 1978a. Reservoir algal productivity. In: A. James (ed.), Mathematical Models in Water Pollution Control. Wiley, New York: 107–135.

Steel, J. A., 1978b. The use of simple plankton models in the management of Thames Valley Reservoirs. DVGW-Schriftenreihe, Wasser 16: 42–59.

Steel, A., 1980. Phytoplankton models. In: E. D. Lecren & R. H. McConnell (eds.), 1980. The Functioning of Freshwater Ecosystems. Cambridge Univ. Press, Cambridge: 221–227.

Stefan, H. G., M. D. Bender, J. Shapiro & D. I. Wright, 1987. Hydrodynamic, design of a metalimnetic lake aerator. Proc. ASCE, J. Envir. Engng Div. 113: 1249–1264.

Stefan, H. G., J. J. Cardoni & A. Y. Fu, 1982. RESQUAL II: A dynamic water quality simulation program for a stratified shallow lake or reservoir: Applications to Lake Chicot, Arkansas. Univ. of Minnesota, St. Anthony Falls Hydraulic Laboratory, Project Report 209, 133 pp.

Stefan, H. G. & M. J. Hanson, 1980. Predicting dredging depths to minimize internal nutrient recycling in shallow lakes. Restoration of lakes and Inland waters. International Symposium on Inland Waters and Lake Restoration, September 8–12, 1980, Portland, Maine. US Envir. Prot. Agency, Washington, D.C.: 79–85.

Straškraba, M., 1973. Limnological basis for modelling reservoir ecosystems. In: W. C. Ackerman, F. G. White & E. B. Worthington (eds.), Man-Made Lakes: Their Problems and Environmental Effects. American Geophysical Union, Washington, D.C., Geophys. Monogr. 17: 517–535.

Straškraba, M., 1974. Sezonální cyklus hloubkového rozšíření ryb v Klíčavské údolní nádrži, studovaný pomocí echolotu. (Seasonal cycle of the depth distribution of fish in Klíčava Reservoir studied by a recording echosounder – In Czech). Živočisná výroba 19: 653–663.

Straškraba, M., 1976. Development of an analytical phytoplankton model with parameters empirically related to dominant controlling variables. Abh. Acad. Wiss. DDR 1974: 33–65.

Straškraba, M., 1979. Mathematische Simulation der Produktionsdynamik in Gewässern und deren Anwendung auf die Produktionssteuerung in Talsperren. Z. f. Wasser- und Abwasser-Forschung 12: 56–64.

Straškraba, M., 1980. The effect of physical variables on freshwater production: Analyses based on models. In: E. D. Lecren & R. H. Lowe-McConnell (eds.), The Functioning of Freshwater Ecosystems. Cambridge Univ. Press, Cambridge: 13–84.

Straškraba, M., 1986. Ecotechnological measures against eutrophication. Limnologica (Berlin) 17: 239–249.

Straškraba, M., 1989. Analysis of freshwater pelagic productivity by dynamic models. Abstracts, XXIV Congress of the International Association of Limnology: 148.

Straškraba, M., 1990. Limnological particularities of multiple reservoir series. Arch. Hydrobiol. Beih. Ergebn. Limnol. 33: 677–678.

Straškraba, M., 1992. Theoretical Models of Population Dynamics and Cybernetic Theory of Ecosystems. In: B. C. Patten (ed.), The George M. Van Dyne Memorial Series in System Ecology. 1 (in print).

Straškraba, M., P. Blažka, Z. Brandl, B. Desortová, J. Komárková, J. Kubečka, L. Procházková & J. Seďa., 1990. A hypothesis on reservoir aging. Arch. Hydrobiol. Beih. Ergebn. Limnol. 33: 803.

Straškraba, M. & A. Gnauck, 1985. Aquatic Ecosystems. Modelling and Simulation. Elsevier, Amsterdam, 309 pp.

Straškraba, M. & G. Hocking, in prep. Simulation model DYRESM for reservoir hydrodynamics used in Czechoslovakia.

Straškraba, M. & P. Javornický, 1973. Limnology of two re-regulation reservoirs in Czechoslovakia. In: J. Hrbáček & M. Straškraba (eds.), Hydrobiological Studies. Academia, Praha, Vol. 2: 249–316.

Straškraba, M. & V. Straškrabová, 1975. Management problems of Slapy Reservoir, Bohemia, Czechoslovakia. Proceedings of a Symposium The Effects of Storage on Water Quality. Reading University, Reading, England: 449–484.

Straškrabová, V., 1976. Self-purification of impoundments. Water Res. 9: 1171–1177.

Straškrabová, V. & K. Šimek, 1984. Total and individual cell uptake of organic substances as a measure of activity of bacterioplankton. Arch. Hydrobiol. Beih. Ergebn. Limnol. 19: 1–6.

Straškrabová, V., Z. Brandl, B. Henderson-Sellers, O. T. Lind, V. Sládeček & J. F. Talling, 1989. Proceedings of the International Conference on Reservoir Limnology and Water Quality. Arch. Hydrobiol. Beih. Ergebn. Limnol. 33, 975 pp.

Strycker, L., 1988. Decaying dam holds tide of trouble. The Register-Guard, Eugene, Oregon 121 (331): 1,8.

Swartzman, G. L. & S. P. Kaluzny, 1987. Ecological Simulation Primer. Macmillan Publishing Company, New York, 370 pp.

Swartzman, G. L., R. T. Haar, D. H. McKenzie & T. Zaret, 1981. Evaluation of usefulness of ecological simulation models in power plant cooling systems. In: W. J. Mitsch, R. V. Bosserman & J. M. Klopatek (eds.), Energy and Ecological Modelling. Proceeding of a Symposium held from 20 to 23 April 1981 at Louisville, Kentucky. Developments in Ecological Modelling 1. Elsevier, Amsterdam: 173–184.

Sylvester, R. O. & R. W. Seabloom, 1965. Influence of site characteristics, on quality of impounded water. J. Amer. Water Works Assoc. 57: 1528–1546.

Symoens, J. J., M. J. Burgis & J. J. Gaudet, 1981. The Ecology and Utilization of African Inland Waters. United Nations Environment Programme, Nairobi, 191 pp.

Symons, J. M., W. H. Irwin, R. M. Clark & G. G. Roebeck, 1967. Management and measurement of DO in impoundments. Proc. ASCE, J. Sanit. Engng Div. 93: 181–209.

Szumiec, M., 1975. The effect of controlled eutrophication on solar radiation penetrating into the ponds. Acta hydrobiol. 17: 149–182.

Talling, J. F., 1957a. The phytoplankton population as a compound photosynthetic system. New Phyt. 56: 133–149.

Talling, J. F., 1957b. Photosynthetic characteristic of some freshwater plankton diatom in relation to underwater radiation. New Phyt. 56: 29–50.

Thérien, N., G. Spiller, 1981. A mathematical model of the decomposition of flooded vegetation in reservoir. Soc. Computer Simulation, California, Simulation Conference Proc. Ser. V9, 2: 87–98.

Thérien, N., G. Spiller & B. Coupal, 1982. Simulation de la decomposition de la matiere végétale et des sols inondés dans les reservoirs de la region de la Baie de James. Can. Water Res. J. 7: 375–396.

Thornton, K. W., B. L. Kimmel & F. F. Payne (eds.), 1990. Reservoir Limnology: Ecological Perspectives. Wiley, New York, 246 pp.

Threlkeld, S. T., 1982. Water renewal effects on reservoir zooplankton, communities. Can. Water Res. J. 7: 151–167.

Triska, F. J., J. R. Sedell & B. Buckley, 1975. The processing of conifer and hardwood leaves in two coniferous forest streams. II. Biochemical and nutrient changes. Verh. Int. Verein. Limnol. 19: 1628–1639.

Tundisi, J. G., 1984. Estratificacao hidraulica em reservatórios e suas conseqüências ecológicas. Cienc. Cult. 36: 1498–1504.

Tundisi, J. G., 1986. Limnologia de represas artificiais. Bol. Hidraul. Sameam 11, 46 pp.

Tundisi, J. G. (ed.), 1988. Limnologia e Manejo de Represas. (Limnology and Management of Reservoirs). Série Monografias em Limnologia. Vol. I. Universidade de Sao Paulo, Escola de Engenharia de Sao Carlos, Centro de Recursos Hídricos e Ecologia Aplicada, Sao Carlos, Tomo 1, 506 pp. Tomo 2, 432 pp.

Tundisi, J. G., 1989. Management of reservoirs in Brasil. In: S. E. Jørgensen & R. A. Vollenweider (eds.), Guidelines for Lake Management. Vol. 1. Principles of Lake Management. United Nations Environmental Programm, International Lake Ecology Society, Otsu, Japan: 155–170.

Tundisi, J. G., 1990a. Key factors of reservoir functioning and geographical aspects of reservoir limnology. Chairman's overview. Arch. Hydrobiol. Beih. Ergebn. Limnol. 33: 645–646.

Tundisi, J. G., 1990b. Perspectives of ecological modelling of tropical and subtropical reservoirs in South America. Ecol. Modelling 52: 7–20.

Tundisi, J. G., 1990c. Spatial distribution, temporal sequence and seasonal cycle of phytoplankton in reservoirs: limiting and controlling factors. Rev. Bras. Biol. 50: 937–955.

Tundisi, J. G. & T. Matsumura-Tunsidi, 1986. Eutrophication processes and trophic state for 23 reservoirs in S. Paulo State, Southern Brazil. Fourth Brazil/Japan Symposium on Science and Technology, Supplementary volume, Publ. Academy of Sciences, Sao Paulo State, 26 pp.

Tundisi, J. G. & T. Matsumura-Tundisi, 1990. Limnology and eutrophication of Barra Bonita Reservoir, S. Paulo State, Southern Brazil, Arch. Hydrobiol. Beih. Ergebn. Limnol. 33: 661–676.

Tundisi, J. G., T. Matsumura-Tundisi, M. C. Calijuri & E. M. L. Novo, 1991. Comparative limnology of five reservoirs in the middle Tiete River, S. Paulo State. Verh. Int. Verein. Limnol. 24: 1489–1496.

Turner, R. R., E. A. Laws & R. C. Harriss, 1983. Nutrient retention and transformation in relation to hydraulic flushing rate in a small impoundment. Freshwater Biol. 13: 113–127.

Tyler, P. A., 1980. Limnological problems in the management of Tasmanian water resources. In: W. D. Williams (ed.), An Ecological Basis for Water Resource Management. Australian National University Press, Canberra, Australia: 43–66.

Uhlmann, D., 1971. Influence of dilution, sinking and grazing rate on phytoplankton populations in hyperfertilized ponds and micro-ecosystems. Mitt. Internat. Verein. Limnol. 19: 100–124.

Uhlmann, D., 1978. The upper limit of phytoplankton production as a function of nutrient load, temperature, retention time of the water, and euphotic zone depth. Int. Revue ges. Hydrobiol. 63: 353–363.

Van Straten, G., 1986. Identification, Uncertainty Assessment and Prediction in Lake Eutrophication. Proefschrift ter verkrijging van de graad van doctor in de technische wetenschappen aan de Universiteit Twente, Twente, 240 pp.

Vannote, R. L., G. W. Minshall, K. W. Cummins, J. R. Sedell & C. E. Cushing, 1980. The river continuum concept. Can. J. Fish. Aquat. Sci. 37: 130–137.

Vincent, T. L. & J. M. Skorkowski (eds.), 1981. Renewable

Resource Management. Proceedings of a Workshop on Control Theory Applied to Renewable Resource Management and Ecology Held in Christchurch, New Zealand, January 7–11, 1980. Lecture Notes in Biomathematics 40, Springer, Berlin.

Vollenweider, R. A., 1965. Calculation models of photosynthesis depth curves and some implications regarding day rate estimates in primary production measurements. Memorie dell' Istituto di Idrobiologia 18 (supplement): 437–457.

Vollenweider, R. A., 1968. Scientific sunfamentals of the eutrophication of lakes and flowing waters, with particular reference to phosphorus and nitrogen as factors in eutrophication. OECD Technical Report DAS/CS1/68. 27, Paris, 159 pp.

Vollenweider, R. A., 1975. Input-output models with special reference to the phosphorus loading concept in limnology. Schweizerische Zeitschrift für Hydrologie 37: 53–84.

Volokhonsky, H., E. Shmain & S. Serruya, 1980. Lake Kinneret water biotopes: a mathematical model of thermal stratification for ecological purposes. Ecol. Modelling 9: 91–120.

Walker, K. F. & G. E. Likens, 1975. Meromixis and a reconsidered typology of lake circulation patterns. Verh. Int. Verein Limnol. 19: 442–458.

Walker, W. W. Jr., 1985. Empirical Methods for Predicting Eutrophication in Impoundments; Report 3, Phase II: Model Refinements. US Army Engineer Waterways Experiment Station, Vicksburg, Mississippi, Technical Report E-81-9, 297 pp.

Walter, H. & S. W. Breckle, 1991. Ökologische Grundlagen in globaler Sicht (Ökologie der Erde, Bd 1). 2, bearbeitete Auflage. Gustav Fischer, Stuttgart, 238 pp.

Ward, J. V., B. R. Davies, C. M. Breen, J. A. Cambray, F. M. Chutter, J. A. Day, F. C. de Moor, J. Geeg, J. H. O'Keefe & K. F. Walker., 1984. Stream regulation. In: R. C. Hart & B. R. Allanson (eds.), Limnological Criteria for Management of Water Quality in the Southern Hemisphere. Inland Water Ecosystem Programmes, Pretoria: 21–63.

Ward, J. V. & J. A. Stanford (eds.), 1979. The Ecology of Regulated Streams. Plenum Press, New York, 389 pp.

Ward, J. V. & J. A. Stanford, 1983. The serial discontinuity concept of lotic ecosystems. In: T. D. Fontaine & S. M. Bartell (eds.), Dynamics of Lotic Ecosystems. Ann Arbor Science, Ann Arbor, Michigan: 29–42.

Ward, R. D., C. Loftis & G. B. McBride, 1986. The "Data-rich but Information-poor" syndrome in water quality monitoring. Environmental Management 10: 291–297.

Welch, E. B., J. P. Michaud & M. A. Perkins, 1982. Alum control of internal phosphorus loading in a shallow lake. Water Res. Bull. 18: 929–936.

Welch, E. B. & C. R. Patmont, 1980. Lake restoration by dillution; Moses Lake, Washington. Water Res. 14: 1317–1325.

Wetzel, R. G., 1990. Reservoir ecosystems: conclusions and speculations. In: K. W. Thornton, B. L. Kimmel & F. E. Payne (eds.), Reservoir Limnology: Ecological Perspectives. Wiley, New York: 227–238.

Whitaker, V. V., 1992. Denitrification in Wetlands. PhD. Thesis, School of Engineering at S. Carlos, Univ. of S. Paulo (in prep.)

Wilhelmus, B., H. Bernhardt & D. Neumann, 1978. Vergleichende Untersuchungen über die Phosphorelimininerung von Vorsperren – Verminderung der Algenentwicklung in Speicherbecken und Talsperren. DFGW-Schriftenreihe Wasser 16: 140–176.

Williams, L. R., V. W. Lambou, S. C. Hearn & R. W. Thoman, 1977. Relationship of productivity and problem conditions to ambient nutrients: National eutrophication findings for 418 eastern lakes. National Eutrophication Survey, US Environmental Protection Agency, Las Vegas.

Windle-Taylor, E., 1964. Rep. Res. bact. chem. biol. Exam. Lond. Wat. 41

Windle-Taylor, E., 1965. Rep. Res. bact. chem. biol. Exam. Lond. Wat. 42, 140 pp.

Windle-Taylor, E., 1967. Forty-third Report of the Metropolitan Water Board: 65–67.

Windle-Taylor, E., 1969. Forty-fourth Report of the Metropolitan Water Board: 86–88.

Wright, J. C., 1968. Effect of impoundment on productivity, water chemistry and heat budgets of rivers. In: Reservoir Fishery Resources Symposium, Proceedings of a Conference of the American Fisheries Society, Athens, Georgia, April 1967. University of Georgia Press, Athens, Georgia: 188–199.

Wrobel, L. C. & C. A. Brebbia, 1991. Water Pollution: Modelling, Measuring, and Prediction. Computational Mechanics Publications, Southampton, 762 pp.

Zalewski, M., B. Brewinska-Zaras, P. Frankiewicz & S. Kalinowski, 1990. The potential for biomanipulation using fry communities in a lowland reservoir: concordance between water quality and optimal recruitment. Hydrobiologia 200/201: 549–556.

Zwirnmann, K. H., 1982. Nonpoint Nitrate Pollution of Municipal Water Supply Sources: Issues of Analysis and Control. International Institute of Applied Systems Analysis, Laxenburg, CP-82-SO4, 302 pp.

Index

Model names in capitals, reservoir names in italics. Only reservoirs mentioned in the text and figures are included.

Acidification 180, 207–208, 236, 237, 267
Adaptation 228, 237, 264, 265
Aeration 269–273
Aging of reservoirs 47–49, 252–257
Algal blooms 148
AQUAMOD 109–110, 236

Bacteria 17, 63, 100–104, 106, 108, 114, 123, 140, 175, 184, 187, 189, 208, 210, 253–255
Balbina Reservoir 25–26, 29, 45, 223, 264
Barra Bonita Reservoir 26, 31, 32, 34, 35, 39, 41, 42, 43, 48, 147, 148, 151, 288
Bautzen Reservoir 33
Belton Reservoir 58, 59, 63, 64
Billings Reservoir 45
Biological diversity 251, 264, 265, 269
Biomanipulation 160–166, 194–195, 266, 271–272
Biomass 2, 15, 22, 24, 26, 40, 45–47, 54, 57, 61, 62, 64, 67, 69, 71–73, 75, 86, 87, 115–118, 122, 124, 129, 130, 133–135, 137, 140–142, 154, 155, 164, 166, 167, 181, 183, 185, 188, 190–195, 201, 210, 217, 222, 229, 247, 253, 254, 256, 261, 265, 270, 272, 274, 283, 285
Blue-greens 34, 48, 51, 65, 69, 72, 95, 107, 127–137, 185, 191, 225, 237, 261, 271
Broa Reservoir 26, 35–37, 41, 43, 46, 277
Brokopondo Reservoir 47
Bystřička Reservoir 158

Calibration 79, 81–83, 131, 139
Catastroph theory 96
CE-QUAL 94, 236
Chaos 96, 277
Chlorophyll-a (CHA) 2, 15, 17–22, 35, 40–42, 58–61, 63, 64, 69, 71–75, 78, 87, 119–122, 124, 128, 140, 147–151, 171, 175, 184, 189, 201, 222, 223, 227, 236, 267, 271
Chlorophyll model 150
Circulation 31, 34, 35, 37, 48, 69, 70, 75, 119, 169, 189, 202, 208, 217, 252, 257, 265
Classification 15–18, 57–67, 94, 129, 151, 164, 166, 180, 192, 238, 250
Clyde Pond 59
Competitive ability 127, 134, 135, 237
Consequences for rivers of building reservoirs 220–223
Copper poisoning 272
Correlation analysis 149, 150
Curua Una Reservoir 214

Dam reservoirs 214–216
Decision support system 275
Density flows 61
Dependence on flow 199–200
Destratification 269–271
Detritus 47, 48, 100–105, 114, 119, 123–124, 147, 222, 253, 255, 256
Differences between natural lakes and impoundments 62
Discriminant analysis 62
Downstream water quality effects 222
DYRESM 75–84, 95, 223, 235, 237–238, 245

Eau Claire Lake 59
Ecological theory 264, 265
Ecotechnological methods 268, 273
Ecotechnology 265, 266, 277
EDD1 95
Eddy diffusion 70, 76, 80, 82, 84, 85, 88, 89, 95, 110
El Cajon Reservoir 214
Epilimnion 70, 80–82, 99–109, 111, 114, 121–124, 184, 232, 270
Eutrophication 1, 2, 15–17, 19, 21–22, 38, 47, 48, 50, 51, 57, 58, 60, 75, 94, 99, 119, 127, 154, 164, 169, 174, 175, 180, 181, 184–186, 213, 229, 234, 236, 237, 246, 251, 257, 261, 263, 264, 267, 268, 273
Eutrophication Management 21, 22
Exchange of phosphorus 99, 106, 115, 116

Factor analysis 71, 73, 74, 131
Feed-back 228, 230, 250, 265, 266
Filling 35, 39, 41, 43–45, 47, 80, 153, 154, 156, 158, 160, 172, 185, 217, 219, 251, 253–257
Fish communities 153, 168
Fish predation 45, 108, 254
Fish stock 153–168, 174, 175, 191, 193–195, 248, 253, 271
Food web 42, 47, 48, 251, 256, 272, 279
Furnas Reservoir 31–32, 37–38
Future research needs 275–278

GIRL OLGA 236
Glebokie Lake 99–104, 110–112, 114, 116
Glen Lake 60
Gorkovskoye Reservoir 155, 157, 159
Growth factor 131

Holistic approach to reservoir science 224
Holistic reservoir limnology 229
Horizontal zonation 35–36, 241
Husinec Reservoir 160
Hydraulic stratification 35, 37, 38, 244

Hypolimnion 31, 39, 51, 70, 75, 76, 81, 82, 87, 88, 99, 104–108, 114, 119, 121–125, 181, 183, 185, 186, 205, 232, 239, 242, 245, 251, 254, 256, 262, 269–271, 273

Impoundments 216–217
Inflow 2, 16, 18, 61, 64, 65, 69, 70, 72, 73–80, 82, 84–88, 108, 110, 129, 140, 170, 172, 173, 176, 183, 185–189, 196, 200–202, 204, 205, 209, 214, 219, 221, 224, 229, 235, 239, 240, 242–245, 253, 255, 259–262, 268–270, 273
Intrusion 69, 76–81, 85, 86, 88, 89
Inulec Lake 99, 100, 116
Itaipu Reservoir 264
Itaparica Reservoir 26, 29

Jesenice Reservoir 158
Jordan Lake 218
Jorzec Lake 99, 100, 106, 107

Kariba Lake 47, 264
Kličava Reservoir 94, 155, 156, 158, 159, 165, 203, 243, 253–254
Kremenchugskoye Reservoir 160
Kuortaneenjärvi Lake 127–134, 140–144

La Concepcion Reservoir 123–125
Lamoille Lake 60
Landštejn Reservoir 156, 158, 159
Light 16, 17, 19, 20, 22, 32, 34, 36, 51, 57, 61, 63, 69, 70–73, 75, 79, 88, 100, 101, 105, 111, 112, 120, 124, 135, 149, 181, 191, 201, 208, 222, 224, 227, 231, 239, 241, 248, 249, 251, 255, 259, 263, 265, 269, 271, 272
Limno-geographical regions 248
Limnological succession 153
Lipno Reservoir 321
Load 48, 63, 73, 75, 79, 87, 88, 99, 104, 111, 127, 128, 130, 132–135, 181–184, 190, 200, 207–209, 237, 244
Longitudinal zonation 63, 64
Lučina Reservoir 155, 159, 165

Macrophyta 39, 40, 46, 51, 63, 217, 223, 251, 255, 257, 269, 273
Major geographical differences in reservoir limnology 248
Management at the watershed level 268
Management of reservoir outflows 272–273
Management within the reservoir 269–272
Mathematical modelling 93–97, 232–237, 264, 273
Mathematical modelling in management 273
Mineralization 119, 121–125, 177, 183, 208
Mingechaurskoye Reservoir 163
MINLAKE 236
Mixing 2, 31, 35, 37, 38, 40, 42, 46, 61–63, 69, 70, 75, 76, 79, 80–82, 84, 86–88, 95, 101, 108, 109, 111, 112, 170, 173, 176, 177, 179, 180, 185, 191, 200, 205, 208, 209, 220, 224, 231, 232, 235, 239–243, 246, 248–250, 255, 259, 270–272
Morávka Reservoir 158
MORDOR 236
Morphometry 16, 26, 31, 32, 34, 36, 49, 51, 86, 216, 217, 219, 224, 231, 237, 240, 259
MS CLEANER 236
Multi-layer model 108–114

Nitrates 132, 175, 182, 186, 199, 202, 205, 207, 208, 222
Nitrogen 3, 15, 16, 18, 20, 38, 40, 48, 49, 58–61, 71, 75, 114, 123, 129–133, 140–142, 145, 174, 181, 182, 191, 199, 201, 207, 222, 235, 267, 269, 270, 273, 276
Nizhnekamskoye Reservoir 158
Normandy Reservoir 64
N:P ratios 18, 20, 21, 135, 191
Numerical experiments 99, 102, 111, 114
Nutrient availability 57, 61, 62, 66, 261, 276
Nutrient recycling 86, 236

One-dimensional models 94
Opatovice Reservoir 157
Optimal timing of sampling 139, 142–145
Organic compounds 182–188, 208, 209, 268, 273
Organic matter 39, 42, 46, 47, 105, 119, 120, 123, 124, 147, 169, 174, 183–185, 188, 189, 201, 208, 222, 235, 236, 238, 239, 246, 249, 251, 253–256, 261, 262, 267
Orlík Reservoir 155, 156, 160, 259, 260, 261
Outflow 33, 46, 47, 51, 65, 69, 70, 72, 75, 76, 80, 84–86, 88, 108, 110, 130, 135, 140, 172, 173, 181, 182, 184, 196, 200, 204, 207, 214–216, 219–224, 229, 232, 236, 239, 240, 244, 259, 268, 269, 272

Parameter estimation 99, 139, 140
Periphyton 40–42, 47, 51, 79, 253, 254
Phosphorus – see total phosphorus
Phosphorus-chlorophyll relationships 19
Phytoplankton 1, 3, 15, 17, 18, 31, 32, 35, 37, 40–44, 58, 61–67, 69–75, 87, 88, 90, 100, 101, 103, 105, 107–114, 119–120, 122–125, 130–135, 141, 143, 147, 148, 150, 169, 174–176, 181–184, 188–193, 201, 202, 205, 207–209, 222–224, 227, 228, 239, 240, 245–250, 252–256, 261–264, 266, 269, 270, 276
Phytoplankton flux 119, 120, 123
Precipitation 37, 38, 40, 65, 69, 70, 80, 84, 85, 87, 88, 95, 128, 140, 172, 181–183, 186, 223, 239, 248–251, 267, 269
Predation 45, 100, 108, 154, 158, 163, 228, 251, 254
Predator 52, 165, 199
Predatory pressure 108, 165
Primary production 17, 31, 32, 40–42, 51, 57, 61, 70–72, 74, 75, 86–88, 101, 111, 114, 119, 123–124, 174, 181, 183, 184, 186, 188, 201, 223, 236, 247, 263, 265, 267, 272
Pulse effects 51, 251
Pump-storage reservoirs 69, 70, 74, 75, 86

Recording 196
Recreation 50, 127, 189, 204, 207–208, 214, 237, 264
Relationship between primary production and sedimentation 119
Remote sensing 51, 147–152, 231, 277
Reservoir cascades 26, 28, 31, 48–51, 257–263
Reservoir management 1, 2, 16, 20–23, 25, 26, 31, 46, 49–51, 57, 58, 65, 66, 71, 93, 96, 127, 136, 139, 147, 154, 165, 166, 169, 172, 194, 205, 208, 213, 215, 220, 221, 229, 230, 232, 235, 236, 252, 263–269, 270–278
Reservoir systems 40, 51, 257–262
Reservoir thermodynamics 95
Reservoir types 170
RESTEMP 243

RESQUAL 236
Retention time (=residence time) 15, 31, 34, 40–42, 46, 48, 51, 57–58, 61–65, 85, 94, 96, 100, 128–130, 170, 172, 176, 178, 180, 182, 184, 185, 188, 191, 200, 202, 204, 207–209, 213, 215–217, 220–223, 227, 236, 239, 240, 241–250, 255, 259, 260–262, 265, 267, 272
Retention time as a key factor 239–248
Řimov Reservoir 155, 156, 158, 159, 160, 165, 184, 204, 245
Rio Bermejo Reservoir 43
Rio Pardo Reservoir 36
Rio Paulo Reservoir 37
River continuum concept 237, 238
Roosevelt Lake 58, 59

Sampling 14, 16, 41, 51, 65, 71, 120, 128, 139–146, 172–176
Sam Rayburn Reservoir 58, 59
Samuel Reservoir 29, 36, 39–40, 44–47, 264
Šance Reservoir 155, 156, 159, 165, 167
Seasonality 31, 51, 124, 198, 231
Seč Reservoir 163, 166
Secchi 3, 15, 19, 20, 22, 32, 34, 35, 41, 58, 60, 63, 64, 149, 150, 177, 236, 251
Sedimentation 72, 74, 75, 87, 88, 99, 100, 101, 105, 106, 108, 110, 112, 114–116, 119–121, 123–125, 188, 189, 191, 205, 208, 209, 223, 245, 247, 248, 255, 262, 265
Sediment inactivation 72
Sediment traps 119–120, 123–124
Sediment-water exchange 115–117
Selforganization 237, 264
Semi-arid region 1, 2, 14–22, 48, 214, 221, 250, 267, 276
Sensitivity analysis 99, 116, 117, 139, 141, 143, 145
Serial discontinuity concept 238, 259
Seston flux 120–123
Simulation 12, 69, 70, 79, 81–83, 86, 88, 93, 94, 99, 101, 103, 104, 106–108, 110–112, 116–117, 127, 130–132, 139–141, 232, 234–238, 243, 244, 267, 274, 275
Slapy Reservoir 155, 156, 160, 164, 165, 178, 186, 187, 197, 203, 205, 223, 232, 233, 237, 238, 241–244, 246, 247, 259–261, 277
Slatersville Reservoir 60
Southern Indiana Lake 264
Spatial scales 93, 231–232
Spatial heterogeneity 66, 251, 269
Spectra 147, 148, 150, 233
Stanovice Reservoir 155, 156, 157, 159, 165, 167
Stillhouse Hollow Reservoir 63, 64
Stratification modelling 95
Succession 41, 46, 47, 51, 121, 153, 154, 157–159, 164–166, 192, 195, 232, 252, 257, 264
Sulejow Reservoir 160, 167

Temporal heterogeneity 57, 63, 230
Temporal scales 23
Thermal model 108, 110, 111
Thermal structure 34, 40, 231, 252
Three-dimensional models 94
Top-down 228, 254, 256, 264, 272
Total phosphorus 2, 15, 17, 19, 20, 21, 58–61, 71, 73–75, 79–81, 99, 106, 175, 183, 190, 191, 201, 209, 227, 245, 246, 254, 261
Transparency 3, 15, 19–22, 32, 34, 46, 57, 58, 61, 64, 74, 174, 175, 177, 180, 181, 270
Trophic-state 20, 21, 57, 58, 61–64, 66, 71, 75, 78, 93, 94, 151, 190
Tropics 25–55, 147–152, 215, 217, 221, 223, 236, 250–251, 263–264, 266–267
Tscheboksarskoye Reservoir 157, 158
Tucuruí Reservoir 26, 29, 32–33, 36, 39–42, 47–49, 263
Turbidity 1, 2, 16, 17, 19, 20, 61–63, 66, 87, 128, 150, 177, 180, 181, 205, 208, 209, 224, 239, 248–250, 255, 263, 267
Two-dimensional models 94
Two-layer model 104

Volta Grande Reservoir 26, 31
Volta Reservoir 263

Waco Reservoir 64, 65
Wägitalersee 69–73, 75–77, 79, 80–87, 257
Waterbury Reservoir 60
Water mass balance 85
Water quality modelling 94, 96, 213
Water quality trends 197–199
Watershed 37, 42, 50, 64–66, 99, 105, 129, 169–172, 177, 181–184, 186, 187, 189, 199, 200, 205–209, 217, 220, 224, 229–231, 235, 250, 253, 255–257, 264, 267–269, 275, 276, 278
Water treatment plants 209
Wellington Reservoir 80, 81, 84
Willow Creek Reservoir 214
Willow Reservoir 60
Wissota Lake 60

Yacireta Reservoir 263

Záskalská Reservoir 157–160
Zonation 35–36, 63–66, 176, 185, 240, 241, 250, 255
Zooplankton 31, 32, 42–45, 48, 69, 72, 100–106, 107–114, 159, 162, 165, 169, 174–176, 180, 181, 188, 189, 192–195, 201, 210, 222, 247, 253–256, 266, 272, 276